Emerging Contaminants and Associated Treatment Technologies

Series Editors

Muhammad Zaffar Hashmi, Institute of Biology and Biotechnology, The University of Lahore, Lahore, Pakistan

Vladimir Strezov, Department of Environmental Sciences, Macquarie University, Sydney, Australia

Emerging Contaminants and Associated Treatment Technologies focuses on contaminant matrices (air, land, water, soil, sediment), the nature of pollutants (emerging, well-known, persistent, e-waste, nanomaterials, etc.), health effects (e.g., toxicology, occupational health, infectious diseases, cancer), treatment technologies (bioremediation, sustainable waste management, low cost technologies), and issues related to economic development and policy. The book series includes current, comprehensive texts on critical national and regional environmental issues of emerging contaminants useful to scientists in academia, industry, planners, policy makers and governments from diverse disciplines. The knowledge captured in this series will assist in understanding, maintaining and improving the biosphere in which we live. The scope of the series includes monographs, professional books and graduate textbooks, edited volumes and books devoted to supporting education on environmental pollution at the graduate and post-graduate levels.

Himanshu Gupta · Sughosh Madhav ·
Soniya Dhiman · Pramod Kumar
Editors

Ubiquitous Polycyclic Aromatic Hydrocarbon Contamination

Recent Perspectives

Springer

Editors
Himanshu Gupta
Department of Chemistry, School of Sciences
IFTM University
Moradabad, Uttar Pradesh, India

Sughosh Madhav
Department of Civil Engineering
Jamia Millia Islamia
New Delhi, Delhi, India

Soniya Dhiman
Helmholtz-Zentrum Dresden-Rossendorf, Helmholtz Institute Freiberg for Resource Technology
Dresden, Germany

Pramod Kumar
Sri Aurobindo College
University of Delhi
New Delhi, Delhi, India

ISSN 2524-6402 ISSN 2524-6410 (electronic)
Emerging Contaminants and Associated Treatment Technologies
ISBN 978-3-032-11763-2 ISBN 978-3-032-11764-9 (eBook)
https://doi.org/10.1007/978-3-032-11764-9

This Springer imprint is published by the registered company Springer Nature Switzerland AG
The registered company address is: Gewerbestrasse 11, 6330 Cham, Switzerland

Contents

1 Overview of Ubiquitous Contaminants Polycyclic Aromatic Hydrocarbons in Different Environmental Matrices

Himanshu Gupta, Sughosh Madhav, and Soniya Dhiman

Abstract

In view of their widespread occurrence in the environmental segments, these are considered as ubiquitous. The incomplete burning of different substances cause their generation and distribution in environment, particularly air. Smaller PAHs such as naphthalene are relatively less toxic compared to higher members such as benzo[a]pyrene, which is carcinogenic in nature. At places, it is reported that PAHs contamination is also present in the drinking water, causing direct harm to the humans. The most toxic effect is their involvement with the DNA and formation of DNA adducts, which results is mutations and sometimes can cause cancer. The dominant PAH pollution mechanism can be differentiated by the application of diagnostic ratios like phenanthrene/anthracene or fluoranthene/pyrene ratios. These ratios suggest the nature of PAH pollution is major from petrogenic or pyrogenic sources. The PAHs from the soil can be transported to roots of the plants and slowly reaches to other parts of the plants including fruits and vegetables. Without any metabolic activation, oxygenated PAHs and nitrogenated PAHs can bind with DNA directly and cause mutations, posing severe threat to living organisms. Various detection and remediation techniques have been applied for the detection of PAHs in various environmental matrices. This chapter gives a basic overview of the ubiquity, sources, distribution, fate, derivatives, transport, detection and remediation of PAHs.

H. Gupta
Department of Chemistry, School of Sciences, IFTM University, Moradabad, Uttar Pradesh, India

S. Madhav
Department of Civil Engineering, Jamia Millia Islamia University, New Delhi, India

S. Dhiman (✉)
Helmholtz-Zentrum Dresden-Rossendorf, Helmholtz Institute Freiberg for Resource Technology, Dresden, Germany
e-mail: s.dhiman@hzdr.de

H. Gupta et al. (eds.), *Ubiquitous Polycyclic Aromatic Hydrocarbon Contamination*, Emerging Contaminants and Associated Treatment Technologies,
https://doi.org/10.1007/978-3-032-11764-9_1

Keywords

PAHs • Ubiquitous • Toxicity • Benzo[a]pyrene • Derivatives • Environmental pollution

Introduction

Various environmental matrices are contaminated with persistent organic pollutants on the global level. Among them, polycyclic aromatic hydrocarbons (PAHs) are reported to be ubiquitous and hazardous to all segments of environment. PAHs comprises of 02 or higher number of benzene rings which are fused to form bigger molecules. Such molecules can be unsubstituted or can be of substituted nature. In view of their widespread occurrence in the environmental segments, these are considered as ubiquitous (Gupta et al. 2017; Gupta and Kumar 2020). PAHs generally have hydrophobic nature and high stability towards temperature, which makes it persistent in the environment and accumulate in different segments such as soil and sediments (Berríos-Rolón et al. 2025). The incomplete burning of different substances cause their generation and distribution in environment, particularly air. The airborne PAHs has a tendency to travel to short as well as long distances. The deposition of airborne PAHs is one of the major cause of PAH pollution in soil and water. The environmental PAHs may undergo degradation through photochemical or microbiological methods. However, accumulation of PAHs is reported to be dominated over the degradation phenomena (Gupta et al. 2017).

Ubiquitous and Toxicological Nature of PAHs

PAHs are produced by incomplete combustion of fuels, wood, paper etc. as well as also released to the environment as a result of forest fires, oil spills, volcanic eruptions etc. (Kumar et al. 2016). As the process of burning is inevitable and present across the globe, the PAHs pollution has became ubiquitous. At longer distances, where the burning process is not in function, the tendency of PAHs to travel longer distances cause pollution. Among a large number of PAHs, 16 PAHs are regarded as priority pollutants by USEPA. The smaller PAHs such as naphthalene are relatively less toxic compared to higher members such as benzo[a]pyrene, which is carcinogenic in nature (Gupta and Gupta 2015). The molecular structure of PAHs is planar and π-electrons are delocalised in the rings, making their higher stability as well as higher boiling or melting points and lower solubility in water (Sahoo et al. 2020). The solubility of smaller PAHs is greater than the higher PAHs (Gupta and Gupta 2016). Due to their accumulative nature, PAHs persists in the environment and reaches to the living organisms through contaminated soil, water or air (Fig. 1.1). PAHs are toxic to different living organisms including human beings (Rusănescu et al. 2025). These get accumulated in the body and interfere

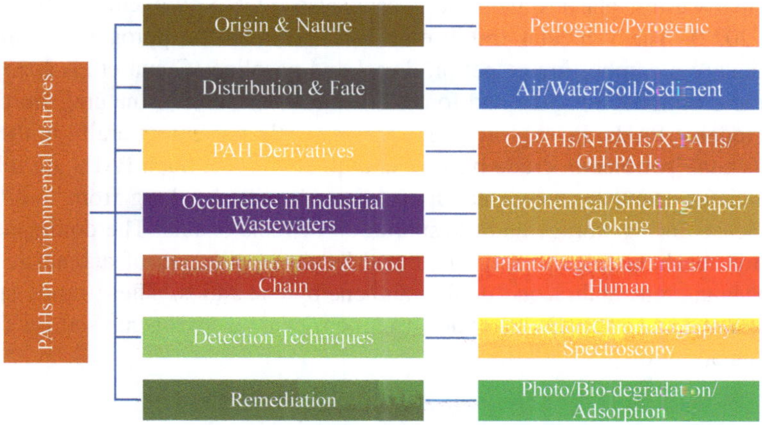

Fig. 1.1 PAHs in various environmental matrices

in the metabolism causing serious damages. The most toxic effect is their involvement with the DNA and formation of DNA adducts, which results is mutations and sometimes can cause cancer (Gupta and Gupta 2015). Their persistence, mutagenic to carcinogenic nature and ubiquitous occurrence necessitates a detailed study on the sources, accumulation, transport mechanism, associated risks and remediation in different environmental matrices.

Origin of PAHs in the Environment

PAH origin in the environment is usually classified by two major methods, namely pyrogenic and petrogenic. It is based on their mechanism of production and initial sources. The study of these origin of PAHs is essential in order to understand the source apportionment, risk assessment, availability and toxicity of PAHs to the living beings and the environment (Gupta and Kumar 2020).

Origin of Pyrogenic and Petrogenic PAHs

The major part of the global PAHs comes in the form of pyrogenic PAHs. These are more common in the areas of heavy traffic and industries. These kind of PAHs originates by incomplete combustion of different materials such as fuels, cow dung cake, wood, plastic, paper, food etc. (Pradhi et al. 2023; Rogers et al. 2024). During the generation of pyrogenic PAHs, generally the larger compounds break down to small fragments and radicals, which recombine, aromatize to form PAHs. The major component of these pyrogenic PAHs is associated with the anthropogenic activities by the humans e.g. energy driven activities and industrial processes. The combustion of coal and wood to generate power produces pyrogenic PAHs. The coking and smelting like industrial processes releases pyrogenic PAHs mainly to the air (Li et al. 2023). The burning/incineration of different wastes including municipal wastes, plastics and other wastes also release pyrogenic PAHs (Cheng et al. 2022). The vehicular traffic emissions are also the pyrogenic sources as these involves combustion of petrol, diesel and gasoline (Gupta et al. 2025). The volcanic eruptions, oil spills and forest fires are considered as natural sources for the pyrogenic PAHs. Petrogenic PAHs are generally present naturally in the fuels such as coal, petroleum, crude oil and natural gas. These. Such PAHs are released to the environment by accidental spills of crude oils, leaching from fossil fuels and release during storage and transportation of fuels or oils. The dominant PAH pollution mechanism can be differentiated by the application of diagnostic ratios like phenanthrene/anthracene or fluoranthene/pyrene ratios. These ratios suggest the nature of PAH pollution is major from petrogenic or pyrogenic sources (Gupta and Kumar 2020).

PAH Distribution and Fate in Environmental Matrices

Once PAHs are released in the environment, these got distributed to different environmental matrices like water, soil, air and sediments. The accumulated PAHs in these matrices transmits to the food chain and reaches to living organisms including humans (Wang and Wang 2006; Ezugwu et al. 2024). The released PAHs to the atmosphere transport to short and long distances, then settles down on the soil and water. The low molecular weight PAHs are reported to be transported to longer distances than high molecular weight PAHs (Gupta and Kumar 2020). The deposition of PAHs occurs by different mechanisms, namely dry deposition, involving gravitation settlement of PAHs and wet deposition, where the PAH molecules are mixed to fog, snow or get settle by rain (Wang et al. 2016). The air pollution due to PAHs, results in secondary pollution to other matrices, water and soil. Other than atmospheric transport, soil get contaminated by spillage of crude oils, coal tar etc. It is also contaminated by landfilling of PAH contaminated wastes and other industrial processes (Liu and Wang 2022). The dermal contact with contaminated soil is harmful to living organisms. The PAHs from the soil can be transported to roots of the plants and slowly reaches to other parts of the plants including fruits and vegetables (Jia et al. 2021). PAHs from the soil can also leach to the groundwater and contaminated it. The availability of PAHs in surface and groundwater is hazardous to the organisms. Water bodies also receives PAHs from agricultural runoffs from PAH contaminated sites, direct discharge of wastewaters by the different industries and other wastewater treatment plants. At places, it is reported that PAHs contamination is also present in the drinking water, causing direct harm to the humans (Ambade et al. 2021). Various studies also reported the presence of PAHs in fishes and other organism in water (Filatova et al. 2023; Uzomah et al. 2021; Damir et al. 2021). The consumption of contaminated fish also result in the PAH toxicity. The PAHs in the water settles down to the sediments and bind with it. The levels of PAHs in sediments may be higher than the water (Xia et al. 2015). Therefore, all environmental matrices comprises of ubiquitous pollutants, PAHs, causing harm to the environment and living beings.

PAH Derivatives in the Environment

Other than simpler PAHs, the different variants of PAHs are also reported to be toxic and referred to as PAH derivatives. These include compounds such as oxygenated PAHs, halogenated PAHs, nitrogenated PAHs and hydroxy PAHs etc. The PAH derivatives are also found to be highly toxic and are responsible for immunotoxicity, genotoxicity, oxidative stress, carcinogenicity as well as neurotoxicity (Peng et al. 2023). The toxicity of hydroxy PAHs was reported higher compared to PAHs in case of sea urchins and fishes (Honda and Suzuki 2020). Biological processes and combustion are primary sources of PAH derivatives. The oxidant and radical mediated photochemical reactions are secondary PAH derivatives sources (Krzyszczak and Czech 2021). Without any metabolic activation, oxygenated PAHs and nitrogenated PAHs can bind with DNA directly and cause mutations, posing severe threat to living organisms (Jeong et al. 2024). Indian ocean has higher levels of halogenated PAHs than the Pacific ocean. Sri Lankan adults are also reported to have potential health risks due to contaminated aquatic species (Wickrama-Arachchige et al. 2021).

PAHs in Industrial Wastewater

The wastewaters of different industries are also reported to be contaminated with PAHs. A few of industries released PAH in the environment includes petrochemical industry, coking (Chen et al. 2020) and smelting industries (Booth and Gribben 2005), paper/pulp industries (Gupta 2018) and municipal wastewater treatment plants (Liu et al. 2021). The discharge of coking plant wastewater in the river increased the levels of PAHs in water (Chen et al. 2020). The co-existence of PAHs and their derivatives was reported in the effluents of wastewater treatment plants (Liu et al. 2021). Greater than 100 PAHs were detected due to smelting processes (Booth and Gribben 2005). Various industries employs different treatment methods for their effluents such as advanced oxidation processes, adsorption, photodegradation and biodegradation, but the removal of PAHs from the wastewaters is usually insufficient causing hazards to the environment.

PAHs in the Foods and Food Chain

The PAHs present in different segments of environment can be transported to the food materials and food chain. Atmospheric PAHs were found to be deposited on food and contaminating them (Bansal et al. 2017). PAHs from soil transport to the roots, stem and finally to the fruits of the plants (Paris et al. 2018). Aquatic PAHs accumulate in the aquatic organisms and transfer to the higher member of the food chain (Honda and Suzuki 2020). Other than the environmental transport of PAHs to foods, various anthropogenic processes also found to be influencing towards the production of PAHs. These include roasting, frying grilling and smoking of the food before eating (Adeyeye and Ashaolu 2022). These processes also produce PAHs, which reaches to the humans directly by ingestion and respiration. PAH contamination due to processed meat can be avoided applying low temperature for long duration, prevent direct contact with flames during barbecue etc. (Adeyeye and Ashaolu 2022). It is also reported that deep frying in different oils releases greater amounts of PAHs in comparison to frying. Among olive, peanut, rapeseed and soyabean oils, the rapeseed oil produces highest levels of PAH and the order of PAH generation in oils is rapeseed > olive > peanut > soyabean (Yao et al. 2015).

Detection and Remediation of PAHs

A number of detection techniques have been applied for the detection of PAHs in various environmental matrices. A few of the techniques involved are solvent extraction, Soxhlet extraction, mechanical agitation, solid phase extraction, solid phase microextraction, microwave-assisted extraction, ultrasonic extraction, supercritical fluid extraction, capillary electrophoresis, gas chromatography, high performance liquid chromatography, UV–Visible spectroscopy, fluorescence spectroscopy, sensors and biosensors etc. (Bansal et al. 2017; Zhang et al. 2020). Different remediation techniques are also being applied in order to treat different PAH-contaminated segments of the environment. The remediation techniques involves the photodegradation and biodegradation of PAHs in soil as well as in water. The degradation of PAHs involve different photochemical as well as biological mechanisms, causing breakdown of PAHs to smaller fragments and sometimes mineralization (Gupta and Dhiman 2023, 2024; Gupta et al. 2025). The study of mechanism of degradation pathways is important as sometimes more toxic compounds may form during the process (Gupta and Gupta 2015). Another major technique applied for the removal of PAHs from water and wastewaters is adsorption (Gupta and Gupta 2016, 2018; Mogashane et al. 2024; Wang et al. 2023). In literature, a variety of adsorbents have been employed to treat contaminated water from PAHs as well as other contaminants. Other remediation methods such as oxidation, advanced oxidation, ozonolysis etc. are also employed to treat PAH contaminated segments.

Conclusion

Polycyclic aromatic hydrocarbons (PAHs) are reported to be ubiquitous and hazardous to all segments of environment. At longer distances, where the burning process is not in function, the tendency of PAHs to travel longer distances cause pollution. Their persistence, mutagenic to carcinogenic nature and ubiquitous occurrence necessitates a detailed study on the sources, accumulation, transport mechanism, associated risks and remediation in different environmental matrices. The study of these origin of PAHs is essential in order to understand the source apportionment, risk assessment, availability and toxicity of PAHs. Biological processes and combustion are primary sources of PAH derivatives. The oxidant and radical mediated photochemical reactions are secondary PAH derivatives sources. The co-existence of PAHs and their derivatives was reported in the effluents of wastewater treatment plants. The order of PAH generation in oils is rapeseed > olive > peanut > soyabean. A number of detection techniques have been applied for the detection of PAHs in various environmental matrices. The remediation techniques involves the photodegradation and biodegradation of PAHs in soil as well as in water. The degradation of PAHs involve different photochemical as well as biological mechanisms. Another major technique applied for the removal of PAHs from water and wastewaters is adsorption. In spite of application of different remediation techniques including advanced and green methods, the abundance of PAHs in the environment is high. Therefore more attention is required to tackle the issue of contamination by PAH and their derivatives.

References

Adeyeye, S.A.O., Ashaolu, T.J.: Polycyclic aromatic hydrocarbons formation and mitigation in meat and meat products. Polycyclic Aromat. Compd. **42**(6), 3401–3411 (2022). https://doi.org/10.1080/10406638.2020.1866039

Ambade, B., Sethi, S.S., Kumar, A., et al.: Health risk assessment, composition, and distribution of polycyclic aromatic hydrocarbons (PAHs) in drinking water of southern Jharkhand, East India. Arch. Environ. Contam. Toxicol. **80**, 120–133 (2021). https://doi.org/10.1007/s00244-020-00779-y

Bansal, V., Kumar, P., Kwon, E.E., Kim, K.H.: Review of the quantification techniques for polycyclic aromatic hydrocarbons (PAHs) in food products. Crit. Rev. Food Sci. Nutr. **57**(15), 3297–3312 (2017). https://doi.org/10.1080/10408398.2015.1116970

Berríos-Rolón, P.J., Cotto, M.C., Márquez, F.: Polycyclic aromatic hydrocarbons (PAHs) in freshwater systems: a comprehensive review of sources, distribution, and ecotoxicological impacts. Toxics **13**(4), 321 (2025). https://doi.org/10.3390/toxics13040321

Booth, P., Gribben, K.: A review of the formation, environmental fate, and forensic methods for PAHs from aluminum smelting processes. Environ. Forensics **6**(2), 133–142 (2005). https://doi.org/10.1080/15275920590952757

Chen, J., Liao, J., Wei, C.: Coking wastewater treatment plant as a sources of polycyclic aromatic hydrocarbons (PAHs) in sediments and ecological risk assessment. Sci. Rep. **10**, 7833 (2020). https://doi.org/10.1038/s41598-020-64835-2

Cheng, K., Li, J., Wang, Y., Ji, W.-Y., Cao, Y.: Characterization and risk assessment of airborne polycyclic aromatic hydrocarbons from open burning of municipal solid waste. Front. Environ. Sci. **10**, 861770 (2022). https://doi.org/10.3389/fenvs.2022.861770

Damir, N.A., Coatu, V., Pantea, E.D., Galațchi, M., Botez, E., Birghilă, S.: Assessment of polycyclic aromatic hydrocarbons content in marine organisms of commercial interest from the Romanian black sea coast. Polycyclic Aromat. Compd. **42**(10), 7595–7606 (2021). https://doi.org/10.1080/10406638.2021.2006243

Ezugwu, A.L., Agbasi, J.C., Egbueri, J.C., et al.: Mechanism, formation and transport of polycyclic aromatic hydrocarbons (PAHs) in fruits, vegetables and fresh fish species in Africa: a systematic review of its health risk. Chem. Afr. **7**, 2321–2344 (2024). https://doi.org/10.1007/s42250-024-00926-1

Filatova, T.S., Abramochkin, D.V.: Physiological effects of polycyclic aromatic hydrocarbons in fish organisms. Moscow Univ. Biol. Sci. Bull. **78**, 115–127 (2023). https://doi.org/10.3103/S0096392523700013

Gupta, H.: PAH determination in effluent and sludge samples of paper industry. Environ. Technol. Innov. **9**, 115–121 (2018)

Gupta, H., Gupta, B.: Photocatalytic degradation of polycyclic aromatic hydrocarbon benzo[a]pyrene by iron oxides and identification of degradation products. Chemosphere **138**, 924–931 (2015)

Gupta, H., Gupta, B.: Adsorption of polycyclic aromatic hydrocarbons on banana peel activated carbon. Desalin. Water Treat. **57**(20), 9498–9509 (2016)

Gupta, H., Gupta, B.: Vehicular tire as potential adsorbent for the removal of polycyclic aromatic hydrocarbons. Polycyclic Aromat. Compd. **38**(4), 354–368 (2018)

Gupta, H., Kumar, R.: Distribution of selected polycyclic aromatic hydrocarbons in urban soils of Delhi, India. Environ. Technol. Innov. **17**, 100500 (2020)

Gupta, H., Kumar, R., Park, H.S., Jeon, B.H.: Photocatalytic efficiency of iron oxide nanoparticles for the degradation of priority pollutant anthracene. Geosyst. Eng. **20**(1), 21–27 (2017). https://doi.org/10.1080/12269328.2016.1218302

Gupta, H., Dhiman, S., Kumar, A., Paliwal, S.K., Khan, G., Singh, H., Mishra, A., Mishra, A.K.: Photo- and bio-degradation of selected persistent organic pollutants. Sep. Purif. Rev. **54**(1), 3–16 (2025). https://doi.org/10.1080/15422119.2024.2332953

Gupta, H., Dhiman, S.: Synthetic nanoparticle-based remediation of soils contaminated with polycyclic aromatic hydrocarbons. In: Shanker, U., Hussain, C.M., Rani, M. (eds.) Handbook of Green and Sustainable Nanotechnology. Springer, Cham (2023). https://doi.org/10.1007/978-3-030-69023-6_102-1

Gupta, H., Dhiman, S.: Emerging contaminants in the environment introduction. In: Shanker, U., Rani, M. (eds.) Occurrence, Distribution and Toxic Effects of Emerging Contaminants, pp. 1–16. CRC Press (2024). https://doi.org/10.1201/9781003335757-1

Honda, M., Suzuki, N.: Toxicities of polycyclic aromatic hydrocarbons for aquatic animals. Int. J. Environ. Res. Public Health **17**(4), 1363 (2020). https://doi.org/10.3390/ijerph17041363

Jeong, J., Kim, G., Lee, J.G.: A review of food contamination with nitrated and oxygenated polycyclic aromatic hydrocarbons: toxicity, analysis, occurrence, and risk assessment. Food Sci. Biotechnol. **33**, 2261–2274 (2024). https://doi.org/10.1007/s10068-024-01653-6

Jia, J., Bi, C., Jin, X., Zeng, Y., Deng, L., Wang, X., Chen, Z.: Uptake, translocation, and risk assessment of PAHs in contaminated soil-air-vegetable systems based on a field simulation experiment. Environ. Pollut. **271**, 116361 (2021). https://doi.org/10.1016/j.envpol.2020.116361

Krzyszczak, A., Czech, B.: Occurrence and toxicity of polycyclic aromatic hydrocarbons derivatives in environmental matrices. Sci. Tot. Environ. **788**, 147738 (2021). https://doi.org/10.1016/j.scitotenv.2021.147738

Kumar, V., Kothiyal, N.C., Saruchi, Vikas, P., Sharma, R.: Sources, distribution, and health effect of carcinogenic polycyclic aromatic hydrocarbons (PAHs)—current knowledge and future directions. J. Chin. Adv. Mater. Soc. **4**(4), 302–321 (2016). https://doi.org/10.1080/22243682.2016.1230475

Li, M., Shi, M., Hu, T., et al.: Geochemical characteristics and behaviors of polycyclic aromatic hydrocarbons (PAHs) in soil, water, and sediment near a typical nonferrous smelter. J. Soils Sediments **23**, 2258–2272 (2023). https://doi.org/10.1007/s11368-023-03476-0

Liu, X., Wang, Y.: Identification and assessment of groundwater and soil contamination from an informal landfill site. Sustainability **14**(24), 16948 (2022). https://doi.org/10.3390/su142416948

Liu, Q., Xu, X., Lin, L., Wang, D.: Occurrence, distribution and ecological risk assessment of polycyclic aromatic hydrocarbons and their derivatives in the effluents of wastewater treatment plants. Sci. Total Environ. **789**, 147911 (2021). https://doi.org/10.1016/j.scitotenv.2021.147911

Mogashane, T.M., Maree, J.P., Mokoena, L.: Adsorption of polycyclic aromatic hydrocarbons from wastewater using iron oxide nanomaterials recovered from acid mine water: a review. Minerals **14**(8), 826 (2024). https://doi.org/10.3390/min14080826

Paris, A., Ledauphin, J., Poinot, P., Gaillard, J.L.: Polycyclic aromatic hydrocarbons in fruits and vegetables: origin, analysis, and occurrence. Environ. Pollut. **234**, 96–106 (2018). https://doi.org/10.1016/j.envpol.2017.11.028

Peng, B., Dong, Q., Li, F., Wang, T., Qiu, X., Zhu, T.: A systematic review of polycyclic aromatic hydrocarbon derivatives: occurrences, levels, biotransformation, exposure biomarkers, and toxicity. Environ. Sci. Technol. **57**(41), 15314–15335 (2023). https://doi.org/10.1021/acs.est.3c03170

Petr, K., Gerhard, L., Alice, D., Jana, K., Andrea, M., Elke, F.: Contamination of Antarctic snow by polycyclic aromatic hydrocarbons dominated by combustion sources in the polar region. Environ. Chem. **7**, 504–513 (2010). https://doi.org/10.1071/EN10066

Pradhi, R., Shukla, P.C., Singh, G.K., Das, D., Gupta, T.: Assessment of entrainment of key PAHs emanating from major combustion sources into the ambient air. Fuel **347**, 128430 (2023). https://doi.org/10.1016/j.fuel.2023.128430

Rogers, K., WaMaina, E., Barber, A., Masood, S., Love, C., Kim, Y.H., Gilmour, M.I., Jaspers, I.: Emissions from plastic incineration induce inflammation, oxidative stress, and impaired bioenergetics in primary human respiratory epithelial cells. Toxicol. Sci. **199**(2), 301–315 (2024). https://doi.org/10.1093/toxsci/kfae038

Rusănescu, C.O., Istrate, I.A., Rusănescu, A.M., Constantin, G.A.: Bioremediation of soil contamination with polycyclic aromatic hydrocarbons—a review. Land **14**(1), 10 (2025). https://doi.org/10.3390/land14010010

Sahoo, B.M., Kumar, B.V.V.R., Banik, B.K., Borah, P.: Polyaromatic hydrocarbons (PAHs): structures, synthesis and their biological profile. Curr. Org. Synth. **17**(8), 625–640 (2020)

Uzomah, A., Lundebye, A.-K., Kjellevold, M., Chuku, F.A., Stephen, O.A.: A review of chemical contaminants in marine and fresh water fish in Nigeria. Foods **10**(9), 2013 (2021). https://doi.org/10.3390/foods10092013

Wang, X., Wang, W.: Bioaccumulation and transfer of benzo(a)pyrene in a simplified marine food chain. Mar. Ecol. Prog. Ser. **312**, 101–111 (2006). https://doi.org/10.3354/meps312101

Wang, Q., Liu, M., Li, Y., Liu, Y., Li, S., Ge, R.: Dry and wet deposition of polycyclic aromatic hydrocarbons and comparison with typical media in urban system of Shanghai, China. Atmos. Environ. **144**, 175–181 (2016). https://doi.org/10.1016/j.atmosenv.2016.08.079

Wang, J., Cao, W., Wei, W., et al.: Adsorption characteristic analysis of PAHs on activated carbon with different functional groups by molecular simulation. Environ. Sci. Pollut. Res. **30**, 32452–32463 (2023). https://doi.org/10.1007/s11356-022-24313-w

Wickrama-Arachchige, A.U.-K., Guruge, K.S., Inagaki, Y., Tani, H., Dharmaratne, T.S., Niizuma, Y., Ohura, T.: Halogenated polycyclic aromatic hydrocarbons in edible aquatic species of two Asian countries: congener profiles, biomagnification, and human risk assessment. Food Chem. **360**, 130072 (2021). https://doi.org/10.1016/j.foodchem.2021.130072

Xia, X., Xia, N., Lai, Y., Dong, J., Zhao, P., Zhu, B., Li, Z., Ye, W., Yuan, Y., Huang, J.: Response of PAH-degrading genes to PAH bioavailability in the overlying water, suspended sediment, and deposited sediment of the Yangtze River. Chemosphere **128**, 236–244 (2015). https://doi.org/10.1016/j.chemosphere.2015.02.011

Yao, Z., Li, J., Wu, B., et al.: Characteristics of PAHs from deep-frying and frying cooking fumes. Environ. Sci. Pollut. Res. **22**, 16110–16120 (2015). https://doi.org/10.1007/s11356-015-4837-4

Zhang, Q., Liu, P., Li, S., Zhang, X., Chen, M.: Progress in the analytical research methods of polycyclic aromatic hydrocarbons (PAHs). J. Liq. Chromatogr. Relat. Technol. **43**(13–14), 425–444 (2020). https://doi.org/10.1080/10826076.2020.1746668

Physical, Chemical and Biological Properties of Polycyclic Aromatic Hydrocarbons

2

Shreya Kotnala, Shalini Tiwari, Richa Saxena, Rashmi Verma, and Mohd Amin Mir

Abstract

Polycyclic aromatic hydrocarbons (PAHs) constitute a significant group of persistent organic pollutants to environmental and human health. PAHs are a family of chemicals with a wide range of natural and anthropogenic origins (i.e., fossil fuel combustion, industrial emissions, and wildfires) that are defined by their hydrophobicity, poor water solubility, and highly binding organic matter. The characteristics enhance their extensive dispersion in air, water and soil and cause their persistence in the environment. Atmospheric transport, deposition and adsorption to particulate matter and sediments operate to regulate the environmental fate of PAHs. Although the degradation of PAHs is enhanced with the involvement of photochemistry, chemicals, and microbes, the high-molecular-weight PAHs are highly resistant to any breakdown leading to persistent contamination. PAHs remain of primary concern to both ecological and human health because of their toxicity, potential to cause cancer and bio accumulative nature. This chapter focuses on the physical, chemical, and

S. Kotnala · S. Tiwari (✉)
Department of Chemistry, School of Basic and Applied Sciences, Shri Guru Ram Rai University, Dehradun, Uttarakhand, India
e-mail: tiwari.shalini100693@gmail.com

R. Saxena
Department of Physics, School of Basic and Applied Sciences, Shri Guru Ram Rai University, Dehradun, Uttarakhand, India

R. Verma
Department of Biotechnology, School of Basic and Applied Sciences, Shri Guru Ram Rai University, Dehradun, Uttarakhand, India

M. A. Mir
Department of Mechanical Engineering, Prince Mohammad Bin Fahd University, Al-Khobar, Saudi Arabia

H. Gupta et al. (eds.), *Ubiquitous Polycyclic Aromatic Hydrocarbon Contamination*, Emerging Contaminants and Associated Treatment Technologies,
https://doi.org/10.1007/978-3-032-11764-9_2

biological character of PAHs and their significance to the environment with the emphasis on the necessity of sophisticated remediation and monitoring solutions that could control the effects of PAHs.

Keywords

Polycyclic aromatic hydrocarbons (PAHs) • Toxicity • Biodegradation • Environmental pollution • Remediation strategies • Persistent organic pollutants

Introduction

One of the major classes of organic compounds which is made up of carbon and hydrogen is polycyclic aromatic hydrocarbons (PAHs) in which these atoms are arranged two or more fused aromatic rings (Gupta and Gupta 2016). They possess distinct ring structure and do not have any heteroatom or functional group due to which these are usually hydrophobic, chemically stable and persistent in environment (Krzyszczak and Czech 2021). On the basis of number of rings, PAHs are classified into two classes: (a) Light PAHs (having two to four rings) and (b) Heavy PAHs (having more than four rings) (Abdel-Shafy and Mansour 2016). Heavy PAHs are more hazardous, less volatile and more stable in comparison to light PAHs (Kuppusamy et al. 2016; Li et al. 2016).

These chemicals are widely present in the environment and the major cause is the incomplete burning of organic resources (such as coal, oil, gas, wood and trash). Natural processes such as volcanic eruptions and forest fires also produce PAHs thereby making them common pollutant that is widely found in air, water and soil (Gupta 2018; Gupta and Gupta 2018; Sakshi et al. 2019; Mallah et al. 2022; Gupta et al. 2025). Various natural and anthropogenic sources of PAHs are depicted in Fig. 2.1. On account of their low water solubility, PAHs tends to bond with sediment and organic matter in the aquatic and the terrestrial environment but these may also exist in vapor phase or be adsorbed onto the particulate matter present in the atmosphere (Kumar et al. 2016). Rather than existing as individual compounds, PAHs are frequently found as complex mixtures that are usually found in soot, tar, crude oil and industrial waste. From the toxicological point of view, PAHs are considered to be carcinogenic, mutagenic and immunosuppressive in nature (Guo et al. 2025; Montano et al. 2025). They may cause major health problems by interfering with enzyme activity and cellular membrane processes. Immunological system negative consequences, such as compromised immunological development and decreased host tolerance, have been linked to exposure to PAHs.

The United State Environmental Protection Agency (USEPA) has designated sixteen PAHs as priority pollutants because of their negative effects on the environment as well as human health (Mojiri et al. 2019). Furthermore, 7 of these PAHs namely, benz[a]anthracene, chrysene, benzo[b]fluoranthene, benzo[k]fluoranthene,

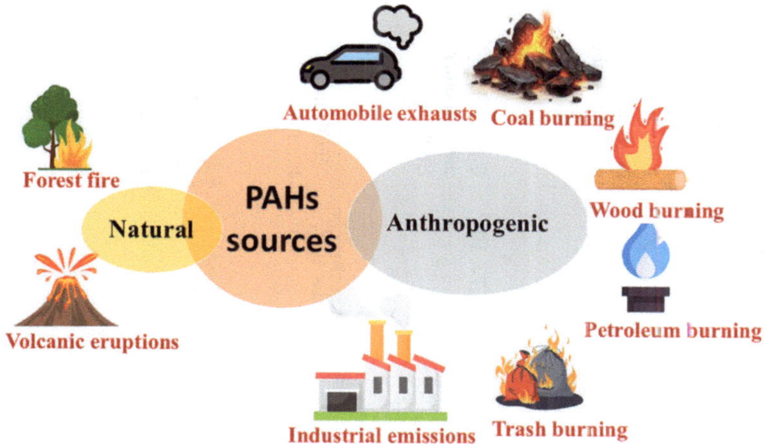

Fig. 2.1 Various sources of PAHs

benzo[a]pyrene, indenol[1,2,2-cd]pyrene, and dibenzo[a,h]anthracene have been designated as carcinogenic by the International Agency for Research on Cancer (Zheng et al. 2016). Benzo[a]pyrene has been designated as human carcinogen due to its genotoxic activity (Sombiri et al. 2024). Various physical and chemical characteristics of PAHs like high melting and boiling point is known to affect their toxicity. The sixteen priority pollutants as per USEPA and their physical and chemical characteristics are summarized in Table 2.1.

Major source of PAHs is the pyrolysis or incomplete combustion of organic materials like coal, petroleum and wood (at temperature ranging from 350 °C to 1200 °C) (Masih et al. 2010; Montano et al. 2025). Repolymerization of hydrocarbon fragments produced during cracking is thought to be a complex step in their production. Natural as well as anthropogenic activities result in their production. Usually, concentration of PAHs in urban areas is high when compared to rural ones (Çabuk et al. 2014; Paris et al. 2018). PAHs can be eliminated from the environment via dry as well as wet depositional mechanisms. Precipitation is involved in wet deposition mechanism for the removal of PAHs whereas gravity settling takes place in dry removal process (Esen et al. 2010; Montano et al. 2025).

PAHs enter and progress via food chain once ingested by the plants and animals. The PAHs are usually transferred from lower to higher trophic levels through the food chain (Sombiri et al. 2024; Wang et al. 2025). PAHs are persistent in nature but they can be decomposed by microbial detraction and photodegradation (Premnath et al. 2021; Liu et al. 2023). Microorganisms can metabolize these compounds in soil (K and Manian 2023).

This chapter gives a detailed insight about PAHs mainly focussing on their environmental impact. Various physical and chemical properties that are important for understanding the nature, environmental behaviour and their interaction with other medias in the environment are discussed in the chapter. This chapter

Table 2.1 List of priority pollutants as per USEPA and their physical and chemical properties (Sverdrup et al. 2002; Balint 2021)

PAH	Structure	Formula	Molecular weight g/mol	Melting point °C	Boiling point °C	Log K_{ow}	Vapour pressure (mm Hg at 25 °C)	Solubility
Acenaphthene	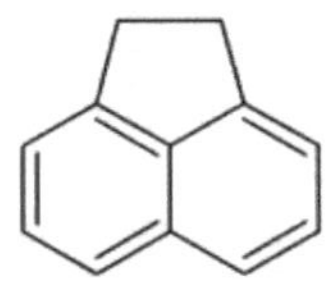	$C_{12}H_{10}$	154.2	93.4 °C	279 °C	4.32	2.2×10^{-3}	Anhydrous acid benzene, $CHCl_3$, petroleum ether, and toluene
Acenaphthylene	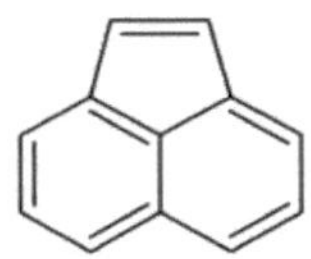	$C_{12}H_8$	152.2	78–82 °C	280 °C	4.08	6.7×10^{-3}	C_6H_6, $CHCl_3$, dietheyl ether, and EtOH
Anthracene	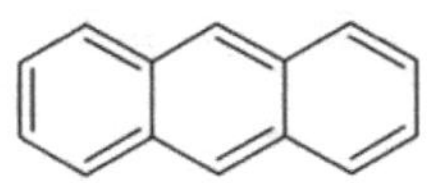	$C_{14}H_{10}$	178.2	218 °C	340 °C	4.45	6×10^{-6}	MeOH and hexane
Benz(a)anthracene	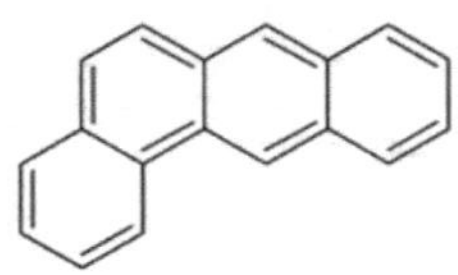	$C_{18}H_{12}$	228.3	158 °C	438 °C	5.6	2.1×10^{-7}	Diethyl ether, 0.00001 g/100 mL in water and acetone

(continued)

Table 2.1 (continued)

PAH	Structure	Formula	Molecular weight g/mol	Melting point °C	Boiling point °C	Log K_{ow}	Vapour pressure (mm Hg at 25 °C)	Solubility
Benzo(a)pyrene	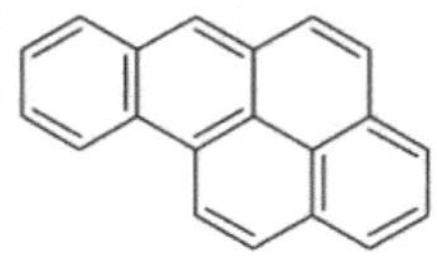	$C_{20}H_{12}$	252.3	179 °C	495 °C	6	5.6×10^{-9}	Benzene, toluene, xylene, and sparingly soluble in alcohol
Benzo(b)fluoranthene	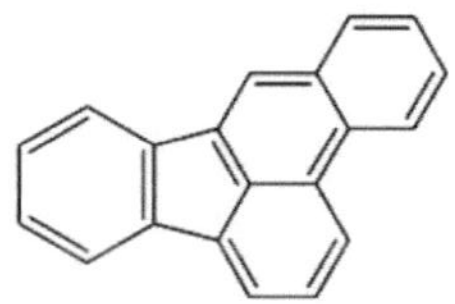	$C_{20}H_{12}$	252.3	168 °C	228.6 °C	6.6	-	Acetone, alcohol, and C_6H_6
Benzo(g,h,i) perylene	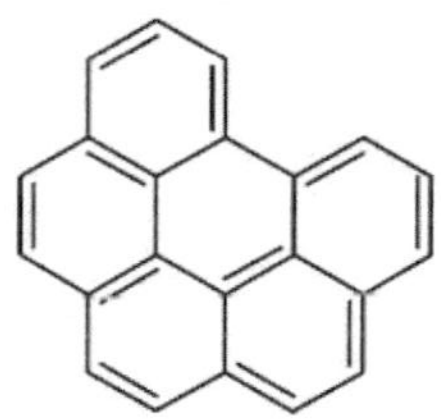	$C_{22}H_{12}$	276.3	278 °C	500 °C	7	1.01×10^{-10}	Acetone, dichloromethane, and 1,4 dioxane

(continued)

Table 2.1 (continued)

PAH	Structure	Formula	Molecular weight g/mol	Melting point °C	Boiling point °C	Log K_{ow}	Vapour pressure (mm Hg at 25 °C)	Solubility
Benzo(k)fluoranthene	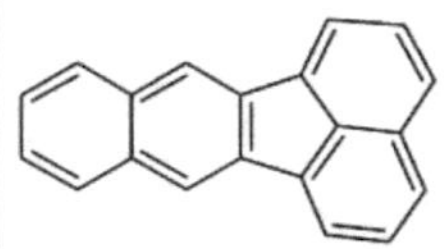	$C_{20}H_{12}$	252.3	217 °C	228.6 °C	6.85	9.6×10^{-11}	Benzene
Chrysene	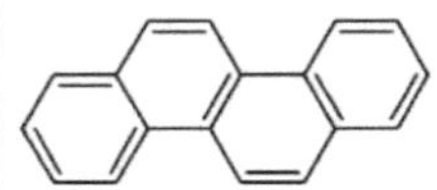	$C_{18}H_{12}$	228.3	254 °C	448 °C	5.6	6.4×10^{-9}	Ethanol
Dibenzo(a,h)anthracene	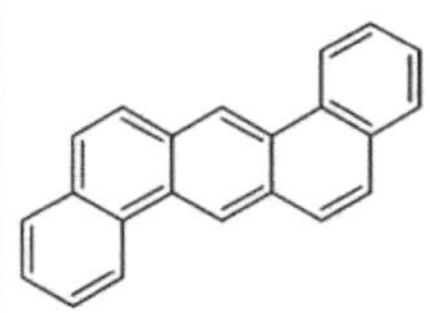	$C_{22}H_{14}$	278.3	262 °C	524 °C	6	-	Benzene, ether, xylene, and slightly soluble in alcohol and ether
Fluoranthene	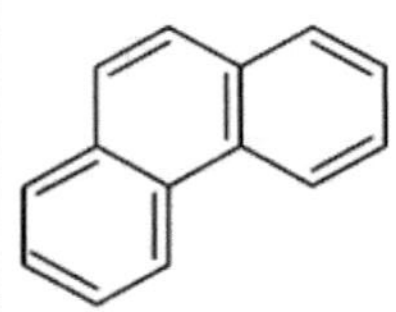	$C_{16}H_{10}$	202.26	375 °C	110.8 °C	5.53	9.2×10^{-4}	Acetic acid, hot alcohol, C_6H_6, ethanol, ethyl ether

(continued)

Table 2.1 (continued)

PAH	Structure	Formula	Molecular weight g/mol	Melting point °C	Boiling point °C	Log K_{ow}	Vapour pressure (mm Hg at 25 °C)	Solubility
Fluorene	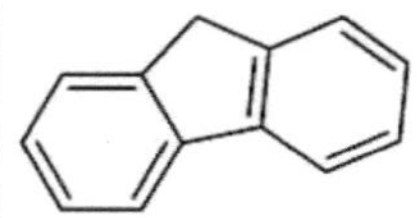	$C_{13}H_{10}$	166.223	295 °C	116–117 °C	4.18	6×10^{-4}	C_6H_6 and ether
Indeno [1,2,3-cd]pyrene	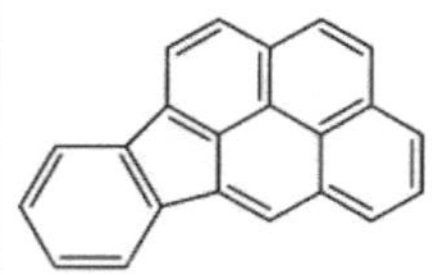	$C_{22}H_{12}$	276.337	162–164 °C	497.101 °C	7.7	–	Benzene
Naphthalene	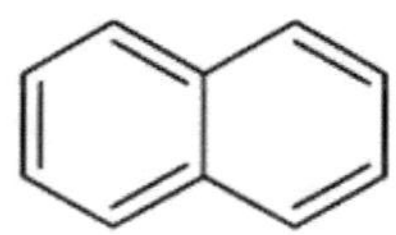	$C_{10}H_8$	128.17	218 °C	80.26 °C	3.36	7.1×10^{-2}	Very soluble in CS_2, $CHCl_3$, and ether; less soluble in C_2H_5OH/CH_3OH; and insoluble in water

(continued)

Table 2.1 (continued)

PAH	Structure	Formula	Molecular weight g/mol	Melting point °C	Boiling point °C	Log K_{ow}	Vapour pressure (mm Hg at 25 °C)	Solubility
Phenanthrene	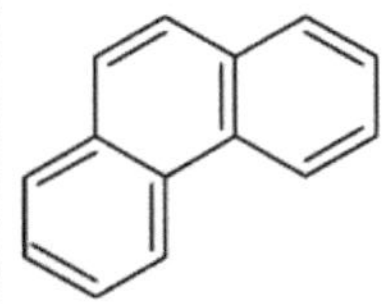	$C_{14}H_{10}$	178.23	101 °C	332 °C	4.46	1.2×10^{-4}	Insoluble in water Soluble in most organic solvents, such as CH_3COOH, C_6H_6, $CHCl_3$, CCl_4, ether, and toluene
Pyrene	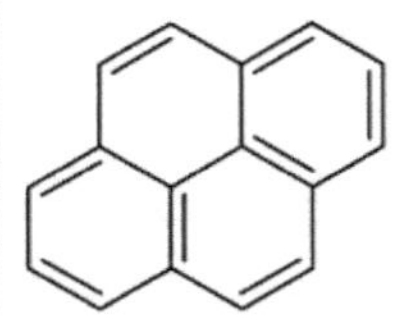	$C_{16}H_{10}$	202.25	145–148 °C	404 °C	5.3	4.5×10^{-6}	Soluble in benzene, ethanol, and ether

further explores the environmental fate and transport of PAHs thereby highlighting the mechanism by which these chemicals persist and move in air, water and soil environment. In addition to this, this chapter also addresses various degradation pathways by which PAHs can be broken down in the environment. Biological impact and toxicity of PAHs is discussed in detail thereby highlighting their carcinogenic, mutagenic and teratogenic effects on living organism. Furthermore, various remediation strategies such as physical, chemical and biological are discussed in detail to mitigate adverse effect of these chemicals on the environment as well as on the human health. The key findings and future perspectives are discussed in the conclusion section thereby emphasizing on the need for the sustainable development and effective methodologies for remediation of the PAHs contaminated environment.

Physical and Chemical Properties

PAHs that are composed of two or more than two fused rings are organic compounds which are usually produced by incomplete combustion of organic resources (like coal, biomass, petroleum products and fossil fuels). In addition to this, they are also found in landfill leachates, exhaust of automobiles, pollutants released from industries and from the kitchen fumes. PAHs are generally produced via both anthropogenic as well as natural processes therefore they are frequently detected in environment (such as soil, water, and air). Generally pure PAHs are white, colourless or pale-yellow solids. They are lipophilic that means they are readily soluble in non-polar solvents such as fats and oils rather than in polar solvents such as water (Sahoo et al. 2020).

Molecular weight and structure of PAHs are known to exhibit substantial impact on the physical and chemical characteristics (Kumar et al. 2016). Low molecular weight PAHs like fluorene and naphthalene consisting of two or three aromatic rings are more volatile and possess high water solubility in comparison to high molecular weight PAHs like benzo[a]pyrene and fluoranthene that possess four or more rings in their structure (Gupta and Gupta 2015). PAHs having low molecular weight are known to exhibit higher vapour pressure in comparison to the ones having high molecular weight. Also, the former ones are less persistent and carcinogenic in comparison to the later ones. PAHs having high molecular weight exhibit less water solubility as well as less vapour pressure (Vikas et al. 2015; Kumar et al. 2016). On account of these characteristics high molecular weight PAHs have high correlation with organic-rich soils, particulate matter, and sediments. It is usually found that with the increase in number of aromatic rings, the vapour pressure sharply decreases.

PAHs usually have high octanol-water partition coefficient (Kow) because they are lipophilic and they exhibit tendency to accumulate in the biological systems (Balint 2021). High soil/sediment partition coefficient (Koc) values indicates that PAH have a great affinity for organic materials found in soils and sediments (Balint 2021). Based on Henry's law constant, these compounds can easily volatize from moist soils and on the basis of vapor pressure, these compounds can easily volatize from dry soils. PAHs present in the soil may undergo microbial breakdown in aerobic conditions but due to their toxicity to microorganisms and slower rate of biological degradation, they usually persist in anaerobic environments (Balint 2021).

The behaviour of PAHs in aquatic media is usually influenced by their density and water solubility. High molecular weight PAHs settle in sediments or mat combine with dissolved organic matter whereas the low molecular weight PAHs are customarily remains dissolved in water (Rabodonirina et al. 2015; Maletić et al. 2019). Some PAHs may sink to the aquifers or may float on the water surface whereas others may volatize into the atmosphere or may remain dissolve based on the density. The air is known to contain huge amount of PAHs which may exist either in gaseous form (lighter PAHs) or may remain suspended to particulate matter (usually heavier PAHs) (Rabodonirina et al. 2015; Maletić et al. 2019). All these substances have the ability to travel both short and long distances in the atmosphere before depositing themselves on the surfaces of soil and water through dry deposition or precipitation (Bozlaker et al. 2008).

Because of their chemical stability, PAHs are categorized as persistent organic pollutants (POPs) (Boethling et al. 2009; Akhtar et al. 2021). Various physical and chemical processes like oxidation, microbial breakdown, and photodegradation are responsible for redistribution and transformation of PAHs in the environment. They pose serious threat to environment and human health on account of their teratogenic, mutagenic, and carcinogenic properties (Drwal et al. 2019; Mallah et al. 2022). Human exposure to PAHs may occur due to ingestion of contaminated food or water, inhalation of contaminated air, dermal contact with contaminated soil or dust particles (Montano et al. 2025). It is evident that the exposure is usually higher in the urban as well as industrial areas due to following activities: (a) hazardous emissions from the traffic, (b) discharge of harmful contaminants from industries and (c) burning of fossil fuels. All these activities result in enhanced concentration of PAHs in the environment. Table 2.2 presents a detailed summary of various physical and chemical characteristics of PAHs.

Due to complex behaviour and diverse physical properties of PAHs, their sampling, extraction and measurement from the environmental samples require careful consideration. (Manousi and Zachariadis 2020; Sajid et al. 2021). Various properties such as vapour pressure, solubility in water and affinity towards particulate matter are known to affect the hazards of exposure, transportation and environmental fate of PAHs. Rapid industrialization, urbanization, population expansion and increased use of fossil fuels have contributed to the increased levels of PAHs in the environment in the developing countries of Asia (Bidleman et al. 2015; Kumar et al. 2016). Thus, it is imperative to know about the physical and chemical characteristics of PAH for determining their effects on the environment, for devising cleanup activities and also for formulating legal requirements for protection of environment as well as public health.

Table 2.2 Physical and chemical properties of PAHs

Property	Description	Reference
Physical state	Pure PAHs are generally white, colourless, or pale yellow-green solids	(Sahoo et al. 2020)
Solubility	Lipophilic in nature; soluble in non-polar solvents (fats, oils); poorly soluble in water	(Sahoo et al. 2020)
Volatility	Low molecular weight PAHs have high vapor pressure and are more volatile; high molecular weight PAHs have low volatility	(Kumar et al. 2016)
Water solubility	Decreases with increasing molecular weight; 2–3 ring PAHs are more water-soluble than those with ≥ 4 rings	(Vikas et al. 2015)
Octanol–water partition Coefficient (Kow)	High Kow indicates strong tendency to bioaccumulate in organisms	(Balint 2021)
Soil/sediment partition coefficient (Koc)	High Koc values reflect strong adsorption to soil and sediment organic matter	(Balint 2021)
Vapor pressure	Decreases significantly with increase in number of aromatic rings	(Kumar et al. 2016)
Environmental phase distribution	LMW PAHs occur in gaseous phase; HMW PAHs associate with particulate matter (solid phase)	(Maletić et al. 2019)
Chemical stability	Chemically stable and resistant to degradation, leading to environmental persistence	(Akhtar et al. 2021)
Photodegradability and biodegradability	Undergo transformation via photodegradation, oxidation, and microbial action, though slower in anaerobic conditions	(Drwal et al. 2019)

Environmental Fate and Transport

PAHs in the environment follow three main factors consisting of their physical and chemical properties combined with source methods and their reactions with ecological living and non-living substances. The distribution and transformation along with degradation pathways which a compound goes through upon environmental release affects both its survival time and total environmental impact (Eldos et al. 2022). The process of burning produces PAHs in atmospheric form as gases along with adsorption to particulate matter. This phenomenon manifests primarily from vehicle exhausts and emissions from industry and biomass fires as well as chemical combustion products of cigarettes. The gaseous part of lighter PAHs predominates in the atmosphere where they either react with sunlight or they evaporate toward remote locations. The heavy PAH compounds react to airborne particulates before they settle by dry and wet deposition which leads to soil and water contamination (Zhang et al. 2022). PAHs stay in soil sediments because they bond tightly to organic matter due to their oil-loving character. The substance adheres firmly to the environment although it restricts animal movement. Aquatic systems maintain PAHs as dissolved chemicals at low concentrations while these contaminants also stick to floating particles and settle down to the sediments. Soil or sediment organic carbon content and grain size determine how much of PAHs will be taken up (Patel et al. 2020).

PAHs experience their environmental fate through two vital degradation pathways consisting of living organisms and non-living environmental factors (depicted in Fig. 2.2). The degradation of biogenic substances by bacteria together with fungi establishes as a fundamental natural process for reduction. Many microorganisms take PAHs as their food source to generate energy under both aerobic and anaerobic environmental conditions (Tesfaye et al. 2025). Microbial breakdown of PAHs faces various barriers including insufficient nutrients and low oxygen concentrations together with the chemical makeup of these compounds. Low molecular weight PAHs break down faster than high molecular weight analogs because they possess simpler arrangements and better aqueous dissolution properties (Ghosal et al. 2016). Leaps in environmental science under abiotic degradation show that photo oxidation joins hydrolysis as well as atmospheric oxidant reactions including ozone, hydroxyl radicals and nitrogen oxides. The atmospheric degradation process of PAHs through photo oxidation leads to the formation of toxic oxygenated derivatives including quinones (Wang et al. 2023).

Pollutants known as PAHs spread beyond surface water into groundwater by means of leaching and runoff. PAHs survive in aquatic ecosystems where they pose extended risks to aquatic life and cause contaminants to build up in food chain

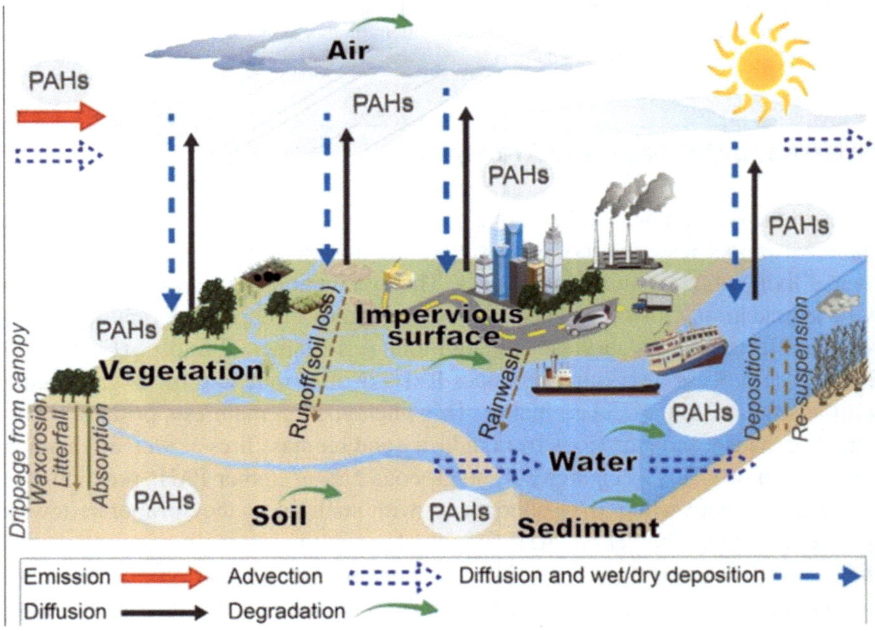

Fig. 2.2 Environmental fate of PAHs (Reprinted from original article by (Nie et al. 2024) after taking permission from Elsevier)

networks. The oil-and-water-loving characteristic of PAHs enables their accumulation in fatty matter of aquatic organisms that builds up toxin levels progressively through successive ecological steps (Singh et al. 2023).

PAHs undergo destiny transformations based on human-initiated processes such as land development changes and waste management operations and soil clean-up procedures. Three main avenues that lead to PAHs entering water bodies are landfill leachates together with wastewater discharges and urban runoff. The precise comprehension of how PAHs interact with the environment and travel through it provides the framework for effective measurement methods and risk evaluation programs together with clean-up solutions (Smol et al. 2017). The environmental distribution and behavioural pattern of PAHs can be predicted using both fugacity models and multimedia environmental fate models including Level III models. These models integrate data on PAH properties, emission sources, environmental parameters, and degradation rates to estimate concentrations and assess exposure risks (Balmer et al. 2019).

In conclusion, the environmental fate and transport of PAHs are determined by a complex interplay of their physical and chemical properties, environmental conditions, and biological interactions. Their persistence and potential for long-range transport necessitate comprehensive management strategies to mitigate ecological and health impacts.

Degradation and Persistence

PAHs exist with different environmental persistence levels due to their unique molecular structures along with the conditions around them. Environmental degradation and ecological risk assessment depend fundamentally on how fast PAHs can degrade in their environmental setting. The environmental behaviour of PAHs depends on three main degradation mechanisms together with sorption and persistence through half-life as well as influential environmental conditions. The Table 2.3.3 summarises the outlines of these essential degradation mechanisms and persistence factors.

Persistence in the Environment: The environmental persistence of PAHs with large molecular weight (HMW) occurs through their stable aromatic bonding structure. These compounds demonstrate high resistance to decay and they preferentially absorb into soil materials as well as underwater sediments and organic substances. Because of their very low water dissolubility and hydrophobic nature these substances stay tightly bound to solid matter yet they cannot move effectively through water. This ability leads to their extended presence in the environment since they continue as long-term contaminants (Yemele et al. 2024). To illustrate an example, a well-known HMW PAH, benzo[a]pyrene, demonstrates half-lives of 0.3–58 years (or more) in anaerobic sediments, depending on the environmental conditions (Abdel-Shafy and Mansour 2016). On the same-way, phenanthrene, a medium molecular weight PAH has soil half-lives of 2.5 days to 5.7 years based

Table 2.3.3 Overview of degradation pathways and environmental persistence of PAHs

Feature	Details	Key compounds	Conditions	Impact	References
Environmental persistence	HMW PAHs are stable and hydrophobic, leading to long-term retention in soil and sediments	HMW PAHs (e.g., benzo[a]pyrene)	Soil, sediments	Persistent and less mobile	Yemele et al. (2024)
Biodegradation	LMW PAHs degrade more easily via microbes, especially under aerobic conditions	LMW PAHs	Aerobic with nutrients	Effective for LMW; slow for HMW	Haritash and Kaushik (2009), Tesfaye et al. (2025)
Photo degradation	UV/visible light breaks down PAHs in air/water, sometimes increasing toxicity	Mainly LMW PAHs	Surface water, air	Can form toxic byproducts	Abdel-Shafy and Mansour (2016)
Chemical degradation	PAHs react with oxidants like ozone and Fenton's reagent, degrading quickly	Various PAHs	Chemical oxidants present	Fast but may form harmful intermediates	Kumari and Kumar (2023)
Hydrolysis	PAHs resist hydrolysis due to lack of suitable functional groups	All PAHs	Water	Minimal degradation through hydrolysis	Winquist et al. (2014)
Sorption and aging	Strongly bind to soil/organic matter, reducing mobility and bioavailability	All PAHs	Organic-rich soils	Persistent and less available to microbes	Ukalska-Jaruga and Smreczak (2020)
Environmental half-life	Degradation varies: days for LMW (e.g., naphthalene), > 1 year for HMW (e.g., benzo[a]pyrene)	Naphthalene, Benzo[a]pyrene	Varies with environment	Half-life ranges from days to over a year	Ukalska-Jaruga and Smreczak (2020)

on the condition of microbial activity and geocomposition of soil (Haritash and Kaushik 2009). The results of field studies indicated that total PAH concentrations reduced at an average of 2–6% in some contaminated soils per year after the rapid decay period (Chi et al. 2024). Such statistics highlight the long-term nature of HMW PAHs and their potential ecological danger, which makes them a major health risk of soil and sediment remediation.

Biodegradation: PAHs undergo natural decomposition thanks to bacteria and fungi microorganisms that perform biodegradation. Mineralization of naphthalene along with phenanthrene occurs more easily than HMW compounds due to their LMW composition. Aerobic biodegradation works better than anaerobic biodegradation because microorganisms use enzymes to add oxygen which produces ring cleavage resulting in mineralization. Microbes under anaerobic environments degrade only simple PAH compounds at a slower pace. The speed of microbial PAH metabolism depends heavily on oxygen availability combined with pH levels and temperatures together with accessible nutrients and specialized PAH-breaking microorganisms existing in the environment (Haritash and Kaushik 2009; Tesfaye et al. 2025). As an example of such variations, naphthalene has half-lives of 2–110 days and phenanthrene can take days to years in soil depending on the nature of the soil (Chi et al. 2024). Aerobic biodegradation occurs much faster than anaerobic, with specialized organisms such as Pseudomonas aeruginosa degrading more than 80–90% of naphthalene and phenanthrene in 10 days (Adebusoye et al. 2007; Karimi et al. 2015). Rates are also dependent on temperature and nutrients, with one study finding an increase in the degradation of naphthalene to 26% at 25 °C, and 98% at 37 °C. The anaerobic degradation is slower and is confined to simpler PAHs, with rates of ~ 0.23 to 0.65 μmol kg^{-1} day^{-1} under sulfate-reducing conditions. Altogether, biodegradation mainly depends on oxygen, temperature, and microbial specialisation requirements (Lawal 2017).

Photo degradation: The absorption of ultraviolet (UV) or visible light by PAHs causes their chemical transformation through the process called photodegradation. Light intensity conditions in atmospheric and surface water environments make this reaction process important. The photochemical reactions of PAHs either on aerosols or water particles generate oxy-PAHs which demonstrate higher toxicity in addition to greater reactivity than their parent constituent compounds. Photodegradation breaks down PAHs in open environments however its rate depends on how much light penetrates and what photosensitizing agents exist (Abdel-Shafy and Mansour 2016). Photosensitizers such as dissolved organic matter and nitrate ions may catalyze degradation of PAHs in aquatic environments by forming reactive forms such as singlet oxygen or hydroxyl radicals (Yang et al. 2024). Experiments revealed that the half-life of benzo[a]pyrene varies between 1.5 and 3.5 h in sunlit surface waters and that the maintains much longer in low-light or turbid waters. Notably, the photochemical chemical reaction of PAHs commonly yields oxygenated PAH (oxy-PAH) which are more water-soluble than their corresponding parent compounds and have potentiated toxicological properties (Marquès et al. 2017). Thus, on one hand, photodegradation as a process contributes to lowering the concentration of PAH in open environments; on the other hand, it presents the potential to emit more reactive and less safe by-products, which determines the importance of sophisticated environmental monitoring (Tesfaye et al. 2025).

Chemical Degradation: PAHs can experience abiotic chemical breakdown following their reaction with strong oxidants such as ozone and hydroxyl radicals (·OH) and reactive oxygen species. Engineered processes that include Fenton oxidation or ozonation and atmospheric environments show regular occurrence of these reactions. The PAH concentration reduction by chemical degradation causes the formation of toxic reaction products that need monitoring in remediation systems (Kumari and Kumar 2023). As an example, research indicates that up to 90% of phenanthrene and 80% of benzo[a]pyrene can be removed through ozonation in 60 min, depending on the pH and the amount of oxidant (Kumari and Kumar 2023). Similarly, Fentons reagent has been shown to extract 85–95% anthracene and Pyrene under optimal conditions (pH ~ 3.5 and Fe^{2+}: $H\ {}_2O_2$ ratio of 1:10) (Das et al. 2025). During the process of such chemical reactions though, intermediate or by-product compounds, such as quinones and epoxides are frequently created and are sometimes more toxic than the parent PAHs. Therefore, the PAH remediation systems require close monitoring and secondary treatment to guarantee that the toxic transformation products have been handled safely and depleted (Idowu et al. 2019).

Hydrolysis: The inadequate number of functional groups in PAH molecules makes hydrolytic degradation an unimportant pathway for their breakdown. The resistance to aqueous solution breakdown through hydrolysis exists because most PAHs contain only stable carbon-carbon bonds which lack hydrolysable functional groups. Hydrolysis operates at virtually none of the levels involved in the environmental transformation processes (Winquist et al. 2014). The resistance to PAH such as naphthalene, anthracene, and benzo[a]pyrene to hydrolysis demonstrated that the half-lives of the latter are greater than 200 years in the neutral pH aquatic system (Lawal 2017). In addition, it is reported that the hydrolysis rate constants of most of the PAHs is $< 10^{-9}\ s^{-1}$ and is taken as insignificant in the determination of environmental fate. Therefore, hydrolysis is not accounted as an important mechanism of PAH degradation under the usual environmental influence and the role of this process inducing the PAH transformations in the soil and water matrices is insignificant in practice (Gundlapalli et al. 2024).

Sorption and Aging Effects: The properties of PAHs regarding their environmental appearance and bioavailability are largely affected by sorption and aging. With time, PAHs become more tightly bound to the particles of soil and sediment, particularly in organic rich matrices, decreasing extractability and access to microbial degradation. (Srogi 2007; Ukalska-Jaruga and Smreczak 2020) describe the persistence of PAHs in aged soils: up to 70–90% of total PAHs can become non-extractable because of assembly to soil organic matter and micropores. As a result of this process, the bioavailability and apparent toxicity of the compounds are lowered, but similar compounds may still present threats in the long term to the environment (Ukalska-Jaruga and Smreczak 2020). Investigation has demonstrated that ageing may decrease the biodegradation rate of PAHs by up to 80% after a year, depending on the soil type and abiotic conditions. Microbial/chemical access is limited by the process of physical trapping in soil micropores, as well as by sorption to humic substances. Although it causes an immediate decrease in ecological threat, ageing PAHs can be washed out of the soil by changes in the environment (e.g. flooding, soil erosions, pH- alterations) refreshing their toxic effect and making post-treatment challenging (Kong et al. 2021).

Environmental Half-Lives: PAH residence times in the environment span widely across compounds and environmental factors affect them differently. LMW (PAHs typically break down within weeks to days when environmental conditions are optimal yet HMW PAHs persist in the environment for hundreds of days and longer durations. Naphthalene decomposes within ten days but benzopyrene exists in soil past one year. Assessing and remediating PAH contamination requires site-specific evaluations because of these differentiating features (Ukalska-Jaruga and Smreczak 2020). As an example, benzo[a]pyrene has a lifespan in the soil of over a year, particularly in anaerobic or under-nutrified environments. This major difference requires location-specific risk characterization in an assessment of the level of contamination and the choice of remediation technologies since degradation rates are grossly dependent (Sushkova et al. 2016).

For a better understanding about the degradation and persistence behaviours of various PAHs under various environmental conditions, the statistical data from previously reported studies have been compiled and is summarised in Table 2.4.

Table 2.4 Comparative analysis of degradation rates, half-lives, and removal efficiencies of selected polycyclic aromatic hydrocarbons (PAHs)

PAH Compound	Degradation method	Reported half-life/removal efficiency	Reference
Naphthalene	Biodegradation (aerobic)	2–110 days	Haritash and Kaushik (2009)
Phenanthrene	Biodegradation (aerobic)	Few days—years	Chi et al. (2024)
Benzo[a]pyrene	Photodegradation	1.5–3.5 h (surface water)	Marquès et al. (2017)
Benzo[a]pyrene	Anaerobic degradation	0.3–58 years (sediment)	Abdel-Shafy and Mansour (2016)
Anthracene	Chemical (Fenton)	85–95% removal (pH ~ 3.5, Fe^{2+}:H_2O_2 = 1:10)	Das et al. (2025)
Phenanthrene	Chemical (Ozonation)	90% in 60 min (pH-dependent)	Kumari and Kumar (2023)

Biological Impact and Toxicity

Rapid industrialization has led to significant contamination of the environment with PAHs emerging as major pollutants on account of their toxicity, mutagenicity and carcinogenicity (Patel et al. 2020; Vijayanand et al. 2023; Suresh et al. 2025). Among all the PAHs, benzo[a]pyrene (BaP) is known to exhibit the highest carcinogenicity and is also classified as a probable human carcinogen (Soares and Botelho 2025). The USEPA has classified seven PAH as probable human carcinogen including benz(a)anthracene, benzo(a)pyrene, benzo(b)fluoranthene, benzo(k) fluoranthene, chrysene, dibenz(ah)anthracene, and indeno(1,2,3-cd) pyrene (Lawal 2017). Through sewage sludge and industrial waste, PAHs infiltrate ecosystems, endangering crops, soil organisms, and human health (Mohanty et al. 2025). Exposure can happen through skin contact, ingestion of tainted food (such grilled meats), or inhalation. Cytochrome P450 enzymes (CYP1A1/1B1) metabolically activate PAHs in the human body, producing reactive intermediates such as o-quinones and diol epoxides that bind DNA, produce adducts, and start the carcinogenesis process (Moorthy et al. 2015). Particularly in urban children, PAHs are linked to respiratory disorders like asthma, liver damage, developmental abnormalities, and lung and oesophageal malignancies (Låg et al. 2020). Despite not being DNA-reactive by nature, PAHs and their metabolites can cause genotoxicity (Montano et al. 2025). PAHs are hence important environmental carcinogens that have a broad biological impact. The mutagenic, carcinogenic, and teratogenic properties of PAHs make them priority toxicants and common environmental contaminants. Exposure to humans is almost inevitable and is associated with bronchitis, lung impairment, and respiratory problems (Utembe and Kamng'ona 2024).

In addition to the particular PAH molecule and individual susceptibility, the method, duration, and degree of exposure all affect how harmful PAHs are. Skin inflammation, nausea, and irritated eyes have all been linked to short-term exposure to PAH combinations (Mallah et al. 2022). Certain PAHs, such as benzo[a]pyrene and anthracene, cause skin irritation and sensitization (Ifegwu and Anyakora 2015). Serious health problems include immunosuppression, respiratory illnesses, liver and kidney damage, and carcinogenesis can arise from prolonged exposure (Vandana et al. 2022). Long-term environmental and occupational exposure has been associated with bladder, gastrointestinal, lung, and skin malignancies (Yari et al. 2018).

Although reactive metabolites of PAHs, such as diol epoxides, bind to DNA and cause mutations including $G > T$ transversions, frameshifts, deletions, and chromosomal abnormalities, PAHs themselves are not genotoxic (Ewa and Danuta 2017). The development of cancer is significantly influenced by these genetic damages. Additionally, smokers and the cord blood of foetuses exposed during pregnancy have been shown to create DNA adducts (Fucic et al. 2017). Another major worry is immunotoxicity, as PAHs impair immunological responses and make people more susceptible to infections and malignancies (Yu et al. 2022). Both immune suppression and potentiation, frequently due to the action of reactive metabolites, might happen depending on the exposure conditions (Cho and Uetrecht 2017).

Low birth weight, early delivery, heart problems, cognitive impairments, and childhood asthma are examples of developmental and embryotoxic consequences. Both human epidemiological data and animal studies have validated these results (Celik Atalay et al. 2025).

Remediation Strategies

The implementation strategy selection for remediating environments contaminated with PAHs depends strongly on properties including physical, chemical, and biological characteristics of the compounds. The remediation of PAHs demands specialized procedures because the compounds are hydrophobic along with their toxic and persistent nature. Various remediation strategies are depicted in Fig. 2.3.

Physical Remediation Strategies

Physical remediation methods leverage the physical properties of PAHs such as low water solubility, volatility (especially for LMW PAHs), and strong sorption to soil and sediments. Soil washing integrates surfactants or solvents into the process to extract PAHs from solid materials thus allowing their removal (Patel et al. 2020). Thermal desorption uses heat to evaporate PAHs from polluted soils which successfully captures both LMW and HMW PAHs. The method of soil vapor

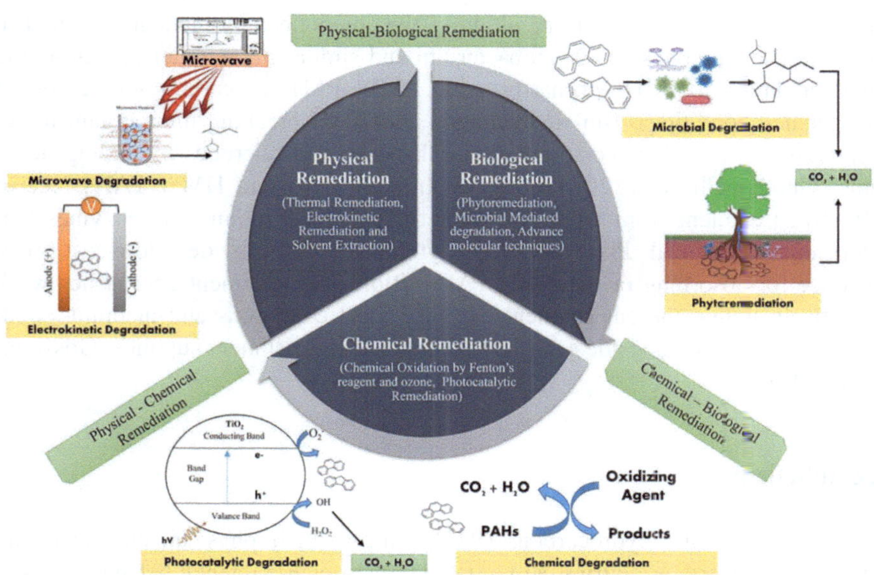

Fig. 2.3 Various remediation strategies for PAHs (Reprinted from original article by (Gupta et al. 2024) after taking permission from Elsevier)

extraction shows great results when extracting volatile and semi-volatile PAHs that exist in the vadose zone. Soil and sediment remediation by these methods does not work effectively for deeply bound compounds and non-volatile substances because their removal warrants chemical and biological treatment approaches (Liu et al. 2022).

Chemical Remediation Strategies

The chemical remediation process uses chemical reactivity's to convert PAHs into either less dangerous substances or compounds which microorganisms can break down. The chemical treatment methods of Fenton's reagent, ozonation and persulfate oxidation generate powerful oxidizing species (hyroxyl radicals) suitable for destroying PAHs by breaking their aromatic rings (Kumari and Kumar 2023). The treatment methods are effective against high molecular weight PAHs when those compounds show resistance to natural degradation processes. UV light photo degradation in combination with titanium dioxide (TiO_2) catalysts proves effective for breaking down contaminants in aquatic settings. Termed efficient but such procedures generate additional by-products that require special attention together with cost considerations (Eldos et al. 2022).

Biological Remediation Strategies

The process of biological remediation produces metabolic changes for PAHs using microorganisms and plants to break down the compounds by biodegradation. The bacteria Pseudomonas and Mycobacterium and similar microorganisms allow the biodegradation of PAHs primarily using LMW PAHs as carbon sources under oxygenated conditions. Bio-augmentation adds PAH-breaking microorganisms to contaminated sites but bio-stimulation enhances local microbes by adding nutrients (Gundlapalli et al. 2024). Overwhelmingly persistent HMW PAHs receive effective treatment from white-rot fungi through their lignin lytic enzymes that work outside the cell. Plants used in phytoremediation provide a low-cost green practice for absorbing or degrading PAHs. Biological treatment approaches work effectively within the environment but need ideal conditions and medium-length treatment periods alongside sufficient oxygen and nourishment supplies (Goswami et al. 2018).

Conclusion

Polycyclic aromatic hydrocarbons (PAHs) due to their physical, chemical, and biological property are intrinsically related to their environmental behavior, persistence and toxicological effect. The tendency to be very stable, low water soluble and attach to organic matter relatively well explains their long term persistence in soils and sediments as persistent environmental pollutants. PAHs are capable of chemical transformation via oxidation, nitration, photo degradation, which tends to produce secondary products that are more toxic or bioavailable than their parent compounds. Biologically, most PAHs are biologically activated into carcinogens, mutagens, or endocrine-disrupting chemicals thereby threatening human health and ecological systems. These basic properties are required to be understood well not only in order to have an accurate environmental risk assessment data not to mention the health impact evaluation but also to arrive at the best remediation strategies. Future study on the behavior and degradation of individual PAHs will therefore play a key role in informing sustainable and science-based strategies aimed at controlling and ameliorating PAH contamination.

References

Abdel-Shafy, H.I., Mansour, M.S.M.: A review on polycyclic aromatic hydrocarbons: source, environmental impact, effect on human health and remediation. Egypt. J. Pet. **25**(1), 107–123 (2016). https://doi.org/10.1016/j.ejpe.2015.03.011

Adebusoye, S.A., Picardal, F.W., Ilori, M.O., Amund, O.O., Fuqua, C., Grindle, N.: Aerobic degradation of di- and trichlorobenzenes by two bacteria isolated from polluted tropical soils. Chemosphere **66**(10), 1939–1946 (2007). https://doi.org/10.1016/j.chemosphere.2006.07.074

Akhtar, A.B.T., Naseem, S., Yasar, A., Naseem, Z.: Persistent Organic Pollutants (POPs): Sources, Types, Impacts, and Their Remediation, pp. 213–246 (2021). https://doi.org/10.1007/978-981-15-5499-5_8

Balint, A.: Physical, chemical and toxicological properties of polycyclic aromatic hydrocarbons (PAHs) in human exposure assessments to contaminated soil and groundwater. MATEC Web Conf. **342**, 03016 (2021). https://doi.org/10.1051/matecconf/202134203016

Balmer, J.E., Hung, H., Yu, Y., Letcher, R.J., Muir, D.C.G.: Sources and environmental fate of pyrogenic polycyclic aromatic hydrocarbons (PAHs) in the Arctic. Emerg. Contam. **5**, 128–142 (2019). https://doi.org/10.1016/j.emcon.2019.04.002

Bidleman, T., et al.: Atmospheric pathways of chlorinated pesticides and natural bromoanisoles in the northern Baltic Sea and its catchment. Ambio **44**(S3), 472–483 (2015). https://doi.org/10.1007/s13280-015-0666-4

Boethling, R., Fenner, K., Howard, P., Klečka, G., Madsen, T., Snape, J.R., Whelan, M.J.: Environmental persistence of organic pollutants: guidance for development and review of POP risk profiles. Integr. Environ. Assess. Manag. **5**(4), 539–556 (2009). https://doi.org/10.1897/IEAM_2008-090.1

Bozlaker, A., Muezzinoglu, A., Odabasi, M.: Atmospheric concentrations, dry deposition and air–soil exchange of polycyclic aromatic hydrocarbons (PAHs) in an industrial region in Turkey. J. Hazard. Mater. **153**(3), 1093–1102 (2008). https://doi.org/10.1016/j.jhazmat.2007.09.064

Çabuk, H., Kılıç, M.S., Ören, M.: Biomonitoring of polycyclic aromatic hydrocarbons in urban and industrial environments of the Western Black Sea Region Turkey. Environ. Monit. Assess. **186**(3), 1515–1524 (2014). https://doi.org/10.1007/s10661-013-3470-2

Celik Atalay, E., Er Demirhan, B., Sagdıcoglu Celep, A.G.: Low-calorie sweeteners and reproductive health: evidence and debates. Curr. Nutr. Food Sci. **21**(3), 309–332 (2025). https://doi.org/10.2174/0115734013315621240802055207

Chi, N.D.T., Huy, D.H., Nguyen, N.T., Nguyen, L.S.P., Ngan, T.A., Takenaka, N., Hien, T.T.: PM2.5-bound polycyclic aromatic hydrocarbons in Ho Chi Minh City, Vietnam: potential sources, health risks, and deposition flux. Aerosol Air Qual. Res. **24**(12), 240127 (2024). https://doi.org/10.4209/aaqr.240127

Cho, T., Uetrecht, J.: How reactive metabolites induce an immune response that sometimes leads to an idiosyncratic drug reaction. Chem. Res. Toxicol. **30**(1), 295–314 (2017). https://doi.org/10.1021/acs.chemrestox.6b00357

Das, N., Kumar, V., Chaure, K., Pandey, P.: Environmental restoration of polyaromatic hydrocarbon-contaminated soil through sustainable rhizoremediation: insights into bioeconomy and high-throughput systematic analysis. Environ. Sci.: Adv. **4**(6), 842–883 (2025). https://doi.org/10.1039/D4VA00203B

Drwal, E., Rak, A., Gregoraszczuk, E.L.: Review: polycyclic aromatic hydrocarbons (PAHs)—action on placental function and health risks in future life of newborns. Toxicology **411**, 133–142 (2019). https://doi.org/10.1016/j.tox.2018.10.003

Eldos, H.I., Zouari, N., Saeed, S., Al-Ghouti, M.A.: Recent advances in the treatment of PAHs in the environment: application of nanomaterial-based technologies. Arab. J. Chem. **15**(7), 103918 (2022). https://doi.org/10.1016/j.arabjc.2022.103918

Esen, F., Tasdemir, Y., Cindoruk, S.S.: Dry deposition, concentration and gas/particle partitioning of atmospheric carbazole. Atmos. Res. **95**(4), 379–385 (2010). https://doi.org/10.1016/j.atmosres.2009.10.010

Ewa, B., Danuta, M.-Š: Polycyclic aromatic hydrocarbons and PAH-related DNA adducts. J. Appl. Genet. **58**(3), 321–330 (2017). https://doi.org/10.1007/s13353-016-0380-3

Fucic, A., Guszak, V., Mantovani, A.: Transplacental exposure to environmental carcinogens: association with childhood cancer risks and the role of modulating factors. Reprod. Toxicol. **72**, 182–190 (2017). https://doi.org/10.1016/j.reprotox.2017.06.044

Ghosal, D., Ghosh, S., Dutta, T.K., Ahn, Y.: Current state of knowledge in microbial degradation of polycyclic aromatic hydrocarbons (PAHs): a review. Front. Microbiol. **7**, 1369 (2016). https://doi.org/10.3389/fmicb.2016.01369

Goswami, L., Manikandan, N.A., Dolman, B., Pakshirajan, K., Pugazhenthi, G.: Biological treatment of wastewater containing a mixture of polycyclic aromatic hydrocarbons using the oleaginous bacterium Rhodococcus opacus. J. Clean. Prod. **196**, 1282–1291 (2018). https://doi.org/10.1016/j.jclepro.2018.06.070

Gundlapalli, M., Sivagami, K., Gopalakrishnan, M., Harshini, P., Janjaroen, D., Ganesan, S.: Biodegradation of low molecular weight polycyclic aromatic hydrocarbons in soil: insights into bacterial activities and bioremediation techniques. Sustain. Chem. Environ. **7**, 100146 (2024). https://doi.org/10.1016/j.scenv.2024.100146

Guo, X., Chen, F., Pan, S.: Pollution, traceability, and health risk of polycyclic aromatic hydrocarbons in traditional Chinese spices. J. Food Compos. Anal. **142**, 107390 (2025). https://doi.org/10.1016/j.jfca.2025.107390

Gupta, H.: Anthracene removal from water onto activated carbon derived from vehicular tyre. Sep. Sci. Technol. **53**(4), 613–625 (2018). https://doi.org/10.1080/01496395.2017.1405038

Gupta, H., Gupta, B.: Photocatalytic degradation of polycyclic aromatic hydrocarbon benzo[a]pyrene by iron oxides and identification of degradation products. Chemosphere **138**, 924–931 (2015). https://doi.org/10.1016/j.chemosphere.2014.12.028

Gupta, H., Gupta, B.: Adsorption of polycyclic aromatic hydrocarbons on banana peel activated carbon. Desalin. Water Treat. **57**(20), 9498–9509 (2016). https://doi.org/10.1080/19443994.2015.1029007

Gupta, H., Gupta, B.: Vehicular tire as potential adsorbent for the removal of polycyclic aromatic hydrocarbons. Polycyclic Aromat. Compd. **38**(4), 354–368 (2018). https://doi.org/10.1080/10406638.2016.1217890

Gupta, N., Banerjee, S., Koley, A., Bharali, P., GhoshThakur, R., Hoque, R.R., Balachandran, S.: Strategies for remediation of polycyclic aromatic hydrocarbons in contaminated soil: a systematic review and bibliometric analysis. Appl. Soil. Ecol. **204**, 105688 (2024). https://doi.org/10.1016/j.apsoil.2024.105688

Gupta, H., Dhiman, S., Madhav, S., Mishra, R.: Emerging contaminants in water and wastewater: an introduction. In: Gupta, H., Madhav, S., Dhiman, S., Mishra, R. (eds) Emerging Contaminants in Water and Wastewater. Environmental Science and Engineering. Springer, Cham (2025). https://doi.org/10.1007/978-3-031-82579-8_1

Haritash, A.K., Kaushik, C.P.: Biodegradation aspects of polycyclic aromatic hydrocarbons (PAHs): a review. J. Hazard. Mater. **169**(1–3), 1–15 (2009). https://doi.org/10.1016/j.jhazmat.2009.03.137

Idowu, O., Semple, K.T., Ramadass, K., O'Connor, W., Hansbro, P., Thavamani, P.: Beyond the obvious: environmental health implications of polar polycyclic aromatic hydrocarbons. Environ. Int. **123**, 543–557 (2019). https://doi.org/10.1016/j.envint.2018.12.051

Ifegwu, O.C. and Anyakora, C.: Polycyclic Aromatic Hydrocarbons. pp. 277–304 (2015). https://doi.org/10.1016/bs.acc.2015.08.001

Karimi, B., Habibi, M., Esvand, M.: Biodegradation of naphthalene using Pseudomonas aeruginosa by up flow anoxic–aerobic continuous flow combined bioreactor. J. Environ. Health Sci. Eng. **13**(1), 26 (2015). https://doi.org/10.1186/s40201-015-0175-1

Kong, L., Song, B., Zhang, T., Gao, K., Liu, J.: Effects of soil organic matter on biochar application in developing the biodegradation potentials of polycyclic aromatic hydrocarbons (PAHs). Appl. Soil. Ecol. **167**, 104046 (2021). https://doi.org/10.1016/j.apsoil.2021.104045

Krzyszczak, A., Czech, B.: Occurrence and toxicity of polycyclic aromatic hydrocarbons derivatives in environmental matrices. Sci. Total Environ. **788**, 147738 (2021). https://doi.org/10.1016/j.scitotenv.2021.147738

Kumar, V., Kothiyal, N.C., Saruchi, Vikas, P., Sharma, R.: Sources, distribution, and health effect of carcinogenic polycyclic aromatic hydrocarbons (PAHs)—current knowledge and future directions. J. Chin. Adv. Mater. Soc. **4**(4), 302–321 (2016). https://doi.org/10.1080/22243682.2016.1230475

Kumari, P., Kumar, A.: Advanced oxidation process: a remediation technique for organic and non-biodegradable pollutant. Results Surf. Interfaces **11**, 100122 (2023). https://doi.org/10.1016/j.rsurfi.2023.100122

Kuppusamy, S., Thavamani, P., Megharaj, M., Naidu, R.: Biodegradation of polycyclic aromatic hydrocarbons (PAHs) by novel bacterial consortia tolerant to diverse physical settings—assessments in liquid- and slurry-phase systems. Int. Biodeterior. Biodegradation **108**, 149–157 (2016). https://doi.org/10.1016/j.ibiod.2015.12.013

Låg, M., Øvrevik, J., Refsnes, M., Holme, J.A.: Potential role of polycyclic aromatic hydrocarbons in air pollution-induced non-malignant respiratory diseases. Respir. Res. **21**(1), 299 (2020). https://doi.org/10.1186/s12931-020-01563-1

Lawal, A.T.: Polycyclic aromatic hydrocarbons. A review. Cogent Environ. Sci. **3**(1), 1339841 (2017). https://doi.org/10.1080/23311843.2017.1339841

Li, T., Wang, Y., Guo, S., Li, X., Xu, Y., Wang, Y., Li, X.: Effect of polarity-reversal on electrokinetic enhanced bioremediation of Pyrene contaminated soil. Electrochim. Acta **187**, 567–575 (2016). https://doi.org/10.1016/j.electacta.2015.11.097

Liu, C., Shi, H., Wang, C., Fei, Y., Han, Z.: Thermal remediation of soil contaminated with polycyclic aromatic hydrocarbons: pollutant removal process and influence on soil functionality. Toxics **10**(8), 474 (2022). https://doi.org/10.3390/toxics10080474

Liu, J., Zhang, C., Jia, H., Lichtfouse, E., Sharma, V.K.: Abiotic transformation of polycyclic aromatic hydrocarbons via interaction with soil components: a systematic review. Crit. Rev. Environ. Sci. Technol. **53**(5), 676–699 (2023). https://doi.org/10.1080/10643389.2022.2083897

Maletić, S.P., Beljin, J.M., Rončević, S.D., Grgić, M.G., Dalmacija, B.D.: State of the art and future challenges for polycyclic aromatic hydrocarbons is sediments: sources, fate, bioavailability and remediation techniques. J. Hazard. Mater. **365**, 467–482 (2019). https://doi.org/10.1016/j.jhazmat.2018.11.020

Mallah, M.A., et al.: Polycyclic aromatic hydrocarbon and its effects on human health: an overeview. Chemosphere **296**, 133948 (2022). https://doi.org/10.1016/j.chemosphere.2022.133948

Manian, R.: Bioremediation of polycyclic aromatic hydrocarbons contaminated soils: recent progress, perspectives and challenges. Environ. Monit. Assess. **195**(12), 1441 (2023). https://doi.org/10.1007/s10661-023-12042-7

Manousi, N., Zachariadis, G.A.: Recent advances in the extraction of polycyclic aromatic hydrocarbons from environmental samples. Molecules **25**(9), 2182 (2020). https://doi.org/10.3390/molecules25092182

Marquès, M., Mari, M., Sierra, J., Nadal, M., Domingo, J.L.: Solar radiation as a swift pathway for PAH photodegradation: a field study. Sci. Total Environ. **581–582**, 530–540 (2017). https://doi.org/10.1016/j.scitotenv.2016.12.161
Masih, J., Masih, A., Kulshrestha, A., Singhvi, R., Taneja, A.: Characteristics of polycyclic aromatic hydrocarbons in indoor and outdoor atmosphere in the North central part of India. J. Hazard. Mater. **177**(1–3), 190–198 (2010). https://doi.org/10.1016/j.jhazmat.2009.12.017
Mohanty, M. et al.: Environmental Hydrocarbon Pollutants: Sources, Transport, and Effect on Human Health, pp. 21–44 (2025). https://doi.org/10.1007/978-3-031-81615-4_2
Mojiri, A., Zhou, J.L., Ohashi, A., Ozaki, N., Kindaichi, T.: Comprehensive review of polycyclic aromatic hydrocarbons in water sources, their effects and treatments. Sci. Total Environ. **696**, 133971 (2019). https://doi.org/10.1016/j.scitotenv.2019.133971
Montano, L., et al.: Polycyclic aromatic hydrocarbons (PAHs) in the environment: occupational exposure, health risks and fertility implications. Toxics **13**(3), 151 (2025). https://doi.org/10.3390/toxics13030151
Moorthy, B., Chu, C., Carlin, D.J.: Polycyclic aromatic hydrocarbons: from metabolism to lung cancer. Toxicol. Sci. **145**(1), 5–15 (2015). https://doi.org/10.1093/toxsci/kfv040
Nie, N., Li, T., Miao, Y., Wei, X., Zhao, D., Liu, M.: Environmental fate and health risks of polycyclic aromatic hydrocarbons in the Yangtze River Delta Urban Agglomeration during the 21st century. J. Hazard. Mater. **465**, 133407 (2024). https://doi.org/10.1016/j.jhazmat.2023.133407
Paris, A., Ledauphin, J., Poinot, P., Gaillard, J.-L.: Polycyclic aromatic hydrocarbons in fruits and vegetables: origin, analysis, and occurrence. Environ. Pollut. **234**, 96–106 (2018). https://doi.org/10.1016/j.envpol.2017.11.028
Patel, A.B., Shaikh, S., Jain, K.R., Desai, C., Madamwar, D.: Polycyclic aromatic hydrocarbons: sources, toxicity, and remediation approaches. Front. Microbiol. **11**, 562813 (2020). https://doi.org/10.3389/fmicb.2020.562813
Premnath, N., et al.: A crucial review on polycyclic aromatic hydrocarbons—environmental occurrence and strategies for microbial degradation. Chemosphere **280**, 130608 (2021). https://doi.org/10.1016/j.chemosphere.2021.130608
Rabodonirina, S., Net, S., Ouddane, B., Merhaby, D., Dumoulin, D., Popescu, T., Ravelonandro, P.: Distribution of persistent organic pollutants (PAHs, Me-PAHs, PCBs) in dissolved, particulate and sedimentary phases in freshwater systems. Environ. Pollut. **206**, 38–48 (2015). https://doi.org/10.1016/j.envpol.2015.06.023
Sahoo, B.M., Ravi Kumar, B.V.V., Banik, B.K., Borah, P.: Polyaromatic hydrocarbons (PAHs): structures, synthesis and their biological profile. Curr. Org. Synth. **17**(8), 625–640 (2020). https://doi.org/10.2174/1570179417666200713182441
Sajid, M., Nazal, M.K., Ihsanullah, I.: Novel materials for dispersive (micro) solid-phase extraction of polycyclic aromatic hydrocarbons in environmental water samples: a review. Anal. Chim. Acta **1141**, 246–262 (2021). https://doi.org/10.1016/j.aca.2020.07.064
Sakshi, Singh, S.K., Haritash, A.K.: Polycyclic aromatic hydrocarbons: soil pollution and remediation. Int. J. Environ. Sci. Technol. **16**(10), 6489–6512 (2019). https://doi.org/10.1007/s13762-019-02414-3
Singh, V., et al.: Polycyclic aromatic hydrocarbons (PAHs) in aquatic ecosystem exposed to the 2020 Baghjan oil spill in upper Assam, India: short-term toxicity and ecological risk assessment. PLoS ONE **18**(11), e0293601 (2023). https://doi.org/10.1371/journal.pone.0293601
Smol, M., Włodarczyk-Makuła, M., Skowron-Grabowska, B.: PAHs removal from municipal landfill leachate using an integrated membrane system in aspect of legal regulations. Desalin. Water Treat. **69**, 335–343 (2017). https://doi.org/10.5004/dwt.2017.20241
Soares, W.M., Botelho, B.G.: Determination of benzo[a]pyrene in barley malt using constant wavelength synchronous fluorescence. J. Food Compos. Anal. **143**, 107611 (2025). https://doi.org/10.1016/j.jfca.2025.107611
Sombiri, S., Balhara, N., Attri, D., Kharb, I., Giri, A.: An overview on occurrence of polycyclic aromatic hydrocarbons in food chain with special emphasis on human health ailments. Discov. Environ. **2**(1), 87 (2024). https://doi.org/10.1007/s44274-024-00121-6

Srogi, K.: Monitoring of environmental exposure to polycyclic aromatic hydrocarbons: a review. Environ. Chem. Lett. **5**(4), 169–195 (2007). https://doi.org/10.1007/s10311-007-0095-0

Suresh, A., Soman, V., K R, A., A, R., Rahman K, H.: Sources, toxicity, fate and transport of polyaromatic hydrocarbons (PAHs) in the aquatic environment: a review. Environ. Forensics **26**(1), 35–57 (2025). https://doi.org/10.1080/15275922.2024.2366801

Sushkova, S., et al.: Analysis of benzo[a]pyrene contamination from an long-term contaminated soil. Am. J. Biochem. Biotechnol. **12**(1), 1–11 (2016). https://doi.org/10.3844/ajbbsp.2016.1.11

Sverdrup, L.E., Nielsen, T., Krogh, P.H.: Soil ecotoxicity of polycyclic aromatic hydrocarbons in relation to soil sorption, lipophilicity, and water solubility. Environ. Sci. Technol. **36**(11), 2429–2435 (2002). https://doi.org/10.1021/es010180s

Tesfaye, E.L., Bogale, F.M., Aragaw, T.A.: Biodegradation of polycyclic aromatic hydrocarbons: the role of ligninolytic enzymes and advances of biosensors for in-situ monitoring. Emerg. Contam. **11**(1), 100424 (2025). https://doi.org/10.1016/j.emcon.2024.100424

Ukalska-Jaruga, A., Smreczak, B.: The impact of organic matter on polycyclic aromatic hydrocarbon (PAH) availability and persistence in soils. Molecules **25**(11), 2470 (2020). https://doi.org/10.3390/molecules25112470

Utembe, W., Kamng'ona, A.W.: Inhalation exposure to chemicals, microbiota dysbiosis and adverse effects on humans. Sci. Total Environ. **955**, 176938 (2024). https://doi.org/10.1016/j.scitotenv.2024.176938

Vandana, Priyadarshanee, M., Mahto, U., Das, S.: Mechanism of toxicity and adverse health effects of environmental pollutants. In: Microbial Biodegradation and Bioremediation, pp. 33–53. Elsevier (2022). https://doi.org/10.1016/B978-0-323-85455-9.00024-2

Vijayanand, M., et al.: Polyaromatic hydrocarbons (PAHs) in the water environment: a review on toxicity, microbial biodegradation, systematic biological advancements, and environmental fate. Environ. Res. **227**, 115716 (2023). https://doi.org/10.1016/j.envres.2023.115716

Vikas, P., Kumar, V., Saruchi.: Toxic equivalency factor and statically analysis of carcinogenic micro air pollutants in urban soil at the northern part of India. J. Chin. Adv. Mater. Soc. **3**(2), 102–118 2015. https://doi.org/10.1080/22243682.2014.964313

Wang, C., Zhang, Z., He, S., Tang, J., Wang, R., Liu, X.: Environmental fate of polycyclic aromatic hydrocarbons (PAHs) in different layers of tar balls in the Bohai Sea, China. J. Clean. Prod. **403**, 136803 (2023). https://doi.org/10.1016/j.jclepro.2023.136803

Wang, H., et al.: Bioaccumulation and potential human health risks of PAHs in marine food webs: a trophic transfer perspective. J. Hazard. Mater. **485**, 136946 (2025). https://doi.org/10.1016/j.jhazmat.2024.136946

Winquist, E., et al.: Bioremediation of PAH-contaminated soil with fungi—from laboratory to field scale. Int. Biodeterior. Biodegrad. **86**, 238–247 (2014). https://doi.org/10.1016/j.ibiod.2013.09.012

Yang, J., Lv, G., Sun, X.: Singlet oxygen and hydroxyl radical in the transformation of acetaminophen in surface water: the mechanism on the degradation pathways based on DFT calculations. J. Environ. Chem. Eng. **12**(6), 114933 (2024). https://doi.org/10.1016/j.jece.2024.114933

Yari, S., Fallah Asadi, A., Nourmohammadi, M.: Occupational and environmental cancer. Asian Pac. J. Environ. Cancer **1**(1), 5–13 (2018). https://doi.org/10.31557/apjec.2018.1.1.5-13

Yemele, O.M., Zhao, Z., Nkoh, J.N., Ymele, E., Usman, M.: A systematic review of polycyclic aromatic hydrocarbon pollution: a combined bibliometric and mechanistic analysis of research trend toward an environmentally friendly solution. Sci. Total Environ. **926**, 171577 (2024). https://doi.org/10.1016/j.scitotenv.2024.171577

Yu, Y., Jin, H., Lu, Q.: Effect of polycyclic aromatic hydrocarbons on immunity. J. Transl. Autoimmun. **5**, 100177 (2022). https://doi.org/10.1016/j.jtauto.2022.100177

Zhang, H., Zhang, X., Wang, Y., Bai, P., Hayakawa, K., Zhang, L., Tang, N.: Characteristics and influencing factors of polycyclic aromatic hydrocarbons emitted from open burning and stove burning of biomass: a brief review. Int. J. Environ. Res. Public Health **19**(7), 3944 (2022). https://doi.org/10.3390/ijerph19073944

Zheng, B., Wang, L., Lei, K., Nan, B.: Distribution and ecological risk assessment of polycyclic aromatic hydrocarbons in water, suspended particulate matter and sediment from Daliao River estuary and the adjacent area, China. Chemosphere **149**, 91–100 (2016). https://doi.org/10.1016/j.chemosphere.2016.01.039

Polycyclic Aromatic Hydrocarbons in the Surface and Groundwater

3

Seeta Dewali, Netra Pal Sharma, Pankaj Bhatt, Deepak Chandra Melkani, Manoj Kumar Arya, Harish Chandra Singh Bisht, Swetapadma Dash, and Satpal Singh Bisht

Abstract

Aquatic habitats are frequently found to contain polycyclic aromatic hydrocarbons (PAHs), a class of persistent and hydrophobic organic contaminants that pose major ecological and public health risks. Both petrogenic and pyrogenic sources contribute to their presence in surface and groundwater. Once released, PAHs can adsorb onto suspended particles, settle in sediments, and gradually infiltrate groundwater systems, making them challenging to remediate. With special attention to the concentration ranges of polycyclic aromatic

S. Dewali (✉) · N. P. Sharma · P. Bhatt · D. C. Melkani · M. K. Arya · H. C. S. Bisht
Department of Zoology, Kumaun University, Nainital, Uttarakhand, India
e-mail: seemadeoli3@gmail.com

N. P. Sharma
e-mail: np_biotech53@rediffmail.com

P. Bhatt
e-mail: bhattpankaj098@gmail.com

D. C. Melkani
e-mail: deepakmelkanintl@gmail.com

M. K. Arya
e-mail: dr.manojkumar19@rediffmail.com

H. C. S. Bisht
e-mail: hcsbishtji@gmail.com

S. Dash
Zoological Survey of India (ZSI), Kolkata, India
e-mail: oasisnainital@gmail.com

S. S. Bisht
Soban Singh Jeena University Almora, Almora, Uttarakhand, India
e-mail: sps.bisht@gnail.com

H. Gupta et al. (eds.), *Ubiquitous Polycyclic Aromatic Hydrocarbon Contamination*, Emerging Contaminants and Associated Treatment Technologies,
https://doi.org/10.1007/978-3-032-11764-9_3

hydrocarbons (PAHs), their main congeners, and the degree of pollution seen in various aquatic ecosystems, this chapter offers a thorough overview of global trends in PAH occurrence. Rapid urbanisation, growing industry, and intensive agricultural practices are major factors influencing PAH profiles in surface and groundwater systems, according to source apportionment studies. The long-term persistence of PAHs presents a significant ecological and public health concern because of their well-established carcinogenic, mutagenic, and teratogenic characteristics. In order to stop more contamination and guarantee the preservation of essential water resources, these worries emphasise the urgent need for comprehensive monitoring frameworks, the implementation of cutting-edge water treatment technology, and the enforcement of strict regulatory procedures.

Keywords

Sediments • Polycyclic aromatic hydrocarbons • Groundwater • Sources • Discharges • Oil spills

Introduction

A family of organic pollutants known as polycyclic aromatic hydrocarbons (PAHs) is made up of two or more fused aromatic benzene rings. They are often divided into two groups according to their molecular size: low molecular weight PAHs (LMW-PAHs), which have two to three rings, and high molecular weight PAHs (HMW-PAHs), which have four or more rings. The hydrophobic and lipophilic properties of the latter group boost their genotoxic, mutagenic, and carcinogenic effects and increase their persistence in the environment, making them more dangerous. As hazardous substances that can cause mutations, cancer, developmental abnormalities, and immune system deficiencies in a variety of organisms, PAHs are generally acknowledged to be harmful. Rapid industrialisation as well as urbanisation have accelerated human activities that emit a variety of pollutants, including PAHs, into the environment (Gupta 2018). Due to their constancy and confrontation to degradation, PAHs persevere for lengthy periods, posing significant biological toxicity and making their remediation a pressing global concern. These pollutants are ubiquitous, occurring in aquatic, terrestrial, and atmospheric environments (Patel et al. 2020; Kumar and Gupta 2020; Gupta et al. 2025; Montano et al. 2025). In soils and sediments, PAH deposition rates are accelerated by their strong hydrophobicity and low aqueous solubility, leading to firm adsorption onto soil particles. As a result, soils act as a significant sink for PAHs. Based on concentration levels, soil PAH contamination can be classified as unpolluted ($\sum$PAH < 200 ng g^{-1}), slightly polluted (200–600 ng g^{-1}), or heavily polluted (> 1000 ng g^{-1}) (Patel et al. 2020).

The causes of PAH effluence are generally anthropogenetic discharge and natural discharge. anthropogenetic discharge, can be further subdivided into four main types: manufacturing, transportable, domestic, and agronomic emissions. Natural discharge, such as volcanic outbreaks, timberland fires, and lightning-triggered upland fires, are often regarded as inconsequential or of lesser significance compared to human-derived emissions (Zhang et al. 2025a, b; Gupta and Kumar 2020).

Three main processes pyrogenic, petrogenic, and biogenic occur in both natural and artificial sources, contributing to the occurrence of PAHs pollutents in the environment. pyrolysis of carbon-based substrates in anoxic environments as during the burning of coal, volcanic eruptions, and other human-caused activities—produces pyrogenic PAHs. Petrogenic PAHs, on the other hand, are mostly linked to crude oil, petroleum derivatives, and related hydrocarbon compounds and are created by the diagenetic alteration of organic matter over geological time (Zhao et al. 2024). PAH Seasonal variation is also evident in concentrations, with winter usually recording the most significant levels, followed by spring, autumn and summer. Increased incomplete combustion of fossil fuels, a higher need for heating in homes, decreased photodegradation, and restricted air dispersion due to weather factors such as low temperatures and calm winds are all associated with sophisticated considerations in the wintertime and spring (Patel et al. 2020). The different kinds of polycyclic aromatic hydrocarbon production causes are depicted in Fig. 3.1

Biogenic PAHs, although comparatively less prevalent, are synthesised by certain plants, microorganisms (including bacteria and fungi), and phytoplankton,

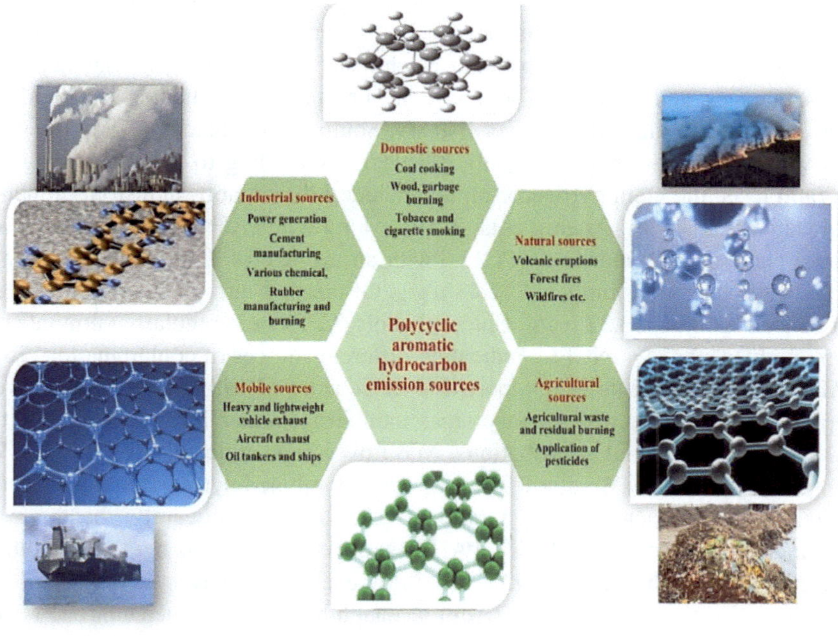

Fig. 3.1 Various types of PAH generation from both natural and man-made sources

typically within specialised ecological niches, without the involvement of diagenetic processes. The identification of PAH sources in environmental matrices is often achieved through the application of diagnostic ratios and molecular marker analyses, which serve as reliable indicators for tracing contamination pathways (Berríos-Rolon et al. 2025).

Surface Water Systems' PAH Sources

Geographical factors, hydrological regimes, land-use patterns, and human-induced activities interact in a complex manner to determine the release and geographical distribution of PAHs in surface waters. Furthermore, by influencing the transit, transformation, and accumulation of PAHs, the physicochemical characteristics of surface water bodies and climatic variability impact the environmental fate of PAHs. Different hydrodynamic and ecological features of surface water systems, including rivers, streams, lakes, wetlands, reservoirs, and glacier-fed waters, influence the relative contributions of various PAH sources and produce contamination profiles unique to a given location.

Rivers

According to Schwanen et al. (2024) and Berrios-Rolon et al. (2025), rivers are important routes for the movement of pollutants, particularly polycyclic aromatic hydrocarbons (PAHs), from upstream catchments into downstream habitats such lakes, marshes, and estuaries. There have been reports of sediment-associated PAH concentrations in the Buffalo River Estuary, South Africa, reaching 22,310 μg/kg, and water samples containing up to 206 μg/L. According to molecular diagnostic ratio tests, the main pyrogenic inputs include urban runoff, industrial effluents, and vehicle emissions (Adeniji et al. 2019). Along with PAHs, sediments also contained phenolic compounds ($\sum$PC11), with an average of 7.95 μg/g and a range from below detection limits to 272.48 μg/g. Of these, 4-nitrophenol had the greatest concentration at 33.32 μg/g. Chlorophenols (CP5) were regularly outnumbered by non-chlorophenols (NCP6), indicating different sources of contamination. Total PAH levels in sediments were reported to range from 12,198 to 26,551 ng/g (average of 22,009 ng/g), with estuary areas exhibiting higher levels because of the Nile River's substantial contributions. MW) over high molecular weight (HMW) congeners (LMW/HMW ratio 0.7–1.18), while combined pyrolytic and petrogenic sources were characterised by a ratio of 0.23–0.39. Significantly, the range of potentially carcinogenic PAHs ($\sum$PAH7) was 5049 to 15,144 ng/g. Compounds like dibenz[a,h]anthracene (DahA), benz[a]pyrene (BaP), phenanthrene (Phe), fluoranthene (Flu), acenaphthylene (Acy), acenaphthene (Ace), phenanthrene (Phe), acenaphthylene (Acy), acenaphthylene (Acy), and acenaphthylene (Ace) were found at concentrations above the effects range median (ERM), indicating a high likelihood of negative biological effects (Abdallah 2025).

TPHs were mostly pyrogenic (anthropogenic) in origin, as indicated by the LMW/HMW ratio. The oil refinery site is considered a high-risk area, and urgent measures are needed to mitigate hydrocarbon pollution and protect the river ecosystem in Iraq (Hassan et al. 2025). The study found notable variations in PAH concentrations across sediment samples. Benzo[a]anthracene was present in high amounts but only in two samples, while fluoranthene showed relatively uniform levels. Benzo[k]fluoranthene had the highest overall concentration, and benzo[a]pyrene the lowest. Atammele sediments were the most carcinogenic (84.23 ng/μL), followed by Isiachala (51.92 ng/μL), whereas the Ube stream had the lowest PAH concentration (11.74 ng/μL). Ecological risk assessments revealed location-specific variations, while human health risk analysis yielded hazard index values ranging from 2.3E–07 to 2.57E–05 and total cancer risks from 4.85E–07 to 5.43E–05 for both adults and children (Woznicki et al. 2025).

Naphthalene showed the highest concentrations in water (1.0564–1.0109 mg/L) and sediments (1.9067–1.8111 mg/L) across all seasons. Health risk assessments revealed hazard indices above safe limits for children and significant carcinogenic risks. PCA identified two primary anthropogenic PAH sources, highlighting contamination control to protect health and the environment (Ezekwe et al. 2025). While post-monsoon concentrations of PAHs ranged from 60.352 to 5663.058 ng/L in water and 2976.770 to 15,238.335 ng/g in sediments, pre-monsoon values ranged from 22.256 to 836.455 ng/L in water and 1,459.941–43,179.243 ng/g in sediments. Sources were identified by PCA and diagnostic ratios as petroleum residues, vehicle emissions, biomass combustion, wastewater, and solid waste burning. High carcinogenic hazards were found by TEF analysis, primarily from benzo[a]pyrene and benzo[a]anthracene (Siddiqui et al. 2024).

Four- to six-ring PAHs were the most common in the Bohai Sea, with values ranging from 26.9 to 50.1 ng/L in surface water, 18.8–44.1 ng/L in bottom water, and 7.4–143.9 ng/g in sediments. Concentrations decreased southeastward, reflecting coastal currents and Yellow River contributions. Correlations with partition coefficients and Henry's constants showed that whereas atmospheric–seawater exchange increased surface concentrations, dilution from riverine silts and waters decreased PAH levels in the south (Cao et al. 2024). According to the Grmasha et al. 2024, pyrolytic and pyrogenic processes accounted for the majority of the possible human-caused sources of PAHs, with pyrogenic sources predominating.

In southern Italy, the Sele River, the region's second-largest by water volume, was studied for PAH distribution and sources, 10 sites across four seasons revealed PAH concentrations ranging from 10.1 to 567.23 ng/L in surface water, 121.23–654.36 ng/L in suspended particulate matter, and 331.75–871.96 ng/g in sediment. About 1808 kg/year of PAHs are discharged into the sea. Analysis indicated that industrial and vehicle emissions influenced PAH levels. Fugacity fraction results showed that PAHs primarily moved from water to sediment (Montuori et al. 2022).

In Nigeria's Ekulu River, surface water PAH levels reached 3.17 mg/L, dominated by high molecular weight (HMW) compounds mainly from combustion-related sources (Umeh et al. 2023). Similarly, HMW PAHs from burning petroleum dominated the water and sediment in Iraq's Euphrates River, with carcinogenic species accounting for up to 55% of the total load (Grmasha et al. 2023). Sewage inputs constitute a significant source of PAHs in the Haihe River, significantly influencing the distribution among sediments and pore water (Liu et al. 2023a, b).

Streams

Streams flowing through urban and woodland catchments often receive significant polycyclic aromatic hydrocarbon (PAH) inputs, primarily from pyrogenic sources associated with combustion activities. Sediments in Suzhou Industrial Park, China, exposed entire PAH values from 180 to 81,000 ng/g. Four-ring PAHs were the maximum prevalent class ($42 \pm 12\%$), and source analysis attributed 18% to vehicle emissions and 61% to the burning of coal (Yuan et al. 2021). Road debris and suspended sediment (SS) from two drainage catchments in Dunedin, New Zealand, were analysed for the occurrence of PAHs. Lead (119–527 µg/g), copper (50–464 µg/g), zinc (241–1325 µg/g), and Σ16PAHs (1.20–11.6 µg/g) were all found in road debris. Similar quantities of contaminants were found in suspended sediments from a primarily rural catchment (20%), but also contained some urban material, suggesting that the primary source was urban traffic litter. Both materials exhibited comparable PAH fingerprints and isomer ratios, indicating a dominant pyrogenic (combustion) origin (Brown and Peake 2006). Streamwater PAHs and quinones (PAQs) were monitored in two creeks with different urbanisation levels in southern Ontario during rainfall events. Seasonal comparisons revealed that PAHs and PAQs had similar sources and were primarily transported by suspended solids. The more urbanised creek had higher concentrations, with PAH loads up to twice as high in cold weather compared to summer. PAH delivery increased with runoff and longer dry periods, highlighting surface flushing as a key process. In the less urbanised creek, PAH levels and behaviour remained relatively consistent across events (Awonaike et al. 2021).

During four sampling occasions, the levels of dissolved organic carbon (DOC) and total and freely dissolved PAHs were assessed in river water from five locations along the River Wyre in the United Kingdom. The flocculation method was used to separate the PAHs that were freely dissolved. In general, dissolved PAHs increased downstream, ranging 2.71–18.9 ng/L and 2.61–16.8 ng/L. In contrast to PAHs with four rings, those with five or more rings exhibited a strong positive correlation with DOC, signifying a communal global source. higher-molecular-weight PAHs mirrored the seasonal trends of DOC, peaking in late summer. A regression model including molecular weight and DOC was used to estimate annual PAH fluxes over two years. Understanding the PAH-DOC relationship can improve the interpretation of monitoring data and support risk assessments for aquatic life and water supplies (Moeckel et al. 2014).

Similarly, sediments in Poland from maintenance reservoirs along the Oliwski and Strzyza streams contained PAHs at concentrations of up to 20.4 mg/kg, dominated by four- and five-ring compounds. The pollution was mainly linked to traffic emissions, domestic heating, and vehicle wear, with some sites posing high ecological risks to benthonic creatures. Broader reviews suggest that wildfires, although diffuse in nature, can be a substantial cause of PAHs to surface waters, with impacts influenced by fire intensity, vegetation composition, and watershed hydrology (Nawrot et al. 2023).

Lakes

Lakes are subject to PAH inputs from a variety of sources, with urban environments being particularly impacted by emissions from coal products, oils etc. combustion. Although regulatory measures in several countries have successfully reduced point-source discharges, diffuse causes, as external runoff and atmospheric deposition, continue to be major pathways for PAHs entering lake ecosystems (Xia et al. 2020).

The geographical and temporal patterns of PAHs and n-alkanes were impacted by changes in land use and basin development. In contrast to urban regions, which had higher levels due to industrial emissions, petroleum inputs, and the burning of fossil fuels, rural areas had lower levels of PAHs and long-chain plant wax n-alkanes. n-Alkane ratios, such as the Terrigenous–Aquatic Ratio and the Carbon Preference Index, revealed changes in the sources of organic matter that were consistent with increased human activity. Three periods were distinguished in the sediment records: natural inputs prior to the 1960s, an increase in anthropogenic influence from the 1960s to the 1990s, and a amalgamation of pyrogenic sources after the 1990s (Enalbes et al. 2025).

HMW PAHs are further likely to adsorb to particles and build up in bottom sediments, LMW PAHs typically stay dissolved in the water column (Berrios-Rolon et al. 2025). Studies of lakes in Ontario, Canada, have shown that pyrogenic activities can contribute more to local-scale atmospheric deposition in urban lake environments than regional sources (Du and Jing 2018). Likewise, PAHs in Lake Baikal have been primarily linked to the burning of fossil fuels, with heavier congeners localised in sediments and lighter PAHs predominantly found in the water column (Semenov et al. 2023). Endosulfans (21.96%), chlordanes (24.6%), and HCHs (24.65%) were the dominant OCPs. DDTs primarily originated from residual technical DDT, dicofol, and antifouling paints used on ships, while HCHs were associated with long-range transport, agricultural use, and industrial products containing lindane (Zhi et al. 2015).

At several locations near Lake Balaton, 16 PAHs were observed. Winter had the largest concentrations of sediment (448.35 ng/g), while summer had the lowest (257.21–465.49 ng/g). LMW PAHs were outnumbered by HMW PAHs. For both adults and children, the incremental lifetime cancer risk (ILCR) through the cutaneous and absorption ways remained increased throughout the year, reaching its highest in winter and then in the spring, summer and autumn (Grmasha et al. 2024). While PAH concentrations in ML increased from 46.8 to 198.25 ng/L, they decreased in LL water from 36.5 to 26.59 ng/L over 43 days. Sediment concentrations declined more gradually, falling from 4812 to 1.04 ng/g in ML over 16.96 years and from 932 to 0.95 ng/g in LL over 13.33 years. For the unrelated case, stabilised concentrations were below both modelled levels and observed values (2170 ng/L in water).

Runoff was identified as the primary mechanism for PAH export by MMBM data. Since ML's output was significantly higher, upstream LL did not increase PAH levels in ML. Sediment characteristics had a significant impact on model results (Zhang et al. 2025a, b). Thirteen out of sixteen target PAHs, primarily pyrene, chrysene, naphthalene, and benzo[b]fluoranthene, were found in 18 fish species from four lakes in China using GC–MS analysis. Site-specific PAH values varied from 26.60 to 47.27 ng/g, with DahA predominating in Yuncheng and BaP in Taihu, Danjiangkou, and Nansi. The lethal equivalents of high-cyclic PAHs were higher than those of low- or mid-cyclic ones ($P < 0.05$). Non-carcinogenic hazards were minimal, although Pleuronichthys cornutus and Lateolabrax japonicus from Yuncheng showed carcinogenic risks ($> 10^{-4}$), primarily from BbF. The primary source of PAHs was found to be combustion, and BaP, DahA, and BbF were designated as priority pollutants (Shi et al. 2024).

Wetlands

Freshwater wetlands are highly productive ecosystems that provide critical ecological functions but are increasingly threatened by polycyclic aromatic hydrocarbon (PAH) contamination from multiple sources. These habitats are susceptible to degradation resulting from petroleum exploitation and other anthropogenic disturbances (Zheng et al. 2022; Amini-Birami et al. 2023). In contrast, wetlands in Northeastern China are primarily impacted by coal combustion and vehicular emissions. Seasonal hydrological dynamics can also alter source profiles, as observed in Wang Lake Wetland, China, where petrogenic PAHs dominate during the wet season.

Constructed and urban wetlands, which play a crucial role in water treatment, are also vulnerable to PAH accumulation. In these systems, contamination often arises from effluent discharges and petroleum combustion, with PAH dispersal patterns influenced by molecular weight. Typically, PAHs present in the water column are of petrogenic origin, while those sequestered in sediments are predominantly pyrogenic, reflecting the complex partitioning behaviour of these compounds in wetland environments (Cheshmvahm et al. 2023). Both pyrogenic and petrogenic sources commonly contaminate wetlands close to oil removal facilities, as the Shadegan Wetland in Iran and the Hoor-Al-Azim Wetland in Lower Mesopotamia. The main contributions are emissions from burning and oil spills (Rokhbar et al.

2023). In Northeast China and Iraq's Al-Hammar marshes—document past contamination from both man-made and natural sources, including industrial processes and wildfires (Salman et al. 2021). Because freshwater wetlands are essential to their ecological processes, they must contend with intricate patterns of PAH pollution from various overlapping sources.

Sediment PAH concentrations demonstrate spatial variability in pollution intensity across global wetlands. Low to moderate contamination levels have been reported in the Anzali Wetland, Iran; Miankaleh International Wetland, Iran; Can Gio Wetland (CGW), Vietnam; Hoor Al-Azim Wetland, Iran, and Shadegan Wetland, Iran (Dhara and Dutta 2025). In contrast, the Ashtamudi Wetland in India predominantly exhibits moderate contamination (Sreedevi and Harikumar 2023) (Table 3.1). These observations highlight the variability in pollution sources among wetlands and emphasise the need for site-specific management strategies to maintain or restore their ecological health. In the Anzali Wetland, sampling sites located near commercial port facilities and areas of intense yacht traffic are particularly vulnerable to contamination from by-products of fossil fuel combustion.

The distribution and transport of polycyclic aromatic hydrocarbons in surface freshwater systems are governed by multiple mechanisms (Fig. 3.2). Physical transport involves advection, where PAHs move downstream with flowing water, diffusion, which is the passive spread from high to low concentration zones, and sediment dynamics, in which PAHs bind to or are released from suspended particles and sediments. Chemical interactions include sorption, the attachment of PAHs to organic-rich particles and sediments, and desorption, their release back into the water column. Biological transport occurs through bioaccumulation and food web transfer, where aquatic organisms incorporate PAHs into tissues, and through microbial activity or sediment bioturbation, which can redistribute these compounds vertically. Environmental modulators play a significant role: hydrology, including rainfall, runoff, and flooding, increases PAH wash-off and sediment transport, while seasonal variations influence dilution and mobilisation patterns. Physicochemical parameters, such as ionic strength, oxidation and reduction potential, etc., affect PAH behaviour by altering sorption and degradation rates (Berrios-Rolon et al. 2025). Additionally, source distribution and land cover determine PAH inputs, with urban and industrial areas contributing higher loads, whereas vegetation and soil surfaces can mitigate transport.

Groundwater-Soil Polycyclic Aromatic Hydrocarbons (PAHs)

Although groundwater is an essential supply of freshwater for household, agricultural, and industrial purposes, contamination by polycyclic aromatic hydrocarbons (PAHs) delivered through surface runoff, leaching, and other transport processes is becoming a greater concern. Hydrological connectivity between surface and groundwater systems facilitates the movement of PAHs, particularly in closely

Table 3.1 Absorption, incidence, source, and degree of contamination by 16 USEPA PAHs in wetland sediments worldwide

S. No.	Study area	Σ16 PAHs concentration (ng/g)	Pollution level*	Dominant PAHs	Sources	References
1	Vembanad wetland, India's west coast	47.1–424.4	Low	Acenaphthylene (Acy) occasional effects	The main causes of oil spills, industrial discharge, home sewage, port operations, vehicle emissions, and urban and agricultural runoff	Suresh and Soman (2025)
2	Ramsar coastal wetland, Kollam district,	0.93–104.8	Low to moderate	Naphthalene	Biomass burning, coal burning (44.8%), petrogenic sources (21%)	Sreedevi and Harikumar (2025)
3	Tamasopo Wetland, Mexico	< 100 ng g^{-1}	Moderate	Pyrene, Benzo[a]pyrene, Dibenz[a,h]anthracene	Combustion of sugarcane, fossil fuels from vehicles and fire lighting	Sandoval-Herazo et al. (2025)
4	In Iran Anzali Wetland,	8.28–806.64	Moderate	NAPH, PHEN and PYR	Traffic, fossil and coal burning	Cheshmvahm et al. (2023)
5	In India Kerala Estuaries,	0.47–126.64	Moderate	NAPH, ACE, ANT, BbF, BkF, BaP, IP, and BghiP	Automobile and surface runoff	Radhakrishnan et al. (2023)
6	In Iran Miankaleh International Wetland,	4.54–196.90	Moderate	NAPH and PHEN	Municipal, industrial wastewater runoff, vehicular exhaust, coal combustion	Rokhbar et al. (2023)
7	In India Ashtamudi Wetland,	158.23–202.81	Moderate	3 rings PAHs	Pipeline and fishing boat oil leaks, port operations, surface runoff, burning of biomass, atmospheric deposition, and urban runoff	Sreedevi and Harikumar (2023)

(continued)

Table 3.1 (continued)

S. No.	Study area	Σ16 PAHs concentration (ng/g)	Pollution level*	Dominant PAHs	Sources	References
8	In India Mahanadi Basin, in	302.6–728.2	Moderate	4 rings PAHs	Combustion of coal, biomass (grass and wood), and petroleum-derived fuels	Kurwadkar et al. (2022)
9	Can Gio Wetland, I n Vietnam	0.73–518	Moderate	2 rings PAHs	Agricultural runoff, industrial discharges, and pollution from intense shipping and boat traffic	Thuy et al. (2021)
10	Diep in South Africa River,	2870–46,910	High	5 rings PAHs	Petroleum-related inputs, along with industrial and agricultural activities	Awe et al. (2020)
11	In Iran Hoor Al-Azim Wetland,	15.78–410.2	Moderate	NAPH, FLU, ANT, BaA, CHRY and PYR	Auto emissions, biomass combustion	Sheikh Fakhradini et al. (2019)
12	In Iran Shadegan Wetland,	10 ± 0.5–317 ± 14.3	Low to moderate	NAPH, PYR, BaA and CHRY	Incomplete burning, gasoline, diesel engine emissions, and fishing boats	Ashayeri et al. (2018)

Abbreviations: PAHs—Polycyclic Aromatic Hydrocarbons, Acy—Acenaphthylene, NAPH—Naphthalene, PHEN—Phenanthrene, ACE—Acenaphthene, ANT—Anthracene BaA—Benzo[a]anthracene, CHRY—Chrysene, PYR—Pyrene, BbF—Benzo[b]fluoranthene, BkF—Benzo[k]fluoranthene, BaP—Benzo[a]pyrene, DbA—Dibenz[a,h]anthracene, IP—Indeno[1,2,3-cd]pyrene, BghiP—Benzo[ghi]perylene and FLU—Fluorene

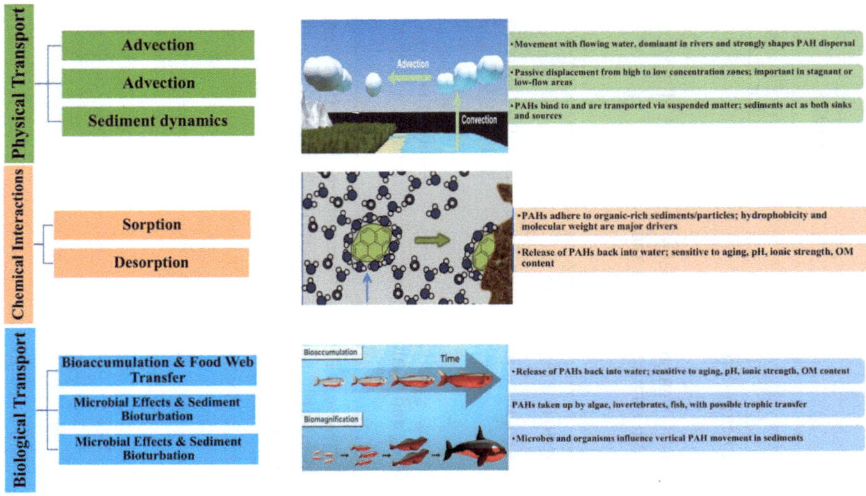

Fig. 3.2 Distribution of PAHs in surface water systems, with key transport mechanisms and influencing factors

interconnected environments. LMW PAHs are more mobile and often exhibit relatively constant concentrations across interconnected water bodies (Berríos-Rolon et al. 2025).

An study evaluated the sources, accumulation, and groundwater–surface water interactions of PAH pollution in the middle Yangtze River. There was no discernible seasonal variation in accumulation; tributary inputs predominated. The principal PAHs were naphthalene and phenanthrene. biomass burning (BB) and coal combustion (CC), with CC concentration increasing by 6.18 ng L^{-1} from the wet to the dry season, indicating an increase in coal consumption during the dry season. Mantel tests and SEM revealed that primary emissions were strongly linked to eutrophication levels, while redox variations influenced nitrogen and phosphorus transformations. Hydrochemical and isotope data ($\delta^{18}O$, $\delta^{2}H$) indicated interaction between ground and surface water during the dry season, as supported by PAH fractionation patterns. PAHs act as effective geochemical tracers, with toxic effects enhanced by such interactions (Liu et al. 2025).

Groundwater PAHs were 3 × (pyrene) to 50 × (indeno [1,2,3-cd]pyrene) above dissolved levels at a coal tar site. While pH-dependent ultrafiltration revealed a humic acid–colloid interaction, similar air–water surface tension values ruled out the presence of cosolvents or surfactants. A pyrene–humic partition coefficient of 10^5 L/kgC explained the enrichment, supporting colloid-facilitated PAH transport. Humic-bound PAHs can significantly enhance groundwater mobility and increase fluxes from tar sources by up to 20 × compared to tar–water partitioning predictions (MacKay et al. 2001). In Chennai, Tamil Nadu, an industrialised coastal region with petrochemical storage tanks, the levels of PAHs measured in the groundwater.

The majority of the 2001 samples were alkaline, and according to pH and EC, 62.5% of them were unfit for human consumption. The confirmation was made through sampling in 2011 and 2012. The prevalence of Na–Cl suggested the influence of seawater. TPH and PAH analyses revealed increased contamination in the eastern part, linked to groundwater flow direction and proximity to coastal storage tanks. Although an underground tank leaking PAHs was closed ~ 50 years ago, pollution has persisted for decades. While the contamination zone has decreased, likely due to natural flushing, the persistence of PAHs highlights the long-term risk from historical petroleum sources (Brindha et al. 2014). Chemical oxidation, biodegradation, and adsorption/desorption are some of the processes that PAHs undergo in soil and groundwater.

Soil Organic Matter (SOM) regulate the sorption behaviour of PAHs when the soil organic carbon (SOC) content exceeds 0.32% (Cheng et al. 2021). Furthermore, different organisms—such as bacteria, fungi, and algae—respond differently in their capacity to degrade PAHs, and the structure and enzymatic activity of microbial communities are critical for this process (Zhang et al. 2020).

Chemical oxidation, including free radical-mediated and photochemical oxidation processes, also plays a vital role in transforming PAHs within these environments (Ghosh and Mukherji 2023). Yet, the soil–groundwater system is inherently complex and highly heterogeneous. Interactions among physical, chemical, and biological factors can be synergistic or antagonistic, impacting key functions as carbon sequestration, nutrient cycling and agricultural productivity.

These changes also affect the behaviour of contaminants, including the transport, alteration, and deprivation of PAHs, heavy metals, and microplastics, thereby influencing environmental quality and human health (Zhang et al. 2023).

Factors Affecting PAH Mobility and Change in Groundwater

Numerous environmental conditions, including dissolved organic matter (DOM), dissolved oxygen (DO), microbial populations, and interactions with surface water, influence the behaviour of PAHs in ground water systems. Among these, DOM, microbial communities, and surface water inputs are particularly influential. Surface water can alter the characteristics of groundwater by contributing to changes in parameters such as oxygen levels, carbon content, and microbial composition through recharge processes. Additionally, fluctuations in water flow and groundwater levels can impact the spatial distribution of PAHs (Hu et al. 2025). Microorganisms play a critical part in reducing the toxicity of PAHs by metabolising them into less harmful or more biodegradable compounds (Zou et al. 2025). DOM, containing reactive groups like hydroxyl and carboxyl, can bind with PAHs through adsorption and complexation, thereby modifying their mobility and degradation rates in the subsurface environment (Salehi 2025).

Role of DOM in Modulating the Mobility and Bioavailability of PAHs in Groundwater

As a natural surfactant in the environment, dissolved organic matter makes polycyclic aromatic hydrocarbons (PAHs) more soluble and, thus, more bioavailable to microbes. Additionally, DOM can help microbial processes that break down and alter PAHs by serving as a source of energy and carbon (Sha et al. 2023). DOM contains diversity including carboxyl, phenolic hydroxyl, aromatic rings, and sulfhydryl groups that facilitate interactions with organic contaminants. For instance, phenolic components in DOM have been shown to stimulate the formation of complexes that enhance the solid-phase adsorption of compounds such as steroid estrogens. the hydroxyl and carboxyl groups in DOM can bind to pollutants through mechanisms such as static interactions and ion exchange, thereby contributing to contaminant removal (Zheng et al. 2021).

In addition to its chemical reactivity, the molecular size and structural complexity of DOM further support its capacity to adsorb pollutants. Interestingly, the impact of DOM on contaminant mobility depends on its concentration. At lower concentrations (below 6 mg/L), DOM can enhance the retention of compounds like naphthalene (Nap) through co-adsorption and cumulative effects, limiting their movement. However, at higher concentrations (above 10 mg/L), DOM begins to behave more like a surfactant, encapsulating Nap within micelles and facilitating its transport via colloidal movement (Hu et al., 2025). DOM doesn't always promote PAH removal. For example, Huang et al. identified competitive adsorption between humic acids (a component of DOM) and pollutants on mineral surfaces, as well as the potential for DOM to interfere with photolytic and microbial degradation pathways. Additionally, a DOM concentration of 10 mg/L decreased ability to adsorb phenanthrene (Phe) by 8.26%, as reported by Meng et al. (2021).

Role of Dissolved Oxygen in Governing PAH Degradation Pathways in Groundwater

The concentration of dissolved oxygen (DO) in groundwater plays a critical role in determining whether PAH degradation occurs under aerobic or anaerobic conditions. Anaerobic biodegradation tends to proceed more slowly and typically follows a three-stage pathway. Initially, under nitrate- and sulfate-reducing conditions, aromatic hydrocarbons are partially broken down into smaller organic acids. These acids can then form complexes with insoluble Fe(III) oxides, thereby mobilising the iron. In the final step, iron-reducing bacteria utilise the activated Fe(III), further promoting the breakdown of aromatic compounds (Hu et al. 2025).

Despite the significance of anaerobic pathways, most research has emphasised aerobic degradation processes, where oxygen is not only a vital reactant but also a limiting factor. Enzymes like monooxygenases and dioxygenases require oxygen to initiate the oxidation. The accessibility of DO in ground water affects the efficiency of PAH biodegradation through oxidative pathways. For example, Gholami et al. demonstrated that increasing DO concentrations significantly enhanced naphthalene (Nap) removal rates—from 52.5% at 100 mg/L of O_2 to about 75.8% at 200 mg/L (Gholami et al. 2018). This improvement is primarily due to the role of DO in supporting microbial activity, which is crucial for PAH degradation. Higher DO levels create more favourable conditions for microbial growth and enzymatic function. For instance, laccase—an enzyme involved in PAH breakdown—showed peak activity at 30% DO, while its activity dropped markedly when DO levels fell below 15% (Beltran-Flores et al. 2023).

Oxygen also enhances the degradation efficiency of PAHs by serving as a terminal electron acceptor, a process that is more effective than anaerobic alternatives. Beyond its direct influence, DO also affects the chemical environment. At concentrations above 1–2 mg/L, iron tends to exist as Fe^{3+}, which promotes PAH adsorption, stimulates microbial populations, and enhances the availability of electron acceptors (Lu et al. 2022). Moreover, elevated DO levels can lead to the creation of reactive species such as singlet oxygen and hydroxyl radicals. These oxidants can alter DOM by breaking down its functional groups, reducing its adsorption capacity, and thereby facilitating increased PAH mobility.

Impact of Surface Water Interactions on Groundwater Microbial Ecology and PAH Fate

Research comparing microbial diversity in these environments has revealed distinct differences: surface waters are typically dominated by phytoplankton, dinoflagellates, ciliates, and algae, whereas groundwater ecosystems are more commonly inhabited by fungi, amoebae, tail algae, and certain ciliates (Capo et al. 2021). The higher biodiversity in surface water is attributed largely to elevated concentrations of carbon-based matter and oxygen, which support the growth of photosynthetic microorganisms. In contrast, microbial life in groundwater is predominantly chemotrophic, relying on chemical energy sources rather than light. Due to the absence of light in subsurface environments, photosynthesis is not a significant driver of microbial diversity in groundwater systems. Through the vadose zone, polycyclic aromatic hydrocarbons migrate surface water to groundwater. This process is influenced by several variables, including soil pore water flow and soil particle detachment. For example, total PAH concentrations in groundwater in Rio de Janeiro, Brazil, have been observed to range from 0.05 to 84.9 μg/L (Do Rego et al. 2007). The migration of PAHs is governed largely by seasonal variations in river water levels, human activities, and geological dynamics associated with river systems (Das et al. 2021).

The most common PAH in the main river channel, tributaries, and groundwater during the wet season was phenanthrene (Phe), which accounted for 30.78%, 29.34%, and 24.06% of all PAHs found, respectively. The main river (22.29 ng/L), groundwater (20.18 ng/L), and tributaries had the most significant average Phe values (24.32 ± 12.50 ng/L) (Liu et al. 2025). Naphthalene (Nap), on the other hand, emerged as the predominant PAH in the dry season, accountancy for

29.19% of the mainstream, 36.48% of tributaries, and 25.26% of groundwater. The corresponding average concentrations in the mainstream were 18.62 ± 7.82 ng/L, 19.48 ± 4.51 ng/L, and 33.14 ± 23.41 ng/L in the tributaries. Augmented runoff and rainfall throughout the wet season, which promotes the deposition and movement of low-molecular-weight PAHs, particularly 3-ring compounds, are likely the cause of this seasonal change (Lu et al. 2023). Conversely, the dry season's higher levels of HMW PAHs may be attributed to warmer temperatures and more intense sunlight, which stimulate microbial activity and abiotic breakdown. The volatility and alteration of HMW-PAHs are further enhanced by climatic features as heat, humidity, and UV light (Zhang and Chen 2017).

Storey et al. revealed additional evidence of this connection when they found by mixing of surface and groundwater might activate inorganic trophic bacteria. These bacteria provide energy by oxidising substances like sulphur, iron, nitrite, and ammonia. Hydrocarbons such as PAHs are then broken down oxidatively with the help of this mechanism (Storey et al. 1999). Krause et al. claim that when oxygenated surface water enters low-oxygen groundwater zones, it initiates a sequence of redox reactions in the soil profile that ultimately lead to the production of methane. The reduction of oxygen, nitrates, iron, manganese, and sulphates is among these processes.

Microbial populations are also significantly shaped by salinity. Hui and Tan claim that areas irrigated with surface water and those irrigated with groundwater have quite different bacterial population compositions. In groundwater-irrigated areas, salt-tolerant taxa, such as Halomonadaceae and Balneolaceae, predominated, and soil salinity had a significant impact on community structure (RV = 0.66, $p < 0.001$) (Hui and Tan 2024). Furthermore, in both surface water and groundwater, significant seasonal relationships between PAHs and essential water chemistry parameters such as bicarbonate (HCO_3^-), calcium (Ca^{2+}), and sodium (Na^+). Additionally, they observed a statistically significant positive correlation ($p < 0.05$, $r > 0.4$) between total phosphorus and PAHs in groundwater during both wet and dry seasons, consistent with findings from surface water samples (Liu et al. 2025).

Microbial Degradation and Adsorption of PAHs by Groundwater Bacterial Communities

Groundwater hosts diverse bacterial communities, several of which exhibit strong capabilities in degrading polycyclic aromatic hydrocarbons (PAHs). Key genera, such as Mycobacterium, Pseudomonas, Sphingomonas, and Sphingobium, are exceptionally proficient in degrading PAHs. Shen et al. observed that strains such as Polaromonas, Pseudomonas, Janthinobacterium, and Methylophilus achieved degradation rates ranging from 6.21% to 16.73% for naphthalene (Nap) and from 13.95% to 24.45% for phenanthrene (Phe) (Zhang et al. 2022). However, the efficiency of degradation tends to decline with increasing groundwater depth. Bacteria play a critical role in metabolising PAHs by using them as carbon sources, thus driving the breakdown process, particularly in subsurface environments. For instance, *Stenotrophomonas maltophilia* was reported to mineralise up to 40% of Phe (Bahr et al. 2015). The susceptibility of PAHs to microbial breakdown is influenced by their chemical structure. Only the low-molecular-weight PAHs (Flu and Phe) shown significant breakdown among the compounds studied, with 65–95% degradation being attained in 72 days. The other chemicals were fluorene (Flu), phenanthrene (Phe), fluoranthene (Flt), and pyrene (Pyr) (Rabodonirina et al. 2019).

Differences in bacterial tolerance to PAHs have also been noted. While most bacterial strains are sensitive to pyrene exposure, some groups, such as Gemmatimonadota, Gaiellales, and Planococcaceae, show notable resistance. the presence of pyrene enhanced cooperation among bacterial strains within the same species, possibly promoting more efficient degradation (Zhang et al. 2022). Even low concentrations of PAHs can disrupt specific groups, such as *Azotobacter*, although the broader microbial community often remains relatively stable (Zhang et al. 2014). In addition to direct biodegradation, bacteria contribute to PAH removal through adsorption mechanisms. Their movement, guided by chemotaxis, helps optimise contact with contaminants, thus improving degradation rates. Moreover, bacterial biofilms can trap PAHs on their surfaces, concentrating them around microbial colonies and influencing the distribution and transport of these compounds within the aquatic system.

Environmental conditions such as pH, temperature, and dissolved oxygen (DO) strongly influence bacterial metabolism and, in turn, the efficiency of PAH degradation in groundwater. Under ideal pH conditions (7.0–9.0) and temperatures (25 °C–40 °C), for example, the YM-6 bacterial strain has been demonstrated to utilise phenanthrene (Phe) concentrations between 100 and 400 mg/L as the carbon source (Liu et al. 2023a, b). Through biodegradation and adsorption, bacterial populations in groundwater primarily affect the mobility and behaviour of PAHs.

Fungal-Mediated Degradation of PAHs in Groundwater

Particularly in aerobic conditions, fungi are very efficient at breaking down PAHs in groundwater. Fungi can mineralise PAHs into carbon dioxide and water through enzymatic activity. To initiate the PAH transformation process, they utilise essential enzymes, such as cytochrome P450 and lignin-degrading enzymes. According to Mao and Guan (2016), for instance, during a 30 day period, fungal strains could break down 82% of benzo[a]pyrene (BaP), 64% of pyrene (Pyr), 62% of fluoranthene (Flt), and 60% of phenanthrene (Phe). Similarly, Mucor fungus broke down 74% of Phe, 81% of BaP, and 63% of Pyr in 14 days, as reported by Zafra et al. (2015).

Extracellular enzymes released by fungi play a important role in this degradation process by dissolving the aromatic rings of PAHs into smaller, easier-to-manage compounds, which are subsequently further broken down inside fungal cells. Necessary enzymes that are especially effective in oxidising high-molecular-weight PAHs are laccases, ligninases, and peroxidases. Some studies have reported degradation rates of up to 99.65% for Phe and 99.58% for Pyr. Cytochrome P450 enzymes can hydroxylate low-molecular-weight PAHs, making them more water-soluble and easier to degrade (Wu et al. 2021). White rot fungi, for instance, produce laccase and peroxidase enzymes that convert PAHs into quinones and facilitate their further oxidation and mineralisation.In addition to their direct degradation capabilities, some fungi and non-ligninolytic molds can modify PAHs through hydroxylation, which improves solubility and enhances subsequent degradation efficiency. Fungi preform supportive role in microbial ecosystems by facilitating the movement of bacteria through saturated or unsaturated environments. Their mycelial networks create physical pathways that enable bacteria to disperse more effectively. Furthermore, fungi can transform PAHs into less toxic, more bioavailable compounds, which bacteria can then further metabolise, demonstrating the benefits of fungal–bacterial cooperation. For instance, while a fungal culture alone achieved a 65% degradation rate of BaP, this increased to 75% in a co-culture with bacteria (Espinosa-Ortiz et al. 2022).

PAH Migration and Transformation in Soil-Groundwater Systems.

Figure 3.3 illustrates the interconnected biological, abiotic, and chemical processes that govern the migration, transformation, and fate of PAHs in soil–groundwater systems. Biological migration occurs through the activity of soil fauna, such as earthworms, which redistribute PAHs via burrowing and selective accumulation—low molecular weight (LMW) PAHs showing greater bioaccumulation potential than high molecular weight (HMW) PAHs. Plants also play a role, with root uptake accounting for approximately 71% of total plant PAH absorption, thereby reducing mobility but enabling transfer into the food chain. Microorganisms, including microalgae, bacteria, and fungi, interact with PAHs via functional groups (e.g., hydroxyl, carboxyl) that facilitate adsorption and biodegradation, although algal contributions are generally smaller than those of macrofauna. Abiotic migration is driven by physical forces, including rainfall infiltration, leaching, and erosion, and is influenced by the molecular weight of PAHs, soil organic matter (SOM), and mineral composition. Minerals like kaolinite and montmorillonite adsorb PAHs through hydrophobic and π–π interactions, while humic acids enhance adsorption at low concentrations but may increase solubility and mobility at higher concentrations. Groundwater migration is influenced by pH, salinity, and dissolved organic matter (DOM), with low pH strengthening DOM–PAH binding and high pH promoting desorption. Competitive ion adsorption (e.g., Cl^-, SO_4^{2-}) can also modify PAH–mineral interactions. DOM acts dually, enhancing retention at low concentrations and facilitating colloidal transport at high concentrations (Meng et al. 2021). Biodegradation processes, mediated by bacterial genera such as *Mycobacterium*, *Pseudomonas*, and *Sphingomonas*, as well as fungal ligninolytic enzymes

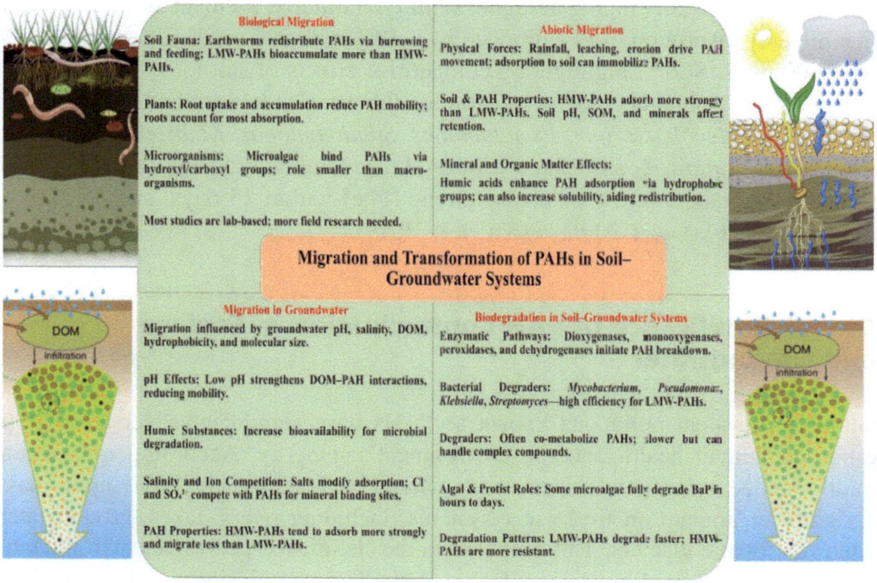

Fig. 3.3 Mechanisms and mechanisms regulating the movement and change of PAHs in soil-groundwater systems

and certain microalgae, further contribute to PAH transformation through enzymatic oxidation, ring cleavage, and mineralisation. Ultimately, the fate of PAHs in soil–groundwater environments depends on the dynamic balance between adsorption, dissolution, transport, and microbial degradation, all of which are modulated by site-specific environmental conditions.

Plants also contribute significantly to the control of PAH migration, primarily through root uptake and accumulation. This phytoaccumulation mechanism reduces the potential mobility of PAHs in the soil. For instance, research by Cao et al. demonstrated that alfalfa could remove 34.7% of LMW-PAHs and 37.3% of HMW-PAHs from polluted soils (Cao et al. 2022). Additionally, Collins et al. (2006) found that while PAH absorption can occur through both roots and leaves, the root system accounted for the majority, approximately 71% of the total uptake.

Both soil parameters, such as pH, SOM, and mineral content, as well as the physicochemical qualities of the PAHs themselves, such as their molecular weight and hydrophobicity, affect how much they migrate through soil. Because they are more hydrophobic, PAHs (HMW-PAHs) tend to be adsorbed onto soil particles more easily than low-molecular-weight PAHs (LMW-PAHs) (Kim et al. 2017).

The interaction of PAHs with soil minerals strongly influences their environmental behavior. Cai et al. (2023) reported that minerals at 0.5 g/L achieved 36.7% adsorption of anthracene (Ant). Groundwater salts also modify adsorption by altering surface charges and competing for binding sites. Yue et al. (2023) observed phenanthrene (Phe) desorption increasing to 34.6 mg/kg at low salinity (< 0.3 wt%)

but decreasing to 20.1 mg/kg at higher salinity. Similarly, Meng et al. (2021) found that mineral adsorption of Phe dropped by 1.38% with 300 mg/L Cl^- and by 14.6% with 300 mg/L SO_4^{2-}, highlighting competitive effects of salt ions.

Among microbial degraders, bacteria are the most widely studied and effective. Species like *Mycobacterium*, *Pseudomonas*, *Sphingomonas*, *Klebsiella*, and *Streptomyces* are commonly associated with PAH degradation by *Sphingomonas* sp. PJ1 shown to degrade 74.32% of phenanthrene (Phe) within 15 days, and 58.18% of fluoranthene (Fla) over 30 days (Xu et al. 2019).

Global Significance of PAH Contamination in Surface and Groundwater

Globally prevalent pollutants known as PAHs have significant adversative properties on human health and the environment's integrity. According to a recent multi-country systematic review, both natural and man-made PAH inputs have a significant impact on freshwater systems, including rivers, lakes, groundwater, and even remote glaciers. This emphasises the chemicals' pervasive environmental presence and risk profile (Berrios-Rolon et al. 2025). Analysing riverine data from around the world reveals remarkable regional variations in PAH contamination, primarily caused by variations in land-use, urbanisation, and industrialisation. The Nile River in Egypt has Σ16 PAH concentrations ranging from 1113 to 4364 ng/L, with an average of 1877.5 ng/L.

In contrast, the Seine estuary in France has comparatively low values, ranging from 4 to 36 ng/L (mean ~ 20 ng/L). The concentrations in the Mississippi River in the United States range from 62.9 to 144.7 ng/L (mean ~ 114.9 ng/L), while the Ganges River in South Asia shows a wide range of 0.05–65.9 ng/L (mean ~ 32.5 ng/L). Similarly, seven major river basins in China exhibit considerably higher readings, ranging from 300 to 7,552 ng/L (mean ~ 1,868 ng/L), while the Danube River in Hungary fluctuates between 25 and 1208 ng/L (mean ~ 122.6 ng/L). These trends indicate that rivers affected by strong industrialisation and intense urban activity, such as the Nile and Chinese basins, can have PAH loads of several thousand ng/L, whereas less affected river systems, like the Seine, often have low PAH loads (<50 ng/L). In addition to directly endangering human populations who depend on these water bodies for agriculture, fishery, and drinking, such high concentrations raise serious ecological issues that impact aquatic biodiversity, bioaccumulation, and sediment quality. All of these data point to PAH contamination as a serious worldwide water quality issue that needs to be addressed through integrated management, pollution control regulations, and international cooperation to reduce the effects on the human health.

PAH pollution is a significant issue in ports and coastal ecosystems due to heavy maritime traffic, industrial emissions, and oil-related activities. The average ΣPAH concentration in port waters is estimated to be 176 ± 178 ng/L, according to global studies, whereas sediments have far greater accumulation levels, averaging 1593 ± 1837 μg/kg (dry weight). With mean ΣPAHs of 268 ± 236 μg/kg, aquatic biota—including economically significant fish species—display notable bioaccumulation, highlighting the risks of trophic transfer and human exposure through seafood consumption (Tulcan et al. 2024).

Regional trends also show sharp disparities: African ports report the highest levels of sediment contamination, which may be related to poor waste management and oil handling procedures, while Asian ports show the highest PAHs concentration in both water and aquatic organisms, reflecting rapid industrialisation and heavy shipping activity (Tulcan et al. 2024). These results demonstrate how vulnerable coastal ecosystems, which are already being strained by habitat loss and climate change, are to long-term PAH deposition, which could have an impact on fisheries output, ecological health, and the livelihoods of coastal communities.

Agricultural areas are dynamic interfaces where irrigation return flows, agrochemical runoff, and infiltration processes frequently cause surface water and groundwater contamination by PAHs to overlap. The North China Plain's Yellow River estuary region has surface waters with PAH concentrations from 11.8 to 393.1 ng/L and groundwater with values ranging from 8.5 to 402.8 ng/L. With surface–groundwater connection promoting pollutant transfer, such similar ranges demonstrate that both compartments are exposed to mild but significant pollution pressures. Seasonal variations, including monsoon-driven recharge and agricultural irrigation practices, further influence the relative distribution and intensity of PAH contamination.

Such results have enormous consequences for human and the environment health. With fish and shellfish from contaminated areas frequently surpassing allowable consumption levels, PAHs bioaccumulate in aquatic creatures, inflowing the food chain and posing long-term health hazards (Mardani 2021). On a global scale, PAHs also act as indicators of human pressure. Petrogenic PAHs (from crude oil and petroleum products) predominate in shipping corridors and oil-producing regions like the Middle East and West Africa. In contrast, pyrogenic PAHs (from coal, biomass, and traffic combustion) are found to be more prevalent in urban-industrial catchments, according to diagnostic ratio studies. Their persistence and transportation highlight their transboundary nature through groundwater recharge, riverine export, and atmospheric deposition, which renders them global pollutants rather than localised ones (Rodrigue 2004).

International objectives for water safety and sustainability are directly threatened by the presence of PAHs in vital water supplies, according to policymakers. Their control is essential to the United Nations' Sustainable Development Goals (SDGs), particularly SDG 6 (Clean Water and Sanitation), SDG 3 (Good Health and Well-being), and SDG 14 (Life Below Water). They are also regulated in drinking water directives throughout the European Union, as well as in US EPA and WHO standards (Tsaridou and Karabelas 2021). This study highlights the critical need for international collaboration in monitoring, risk assessment, and mitigation techniques by offering a comparative framework of PAH contamination across continents.

Studying PAHs in surface and groundwater is essential globally because it can link local pollution trends to global water security issues. The widespread threats posed by PAHs to ecosystems and human populations are highlighted by the persistent findings of high concentrations in industrialised basins, their persistence in sediments, and exceeding health-based limits in some hotspots. Therefore, addressing PAH contamination is a global health concern that necessitates integrated monitoring, stricter emission restrictions, and advanced remediation techniques worldwide. It is not only a localised environmental issue.

Conclusion

Seasonality, pollution sources, and watershed characteristics all have a influence on the amounts of polycyclic aromatic hydrocarbons (PAHs) in surface and groundwater. While some Indian wetlands, such as Vembanad, record ~ 47 ng/g in sediments, large river basins often have seasonal water averages of 225–366 ng/L, which are dominated by high-molecular-weight congeners. Σ16PAH levels in surface waters usually range from approximately 100 to 3800 ng/L. Groundwater typically remains significantly less contaminated (0–5 ng/L in uncontaminated aquifers), despite localised industrial and petroleum-affected regions recording tens to hundreds of ng/L, with occasionally high readings surpassing 1000 ng/L in hotspot areas. Despite frequently being below detection limits, carcinogenic benzo[a]pyrene (BaP) has been detected in impacted sources at concentrations as high as 0.30 μg/L, exceeding the drinking water limitations set by the US EPA, Health Canada and the WHO (0.7 μg/L). While petrogenic signals are discernible around maritime and petroleum zones, composition patterns in urbanised catchments exhibit pyrogenic dominance from traffic and biomass/coal combustion; diagnostic ratios consistently identify industrial and vehicle emissions as the primary sources. Sediments often store more ΣPAHs than the surrounding waters because they are long-term sinks, and fluxes indicate net deposition. Despite the low background levels of groundwater, these data demonstrate that surface waters in urban-industrial areas and contaminated aquifers present significant ecological and health hazards. Therefore, it is essential to keep PAHs under safe regulatory criteria by utilising effective treatment technologies, implementing more stringent source-reduction strategies, and maintaining continuous monitoring.

References

Abdallah, M.A.M.: Distribution and risk assessment of phenols and polycyclic aromatic hydrocarbons in sediments from the continental shelf adjacent to Nile River estuaries, Egypt. Mar. Pollut. Bull. **212**, 117478 (2025)

Adeniji, A.O., Okoh, O.O., Okoh, A.I.: Levels of polycyclic aromatic hydrocarbons in the water and sediment of Buffalo River Estuary, South Africa and their health risk assessment. Arch. Environ. Contam. Toxicol. **76**(4), 657–669 (2019)

Amini-Birami, F., Keshavarzi, B., Esmaeili, H.R., Moore, F., Busquets, R., Saemi-Komsari, M., Zarei, M., Zarandian, A.: Microplastics in aquatic species of Anzali wetland: an important freshwater biodiversity hotspot in Iran. Environ. Pollut. **330**, 121762 (2023)

Ashayeri, N.Y., Keshavarzi, B., Moore, F., Kersten, M., Yazdi, M., Lahijanzadeh, A.R.: Presence of polycyclic aromatic hydrocarbons in sediments and surface water from Shadegan wetland–Iran: a focus on source apportionment, human and ecological risk assessment and sediment-water exchange. Ecotoxicol. Environ. Saf. **148**, 1054–1066 (2018)

Awe, A.A., Opeolu, B.O., Olatunji, O.S., Fatoki, O.S., Jackson, V.A., Snyman, R.: Occurrence and probabilistic risk assessment of PAHs in water and sediment samples of the Diep River, South Africa. Heliyon **6**(6) (2020)

Awonaike, B., Lei, Y.D., Parajulee, A., Mitchell, C.P., Wania, F.: Polycyclic aromatic hydrocarbons and quinones in urban and rural stormwater runoff: effects of land use and storm characteristics. ACS ES&T Water **1**(5), 1209–1219 (2021)

Bahr, A., Fischer, A., Vogt, C., Bombach, P.: Evidence of polycyclic aromatic hydrocarbon biodegradation in a contaminated aquifer by combined application of in situ and laboratory microcosms using 13C-labelled target compounds. Water Res. **69**, 100–109 (2015)

Beltran-Flores, E., Tayar, S., Blanquez, P., Sarra, M.: Effect of dissolved oxygen on the degradation activity and consumption capacity of white-rot fungi. J. Water Process Eng. **55**, 104105 (2023)

Berríos-Rolon, P.J., Cotto, M.C., Márquez, F.: Polycyclic aromatic hydrocarbons (PAHs) in freshwater systems: a comprehensive review of sources, distribution, and ecotoxicological impacts. Toxics **13**(4), 321 (2025)

Berrios-Rolon, P.J., Cotto, M.C., Márquez, F.: Polycyclic aromatic hydrocarbons (PAHs) in freshwater systems: a comprehensive review of sources, distribution, and ecotoxicological impacts. Toxics **13**(4), 321 (2025)

Brindha, K., Elango, L.: PAHs contamination in groundwater from a part of metropolitan city, India: a study based on sampling over a 10-year period. Environ. Earth Sci. **71**(12), 5113–5120 (2014)

Brown, J.N., Peake, B.M.: Sources of heavy metals and polycyclic aromatic hydrocarbons in urban stormwater runoff. Sci. Total Environ. **359**(1–3), 145–155 (2006)

Cai, Z., Yang, F., Song, Y., Liu, Y., Liu, W., Wang, Q., Sun, X.: Semiconducting mineral induced photochemical conversion of PAHs in aquatic environment: mechanism study and fate prediction. Sci. Total Environ. **860**, 160382 (2023)

Cao, X., Cui, X., Xie, M., Zhao, R., Xu, L., Ni, S., Cui, Z.: Amendments and bioaugmentation enhanced phytoremediation and micro-ecology for PAHs and heavy metals co-contaminated soils. J. Hazard. Mater. **426**, 128096 (2022)

Cao, Y., Wang, J., Xin, M., Wang, B., Lin, C.: Spatial distribution and partition of polycyclic aromatic hydrocarbons (PAHs) in the water and sediment of the southern Bohai Sea: Yellow River and PAH property influences. Water Res. **248**, 120873 (2024)

Capo, E., Ninnes, S., Domaizon, I., Bertilsson, S., Bigler, C., Wang, X.R., Bindler, R., Rydberg, J.: Landscape setting drives the microbial eukaryotic community structure in four swedish mountain lakes over the Holocene. Microorganisms **9**(2), 355 (2021)

Cheng, J., Ye, Q., Lu, Z., Zhang, J., Zeng, L., Parikh, S.J., Ma, W., Tang, C., Xu, J., He, Y.: Quantification of the sorption of organic pollutants to minerals via an improved mathematical model accounting for associations between minerals and soil organic matter. Environ. Pollut. **280**, 116991 (2021)

Cheshmvahm, H., Keshavarzi, B., Moore, F., Zarei, M., Esmaeili, H.R., Hooda, P.S.: Investigation of the concentration, origin and health effects of PAHs in the Anzali wetland: the most important coastal freshwater wetland of Iran. Mar. Pollut. Bull. **193**, 115191 (2023)

Collins, C., Fryer, M., Grosso, A.: Plant uptake of non-ionic organic chemicals. Environ. Sci. Technol. **40**(1), 45–52 (2006)

Das, P., Mukherjee, A., Hussain, S.A., Jamal, M.S., Das, K., Shaw, A., Layek, M.K., Sengupta, P.: Stable isotope dynamics of groundwater interactions with Ganges river. Hydrol. Process. **35**(1), e14002 (2021)

Dhara, A., Dutta, R.: A review on sources and distribution of polycyclic aromatic hydrocarbons (PAHs) in wetland ecosystem: Focusing on plant-biomonitoring and phytoremediation. Environ. Sci. Pollut. Res. 1–23 (2025)

Do Rego, E.C.P., Pereira Netto, A.D.: PAHs and BTEX in groundwater of gasoline stations from Rio de Janeiro City, Brazil. Bull. Environ. Contam. Toxicol. **79**(6), 660–664 (2007)

Du, J., Jing, C.: Anthropogenic PAHs in lake sediments: a literature review (2002–2018). Environ. Sci. Process. Impacts **20**(12), 1649–1666 (2018)

Enalbes, C.J., Njagi, D.M., Luo, C., Olago, D., Routh, J.: Reconstructing historical land use and anthropogenic inputs in lake Victoria basin: insights from PAH and n-Alkane trends. Toxics **13**(2), 130 (2025)

Espinosa-Ortiz, E.J., Rene, E.R., Gerlach, R.: Potential use of fungal-bacterial co-cultures for the removal of organic pollutants. Crit. Rev. Biotechnol. **42**(3), 361–383 (2022)

Ezekwe, C., Onwudiegwu, C.A., Uzoekwe, S.A.: Risk assessment and distribution of polycyclic aromatic hydrocarbons in water and sediments along Ibelebiri Axis of Kolo Creek, Ogbia Local Government Area, Niger Delta region of Nigeria. Chem. Africa **8**(3), 1199–1213 (2025)

Gholami, F., Shavandi, M., Dastgheib, S.M.M., Amoozegar, M.A.: Naphthalene remediation from groundwater by calcium peroxide (CaO_2) nanoparticles in permeable reactive barrier (PRB). Chemosphere **212**, 105–113 (2018)

Ghosh, P., Mukherji, S.: Fate, detection technologies and toxicity of heterocyclic PAHs in the aquatic and soil environments. Sci. Total Environ. **892**, 164499 (2023)

Grmasha, R.A., Abdulameer, M.H., Stenger-Kovács, C., Al-Sareji, O.J., Al-Gazali, Z., Al-Juboori, R.A., Meiczinger, M., Hashim, K.S.: Polycyclic aromatic hydrocarbons in the surface water and sediment along Euphrates River system: occurrence, sources, ecological and health risk assessment. Mar. Pollut. Bull. **187**, 114568 (2023)

Grmasha, R.A., Al-Sareji, O.J., Meiczinger, M., Al-Juboori, R.A., Stenger-Kovács, C., Lengyel, E., Majdi, H.S., AlKhaddar, R., Mohammed, S.J., Hashim, K.S.: Seasonal variation and concentration of PAHs in Lake Balaton sediment: a study on molecular weight distribution and sources of pollution. Mar. Pollut. Bull. **202**, 116333 (2024)

Gupta, H.: PAH determination in effluent and sludge samples of paper industry. Environ. Technol. Innov. **9**, 115–121 (2018)

Gupta, H., Kumar, R.: Distribution of selected polycyclic aromatic hydrocarbons in urban soils of Delhi, India. Environ. Technol. Innov. **17**, 100500 (2020). https://doi.org/10.1016/j.eti.2019.100500

Gupta, H., Dhiman, S., Kumar, A., Paliwal, S.K., Khan, G., Singh, H., Mishra, A., Mishra, A.K.: Photo- and bio-degradation of selected persistent organic pollutants. Sep. Purif. Rev. **54**(1), 3–16 (2025). https://doi.org/10.1080/15422119.2024.2332953

Hassan, F.M., Al-Obaidy, A.H.M.J., Noor, A.R., Alansari, N.: Pathway of total petroleum hydrocarbon in a lotic ecosystem: water and sediment of Tigris River, Iraq. Appl. Water Sci. **15**(8), 1–14 (2025)

Hu, W., Zhang, J., Li, D., Yuan, Y., Tang, Y., Hui, K., Jiang, Y., Tan, W., 2025. Study on factors influencing the transport and transformation of polycyclic aromatic hydrocarbons in soil–groundwater systems. Emerg. Contam. 100472

Hui, R., Tan, H.: Divergence of nutrients, salt accumulation, bacterial community structure and diversity in soil after 8 years of flood irrigation with surface water and groundwater. BMC Microbiol. **24**(1), 477 (2024)

Kim, P.G., Roh, J.Y., Hong, Y., Kwon, J.H.: Effects of soil water saturation on sampling equilibrium and kinetics of selected polycyclic aromatic hydrocarbons. Chemosphere **184**, 86–92 (2017)

Kumar, A., Gupta, H.: Activated carbon from sawdust for naphthalene removal from contaminated water. Environ. Technol. Innov. **20**, 101080 (2020)

Kurwadkar, S., Sethi, S.S., Mishra, P., Ambade, B.: Unregulated discharge of wastewater in the Mahanadi River Basin: risk evaluation due to occurrence of polycyclic aromatic hydrocarbon in surface water and sediments. Mar. Pollut. Bull. **179**, 113686 (2022)

Liu, Y., He, Y., Han, B., Liu, H., Tao, S., Liu, W.: Sewage discharge and organic matter affect the partitioning behaviors of different polycyclic aromatic hydrocarbons in a river surface sediment-pore water system. J. Hazard. Mater. **446**, 130757 (2023a)

Liu, Y., Li, X., Zhou, W., He, R., Zhang, Y., Zhao, N.: Electrical stimulation accelerated phenanthrene biodegradation coupling with nitrate reduction in groundwater. Sep. Purif. Technol. **309**, 123019 (2023b)

Liu, W., Xing, X., Zou, Y., Li, X., Gao, Y., Liu, Y., Zhu, X., Qi, S.: Novel insights into PAHs accumulation and multi-method characterisation of interaction between groundwater and surface water in middle Yangtze River: hydrochemistry, isotope hydrology and fractionation effect. Sci. Total Environ. **958**, 178023 (2025)

Lu, J., Guo, Z., Pan, Y., Li, M., Chen, X., He, M., Wu, H., Zhang, J.: Simultaneously enhanced removal of PAHs and nitrogen driven by Fe^{2+}/Fe^{3+} cycle in constructed wetland through automatic tidal operation. Water Res. **215**, 118232 (2022)

Lu, Z., Guo, Y., Xu, D., Xiao, H., Dai, Y., Liu, K., Chen, L., Wang, H.: Developmental toxicity and programming alterations of multiple organs in offspring induced by medication during pregnancy. Acta Pharmaceutica Sinica B **13**(2), 460–477 (2023)

MacKay, A.A., Gschwend, P.M.: Enhanced concentrations of PAHs in groundwater at a coal tar site. Environ. Sci. Technol. **35**(7), 1320–1328 (2001)

Mao, J., Guan, W.: Fungal degradation of polycyclic aromatic hydrocarbons (PAHs) by Scopulariopsis brevicaulis and its application in bioremediation of PAH-contaminated soil. Acta Agric. Scand. Sect. B—Soil Plant Sci. **66**(5), 399–405 (2016)

Mardani, A.: Health Impacts of Consuming Shellfish Contaminated with Polycyclic Aromatic Hydrocarbons (PAHs), Harvested from Waters Polluted by Petroleum and Industrial Activities. Doctoral dissertation, University of British Columbia (2021)

Meng, X., Wang, X., Zhang, C., Yan, S., Zheng, G., Zhou, L.: Co-adsorption of As (III) and phenanthrene by schwertmannite and Fenton-like regeneration of spent schwertmannite to realise phenanthrene degradation and As (III) oxidation. Environ. Res. **195**, 110855 (2021)

Moeckel, C., Monteith, D.T., Llewellyn, N.R., Henrys, P.A., Pereira, M.G.: Relationship between the concentrations of dissolved organic matter and polycyclic aromatic hydrocarbons in a typical UK upland stream. Environ. Sci. Technol. **48**(1), 130–138 (2014)

Montano, L., Baldini, G.M., Piscopo, M., Liguori, G., Lombardi, R., Ricciardi, M., Esposito, G., Pinto, G., Fontanarosa, C., Spinelli, M., Palmieri, I.: Polycyclic aromatic hydrocarbons (PAHs) in the environment: occupational exposure, health risks and fertility implications. Toxics **13**(3), 151 (2025)

Montuori, P., De Rosa, E., Di Duca, F., De Simone, B., Scippa, S., Russo, I., Sarnacchiaro, P., Triassi, M.: Polycyclic aromatic hydrocarbons (PAHs) in the dissolved phase, particulate matter, and sediment of the Sele River, Southern Italy: a focus on distribution, risk assessment, and sources. Toxics **10**(7), 401 (2022)

Nawrot, N., Pouch, A., Matej-Łukowicz, K., Pazdro, K., Mohsin, M., Rezania, S., Wojciechowska, E.: A multi-criteria approach to investigate spatial distribution, sources, and the potential toxicological effect of polycyclic aromatic hydrocarbons (PAHs) in sediments of urban retention tanks. Environ. Sci. Pollut. Res. **30**(10), 27895–27911 (2023)

Patel, A.B., Shaikh, S., Jain, K.R., Desai, C., Madamwar, D.: Polycyclic aromatic hydrocarbons: sources, toxicity, and remediation approaches. Front. Microbiol. **11**, 562813 (2020)

Rabodonirina, S., Rasolomampianina, R., Krier, F., Drider, D., Merhaby, D., Net, S., Ouddane, B.: Degradation of fluorene and phenanthrene in PAHs-contaminated soil using Pseudomonas and Bacillus strains isolated from oil spill sites. J. Environ. Manage. **232**, 1–7 (2019)

Radhakrishnan, K., Prakasheswar, P., Pradhap, D., Akramkhan, N., Rajkumar, A., Maheswaran, S.P., Hussain, S.M., Gandhi, K.S., Krishnakumar, S.: A baseline study on polycyclic aromatic hydrocarbons (PAHS) in surface Sediments, Kerala Estuaries, west coast of India. Total Environ. Res. Themes **6**, 100055 (2023)

Rodrigue, J.P.: Straits, passages and chokepoints: a maritime geostrategy of petroleum distribution. Cahiers De Géographie Du Québec **48**(135), 357–374 (2004)

Rokhbar, M., Keshavarzi, B., Moore, F., Zarei, M., Hooda, P.S., Risk, M.J.: Occurrence and source of PAHs in Miankaleh international wetland in Iran. Chemosphere **321**, 138140 (2023)

Salehi, M.: Surface complexation at charged organic surfaces. Rev. Mineral. Geochem. **91**(1), 149–173 (2025)

Salman, N.A., Al-Saad, H.T., Al-Imarah, F.J.: The status of pollution in the southern marshes of Iraq: a short review. South. Iraq's Marshes: Their Environ. Conserv. 505–516 (2021)

Sandoval-Herazo, E.J., Alfaro-De la Torre, M.C., Flores-Ramírez, R., Rodríguez-Torres, I., Lizardi-Jiménez, M.A., Berumen-Rodríguez, A.A., Espinosa-Reyes, G.: Polycyclic aromatic hydrocarbons emissions and sugarcane industry impacts in the Tamasopo Wetland, San Luis Potosi, Mexico. Int. J. Environ. Sci. Technol. 1–14 (2025)

Schwanen, C.A., Kronsbein, P.M., Balik, B., Schwarzbauer, J.: Dynamic transport and distribution of organic pollutants in water and sediments of the Rur River. Water Air Soil Pollut. **235**(1), 9 (2024)

Semenov, M.Y., Marinaite, I.I., Silaev, A.V., Begunova, L.A.: Composition, concentration and origin of polycyclic aromatic hydrocarbons in waters and bottom sediments of Lake Baikal and its tributaries. Water **15**(13), 2324 (2023)

Sha, H., Liu, Z., Sun, Y., Wang, Y., Wang, X., Zheng, J., Ma, Y., He, X.: Leachate leakage enhances the microbial diversity and richness but decreases Proteobacteria and weakens stable microbial ecosystem in landfill groundwater. Water Res. **243**, 120321 (2023)

Sheikh Fakhradini, S., Moore, F., Keshavarzi, B., Lahijanzadeh, A.: Polycyclic aromatic hydrocarbons (PAHs) in water and sediment of Hoor Al-Azim wetland, Iran: a focus on source apportionment, environmental risk assessment, and sediment-water partitioning. Environ. Monit. Assess. **191**(4), 233 (2019)

Shi, Q., Qin, H., Huo, X., Lu, H., Lu, S.: Polycyclic aromatic hydrocarbons in fish from four lakes in central and eastern China: bioaccumulation, pollution characteristics, sources, and health risk assessment. Sci. Total Environ. **951**, 175530 (2024)

Siddiqui, M.H., Bhanbhro, U., Shaikh, K., Ahmed, S., Begum, R., Shoukat, S., Khokhar, W.A.: Seasonal occurrence and ecological risk assessment of polycyclic aromatic hydrocarbons in the sediments and water in the left-bank canals of Indus River, Pakistan. Environ. Monit. Assess. **196**(12), 1165 (2024)

Sreedevi, M.A., Harikumar, P.S.: Occurrence, distribution, and ecological risk of heavy metals and persistent organic pollutants (OCPs, PCBs, and PAHs) in surface sediments of the Ashtamudi wetland, south-west coast of India. Regional Stud. Mar. Sci. **64**, 103044 (2023)

Sreedevi, M.A., Harikumar, P.S.: EPA-Priority polycyclic aromatic hydrocarbons in Ashtamudi Ramsar wetland, southwest coast of India: perspectives from the PMF modelling, health risk evaluation, and ERA technique. Regional Stud. Mar. Sci. **81**, 103940 (2025)

Storey, R.G., Fulthorpe, R.R., Williams, D.D.: Perspectives and predictions on the microbial ecology of the hyporheic zone. Freshw. Biol. **41**(1), 119–130 (1999)

Suresh, A., Soman, V.: Polycyclic aromatic hydrocarbons (PAHs) in surficial sediments of Vembanad wetland ecosystem, southwest coast of India: an integrated assessment of dissemination, source apportionment, ecological and public health implication. Mar. Pollut. Bull. **220**, 118369 (2025)

Thuy, H.T.T., Loan, T.T.C., Luu, P.T., Van Dong, N., Bao, L.D., Phuong, T.H., Khanh, N.T., Yen, T.T.H., Huy, D.X.: Spatial and temporal variations of PAHs in surface sediments of estuarine and coast of CanGio wetland, Vietnam. Environ. Sci. Poll. Res. **28**(10), 11962–11975 (2021)

Tsaridou, C., Karabelas, A.J.: Drinking water standards and their implementation—a critical assessment. Water **13**(20), 2918 (2021)

Tulcan, R.X.S., Liu, L., Lu, X., Ge, Z., Rojas, D.Y.F., Silva, D.M.: PAHs contamination in ports: status, sources and risks. J. Hazard. Mater. **475**, 134937 (2024)

Umeh, C.T., Nduka, J.K., Omokpariola, D.O., Morah, J.E., Mmaduakor, E.C., Okoye, N.H., Lilian, E.E.I., Kalu, I.F.: Ecological pollution and health risk monitoring assessment of polycyclic aromatic hydrocarbons and heavy metals in surface water, southeastern Nigeria. Environ. Anal. Health Toxicol. **38**, e2023007 (2023)

Woznicki, S.A., Barber, J., Butcher, J.B., Essoka, J., Harris, M., Mehaffey, M., Pluta, B., Shabani, A., Whung, P.Y.: Compound Impacts of Fluvial Flooding and Sea-Level Rise on Benzo[a]pyrene Transport in the Lower Darby Creek Area Superfund Site, Pennsylvania, USA. ACS ES&T Water (2025)

Wu, P., Wang, Z., Bhatnagar, A., Jeyakumar, P., Wang, H., Wang, Y., Li, X.: Microorganisms-carbonaceous materials immobilised complexes: synthesis, adaptability and environmental applications. J. Hazard. Mater. **416**, 125915 (2021)

Xia, Y., Zhang, M., Tsang, D.C., Geng, N., Lu, D., Zhu, L., Igalavithana, A.D., Dissanayake, P.D., Rinklebe, J., Yang, X., Ok, Y.S.: Recent advances in control technologies for non-point source pollution with nitrogen and phosphorous from agricultural runoff: current practices and future prospects. Appl. Biol. Chem. **63**(1), 8 (2020)

Xu, X., Zhou, H., Chen, X., Wang, B., Jin, Z., Ji, F.: Biodegradation potential of polycyclic aromatic hydrocarbons by immobilised Klebsiella sp. in soil washing effluent. Chemosphere **223**, 140–147 (2019)

Yuan, Z., He, B., Wu, X., Simonich, S.L.M., Liu, H., Fu, J., Chen, A., Liu, H., Wang, Q.: Polycyclic aromatic hydrocarbons (PAHs) in urban stream sediments of Suzhou Industrial Park, an emerging eco-industrial park in China: occurrence, sources and potential risk. Ecotoxicol. Environ. Saf. **214**, 112095 (2021)

Yue, R., An, C., Ye, Z., Li, X., Li, Q., Zhang, P., Qu, Z., Wan, S.: A pH-responsive phosphoprotein washing fluid for the removal of phenanthrene from contaminated peat moss in the cold region. Chemosphere **313**, 137389 (2023)

Zafra, G., Moreno-Montaño, A., Absalón, A.E., Cortés-Espinosa, D.V.: Degradation of polycyclic aromatic hydrocarbons in soil by a tolerant strain of Trichoderma asperellum Environ. Sci. Pollut. Res. **22**(2), 1034–1042 (2015)

Zhang, P., Chen, Y.: Polycyclic aromatic hydrocarbons contamination in surface soil of China: a review. Sci. Total Environ. **605**, 1011–1020 (2017)

Zhang, X.Z., Xie, J.J., Sun, F.L.: Effects of three polycyclic aromatic hydrocarbons on sediment bacterial community. Curr. Microbiol. **68**(6), 756–762 (2014)

Zhang, M., Tao, S., Wang, X.: Interactions between organic pollutants and carbon nanomaterials and the associated impact on microbial availability and degradation in soil: a review. Environ. Sci. Nano **7**(9), 2486–2508 (2020)

Zhang, L., Yi, M., Lu, P.: Effects of pyrene on the structure and metabolic function of soil microbial communities. Environ. Pollut. **305**, 119301 (2022)

Zhang, P., Yuan, Y., Zhang, J., Wen, T., Wang, H., Qu, C., Tan, W., Xi, B., Hui, K., Tang, J.: Specific response of soil properties to microplastics pollution: a review. Environ. Res. **232**, 116427 (2023)

Zhang, S., Xing, X., Yu, H., Du, M., Zhang, Y., Li, P., Li, X., Zou, Y., Shi, M., Liu, W., Qi, S.: Fate of polycyclic aromatic hydrocarbon (PAHs) in urban lakes under hydrological connectivity: a multi-media mass balance approach. Environ. Pollut. **366**, 125556 (2025a)

Zhang, Y., Guo, Z., Peng, C., Zhang, C., Shen, M., Li, J.: An integrated framework for identifying primary risk contributors and risk zones of combined soil pollution in an industrial city. J. Environ. Chem. Eng. **13**(2), 115607 (2025b)

Zhao, X., Yang, F., Li, Z., Tan, H.: Formation and emission characteristics of PAHs during pyrolysis and combustion of coal and biomass. Fuel **378**, 132935 (2024)

Zheng, Y., Cao, T., Zhang, Y., Xiong, J., Dzakpasu, M., Yang, D., Yang, Q., Liu, Y., Li, Q., Liu, S., Wang, X.: Characterisation of dissolved organic matter and carbon release from wetland plants for enhanced nitrogen removal in constructed wetlands for low C-N wastewater treatment. Chemosphere **273**, 129630 (2021)

Zheng, X., Wang, H., Tao, Y., Kou, X., He, C., Wang, Z.: Community diversity of soil meso-fauna indicates the impacts of oil exploitation on wetlands. Ecol. Ind. **144**, 109451 (2022)

Zhi, H., Zhao, Z., Zhang, L.: The fate of polycyclic aromatic hydrocarbons (PAHs) and organochlorine pesticides (OCPs) in water from Poyang Lake, the largest freshwater lake in China. Chemosphere **119**, 1134–1140 (2015)

Zou, J.J., Dai, C., Hu, J., Tong, W.K., Wang, J., Gao, M.T., Zhang, Y., Leong, K.H., Zhou, L.: Nanobubbles promote microbial degradation of polycyclic aromatic hydrocarbons: enhanced detoxification and optimised degradation pathway. Environ. Res. 122368 (2025)

Effects and Risk Assessment of Polycyclic Aromatic Hydrocarbons in the Soil Ecosystems

4

Madhupriya, Sushil Kumar, Rakesh Kumar, Chander Parkash, and Srinivasan Balachandran

Abstract

Polycyclic aromatic hydrocarbons (PAHs) represent as a long-lasting and stable environmental organic pollutant composed of hydrocarbons that threaten both ecology and individual well-being with long-term stability, mutagenic, carcinogenic, and teratogenic properties. Soil PAHs are dependent on industrial processes, transportation, automobile exhaust, and physico-chemical characteristics, including hydrophobicity and non-polar nature. Low solubility and strong sorption will lead PAHs to remain intact in soil and threaten both ecosystems and human health for a long time. PAHs ubiquitous occurrence in various environmental components (air, water, soil, biological organisms) and associated effects are future need using recent advanced detection techniques and improves the understanding of health risk assessment. The present chapter discusses the multiplicity of PAHs routes, associated risk assessment and reviews the recent practices to evaluate the PAH related risks and advance sustainable soil management practices and future perspectives.

Keywords

Polycyclic aromatic hydrocarbons • Organic pollutants • Risk assessment • Soil ecosystem

Madhupriya · S. Kumar (✉) · S. Balachandran
Department of Environmental Studies, Siksha-Bhavana (Institute of Science), Visva-Bharati, Santiniketan, West Bengal, India
e-mail: sushil.kumar@visva-bharati.ac.in

R. Kumar
Department of Environmental Sciences, University of Jammu (J&K), Jammu, India

C. Parkash
Department of Chemistry, I.K. Gujral Punjab Technical University, Kapurthala, Punjab, India

H. Gupta et al. (eds.), *Ubiquitous Polycyclic Aromatic Hydrocarbon Contamination*, Emerging Contaminants and Associated Treatment Technologies,
https://doi.org/10.1007/978-3-032-11764-9_4

Introduction

PAHs are one of the contaminants that are continuously added to the environment with the rising scale of urbanization, industrialization and lifestyle changes (Zhang and Chen 2017; Hussain et al. 2018). They are family of hydrocarbon-based pollutants, having multiple (at least two or more) benzoic rings, either arranged as a single chain, clustered or angular patterns composed solely of hydrocarbons (carbon and hydrogen). It including unsubstituted parent PAHs and their alkyl-substituted derivatives. However, the broader term 'polycyclic aromatic compounds also comprises of functional derivatives such as nitro-PAHs and heterocyclic analogs like aza-arenes (Abdel-Shafy and Mansour 2016; Hussain et al. 2018). PAHs are a serious global concern owing to their extensive distribution, long-term environmental persistence, high lipid solubility and bioaccumulation, human toxicity (carcinogenic, teratogenic, and mutagenic) etc. It is known for their toxicity and exhibit long-term persistence in surroundings as a result of low vapour pressure, weak volatility and persistence against degradation (Brimo et al. 2018; Patel et al. 2020; Montano et al. 2025). The rising occurrence of PAHs in soil is widely emitted from natural and human interventions. The anthropogenic sources include car exhaust fumes, cigarette smoke, gas and oil spills, industrial and wastewater discharge, coal, and burning biomass and food while natural sources encompass wildfire, volcanic outbursts, and biological processes (Venkatraman et al. 2024; Wang et al. 2024b, c). To address the growing concern of pollutants, PAH are categorised as priority contaminants by UNEPA, posing toxicological significance affecting individual health and the ecosystem (Wang et al. 2024b, c).

PAHs are distributed across different environmental settings however, soil is intended as the dominant sink and reservoir (Hussain et al. 2018; Maletic et al. 2019; Kumar et al. 2021). In general, they are usually divided into two distinct groups depending on their molecular structure, different classes and sources of emission: (a) Category-I includes Low Molecular Weight (LMW) with four or less than four benzoic rings, they are emitted in gaseous phase due to their Low Molecular Weight and (b) Category-II High Molecular Weight (HMW) containing five or more aromatic rings, considering their molecular mass they are emitted as particulate PAHs. Based on sources of emission, it is classified as petrogenic PAHs emitted from petroleum products, biogenic emitted from biological activities, and pyrogenic emitted from combustion processes (Yebra-Pimentel et al. 2015; Manaswini et al. 2024). Since 1983, sixteen PAHs are listed as priority hazardous compounds by UNEPA and EU owing to their occurrence in the environmental matrices in highest concentration, hazardous properties, greater exposure, toxicity, and recalcitrant properties, listed as Naphthalene having two rings, Acenaphthene, Acenaphthylene, Fluorene, Anthracene, Phenanthrene contain three aromatic rings, these above comes under low molecular weight PAHs. Fluoranthene, Pyrene, Benzo[a]anthracene, Chrysene, have four benzoic

Table 4.1 The key physical properties of sixteen (16) prioritised PAHs listed by the USPEA

PAH	Molecular formula	Molecular weight	Water solubility (mg/L)	Phase distribution
Acenaphthene	$C_{12}H_{10}$	154.2	3.8	Particle
Acenaphthylene	$C_{12}H_8$	152.1	16.1	Gas
Anthracene	$C_{14}H_{10}$	178.2	0.045	Particle gas
Benz[a]anthracene	$C_{18}H_{12}$	228.3	0.011	Particle
Benzo[b]fluoranthene	$C_{20}H_{12}$	252.3	0.0008	Particle
Benzo[k]fluoranthene	$C_{20}H_{12}$	252.3	0.0015	Particle
Benzo[a]pyrene	$C_{20}H_{12}$	252.3	0.0038	Particle
Benzo[ghi]perylene	$C_{22}H_{12}$	276.3	0.00026	Particle
Chrysene	$C_{18}H_{12}$	228.2	0.0015	Particle
Dibenz[a,h]anthracene	$C_{22}H_{14}$	278.3	0.0005	Particle
Fluoranthene	$C_{16}H_{10}$	202.2	0.26	Particle gas
Fluorene	$C_{13}H_{10}$	166.2	1.9	Gas
Indeno[1,2,3 cd] pyrene	$C_{22}H_{12}$	276.3	0.062	Particle
Naphthalene	$C_{10}H_8$	128.1	31	Particle
Phenanthrene	$C_{14}H_{10}$	174.2	1.1	Particle gas
Pyrene	$C_{16}H_{10}$	202–2	0.132	Particle gas

Source Yebra-Pimentel et al. (2015); Alegbeleye et al. (2017); Hussain et al. (2018); Vijayanand et al. (2023)

rings, Benzo[b]fluoranthene, Benzo[k]fluoranthene, Dibenzo[ah]anthracene possess five rings, and Benzo[ghi]prylene, and Indene[1,2,3-cd] pyrene contain six rings belongs to high molecular weight PAHs (Table 4.1) (Hussain et al. 2018; Rabieimesbah et al. 2024).

Soil is known to be both a source as well as a major sink/reservoir for the atmospheric PAHs, with atmospheric dry and wet deposition, particularly through precipitation, is a key pathway of PAHs entry into the soil (Ali et al. 2024). Human exposure to PAHs via soil is more noticeable than water and air (Hussain et al. 2018). The occurrence of hydrophobicity and insolubility in water, they have high tendency of absorption on the surface soil rich in organic matter. The adsorbed PAHs on particles become mobile after deposition on the soil surface over a prolonged duration, worsening the biological condition of the soil, affects the biological communities, degrades the soil quality, and ultimately transfer to food supply chain and affect human health, ecosystem (Saeedi et al. 2020; Kumar et al. 2021; Aparicio et al. 2022).

The present chapter provides an outline of the multiplicity of PAHs sources in soil systems, general physicochemical properties of various polyaromatics, distribution patterns in environmental components, and recent advances for the evaluation of risks to individual health and ecosystem.

Multiplicity of Origin and Properties of PAHs

In soils, PAH originate from a range of natural and anthropogenically derived sources, mainly derived from process of incomplete combustion of non-renewable products containing hydrocarbons (Abdel-Shafy and Mansour 2016; Wiłkomirski et al. 2018). Based on their origin, PAHs sources are classified into biogenic, petrogenic, and pyrogenic. Among them, forest and grassland fires, oil spills, chlorophyllous vegetation, volcanic eruptions, and microbial activities (e.g., fungi and bacteria) are a few examples of PAH additions from natural sources (Maletic et al. 2019; Patel et al. 2020).

In contrast, anthropogenic sources include industrial emissions, in the form of incomplete combustion activities such as waste incineration, during the conversion of coal tar into pitch, production of gas from coal, industrial production of cement, asphalt production, dye manufacturing, tire-rubber production, insecticides and fungicides, exhaust from refineries and power plants, diesel and gasoline engines, aluminium manufacturing are key contributors of PAHs (Lawal 2017; Maletic et al. 2019). Besides the above-cited sources, vehicular exhaust, burning of agricultural stubble, domestic emissions from garbage and wood burning, cooking on kerosene/oil/coal stoves contribute a significant amount of PAHs across the surroundings (Fig. 4.1) (Hussain et al. 2018; Swit et al. 2023). Moreover, in the recent times, the open burning of electronic waste (e-waste) for the recovery of precious metals has emerged to be a primary source of PAHs into soils due to a lack of a proper disposal mechanism. For instance, in Vietnam, the PAHs concentrations in soil were found to range between 25 and 37,000 ng/g dw (Hao et al. 2020; Barathi et al. 2023). Entire world generates 510,000 t a^{-1} PAHs from different sources in which China alone emits 125,000 t a^{-1}. China emits 56,000 t a^{-1} PAHs from boiler installed in industries, production of coke and coal burning (Shen 2016; Zhang et al. 2021). To identify PAHs source in the soil ecosystem various studies relied on analysing the diagnostic ratio. In this method the ratios of different PAHs like ANT to (ANT + PHE), FL-2 to (FL-2 + PYR), BEN-a to (BEN-a + CHR), IDP/(IDP + BEN-g) and BEN-a/CHR suggest possible sources. If the ratio of FL-2/(FL-2 + PYR) is less than 0.4 is shows petroleum or liquid fuel sources and if FL-2 to (FL-2 + PYR) is greater than 0.5 shows the sources from petroleum and biofuel burning. If the ratio of IDP/ (IDP + BEN-g) falls between 0.2 and 0.5 it suggests sources like biomass, coal burning and petroleum combustion (Liu et al. 2019; Qu et al. 2020; Zhang et al. 2021; Shukla et al. 2022; Ambade et al. 2023). Shukla et al. (2022) in their study found the ratio of IDP/ (IDP + BEN-g) between 0.2 and 0.35 suggesting incomplete burning of carbon based fuels in 60% of the selected sites. Zhang et al. (2021) study shows the ratio of IDP/(IDP + BEN-g) between 0.2 and 0.6 indicating petroleum and other sources.

Fig. 4.1 Showing PAHs major sources in different environmental matrices

Effects of PAHs on Human Health

The health-related impact of PAHs is a function of exposure pathways and concentration in environmental components (air, water, and soil). PAHs contamination within soil matrix is a serious problem to both environmental matrices and human health as they are highly toxic, have high bioaccumulation potential, and are recalcitrant to ecosystems and living organisms (Balcıoglu 2016; Barbosa et al. 2023; Wang et al. 2024b, c). In general, PAHs exhibit toxicity by binding to their reactive metabolites (e.g., Dihydrodiols and Epoxides) to the nucleic acid and cellular polypeptides, which leads to developmental malfunction, alteration in gene and development of cancer due to cell damage and biochemical disruption (Mishra et al. 2022; Sankar et al. 2023; Montano et al. 2025). The adsorption, dispersion and toxic effects of PAH in humans is governed by multiple conditions such as exposure pathway, physicochemical properties of chemical species, and concentration (Mallah et al. 2022; Shukla et al. 2022). In general, people come in contact with PAHs, which include skin contact, the respiratory tract (inhalation of contaminated air), and the gastrointestinal tract (consuming food exposed to PAHs, smoking tobacco) (Hussain et al. 2018; Mallah et al. 2022). The severity of health risks depends upon the rate of exposure, exposure time, level of PAH, and the innate toxicity of individual PAHs. In addition, other factors includes pre-existing health conditions, age, a person's habits, individual traits, presence of other pollutants etc. (Kumar et al. 2014). For instance, approximately 70% of PAHs enter their body through diet, including cereals, vegetables and oils, even in a non-smoking person. Consuming fire-cooked (over an open flame) food is primary contributor of PAH within human diet. Smoking of a single cigarette leads to uptake of BaP ranges from 20 to 40 ng and also contains 40 suspected human carcinogens (Hussain et al. 2018; Berlinger et al. 2024). Tongo et al. (2017) worked on fishes and other meat products in which concentration of $\sum$16 PAH is found 3585 $\mu g\ kg^{-1}$ in S. scombrus fish species (Tongo et al. 2017), 154.5 $\mu g\ kg^{-1}$ in roasted pork (Ofori et al. 2020), and 0.4 to 2290.32 $\mu g\ kg^{-1}$ in meat (Kalteh et al. 2024). Gholami-Borujeni et al. (2025) observed that meat kebabs either barbecued, grilled or smoked contains 359.32 $\mu g\ kg^{-1}$ pyrene, 263.20 $\mu g\ kg^{-1}$ naphthalene. PAHs contaminate water through solubilizing chemical compounds present in soil, through accidental spillage of oil during shipment, from industrial effluents (Yebra-Pimentel et al. 2015). The key problem regarding PAH to human that it can cause cancer (lung, skin, bladder), effects reproduction (developmental abnormalities, low birth weight, decrease fertility), effects respiratory systems (can cause asthma, bronchitis, hamper lung function), weakens immune system which make person prone to infections, effects nervous system (leads to behavioural modification, mental deterioration), also can cause dermatitis, skin irritation, rashes, patches when PAHs comes in contact with skin (Sombiri et al. 2024; Montano et al. 2025). The effects of PAHs on humans and others are provided in Table 4.2.

Effects of PAHs on Soil Health

Soil contamination with PAHs exhibits significant cancer-causing potential, hazard potential, and mutagenicity in fauna, flora, and humans, resulting in the loss of biodiversity, heterogeneous distribution of soil microorganisms, and degraded soil quality, and hampers soil physical and chemical properties. This in turn can affect the enzymatic activities of soil microbiota, which severely affects organic carbon and pH of soil and sediments, and binds with soil organic matter, and acts as a depositional reservoir of PAH (Hussain et al. 2018; Li et al. 2024a, b). Organic pollutants (PAHs) associated with soil percolate downward and pollute the groundwater, putting human well-being and ecological system at risk. They are easily transported from contaminated sites to the hierarchy of energy levels through food chain, affects the organisms of the respective trophic level and higher organisms (Eldos et al. 2022; Barathii et al. 2023). Further, accumulation of contaminants in organisms loaded with Pyrene negatively impacts the dehydrogenase enzymatic activities of soil and severely affects the microbes of phyla such as *Chlorflexi, Alphaproteobacteria, Actinobacteria, Deltapro teobacteria* and *Crenarchaeota* shows decreasing trends of population. In contrast, phyla belonging to *Acidobacteria, Betaproteobacteria and Gamaproteobacteria* exhibit growth in their population (Lipinska et al. 2014; Ren et al. 2015; Alegbeleye et al. 2017). In addition, exposure to Anthracene, Fluorene, Naphthalene, and Phenanthrene results in reduction of phosphatase reaction in soil (Lipinska et al. 2019). Phosphatase reaction in soil refers to the ability of enzymes to convert organically bound phosphorus into inorganic phosphate, enhancing its availability to plants. This mechanism is crucial for the phosphorus cycle in soils and is particularly important in phosphorus-deficient soils. Similarly, the PAHs like Phenanthrene

Table 4.2 The summary of the impact on humans, plants, other organisms, and mechanisms of PAHs exposure

Catergory	Effects of PAHs	Mechanisms	References
Humans	Carcinogenic effects (e.g., lung, bladder and gastro-intestinalcancers	Rate of exposure, time, amount, and the inherent toxic properties of each PAH	Balcıoglu (2016) Barbosa Jr et al. (2023)
	Eye inflammation, jaundice, nausea, muscle spasms, Respiratory problems (e.g., asthma, lung dysfunction, abnormalities, chronic bronchitis, kidney and liver dysfunction	Occurs via inhalation, ingestion and dermal	Sankar et al. (2023)
	Developmental and Reproductive toxicity		Montano et al. (2025)
	DNA damage, mutation		
	Abnormal development of foetus, development of malignant tumours, reduce weight of newborn		
Plants	Hampers growth, development and reproduction	Adsorb on leaf surface blocks stomata pore supress chlorophyll synthesis	Tarigholizadeh et al. (2023)
	Obstruct respiration, photosynthesis, and nutrient uptake	It can enter from contaminated soil via roots	Wei et al. (2024)
	Increases chlorophyll gas exchange, induces oxidative stress resulting in damage to plant cell membrane, DNA, and proteins		
Aquatic organisms	Carcinogenic, DNA mutagens and endocrine disruptors	Enters through ingestion, inhalation and skin contact react with other pollutants lead to organ toxicity	Alegbeleye et al. (2017)
	Reduced abundance of benthic invertebrates		Li et al. (2024)
	Behavioural and physiological changes		
	Alter egg production in fish, hampers the development of embryonic yolk sac		
	Accumulation of unwanted fluid around the heart, haemorrhaging	Embryonic exposures with PAHs	

(continued)

Table 4.2 (continued)

Catergory	Effects of PAHs	Mechanisms	References
	Hampers functioning of heart, loss of nerve cells and reduce movement		
	Bioaccumulation, biomagnification		
	Reduced enzymatic activities in reptiles	PAHs get easily absorb as they have low digestion rate	Wang et al. (2024b, c)
Other Organisms	Results in loss of some species of microbes	More toxic to sensitive species, accumulates in soil and sediments impacting food chain	Alegbeleye et al. (2017)
	Abundance of hydrocarbon clastic bacteria in soil		
	Reduce biodiversity and evenness of many species		

and Naphthalene obstruct oxidation–reduction activity and urea hydrolysis in soil (Lipinska et al. 2013, 2014). In soils, the microbial function and enzymatic activities represent the health of the soil, any obstruction in their activity can render soil unproductive for plant growth and in long term cause damage to ecosystems. Humans and other organisms are directly or indirectly get exposed by the accumulated PAHs in soil during the uptake of vegetables, plant-based product, and food chain, as plants absorb PAHs from contaminated sites through roots and transport them to other parts of the plant. From soils, the PAHs could be transported to groundwater and atmosphere through resuspension, downward leaching, and evaporation (Alegbeleye et al. 2017; Hussain et al. 2018). Based on the abundance of PAHs in soils, the degree of contamination has been classified into three categories, (a) unpolluted having PAHs concentration less than 200 ng/g, (b) weakly polluted having PAHs concentration between 200–600 ng/g and (c) heavily polluted having concentration of more than 600 ng/g (Bortey-Sam et al. 2014). For effective implementation of on-field remediation of contaminated soils, a thorough knowledge about the composition of PAHs with biochemical markers of soil is required (Sui et al. 2021; Das et al. 2025). Strong perception of pollution dynamics and their correlation with microbial profile of soil will be effective in choosing remediation techniques (Davletgildeeva and Kuznetsov 2024). The effects of PAHs on huma health and soil ecosystem are depicted in Fig. 4.2.

Effects of PAHs on Plants Health

Plant health in the terrestrial environment is directly influenced via the contamination status of soil and the surrounding environment. The level of toxicity depends upon factors like concentration of PAHs, type of PAHs, exposure duration, type of

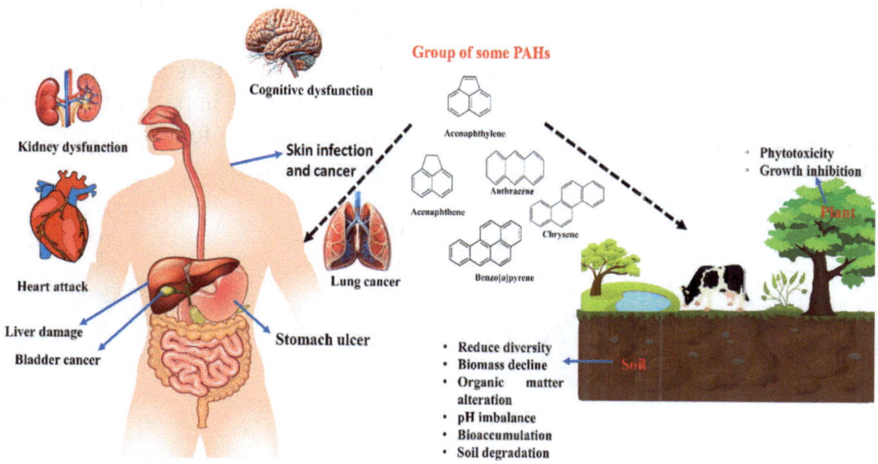

Fig. 4.2 Showing the Effects of PAHs on human health and soil ecosystem

plant species and state of the environment (Barathi et al. 2023; Tarigholizadeh et al. 2023). The rising concentration level of soil bound PAH, adversely alter growth, development and reproduction in vegetation. PAHs directly hamper plants respiration, photosynthesis, and nutrient uptake, increase chlorophyll gas exchange and growth inhibition hormones, and promote oxidative imbalance, resulting in harm to plant cell membranes, DNA, and proteins. PAHs directly transfer from root to grain. Wang et al. (2020) in their investigation on wheat plant found the concentration of PAHs in soil (1081 μg kg − 1 DM) transferred to root (464 μg kg − 1 DM), stems (365 μg kg − 1 DM), and leaves (323 μg kg − 1 DM) showing transportation of PAH from root system to shoots. Zhao et al. (2022) detected PAH at a concentration of 274.66 ng/g in plant tissue. Sushkova et al. (2021) measured the level of $\sum$16PAH in roots of plant Phragmites australis Cav. was 327 μg kg^{-1}. Alteration in soil microbial species, reduced soil fertility, nutrient availability, increased soil acidity, and affect soil's water holding capacity, impacting plants by restricting their growth and efficiency (Schwab and Dermody 2021; Tarigholizadeh et al. 2023; Wei et al. 2024). Plants absorb PAHs via roots and move them to different tissues, and accumulate them in the edible parts of the plant, thus directly threatening living organisms in successive trophic levels in the food chain (Alagic et al. 2015; Hunt et al. 2019).

Ecological Risk Assessment

Risk assessment and risk indices of single and mixture of PAHs relies on total concentration of observed PAHs estimated from different environmental media including soil, air and water. Severity of threats are calculated in the form of either non-threshold or threshold effects based on the duration of exposure to the respective pollutants (PAHs). The calculation of potential ecological impacts resulting from deposition of PAHs in soil, aquatic bodies and plant tissues are based on the estimated mass concentration. For assessing ecological impacts of PAHs in soil, using the risk quotient (RQ) will be calculated using the following equations (Abdel-Shafy et al. 2016; Feng et al. 2025; Rigi et al. 2025)

$$RQ = \frac{CPAH_S}{CQV} \tag{4.1}$$

$$RQNC_S = \frac{CPAH_S}{CQV(NC)_S} \tag{4.2}$$

$$RQMPC_S \frac{CPAH_S}{CQV(MPC)_S} \tag{4.3}$$

$$RQ \sum PAH_S = \sum_{i=1}^{16} RQ_i \quad RQ_i \geq 1 \tag{4.4}$$

$$RQ \sum PAH_S(NC)_S = \sum_{i=1}^{16} RQ_i(NC)_S \quad RQ_i(NC_S) \geq 1 \tag{4.5}$$

$$RQ \sum PAH_S(MPC_S) = \sum_{i=1}^{16} RQ_i(MPC_S) \quad RQ_i(MPC_S) \geq 1 \tag{4.6}$$

Here *CPAHs* represent the concentration of individual PAH, while *CQV* refers to equivalent grade estimate of specific PAH. NCs stands for the concentration of PAHs where effects are very small and MPCs stands for the concentration at which the effects are unbearable. The *CQV(NCs) and CQV (MPCs)* represents these thresholds in soil ecosystem. *RQ∑PAHs (NCs) and RQ∑PAHs (MPCs)* are individual-level calculated as RQNCs and RQMPCs (Younis et al. 2023; Rabieimesbah et al. 2024). Categories linked to ecological risk are provided in Table 4.3.

Human Health Risk Assessment

Health risk evaluation is an empirical approach to understand the effect of PAHs on different age groups of humans to evaluate carcinogenic and non-carcinogenic effects was formulated by the United States Environmental Protection Agency (USEPA). In order to evaluate the risk Toxicity Equivalency Factor (TEF) method is used for assessing the health effects of PAH as it can harm individual health through respiratory, oral and skin exposure. The calculation of exposure is based on the value provided for exposure time, exposure rate, body weight, and age to examine possible health impacts. The carcinogenic toxicity of an individual PAHs

Table 4.3 Classification of the Ecological risk of specific and total PAHs with different risk categories

PAH type	Risk level	RQ (NCs)	RQ (MPCs)
Individual PAHs	Minimum	< 1	< 1
Individual PAHs	Moderate	≥ 1 and < 1	< 1
Individual PAHs	Severe	≥ 1	≥ 1
Combined PAHs (∑PAHs)	Risk free	< 1	< 1
Combined PAHs (∑PAHs)	Low	≥ 1, < 800	< 1
Combined PAHs (∑PAHs)	Moderate 1	≥ 800	< 1
Combined PAHs (∑PAHs)	Moderate 2	< 800	≥ 1
Combined PAHs (∑PAHs)	High	≥ 800	≥ 1

Source Rabieimesbah et al. (2024); Younis et al. (2023)

to the toxicity of Benzo[a]Pyrene taken as reference while cancer risk is estimated by aggregating individual PAH concentrations and multiplying with TEF values (Samburova et al. 2017; Hussain et al. 2018; Qu et al. 2020; Chebykina et al. 2021; Rabieimesbah et al. 2024).

$$Total\ BaPeq/TEQ = \sum_{i=1}^{n} C_i \times TEF_i \tag{4.7}$$

Here, C_i represents specific level of PAH, Toxic equivalency factor denoted by TEF_i of the specific PAHs.

The method in which Benzo[a]Pyrene is used to determine the relative hazardous properties of a single PAH is known as Toxicity Equivalency Quotient (*TEQ*) (Samburova et al. 2017; Agapkina et al. 2018). Based on carcinogenic risk from contaminated soil, the Canadian Soil Quality Guidelines set Maximum Acceptable Concentration (MAC) for carcinogenic chemicals in soil. It was calculated as ILCR 10^{-5} for MAC value of 5.3 ng/g of BaP or 10^{-6} for MAC value of 0.6 ng/g of BaP (CCME 2010; Kumar et al. 2015; Hussain et al. 2018). ILCR through ingestion, inhalation and dermal can be calculated using Eqs. 4.8, 4.9, 4.10 and 4.11.

$$ILCRS(ingestion) = \frac{TEQ \times (CSF(ingestion) \times \sqrt[3]{BW \div 70} \times IRSoil \times EF \times ED}{BW \times AT \times Cf} \tag{4.8}$$

$$ILCRS(inhalation) = \frac{TEQ \times (CSF(inhalation) \times \sqrt[3]{BW \div 70} \times IRair \times EF \times ED}{BW \times AT \times PEF \times Cf} \tag{4.9}$$

$$ILCRS(dermal) = \frac{TEQ \times (CSF(dermal) \times \sqrt[3]{BW \div 70} \times SA \times AF \times ABS \times EF \times ED}{BW \times AT \times Cf} \quad (4.10)$$

$$ILCRs = \sum (ILCR_{(ing)} + ILCR_{(inh)} + ILCR_{(der)}) \quad (4.11)$$

where CSF refers to the Cancer Slope Factor, CSF (ingestion) = 7.30, CSF (inhalation) = 3.85, CSF (dermal) = 25.0 mg kg^{-1} day^{-1}. BW (Body Weight) = 70 kg for adult and 15 kg for child; EF (Exposure frequency) = 365 days $year^{-1}$ for adult and 180 days for child; ED (Exposure duration) = 30 years for adult and 6 years for child; AT (Average life span) = 70 years; Conversion factor (Cf) is equals to 10^{-6}; PEF (Soil emission factor) = 1.32×10^9 m^3 kg^{-1}; SA denotes the Skin surface area assigned as 0.5 m^2 for adult and 0.28 m^2 for child; AF, representing soil- skin adherence factor for adult and child equals 10^{-5} mg cm^{2-1} and 0.2 kg cm^{2-1}; ABS (Dermal adsorption factor) = 0.1; IRSoil (Soil intake rate) = 10^{-4} kg day^{-1}; Inhalation rate (IRair) is taken as 20 m^3 day^{-1} for adult and 10 m^3 day^{-1} (Agapkina et al. 2018; Shi et al. 2021). In the soil ecosystem, concentration of PAH, benzo(a)pyrene, and corresponding ILCR values are provided in Table 4.4.

Table 4.4 A comparison of PAH concentrations, benzo(a)pyrene equivalent, and the associated ingested incremental Lifetime Cancer Risk (ILCRs) in soil

Location	∑16PAHs concentration (μg kg^{-1})	∑16PAHs (BaPeq) toxicity	∑16PAHs ILCRs (Adult)	∑16PAHs ILCRs (Children)	References
Shanghai, China	807 μg kg^{-1}	146 μg kg^{-1}	4.17×10^{-6}	NA	Wang et al. (2015)
Nanjing, China	3330 ng g^{-1}	445 ng g^{-1}	NA	NA	Wang et al. (2015)
Shanxi, China	2780.42 ng g^{-1}	458.9 μg kg^{-1}	NA	NA	Liu et al. (2016)
Niger Delta, Nigeria	684 μg kg^{-1}	166 μg kg^{-1}	1.52×10^{-3}	2.19×10^{-3}	Iwegbue et al. (2016)
Himalayas, India	458 ng g^{-1}	1253 ng g^{-1}	NA	NA	Devi et al. (2016)
Gujarat, India	345 μg kg^{-1}	NA	NA	NA	Dudhagara et al. (2016)

(continued)

Table 4.4 (continued)

Location	$\sum$16PAHs concentration (μg kg^{-1})	$\sum$16PAHs (BaPeq) toxicity	$\sum$16PAHs ILCRs (Adult)	$\sum$16PAHs ILCRs (Children)	References
Yangtze River Delta region, China	471.3 μg kg^{-1}	NA	9.17×10^{-4}	2.19×10^{-4}	Wang et al. (2017)
Ningde, China	489 ng g^{-1}	93.2 ng g^{-1}	1.19×10^{-3}	NA	Zheng et al. (2019)
Jilin, China	740.21 ng g^{-1}	133 ng g^{-1}	1.61×10^{-6}	1.36×10^{-6}	Chen et al. (2018)
Hebei, China	225 μg kg^{-1}	41.23 μg kg^{-1}	1.42E-07	3.99E-07	Liang et al. (2019)
Stockholm, Sweden	4836 ng g^{-1}	NA	4.06×10^{-5}	1.46×10^{-5}	Dreij et al. (2020)
Lagos, Nigeria	2772 μg kg^{-1}	523 μg kg^{-1}	6.73×10^{-3}	8.77×10^{-2}	Ehigbor et al. (2020)
Pljevlja, Montenegro	271.49 μg kg^{-1}		1.59E-05	1.16E-05	Bigovic et al. (2022)
Hubei, China	1678 μg kg^{-1}	NA	NA	NA	Wang et al. (2024b, c)
Hamedan, Iran	1806 μg kg^{-1}	284.63 μg kg^{-1}	1.59×10^{-6}	$7/51 \times 10^{-7}$	Rabieimesbah et al. (2024)

Note NA, not available; PAH, Polycyclic aromatic hydrocarbons; BaPeq, benzo(a)pyrene equivalent; ILCR, Ingested incremental Lifetime Cancer Risk

Conclusion

Soil contamination is a growing threat to human health and the ecosystem due to the rapid growth of industries, vehicles, and agricultural activities. Recalcitrant pollutant like PAHs enters the soil ecosystem through natural or anthropogenic sources and threaten the environment and human health. In the soil ecosystem, the ecological risk evaluation of PAHs poses multifaceted challenges that demand integrated approaches in combination with analytical chemistry, soil science, and environmental toxicology. The behaviour of PAHs depends upon soil characteristics like pH, organic matter content, porosity, microbial activity and other physicochemical properties. Although remediation initiatives are effective, particularly when specialised to site-specific conditions, but demand continual refinement and cost-effectiveness. Eventually, safeguarding environmental health and assuring sustainable land use requires a comprehensive understanding of PAHs dynamics in soil promoted by interdisciplinary cooperation and policy integration.

Furthermore, researches are required to understand the movement of PAHs, risk assessment models and protocols focusing on spatial variability and continuous monitoring for different topographical and environmental settings, along with promoting innovative remediation technologies for eliminating PAH contamination in soil.

Acknowledgements The author (Madhupriya) is thankful to the University Grant Commission (UGC) for providing the Junior Research Fellowship (JRF) during her PhD.

References

Abdel-Shafy, H.I., Mansour, M.S.: A review on polycyclic aromatic hydrocarbons: source, environmental impact, effect on human health and remediation. Egypt. J. Pet. **25**(1), 107–123 (2016)

Agapkina, G.I., Brodskii, E.S., Shelepchikov, A.A.: The benzo(a)pyrene total toxicity equivalent of polycyclic aromatic hydrocarbons in soils of Moscow and assessment of carcinogenic risk to human beings. Mosc. Univ. Soil Sci. Bull. **73**, 186–194 (2018)

Alagic, S.C., Maluckov, B.S., Radojicic, V.B.: How can plants manage polycyclic aromatic hydrocarbons? May these effects represent a useful tool for an effective soil remediation? Rev. Clean Technol. Environ. Policy **17**, 597–614 (2015)

Alegbeleye, O.O., Opeolu, B.O., Jackson, V.A.: Polycyclic aromatic hydrocarbons: a critical review of environmental occurrence and bioremediation. Environ. Manage. **60**, 758–783 (2017)

Ali, M., Xu, D., Yang, X., Hu, J.: Microplastics and PAHs mixed contamination: an in-depth review on the sources, co-occurrence, and fate in marine ecosystems, Water Res. 121622 (2024)

Ambade, B., Sethi, S.S., Chintalacheruvu, M.R.: Distribution, risk assessment, and source apportionment of polycyclic aromatic hydrocarbons (PAHs) using positive matrix factorization (PMF) in urban soils of East India. Environ. Geochem. Health **45**(2), 491–505 (2023)

Aparicio, J.D., Raimondo, E.E., Saez, J.M., Costa-Gutierrez, S.B., Alvarez, A., Benimeli, C.S., Polti, M.A.: The current approach to soil remediation: a review of physicochemical and biological technologies, and the potential of their strategic combination. J. Environ. Chem. Eng. **10**(2), 107141 (2022)

Balcıoglu, E.B.: Potential effects of polycyclic aromatic hydrocarbons (PAHs) in marine foods on human health: a critical review. Toxin Rev. **35**(3–4), 98–105 (2016)

Barathi, S., Gitanjali, J., Rathinasamy, G., Sabapathi, N., Aruljothi, K.N., Lee, J., Kandasamy, S.: Recent trends in polycyclic aromatic hydrocarbons pollution distribution and counteracting bioremediation strategies. Chemosphere **337**, 139396 (2023)

Barbosa Jr, F., Rocha, B.A., Souza, M.C., Bocato, M.Z., Azevedo, L.F., Adeyemi, J.A., Santana, A., Campiglia, A.D.: Polycyclic aromatic hydrocarbons (PAHs): updated aspects of their determination. Kinetics in the human body and toxicity. J. Toxicol. Environ. Health Part B **26**(1), 28–65 (2023)

Berlinger, B., Fehérvári, P., Kovago, C., Lanyi, K., Matis, G., Mackei, M., Konyves, L.: There is still a need for a comprehensive investigation of the health consequences of exposure to urban air with special regard to particulate matter (PM) and cardiovascular effects. Atmosphere **15**(3), 296 (2024)

Bigovic, M., Durovic, D., Nikolic, I., Ivanovc, L., Bajic, B.: Profile, sources ecological and health risk assessment of PAHs in agricultural soil in a pljevlja municipality. Int. J. Environ. Res. **16**(5), 90 (2022)

Bortey-Sam, N., Ikenaka, Y., Nakayama, S.M, Akoto, O., Yohannes, Y.B., Baidoo, E., Mizukawa, H., Ishizuka, M.: Occurrence, distribution, sources and toxic potential of polycyclic aromatic hydrocarbons (PAHs) in surface soils from the Kumasi Metropolis, Ghana. Sci. Total Environ. **496**, 471–478 (2014)

Brimo, K., Ouvrard, S., Houot, S., Lafolie, F., Garnier, P.: Modelling the fate of PAH added with composts in amended soil according to the origin of the exogenous organic matter. Sci. Total. Environ. **616**, 658–668 (2018)

Canadian Council of Ministers of the Environment (CCME): Canadian Soil Quality Guidelines for Carcinogenic and Other Polycyclic Aromatic Hydrocarbons (Environmental and Human Health Effects), Scientific Criteria Document (Rev. 2010) (2010). Available at https://ccme.ca/en/res/polycyclic-aromatic-hydrocarbons-2010-canadian-soil-quality-guidelines-for-the-protection-of-environmental-and-human-health-en.pdf. Accessed 24 June 2025

Chebykina, E., Abakumov, E., Shamilishvilly, G.: Polycyclic aromatic hydrocarbons accumulation and toxic equivalency factors (TEFs) in postpyrogenic soils, In: IOP Conference Series: Earth and Environmental Science vol. 862, no. 1, p. 012094. IOP Publishing (2021)

Chen, Y., Zhang, J., Zhang, F., Liu, X., Zhou, M.: Contamination and health risk assessment of PAHs in farmland soils of the Yinma River Basin, China. Ecotoxicol. Environ Saf. **156**, 383–390 (2018)

Das, N., Kumar, V., Chaure, K., Pandey, P.: Environmental restoration of polyaromatic hydrocarbon-contaminated soil through sustainable rhizoremediation: insights into bioeconomy and high-throughput systematic analysis. Environ. Sci.: Adv. (2025)

Davletgildeeva, A.T., Kuznetsov, N.A.: Bioremediation of polycyclic aromatic hydrocarbons by means of bacteria and bacterial enzymes. Microorganisms **12**(9), 1814 (2024)

Devi, N.L., Yadav, I.C., Shihua, Q., Dan, Y., Zhang, G., Raha, P.: Environmental carcinogenic polycyclic aromatic hydrocarbons in soil from Himalayas, India: Implications for spatial distribution, sources apportionment and risk assessment. Chemosphere **144**, 493–502 (2016)

Dreij, K., Lundin, L., Le Bihanic, F., Lundstedt, S.: Polycyclic aromatic compounds in urban soils of Stockholm City: occurrence, sources and human health risk assessment. Environ. Res. **182**, 108989 (2020)

Dudhagara, D.R., Rajpara, R.K., Bhatt, J.K., Gosai, H.B., Sachaniya, B.K., Dav, B.P.: Distribution, sources and ecological risk assessment of PAHs in historically contaminated surface sediments at Bhavnagar coast, Gujarat, India. Environ. Pollut. **213**, 338–346 (2016)

Ehigbor, M.J., Iwegbue, C.M., Eguavoen, O.I., Tesi, G.O., Martincigh, B.S.: Occurrence, sources and ecological and human health risks of polycyclic aromatic hydrocarbons in soils from some functional areas of the Nigerian megacity, Lagos. Environ. Geochem. Health **42**(9), 2895–2923 (2020)

Eldos, H.I., Zouari, N., Saeed, S., Al-Ghouti, M.A.: Recent advances in the treatment of PAHs in the environment: Application of nanomaterial-based technologies. Arab. J. Chem. **15**(7), 103918 (2022)

Feng, Y., Li, Z., Li, W.: Polycyclic aromatic hydrocarbons (PAHs): environmental persistence and human health risks. Nat. Prod. Commun. **20**(1), 1934578X241311451 (2025)

Gholami-Borujeni, F., Mortezazadeh, F., Hoseinvandtabar, S., Mohseni-Bandpei, A., Niknejad, H.: Health risks assessment of polycyclic aromatic hydrocarbons (PAHs) in meat kebabs through meta-analysis and monte carlo simulation. MethodsX 103502 (2025)

Hoa, N.T.Q., Anh, H.Q., Tue, N.M., Trung, N.T., Da, L.N., Van Quy, T., Huong, N.T.A., Suzuki, G., Takahashi, S., Tanabe, S., Thuy, P.C.: Soil and sediment contamination by unsubstituted and methylated polycyclic aromatic hydrocarbons in an informal e-waste recycling area, northern Vietnam: occurrence, source apportionment, and risk assessment. Sci. Total Environ. **709**, 135852 (2020)

Hunt, L.J., Duca, D., Dan, T., Knopper, L.D.: Petroleum hydrocarbon (PHC) uptake in plants: a literature review. Environ. Pollut. **245**, 472–484 (2019)

Hussain, K., Hoque, R.R., Balachandran, S., Medhi, S., Idris, M. G., Rahman, M., Hussain, F.L.: Monitoring and risk analysis of PAHs in the environment. Handb. Environ. Mater. Manag. **10**, 978–3 (2018)

Iwegbue, C.M., Obi, G., Aganbi, E., Ogala, J.E., Omo-Irabor, O.O., Martincigh, B.S.: Concentrations and health risk assessment of polycyclic aromatic hydrocarbons in soils of an urban environment in the Niger Delta, Nigeria. Toxicol. Environ. Health Sci. **8**(3), 221–233 (2016)

Kalteh, S., Ahmadi, E., Ghaffari, H., Yousefzadeh, S., Abtahi, M., Dobaradaran, S., Saeedi, R.: Occurrence of polycyclic aromatic hydrocarbons in meat and meat products: systematic review, meta-analysis and probabilistic human health risk. Int. J. Environ. Anal. Chem. **104**(15), 3533–3547 (2024)

Kumar, S.N., Verma, P., Bastia, B., Jain, A.K.: Health risk assessment of polycyclic aromatic hydrocarbons: a review. Jakraya J. Pathol. Toxicol. **1**(1), 16–30 (2014)

Kumar, B., Verma, V.K., Sharma, C.S., Akolkar, A.B.: Estimation of toxicity equivalency and probabilistic health risk on lifetime daily intake of polycyclic aromatic hydrocarbons from urban residential soils. Human Ecol. Risk Assess.: Int. J. **21**(2), 434–444 (2015)

Kumar, M., Bolan, N.S., Hoang, S.A., Sawarkar, A.D., Jasemizad, T., Gao, B., Keerthanan, S., Padhye, L.P., Singh, L., Kumar, S., Vithanage, M.: Remediation of soils and sediments polluted with polycyclic aromatic hydrocarbons: to immobilize, mobilize, or degrade? J. Hazard. Mater. **420**, 126534 (2021)

Lawal, A.T.: Polycyclic aromatic hydrocarbons: a review. Cogent Environ. Sci. **3**(1), 1339841 (2017)

Li, X., Zhang, S., Guo, R., Xiao, X., Liu, B., Mahmoud, R.K., Abukhadra, M.R., Qu, R., Wang, Z.: Transformation and degradation of PAH mixture in contaminated sites: clarifying their interactions with native soil organisms. Toxics **12**(5), 361 (2024)

Liang, M., Liang, H., Rao, Z., Hong, X.: Characterization of polycyclic aromatic hydrocarbons in urban-rural integration area soil, North China: spatial distribution, sources and potential human health risk assessment. Chemosphere **234**, 875–884 (2019)

Lipinska, A., Kucharski, J., Wyszkowska, J.: The effect of polycyclic aromatic hydrocarbons on the structure of organotrophic bacteria and dehydrogenase activity in soil. Polycyclic Aromat. Compd. **34**(1), 35–53 (2014)

Lipinska, A., Kucharski, J., Wyszkowska, J.: Activity of phosphatases in soil contaminated with PAHs. Water Air Soil Pollut. **230**, 1–15 (2019)

Lipinska, A., Kucharski, J., Wyszkowska, J.: Urease activity in soil contaminated with polycyclic aromatic hydrocarbons. Polish J. Environ. Stud. **22**(5) 2013

Liu, G., Niu, J., Guo, W., An, X., Zhao, L.: Ecological and health risk-based characterization of agricultural soils contaminated with polycyclic aromatic hydrocarbons in the vicinity of a chemical plant in China. Chemosphere **163**, 461–470 (2016)

Liu, J., Liu, Y.J., Liu, Z., Zhang, A., Liu, Y.: Source apportionment of soil PAHs and human health exposure risks quantification from sources: the Yulin national energy and chemical industry base, China as case study. Environ. Geochem. Health **41**(2), 617–632 (2019)

Maletic, S.P., Beljin, J.M., Roncevic, S.D., Grgic, M.G., Dalmacija, B.D.: State of the art and future challenges for polycyclic aromatic hydrocarbons is sediments: sources, fate, bioavailability and remediation techniques. J. Hazard. Mater. **365**, 467–482 (2019)

Mallah, M.A., Changxing, L., Mallah, M.A., Noreen, S., Liu, Y., Saeed, M., Xi, H., Ahmed, B., Feng, F., Mirjat, A.A., Wang, W.: Polycyclic aromatic hydrocarbon and its effects on human health: an overeview. Chemosphere **296**, 133948 (2022)

Manaswini, G., Sivagami, K., Gopalakrishnan, M., Harshini, P., Janjoren, D., Ganesan, S.: Biodegradation of low molecular weight polycyclic aromatic hydrocarbons in soil: insights into bacterial activities and bioremediation techniques. Sustain. Chem. Environ. 100146 (2024)

Mishra, A., Kumari, M., Kumar, R., Iqbal, K., Thakur, I.S.: Persistent organic pollutants in the environment: Risk assessment, hazards, and mitigation strategies. Bioresour. Technol. Rep. **19**, 101143 (2022)

Montano, L., Baldini, G.M., Piscopo, M., Liguori, G., Lombardi, R., Ricciardi, M., Esposito, G., Pinto, G., Fontanarosa, C., Spinelli, M., Palmieri, I.: Polycyclic aromatic hydrocarbons (PAHs) in the environment: occupational exposure, health risks and fertility implications. Toxics **13**(3), 151 (2025)

Ofori, S.A., Cobbina, S.J., Doke, D.A.: The occurrence and levels of polycyclic aromatic hydrocarbons (PAHs) in African environments—a systematic review. Environ. Sci. Pollut. Res. **27**(26), 32389–32431 (2020)

Patel, A.B., Shaikh, S., Jain, K.R., Desai, C., Madamwar, D.: Polycyclic aromatic hydrocarbons: sources, toxicity, and remediation approaches. Front. Microbiol. **11**, 562813 (2020)

Qu, Y., Gong, Y., Ma, J., Wei, H., Liu, Q., Liu, L., Wu, H., Yang, S., Chen, Y.: Potential sources, influencing factors, and health risks of polycyclic aromatic hydrocarbons (PAHs) in the surface soil of urban parks in Beijing, China. Environ. Pollut. **260**, 114016 (2020)

Rabieimesbah, A., Sobhanardakani, S., Cheraghi, M., Lorestani, B.: Concentrations, source identification and potential ecological and human health risks assessment of polycyclic aromatic hydrocarbons (PAHs) in agricultural soils of hamedan county, west of Iran. Soil Sediment Contam.: Int. J. **33**(4), 482–506 (2024)

Ren, G., Ren, W., Teng, Y., Li, Z.: Evident bacterial community changes but only slight degradation when polluted with pyrene in a red soil. Front. Microbiol. **6**, 22 (2015)

Rigi, P., Kamani, H., Ansari, H., Mohammadi, L., Dargahi, A.: Health risk assessment of polycyclic aromatic hydrocarbon compounds (PAHs) in grilled meats in Zahedan city of Iran. Sci. Rep. **15**(1), 24267 (2025)

Saeedi, M., Li, L.Y., Grace, J.R.: Effect of co-existing heavy metals and natural organic matter on sorption/desorption of polycyclic aromatic hydrocarbons in soil: a review. Pollution **6**(1), 1–24 (2020)

Samburova, V., Zielinska, B., Khlystov, A.: Do 16 polycyclic aromatic hydrocarbons represent PAH air toxicity? Toxics **5**(3), 17 (2017)

Sankar, T.K., Kumar, A., Mahto, D.K., Das, K.C., Narayan, P., Fukate, M., Awachat, P., Padghan, D., Mohammad, F., Al-Lohedan, H.A., Soleiman, A.A.: The health risk and source assessment of polycyclic aromatic hydrocarbons (PAHs) in the soil of industrial cities in India. Toxics **11**(6), 515 (2023)

Schwab, A.P., Dermody, C.L.: Pathways of polycyclic aromatic hydrocarbons assimilation by plants growing in contaminated soils. Adv. Agron. **169**, 193–250 (2021)

Shen, H.: Polycyclic Aromatic Hydrocarbons: Their Global Atmospheric Emissions, Transport, and Lung Cancer Risk. Springer (2016)

Shi, M., Hu, T., Mao, Y., Cheng, C., Liu, W., Tian, Q., Chen, Z., Xing, X., Qi, S.: Polycyclic aromatic hydrocarbons in atmospheric PM2. 5 during winter haze in Huang-gang, central China: characteristics, sources and health risk assessment. Aerosol Air Qual. Res. **21**(5), 200337 (2021)

Shukla, S., Khan, R., Bhattacharya, P., Devanesan, S., AlSalhi, M.S.: Concentration, source apportionment and potential carcinogenic risks of polycyclic aromatic hydrocarbons (PAHs) in roadside soils. Chemosphere **292**, 133413 (2022)

Sombiri, S., Balhara, N., Attri, D., Kharb, I., Giri, A.: An overview on occurrence of polycyclic aromatic hydrocarbons in food chain with special emphasis on human health ailments. Discov. Environ. **2**(1), 87 (2024)

Sui, X., Wang, X., Li, Y., Ji, H.: Remediation of petroleum-contaminated soils with microbial and microbial combined methods: advances, mechanisms, and challenges. Sustainability **13**(16), 9267 (2021)

Sushkova, S., Minkina, T., Tarigholizadeh, S., Rajput, V., Fedorenko, A., Antonenko, E., Dudnikova, T., Chernikova, N., Yadav, B.K., Batukaev, A.: Soil PAHs contamination effect on the cellular and subcellular organelle changes of Phragmites australis Cav. Environ. Geochem. Health **43**(6), 2407–2421 (2021)

Swit, P., Orzeł, J., Maslanka, S.: Investigation of the origin and concentration of polycyclic aromatic hydrocarbon with improved accuracy by the use of a multi-component integrated calibration method in the Katowice region, Poland. J. Hazard. Mater. **448**, 130871 (2023)

Tarigholizadeh, S., Sushkova, S., Rajput, V.D., Ranjan, A., Arora, J., Dudnikova, T., Barbashev, A., Mandzhieva, S., Minkina, T., Wong, M.H.: Transfer and degradation of PAHs in the soil–plant system: a review. J. Agric. Food Chem. **72**(1), 46–64 (2023)

Tongo, I., Ogbeide, O., Ezemonye, L.: Human health risk assessment of polycyclic aromatic hydrocarbons (PAHs) in smoked fish species from markets in Southern Nigeria. Toxicol. Rep. **4**, 55–61 (2017)

Venkatraman, G., Giribabu, N., Mohan, P.S., Muttiah, B., Govindarajan, V.K., Alagiri, M., Rahman, P.S.A., Karsani, S.A.: Environmental impact and human health effects of polycyclic aromatic hydrocarbons and remedial strategies: a detailed review. Chemosphere **351**, 141227 (2024)

Vijayanand, M., Ramakrishnan, A., Subramanian, R., Issac, P.K., Nasr, M., Khoo, K.S., Rajagopal, R., Greff, B., Azelee, N.I.W., Jeon, B.H., Chang, S.W.: Polyaromatic hydrocarbons (PAHs) in the water environment: a review on toxicity, microbial biodegradation, systematic biological advancements, and environmental fate. Environ. Res. **227**, 115716 (2023)

Wang, C., Wu, S., Zhou, S., Wang, H., Li, B., Chen, H., Yu, Y., Shi, Y.: Polycyclic aromatic hydrocarbons in soils from urban to rural areas in Nanjing: concentration, source, spatial distribution. and potential human health risk. Sci. Total. Environ. **527**, 375–383 (2015)

Wang, J., Zhang, X., Ling, W., Liu, R., Liu, J., Kang, F., Gao, Y.: Contamination and health risk assessment of PAHs in soils and crops in industrial areas of the Yangtze River Delta region, China. Chemosphere **168**, 976–987 (2017)

Wang, J., Zhang, H., Bao, H., Li, J., Li, J., Xing, W., Hong, H., Wu, F.: Dynamic distribution and accumulation of PAHs in winter wheat during whole plant growth: field investigation. Ecotoxicol. Environ. Saf. **202**, 110886 (2020)

Wang, X.T., Chen, L., Wang, X.K., Lei, B.L., Sun, Y.F., Zhou, J., Wu, M.H.: Occurrence, sources and health risk assessment of polycyclic aromatic hydrocarbons in urban (Pudong) and suburban soils from Shanghai in China. Chemosphere **119**, 1224–1232 (2015)

Wang, Liu, D., Lv, Y., Wang, W., Wu, Q., Huang, L., Zhu, L.: Ecological and health risk assessments of polycyclic aromatic hydrocarbons (PAHs) in soils around a petroleum refining plant in China: a quantitative method based on the improved hybrid model. J. Hazard. Mater. **461**, 132476 (2024a)

Wang, S., Li, C., Zhang, L., Chen, Q., Wang, S.: Assessing the ecological impacts of polycyclic aromatic hydrocarbons petroleum pollutants using a network toxicity model. Environ. Res. **245**, 117901 (2024b)

Wang, Z., Meng, Q., Sun, K., Wen, Z.: Spatiotemporal distribution bioaccumulation, and ecological and human health risks of polycyclic aromatic hydrocarbons in surface water: a comprehensive review. Sustainability **16**(23), 10346 (2024c)

Wei, X., Xie, B., Wan, C., Song, R., Zhong, W., Xin, S., Song, K.: Enhancing soil health and plant growth through microbial fertilizers: mechanisms, benefits, and sustainable agricultural practices. Agronomy **14**(3), 609 (2024)

Wiłkomirski, B., Jabbarov, Z.A., Abdrakhmanov, T.A., Vokhidova, M.B., Jabborov, B.T., Fakhrutdinova, M.F., Okolelova, A.A., Kholdorov, S.M., Abdullayeva, Y.D.: Polycyclic aromatic hydrocarbons (PAHs) in natural and anthropogenically modified soils (a review). Biogeosystem Tech. **5**, 229–243 (2018)

Yebra-Pimentel, I., Fernandez-Gonzalez, R., Martínez-Carballo, E., Simal-Gandara, J.: A critical review about the health risk assessment of PAHs and their metabolites in foods. Crit. Rev. Food Sci. Nutr. **55**(10), 1383–1405 (2015)

Younis, A.M., Hanafy, S., Elkady, E.M., Ghandour, M.A., El-Sayed, A.A.Y., Alminderej, F.M.: Polycyclic aromatic hydrocarbons (PAHs) in Egyptian red sea sediments: seasonal distribution, source identification, and toxicological risk assessment. Arab. J. Chem. **16**(9), 104999 (2023)

Zhang, P., Chen, Y.: Polycyclic aromatic hydrocarbons contamination in surface soil of China: a review. Sci. Total. Environ. **605**, 1011–1020 (2017)

Zhang, Q., Meng, J., Su, G., Liu, Z., Shi, B., Wang, T.: Source apportionment and risk assessment for polycyclic aromatic hydrocarbons in soils at a typical coking plant. Ecotoxicol. Environ. Saf. **222**, 112509 (2021)

Zhao, Z., He, W., Wu, R., Xu, F.: Distribution and relationships of polycyclic aromatic hydrocarbons (PAHs) in soils and plants near major lakes in Eastern China. Toxics **10**(10), 577 (2022)

Zheng, H., Qu, C., Zhang, J., Talpur, S.A., Ding, Y., Xing, X., Qi, S.: Polycyclic aromatic hydrocarbons (PAHs) in agricultural soils from Ningde, China: levels, Sources and human health risk assessment. Environ. Geochem. Health **41**(2), 907–919 (2019)

Polycyclic Aromatic Hydrocarbons and Their Derivatives in the Environment

5

Hariom Nagar, Jaswant Raj, Lokesh Kumar Kumawat, Jai Prakash, Shyam Raj Yadav, and Sujeet Kumar

Abstract

Polycyclic aromatic hydrocarbons are harmful and cancer-causing compounds that come from two primary processes: pyrogenic and petrogenic. Oil extraction and drilling operations, such as oil spills, pollution from refineries and industrial regions, and most significantly, emissions from automobile traffic, are the causes of the petrogenic component. The pyrogenic component, on the other hand, is created by things like incineration, volcanic eruptions, and wildfires. Polycyclic aromatic hydrocarbons (PAHs take a long time to degrade, and new research shows that they significantly accumulate in soil, water, and the environment). Smog clouds, which carry pollutants from air to soil and water sources, and eventually to people, contribute to the increase in pollution and its migration as winter approaches. This chapter presents the most recent PAH measurements and concentrations worldwide, with a focus on areas that have been most heavily affected, such as China, India, and former Eastern European countries. Monitoring and assessing the distribution of PAHs is crucial for environmental scientists due to the toxic and carcinogenic properties of many PAHs. In recent years, significant attention has been focused on identifying PAHs in various environmental segments. PAHs derivatives consist of compounds where hydrogen atoms present in the aromatic ring have been replaced by reactive groups, nitro, hydroxyl, and carbonyl (N-PAH, OH-PAH and O-PAH, respectively), or carbon atom replaced by sulfur atom in the aromatic

H. Nagar (✉) · J. Prakash · S. R. Yadav · S. Kumar
Department of Chemistry, S.P. Jain College (VKSU, Ara), Sasaram, Bihar, India
e-mail: hariomnagariitr@gmail.com

J. Raj
Department of Chemistry, University of Rajasthan, Jaipur, Rajasthan, India

L. K. Kumawat
Global Process Development Centre, Viatris, Little Island, Cork, Ireland

H. Gupta et al. (eds.), *Ubiquitous Polycyclic Aromatic Hydrocarbon Contamination*, Emerging Contaminants and Associated Treatment Technologies,
https://doi.org/10.1007/978-3-032-11764-9_5

structure of PAHs. PAHs, as well as their derivatives, as primary pollutants, can either enter the atmosphere directly or be produced through homogeneous and heterogeneous oxidation reactions. However, it is crucial to remember that PAH determination methods serve the ultimate goal of assessing the health risks that are associated with suspended particulate matter in the air.

Keywords

Suspended particulate matter • Polycyclic aromatic hydrocarbons • Gas chromatography • High performance liquid chromatography • Toxicological effects

Introduction

Smog poses a significant issue in numerous cities worldwide. The English terms smoke and fog are combined to form the phrase "smog" (Schwartz 1997). Windless conditions and artificial fog composed of remnants from the incomplete burning of different fuels are the leading causes of this unnatural occurrence (Abbasi and Keshavarzi 2019). Suspended particulate matter (SPM) made up of mainly by component of smog; aerosols of liquid droplets or solid particles. These SPM particles may be artificial (such as dust from combustion operations) or natural (such as sea salt) (Reizer 2016). In the latter case, SPM, which is classified by size, is a measure of air pollution. Particles having diameter less than 2.5 and 10 μm are called as $PM_{2.5}$, and PM_{10}, respectively. Additionally, various studies reference PM_4 for particles measuring 4.0 μm and PM_1 for those measuring 1.0 μm (Kic 2018).

There are two ways that SPM affects health. The respiratory system may be at risk from tiny SPM particles. Because of the possible harm it could do to the heart and bronchi, the PM_{10} fraction is known as the "chest fraction" (Brown et al. 2013). The $PM_{2.5}$ fraction is referred to as the "respiratory fraction" because it is much smaller and can enter the pulmonary alveoli (Boldo et al. 2006). Additionally, SPM is a carrier of numerous dangerous substances that are created when different fuel processes burn incompletely. $PM_{2.5}$ provides a surface on which heavy metals (for eg., Ni and Cd) are adsorbed (Deng et al. 2006). Furthermore, $PM_{2.5}$ support their insertion into brain which results in illnesses associated with metal exposure.

PAHs are molecules made up of aromatic units, with benzene rings at their core. This group has drawn significant attention, particularly as PAHs adhere to SPM. More than 10,000 compounds (Kurnia et al. 2018) are classified as PAHs. Their subtypes based on their molecular weight are as follows; low molecular weight (LMW) and high molecular weight (HMW). Unlike HMW PAHs, which have four or more rings (Saldarriaga-Noreña et al. 2018), LMW PAHs usually have two or three rings. Numerous studies demonstrate their mutagenic, poisonous, and carcinogenic qualities. Carcinogenicity (Hayakawa 2018; Bandowe et al. 2019) rises with molecular weight: LMW PAHs typically show increased toxicity (Bandowe et al. 2014), whilst HMW PAHs are frequently more carcinogenic. The presence of numerous PAHs in the air poses greater risks (Keith 2015). Additionally, under favourable conditions, these compounds can transform into more hazardous oxidised forms (Mueller et al. 2019). Despite current studies into their carcinogenicity, oxy- and nitro-PAHs (because of their direct carcinogenic effects) are considered as more dangerous than related PAHs (Ringuet et al. 2012; de Oliveira Galvao et al. 2018). Additionally, the acute toxicity of the original PAHs and their derivatives differs; for instance, fluorene has an LD_{50} of 16,000 mg/kg (oral, rat), compared to 11,600 mg/kg for nitrofluorene (Zaciera 2007). Because PM has such negative consequences on human health, research into it is becoming more and more important. Beyond the morphological characteristics of particles, their chemical composition has a significant influence on the adverse health impacts. Most low-vapor-pressure PAHs in the atmosphere adhere to PM, as these compounds are highly lipophilic despite their low water solubility (Walgraeve et al. 2010).

PAHs that have been adsorbed onto PM or evenly dispersed in water may get decomposed on sun (UV) irradiation. When they interact with pollutants like O_3, SO_2, and nitrogen oxides, they can produce nitro-dinitro-PAHs, diones, and sulfonic acids in the environment. Furthermore, PAHs may be broken down by some soil bacteria (Srogi 2007).

However, PAHs (during the synthesis) and their presence in soil, water and air various chemical, photochemical and physical processes can create variants that contain atoms other than carbon, nitrogen, oxygen, or sulfur. As a result, various compounds, like nitrogenated-PAHs (N-PAHs), oxygenated-PAHs (O-PAHs), hydroxyl-PAHs (OH-PAHs), phosphogenated-PAHs (PASHs), or azarenes (AZAs), can form. All these compounds are generally categorised as thermally stable. Typically, the solubility of PAH derivatives in water decreases as the number of rings increases (Zhang et al. 2011). PAHs as well as derivatives can be discovered in sediments and are associated with geological processes, tobacco smoke, and biogenic precursors such as plant terpenes (Miet et al. 2009).

Days and seasons had a substantial impact on the concentrations of individual N- and O- derivatives of PAHs in both gas as well as particle phases (Bamford et al. 2003). Results showed that total suspended particle matter (TSP) in samples obtained from Chiang Mai, Thailand, contained significantly more 19 N-PAHs, especially 9-nitroanthracene (9-NA), in the period of dry season (Chuesaard et al. 2014). In winter season O-PAH concentrations was found higher than the same in other seasons. This discrepancy could be caused by a tendency for O-PAHs to stay in the particle phase throughout the winter instead of moving into the gas phase, a decrease in mixing height during the colder months, or an increase in O-PAH emissions. Perinaphthenone was the most abundant among these compounds. Parent PAHs and O-PAHs might come from similar sources, with a significant proportion of O-PAHs formation likely stemming from primary emissions rather than atmospheric reactions. Increased radiative forcing has led to enhanced photolysis and a subsequent reduction in O-PAHs levels during summer (Lee et al. 2018). They can also be formed when singlet molecular oxygen photo-oxidises PAHs (Barbas

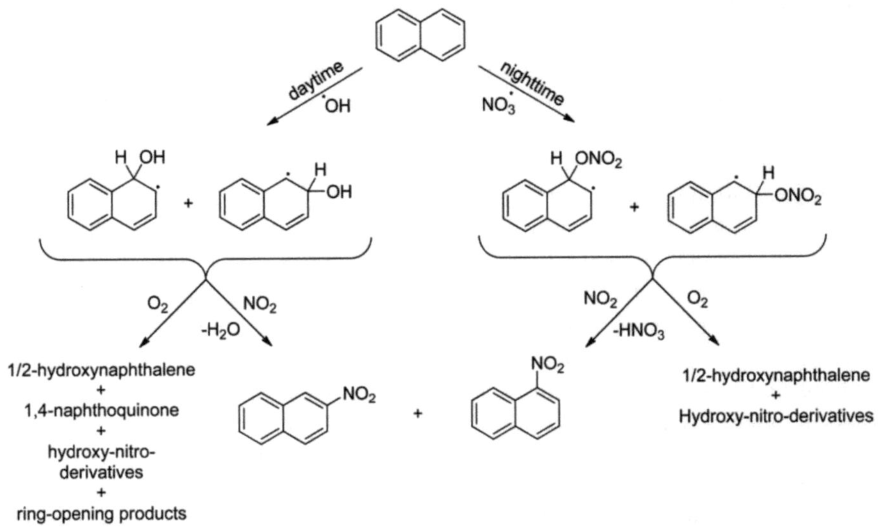

Fig. 5.1 Reactions of naphthalene and radical oxidants (present in the atmosphere) both in daytime and nighttime "adapted from original article by Vione et al. (2004) after taking permission from WILEY-VCH VERLAG GMBH & CO"

et al. 1996). PAH derivatives (contains oxygen) are the most mutagens among human-cell mutagens found in airborne particles in the northeastern US (Pedersen et al. 2004). A target yearly average value of 1.0 ng m^{-3} for benzo[a]pyrene contained in PM_{10} is established by European Directive 107/2009/EC (EU 2005), which designates benzo[a]pyrene as an indicator for the marking of carcinogenic effects associated with PAHs. Day time and night time reactions of naphthalene with radical oxidants (present in atmosphere; e.g., OH, NO_3, SO_2, O_3, and NH_2) are shown in Fig. 5.1 (Vione et al. 2004).

However, challenges still exist, particularly during the sample preparation phase. Measuring PAHs can be difficult and expensive because of their diverse physical and chemical characteristics. PAHs can be determined using flow chart as shown in Fig. 5.2 (Bandowe and Meusel 2017).

For one to accurately measure current concentration levels, evaluate the impact of upcoming control measures, and improve operations by standardising and valuing the effects on human health, it has significant role to standardise and validate sampling, analysis, and emission estimating processes. To enhance current techniques and technical requirements, ongoing research is essential. Traditional solid-phase extraction (SPE) procedures are modified by several of these novel approaches, and liquid-liquid extraction (LLE) methods are also recognised.

Significant progress has been made in creating sorbents for micro- and miniaturised SPE techniques.

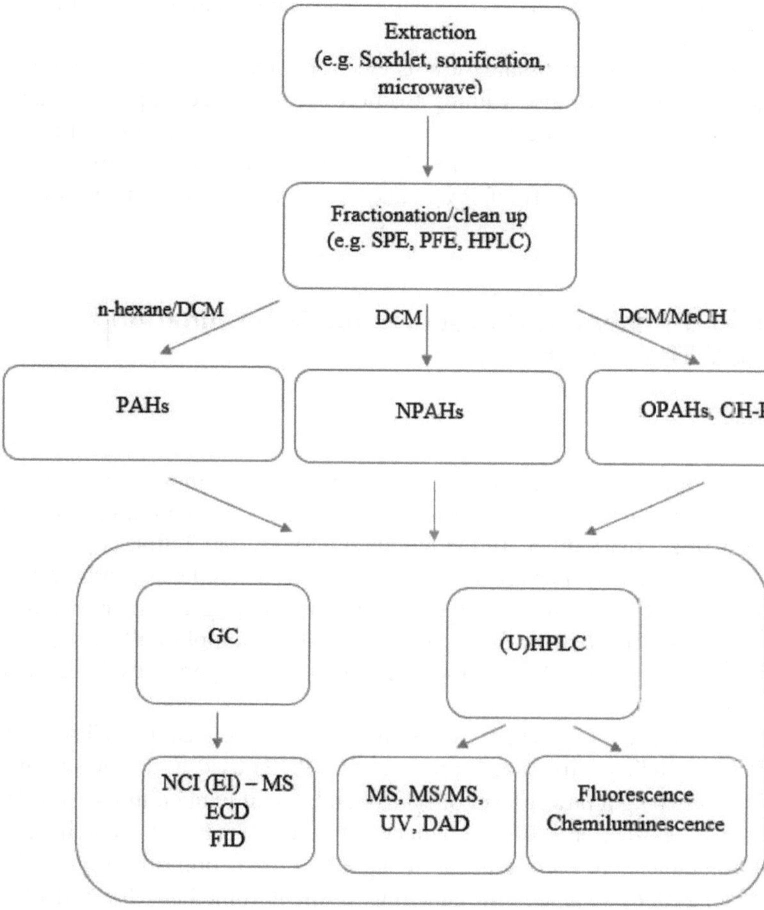

Fig. 5.2 Flow chart to determine PAHs and their derivatives "adapted from original article by Bandowe and Meusel (2017) after taking permission from Elsevier"

The choice of extraction technique and sorbent should align with the analytical requirements, as well as the available laboratory equipment and selectivity. Due to the ease of sample handling, dispersive and matrix solid-phase extraction (d-SPE and MSSPE, respectively) have garnered significant attention as these two methods are also simple to perform, fast and environmental friendly. Their application to assess wise variety of sorbents grown rapidly.

Additionally, innovative miniaturised extraction methods, such as solid-phase microextraction (SPME), packed-tube SPE (PT-SPE), and flow-probe SPE (FPSE), have gained popularity in recent years. In comparison, these methods have not been evaluated as thoroughly as SPE, MSPE, and d-SPE; their operational simplicity positions them as a valuable substitute that enhances the repertoire of chemists and analytical specialists (Jiping et al. 2020; Sun et al. 2020).

Recently, special attention has been given to research related to latest detection techniques for nitro-PAHs.

Separation devices encompassing various electrophoresis forms and microchip capillary electrophoresis are also gaining traction. All these techniques are anticipated to have significant applications in real-time analysis or on-site settings of nitro-PAH (Lisowski and Zarzycki 2013). This chapter also seeks to explore the latest advancements in the study of PAHs and their oxidised derivatives sourced from SPM samples.

Sampling and Analytical Procedures for Determination of PAHs and Their Derivatives

PAHs

A variety of sampler types, including passive samplers and those with low and high volume sampling rates, can be used for sampling. The sampling materials comprise glass fiber, quartz fiber, Teflon membrane filters, and Teflon-coated glass fiber (Szulejko et al. 2014). For few hours, these materials are heated at high temperature (400–550 °C) and then weighed before to sampling in order to reduce the blank levels associated with fresh filters. Following sample collection, prior weighed filters are reweighed, preserved, and kept at temperatures lower than 15 °C.

The techniques used to prepare the samples differ noticeably. With a 40-min temperature gradient program, Barrado et al. (2012) used a microwave oven (MWE) to extract using a power level of 900 W and a final temperature of 120 °C. They reported an extraction solvent efficiency of above 80% and recommended using dichloromethane. In their sonication-supported extraction approach, Crimmins and Baker (2006) sonicated the materials in dichloromethane for half of an hour after that kept at − 20 °C for 48 h. The extract was concentrated to 200 μL under a nitrogen flow prior to analysis using gas chromatography combined with mass spectrometry (GC/MS).

For HPLC analysis using a fluorescence detector, Jakovljević et al. (2018) used a toluene-cyclohexane (7:3) mixture for extraction through sonicator, followed by evaporation to dryness and redissolving in acetonitrile. Using dichloromethane (DCM), Umbuzeiro et al. (2008) carried out Soxhlet extraction, using identical protocols described by Lammel (2015). The solvent was almost completely evaporated throughout the 20-h process at 40 °C, and then sample was redissolved in a little volume of DCM.

Nitro-PAHs

Nitro-PAHs from environment enter humans and animals body primarily by inhalation, oral ingestion, and skin contact. Nitro-PAHs' carcinogenic and mutagenic properties in atmospheric media, including sediments (soil) and air particles, have been evaluated using a variety of experimental techniques. These techniques include biological molecules (like DNA), fish and mammalian cell lines, bacterial strains (like Salmonella used in the Ames test), and in vivo research on plants, rats, and mice.

Numerous nitro-PAHs have been categorised into carcinogenic classes based on scientific research and the International Agency for Research on Cancer's (IARC) working on the assessment of carcinogenic risks to humans. Compared to benzo[a]pyrene, several nitro-PAHs are substantially more hazardous (Nowakowski et al. 2022). The techniques previously described for PAHs is also applicable to simultaneous collection of PAHs and nitro-PAHs. GC/HPLC (with a variety of detectors) can also be used for analysis of nitro-PAHs, like other PAHs.

Numerous ionisation techniques, such as atmospheric pressure chemical ionisation (APCI), negative ion chemical ionisation (NICI), electron impact (EI), and atmospheric pressure photoionisation (APPI), can be employed. They all operate in modes; multiple reaction monitoring or selective ion monitoring. In addition to mass spectrometry, other detection methods that can be applied include fluorescence, chemiluminescence, and photodiode arrays. Because nitro-PAHs may break down at high temperature, Sun et al. (2020) suggest HPLC as the best analytical technique. Because nitro-PAH concentrations are usually low, sample preparation is essential regardless of the analytical method employed.

Zhao et al. (2020) explain the preparation process, which includes the extraction of filters with DCM using Soxhlet at 45 °C. Through evaporation, the extracted samples are concentrated to about 1.0 mL and purified using silica gel and alumina columns. The solvent composition; hexane and hexane/DCM (1:1 v/v) is then used to elute them. The samples are then subjected to GC/MS analysis utilising EI ionisation after the eluates have been evaporated once more. A similar approach was also used by Zhao et al. (2020). GC-MS (Oliveira et al. 2019; Kitanovski et al. 2020) and similar methods was recommended for the analysis of both nitro- as well as oxy-PAHs (Mueller et al. 2019; Gao et al. 2018; Dos Santos et al. 2019). Nitro-PAHs are more stable at room temperature, HPLC can perform measurements, which is the primary benefit of HPLC over GC.

Smaller volatile molecules can be separated from larger and unstable ones by using C-18 columns (Schauer et al. 2004; Mirivel et al. 2010; Nyiri et al. 2016) or a phenylhexyl column (García-Alonso et al. 2012). Certain sample preparation methodologies were discussed, including traditional Soxhlet extraction, pressurised fluid extraction (PFE), and sonication, which were highlighted by Zielinska and Samy (2006). They stated that commonly used solvents include benzene, toluene, and dichloromethane. PFE is an efficient, time-saving method that reduces solvent consumption. For sample preparations, Ringuet et al. (2012) and Bamford et al. (2003) also employed PFE. PAHs and nitro-PAHs were extracted simultaneously by using following solvents; PFE and DCM (Bandowe et al. 2014). Han et al. (2019) described GC-NICI-MS method for analysing PAHs and slightly modified method for nitro-PAHs. HPLC with fluorescent detector was used for analysis of PAHs and nitro-PAHs (Pham et al. 2019). Schauer et al. (2004) stressed the necessity of separating PAHs from nitro-PAHs before analysis, achieving this separation with a silica glass column, where PAHs were eluted using a cyclohexane/dichloromethane (1:1) mixture.

O-PAHs

For effective description of analytical techniques for O-PAHs, it is crucial to classify these into hydroxy-PAHs and oxy-PAHs are two classes of O-PAHs. Special processing is required for hydroxyl group containing molecules prior to analysis. Before performing GC-MS analysis on hydroxy-PAHs, sample derivatisation is essential, according to Cochran et al. (2012). A combination of trimethylchlorosilane (TMCS) and bis(trimethylsilyl)trifluoroacetamide (BSTFA) were used as derivatising agent at 70 °C for six hours, completely transforming the hydroxyl group. Richter-Brockmann et al. (2019) reported employing sodium hydroxide, methyl iodide, and dimethyl sulfoxide for the formation of methyl ether derivative of OH-PAH (3-OH-BaP). Furthermore the formed ether was extracted by SPE using cyclohexane, and GC-APLI-MS analysis was then suggested. It is not necessary to derivatize hydroxy-PAHs when using the LC-MS technique (Han et al. 2019). Methanol was used as the extraction solvent in a low-temperature ultrasonication process by Barrado et al. (2012). O'Connell et al. (2013) suggested an ethyl acetate-based extraction process, which was followed by the quantification of 24 O-PAHs using GC and LC-MS techniques. Analysis of nitro-PAHs, PAHs without derivatisation, and oxy-PAHs can be done together (Albinet et al. 2006; Han et al. 2019). Compared to PAHs, oxy-PAHs demand slightly different temperature settings for analytical studies (Souza et al. 2014). The collected O-PAH analytical methods include similar studies on soil, sediment, and urine analysis in addition to PM-related air contaminations.

PAH Derivatives in the Air

There are two phases (gas and particle) for research on atmospheric PAH derivatives, although the predominant focus is on the particle phase. Currently, studies on PAH derivative concentrations are relatively sparse compared to those of parent PAHs. In terms of pollution levels, PAHs more pollute the atmosphere while PAH derivatives pollute lesser. The increased amount of PAH derivatives in the particle phase is our main worry. Nigeria's air is most concentrated (17.56 μg/L) by PAHs. In air samples, $PM_{2.5}$-bound PAH concentrations vary from 10^{-1} to 102 ng/m^3; the highest concentration was reported in Taiyuan, China, at 314.55 ng/m^3 (Song et al. 2020). In general, N-, A-, and OH-PAHs concentration is lower than O-PAH. The largest concentration of $PM_{2.5}$-bound O-PAHs was found in Estarreja, Portugal, at 142.6 ng/m^3 (Alves et al. 2023), followed by Xingtai, Hebei Province,

China, at 62.7 ng/m^3 (Li et al. 2022), while Thessaloniki, Greece air sample was very dilute (0.09 ng/m^3) (Besis et al. 2022). In terms of $PM_{2.5}$-bound N-PAHs, the most (27.24 ng/m^3) and least (0.0125 ng/m^3) concentrated air samples were from Xian, China (Niu et al. 2019) and Singapore (Wang et al. 2022), respectively.

There is little research on A-PAH and OH-PAH levels in the atmosphere. With $PM_{2.5}$-bound OH-PAH concentrations in the range of 0.045–10.95 ng/m^3 in China and other countries, the available data indicate that China's environment is more concentrated (by $PM_{2.5}$-bound A-PAHs and OH-PAHs) than other countries. Studies have shown that PAH derivatives can coexist with gas/particle phases (Ma et al. 2020). According to reports, 69–79% of PAHS is found in the gas phase, whereas 42–46%, 34–40%, and 19–24%, respectively, of A-PAHS, N-PAHS, and O-PAHS (Zhang et al. 2021). Therefore, although their affinities vary, PAHs and their derivatives can adsorb onto respirable particles. Numerous investigations have demonstrated that distribution of PAH derivatives in gas and particle phases significantly affected by molecular mass and vapour pressure (Huang et al. 2014). Current research suggests that PAHs with LMW are mainly present in the gaseous form, while those with HMW tend to exist in particulate form. Additionally, temperature affects the distribution of gases and particles. Due to the lower ambient temperatures in the spring, the results show that the combined proportions of PAHs, O-PAHs, and N-PAHs with TSP are higher (Wei et al. 2021). Figure 5.3 showcasing phase distribution of particulate phase PAHs and derivatives in Brazilian urban area whereas Fig. 5.4 shows the annual means of gas phase PAHs, nitro-PAHs, and oxy-PAHs concentrations for the sampling site in the dry and rainy months (Dos Santos et al. 2020). Figure 5.5 depicted the varying % percentage concentrations of PAHs derivatives (during August 2003 and March 2004) in samples collected in urban air of the site Athinas St. (Andreou and Rapsomanikis 2009).

PAH Derivatives in the Water and Sediment

Derivatives of PAHs have been analysed quantitatively in water as well as sediment because of their detrimental impacts on living things. Although, studies conducted on PAH derivatives in air environment are comparatively greater than those in ecosystems and sediments. Moreover, in Asia most studies are carried out for quantitative analysis of PAH derivatives in aquatic environments. Compared to PAHs, the concentrations of derivatives in marine ecosystems are very modest. However, the lowest concentration of N-PAHs (5.56 ng/L) was found in sea-surface micro-layer of water surface of the Mediterranean Sea off the shore of Alexandria (Nassar 2017).

The largest quantity, however, is seen in groundwater samples taken close to gas stations (8.96 μg/L) (Santos et al. 2017). Studies on O-PAH concentrations in water are also part of the research; surface water from the Chaobai River System has the highest amounts, measuring 321 ng/L (Qiao et al. 2020). Additionally, some wastewater treatment plants' receiving rivers have been discovered to contain

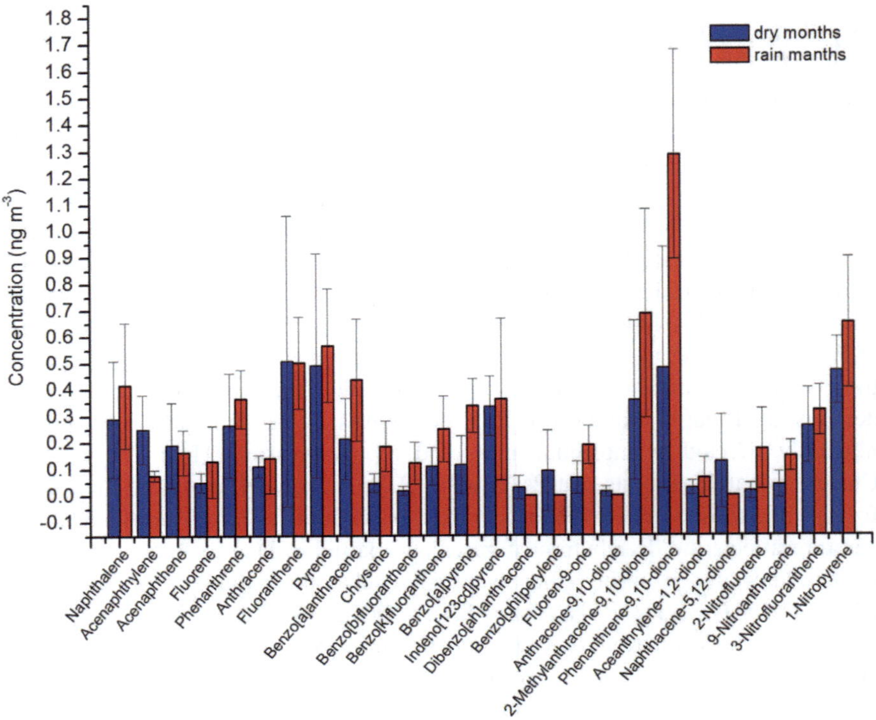

Fig. 5.3 Mean concentration (ng/m^3) of particulate phase PAHs, nitro-PAHs and oxy-PAHs ($PM_{2.5}$) in ambient air samples collected from May 2017 to April 2018 (n = 93). Dry months: April to September, rainy months: October to March "adapted from original article by Dos Santos et al. (2020) after taking permission from Elsevier"

hundreds of ng/L of O-PAHs (Qiao et al. 2018). On the other hand, the Yangtze River had a significantly lower O-PAH value (20.2 ± 7.49 ng/L) (Qiao et al. 2019). The quantitative analyses of A-PAHs and OH-PAHs in river water, however, have not been thoroughly studied. Sediments' seasonal changes in PAH derivatives mirrored those in the aquatic environment. N-PAHS and OH-PAHS concentrations were found to be greater during dry seasons, most likely due to rainy seasons diluting the sediments (Zhu et al. 2022).

PAH Derivatives in the Soil

Pollutants derived from PAHs are significantly absorbed by soil. A lot of reports now indicate that how much PAH derivative is present in soil. China's PAH pollution levels are lower than those of other nations. Soil samples contain more concentrations of PAH than its derivatives; in China's topsoil, PAH concentrations

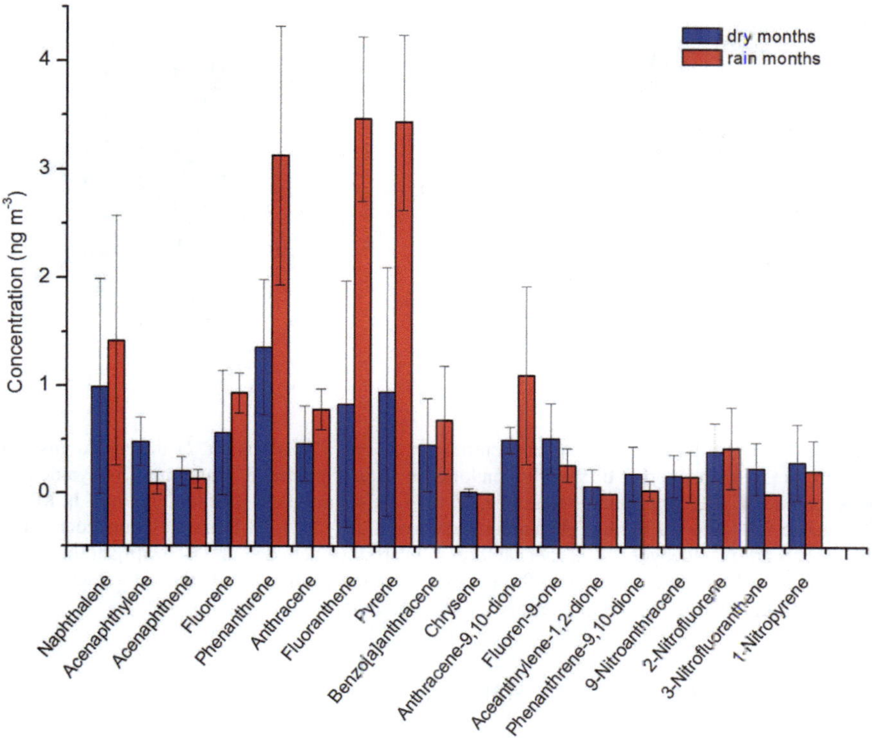

Fig. 5.4 Mean concentration (ng/m^3) of gas phase PAHs, nitro-PAHs and oxy-PAHs (PUF) in ambient air samples (n) collected from May 2017 to April 2018 (n = 93). Dry months: April to September, rainy months: October to March "adapted from original article by Dos Santos et al. (2020) after taking permission from Elsevier"

can reach several hundred ng/g. In general, N-PAHs and OH-PAHs are less abundant in soil than PAHs, O-PAHs, and A-PAHs. Notably, studies show that PAH levels in Sicily can reach up to 7804.3 ng/g (De Guidi et al. 2017), whereas high concentrations of PAHs (550 ng/g) (Ma et al. 2014) have been observed in China's Liao River Delta. According to present analysis, urban soil was found more concentrated than rural soil, while forest soil was more concentrated than agricultural soil by PAHs and their derivatives (Bandowe et al. 2019). Additionally, a different study found that Delta of Liao River, Northeast China contains A-PAHs, was observed as 517 ± 838 ng/g, indicating that burning biomass, air-soil exchange, and petroleum could be the sources. A different Chinese study that examined soil samples from 26 provinces in Eastern China's agricultural region found that the total amounts of N-PAH and O-PAH were 50.0 ng/g and 9.0 ng/g, respectively (Sun et al. 2017).

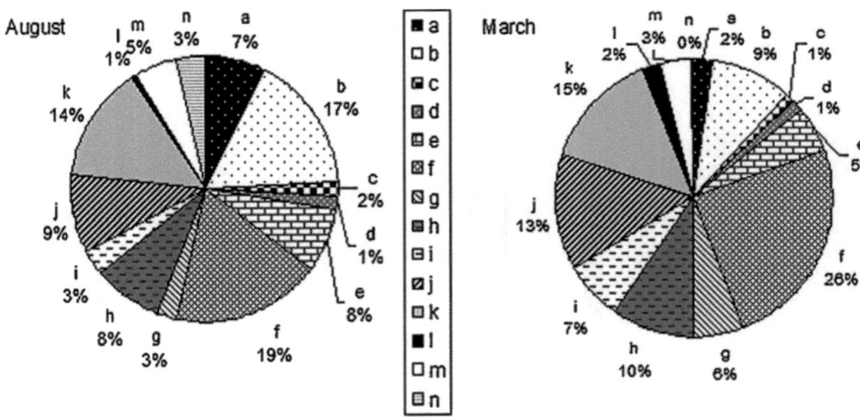

Fig. 5.5 % Contribution of mean concentrations of oxy PAHs (ng m^{-3}) during August and March. **a** Benzophenone, **b** 9H-Fluoren9one, **c** Athracenone, **d** Fluorenecarboxyaldehyde, **e** Xanthone, **f** 9,10-phenanthrenequinone, **g** 4H-Cyclopenta[d,e,f] phenanthrenone, **h** Methyl Athracene 9,10-dione, **i** Methyl phenanthrenecarboxyaldehyde, **j** 7H-Benz[d,e]athnracene-7-one, **k** Pyrenecarboxyaldehyde, **l** C2phenanthrene-9-carboxyaldehyde, **m** Benza[a]anthracenone-7,12-dione, **n** Benza[a]anthracenone-9,10-dione] "adapted from original article by Andreou and Rapsomanikis (2009) after taking permission from Elsevier"

PAH Derivatives in Other Matrices

Qiao et al. (2014) synthesised PAHs and their derivatives, which need to be examined in a variety of matrices because of the harmful effects they have on living things. In river samples, the levels of 16 parent PAHs, methyl-, N-, and O-PAHs were investigated. Because of photochemical degradation in aquatic environments, they discovered that N-PAHs were below the detection limit. On the other hand, O-PAHs were found in higher concentrations than the parent PAHs. Additionally, N-PAHs are more likely to undergo photo-oxidation, which transforms them into O-PAHs, whereas the other derivatives do not undergo this transformation. The researchers claim that when PAHs are altered, O-PAHs are created. Furthermore, wastewater treatment plants are the main source of PAHs, O-PAHs, and methyl-PAHs taken from rivers.

PAHs and their derivatives shows detrimental haze (as per their distribution), particularly in terms of particle size. Furthermore, Cao et al. (2020) revealed that four to six rings containing PAHs predominated in every sample. The levels of PAHs and their hydroxyl, carboxyl, oxygen and certain other derivatives were analysed in edible marine fish species. The findings demonstrated that mostly low-molecular-weight PAHs concentrations have been found greater in the liver than in the muscle. It means the level of PAHs is controlled by the fish's lipid composition (Jafarabadi et al. 2019). Moreover, PAHs and their derivatives are abundant in coal gasification and its byproducts, including ash, tar, and char; char has up to 56.2% of 3-ring PAHs (Ütnü et al. 2020).

Case Studies: Health Impact of PAH Derivatives on Humans and Other Species

Multiple, frequently simultaneous pathways, including inhalation, direct touch, and nutritional consumption, expose humans and other species to PAH derivatives. The particular organs and systems that these various exposure pathways impact are those that target the PAH derivatives that enter the body. Furthermore, after entering the body, PAH derivatives undergo metabolism and can damage almost every organ and system. Variations in toxicity are partly caused by the biological accessibility of PAH derivatives to the relevant organs, which varies depending on the exposure route. This implies that the exposure pathway has a major impact on the toxicity of PAHs to particular organs. This section of the chapter compiles case studies on the harmful effects of PAH compounds on different organs in humans and animals. Figure 5.6 exploring the health impact of PAHs and their derivatives on various organs.

Respiratory System

This system includes respiratory tract and the lungs. People usually breathe in a variety of PAH compounds found in the air. The lungs are where these breathed PAH derivatives mostly build up and potentially harm the lungs. PAH derivatives are frequently found in airborne particles, particularly $PM_{2.5}$. When inhaling, tiny particles may enter the alveoli and release toxic PAH compounds into the respiratory system. These chemicals mainly affect respiratory health and can cause

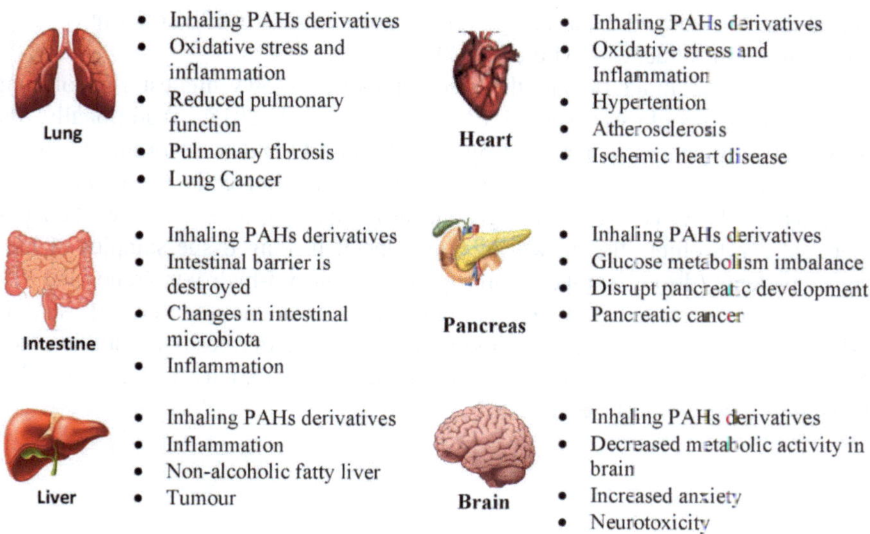

Fig. 5.6 Inhaled PAHs and their derivatives, and their health impact on various organs

respiratory disorders when inhaled. Epidemiological and toxicological studies are the primary focus of current research on the respiratory dangers associated with PAH compounds. The risk of respiratory disorders is increased by breathing PAH compounds, according to numerous epidemiological studies. Following exposure to Naphthalene, the incidence of shortness of breath, chronic cough, and ongoing headaches increased considerably (Bortey-Sam et al. 2017).

According to recent research, children's asthma symptoms' development, progression, and aggravation are all correlated with their exposure to PAHs (Karimi et al. 2015). Similar findings were observed about PAH exposure, with methylphenanthrene exposure raising children's chance of developing asthma (Jung et al. 2014).

Additionally, exposure to low concentrations of PAHs causes a dose-dependent decline in young people's lung functioning (Alhamdow et al. 2021). Even more concerning, human lung cancer risk can be increased by inhaling PAHs (Tsay et al. 2013). A study carried out in Xuanwei, Yunnan Province, China, found that the greatest concentrations of total PAHs (TPAHs) were likewise found in areas where lungs cancer mortality rate was highest (Lui et al. 2017). These findings clearly show a connection between inhaling PAHs and developing lung cancer in humans.

In A549 cells, PAHs can cause inflammation and oxidative stress (Guo et al. 2021). Additionally, it has been shown that inflammation is caused by N-PAHs and O-PAHs. According to recent studies, BEAS-2B cells exhibit inflammatory responses to 1-nitropyrene, 3-nitrofluoranthene, and 3-nitrobenzanthrone (Øvrevik et al. 2010). Additionally, it has been demonstrated that 1-aminopyrene and phenanthrenequinone raise the amounts of proinflammatory proteins in these cells (Koike et al. 2014). Based on experimental results Hu et al. (2020) suggested that lungs experience severe inflammation on its acute exposure to 1-nitropyrene, which activates many signaling pathways. Even more concerning, it has been demonstrated that 1-nitropyrene can induce lung fibrosis in male C57BL6/J/J mice when administered intratracheally (Fu et al. 2021).

The coal/coke industry and the steel industries shows highest risk of lung cancer among people exposed to PAHs (Singh et al. 2018). Additionally, the growth of lung cancer is triggered by N-PAHs and O-PAHs, according to experimental study. For example, human bronchial epithelial cells (Li et al. 2019) and A549 cells tested with varying concentrations of 9-fluorenone resulted that chemical can promote the growth of lung cancer. In lung tissue samples of lung cancer patients (do not smoke), 1-nitropyrene, 1-nitro-3-hydroxypyrene, and 1,3-dinitropyrene were found, which further supports the notion that exposure to N-PAH can induce lung cancer (Tokiwa et al. 1993). Furthermore, studies have shown that 9-nitroanthracene contributes to lung genotoxicity by causing DNA damage in lung tissue (Li et al. 2017).

Cardiovascular System

Heart, arteries, veins, and capillaries are constituents of cardiovascular system (often known as "circulatory system"). Studies have connected exposure to PAHs to heart-related issues include ischemic heart disease, high blood pressure, and irregular heartbeats (Leachi et al. 2020).

Inhaled PAHs and their variants have the ability to reach quickly the circulatory system through inhalation. Absorbed through the alveoli, inhaled B[a]P primarily enters the bloodstream, with the majority reaching the cardiovascular system in its unmetabolised state (Gerde et al. 2001). Therefore, damages the cardiovascular system in addition to the pulmonary system. According to epidemiological studies, diesel particulate matter shows a positive dose-dependent correlation with young women's hypertension, with PAHs in diesel particulate matter being a major contributing factor (Bangia et al. 2015). Air particle matter pollution significantly altered pulse pressure in older persons, where PAHs specifically correlated with higher systolic and pulse pressure (Jacobs et al. 2012).

Studies found that pro-inflammatory cytokines were positively connected with elevated levels of OH-PAH of urine samples taken from taxi drivers exposed to air pollution (Brucker et al. 2013). These results were corroborated by toxicological investigations, which showed that substances such as pyrene, fluoranthene, and phenanthrene caused endothelial dysfunction and inflammation in BALB/c mice when administered as nasal drops (Rojas et al. 2022). Atherosclerosis and other cardiovascular disorders are significantly influenced by endothelial dysfunction. Certain in vitro studies have shown that endothelial dysfunction can be brought on by PAHs, N-PAHs, and O-PAHs. Compounds including 9-fluorene, 9,10-anthraquinone, 1-nitronaphthalene, and 9-nitroanthracene block nitric oxide synthase in human endothelial (Wu et al. 2022), a prior indicator for endothelial dysfunction. Additionally, anthracene and its derivatives weakened the endothelial barrier, which made HUVEC monolayers more permeable (Wu et al. 2022). A link to endothelial dysfunction is suggested by the decrease in nitric oxide generation by N-PAHs and O-PAHs. These results suggest that endothelial dysfunction brought on by PAHs and their derivatives may be a contributing factor to cardiovascular disease. Another important contributing element to the emergence of cardiovascular disorders is DNA damage. The cardiovascular system may be harmed by PAHs and their derivatives when all three exposure pathways are present. In zebrafish, the combination of B[a]P and O-PAHs has been connected to elevated cardiovascular toxicity and embryonic development (Cunha et al. 2020). Similarly, zebrafish embryos exposed systemically to A-PAHs showed cardiovascular damage (Jung et al. 2013).

Digestive System

Consumption of PAHs may cause the adverse gastrointestinal reactions. Research indicates that PAH consumption can lead to stomach cancer (Wester et al. 2012).

Upon reaching the intestines, PAHs may harm intestinal health by triggering inflammation and altering intestinal permeability. The intestinal mucosa preserves the homeostasis of intestinal cells by serving as a barrier against pathogens (Chen et al. 2013). Several intestinal disorders are closely linked to the deterioration of the mucosal barrier. In female mice and their progeny, dietary exposure to 3-methylcholanthrene has been demonstrated to disrupt gut barrier function and intrahepatic circulation (Xu et al. 2022). If the intestinal barrier is compromised, toxic substances could get into the bloodstream and cause more damage to the body. In C57BL/6 mice, oral exposure to benzo[a]pyrene (B[a]P) significantly altered the quantity and makeup of gut microbiota, causing mild inflammation of the mucosa of the colon and ileum (Ribière et al. 2016). These findings demonstrate the potential harm to intestine function from ingesting PAHs and their derivatives. Consuming PAHs may also have an impact on pancreatic development and increase the risk of developing pancreatic cancer. The single hormone that lowers blood glucose levels to maintain homeostasis, insulin, is secreted by islet β-cells. Pregnant C57BL/6 mice gavaged with PAHs exhibited decreased β-cell growth in their adult male offspring, according to a recent study (Ou et al. 2022). This implies that inhalation of PAHs may be the cause of more pancreatic dysplasia and diabetes, in pregnant women. Additionally, pancreatic cancer have been correlated with consumption of PAHs; a population-based case-control study found that higher dietary intake of B[a]P causes higher risk of pancreatic cancer (Anderson et al. 2005).

Furthermore, liver health may suffer if PAHs and their compounds are consumed. In the digestive tract, the liver is essential to the synthesis of bile. According to several studies done, on female rats, their livers faces inflammation and oxidative stress when phenanthrene has been given them orally (Ma et al. 2020). According to a study on healthy males, dietary PAHs also cause oxidative stress in liver cells (Elhassaneen and El-Badawy 2013). Nonalcoholic fatty liver disease may be the result of triglycerides accumulation in liver due to its exposure to PAHs. Oral ingestion of PAHs (as carcinogens) may also lead to liver carcinogenicity. For instance, liver cell carcinoma has been developed in Wistar rats after oral ingestion of B[a]P, for two years (Uno et al. 2018).

Neurotoxicity

Brain endothelial cells have a lower permeability to chemicals, however, blood-brain barrier can be penetrated by PAHs and their derivatives. PAHs and B[a]P are neurotoxic (Olasehinde and Olaniran 2022). According to earlier research, ingesting PAHs and related derivatives can result in neurotoxicity. For example, adult male rats exposed to PAHs in the mother's prenatal diet showed decreased metabolic activity in many parts of the brain and increased anxiety-related behaviors (Crépeaux et al. 2012). It's interesting to note that a recent study, which used intradermal administration to expose animals to ambient PAH concentrations, discovered that low doses of PAHs did not cause appreciable neurotoxicity (Kuang et al. 2022).

Immunotoxicity

The harmful effects that pollutants have on the immune system are referred to as immunotoxicity. Both humans and animals have shown immunotoxic reactions to PAHs. According to a study done in Gansu, China, there is a strong link between increased exposure to PAHs and both oxidative DNA damage and the suppression of Treg cell immunological activities (Peiffer et al. 2016). Research on PAH derivatives is somewhat lacking, even though the immunotoxicity of PAHs is well recognised. According to recent studies, sea bass exposed to a combination of 41 parent and alkyl PAHs for 21 days showed signs of both specific and general immune function impairment, including leukopenia, increased leukocyte mortality, and decreased phagocytosis (Peiffer et al. 2013). Furthermore, it has been suggested that N-PAHs have immunotoxic properties. 1-Nitropyrene, for example, forms hydrogen bonds and does π–π interactions with the aryl hydrocarbon receptor, reduces cytokine levels, activates CYP1A1 transcription, and can strongly inhibit phagocytosis (Yao et al. 2019). Moreover, 1-nitropyrene can induce macrophage apoptosis by generating reactive oxygen species and nuclear translocation of AIF (Danion et al. 2011).

Reproduction Toxicity

It has been shown that PAHs and their derivatives affects both men's and women's reproductive health. These substances have the potential to damage the female reproductive system by interfering with endocrine signals. For example, ovarian follicle pool of offspring mice was affected by exposure to PAHs (Wang et al. 2017). Similar effects are seen in the male reproductive system; research indicates a connection between increased sperm DNA damage and environmental PAH exposure (Han et al. 2011). Reproductive toxicity is another documented effect of ingesting PAHs and their derivatives. Fetal growth restriction was observed in pregnant mice given 1-nitropyrene by gavage every day throughout the third trimester (Li et al. 2018). Additionally, following two weeks of oral therapy with B[a]P and pyrene, C57BL/6 mice displayed damage to their testicular tissue (Zhang et al. 2022). After 90 days of oral intake of B[a]P, adult male [Sprague-Dawley] rats showed testicular shrinkage and lower sperm quality in the epididymis (Chung et al. 2011).

Conclusion

The studies presented in this chapter demonstrate that smog and suspended particulate matter (SPM) pose a significant environmental and public health issue in cities worldwide. The most dangerous components of SPM are PAHs and their oxidised derivatives (e.g. nitro PAHs, O-PAHs, OH-PAHs), which are produced mainly by human combustion. These compounds usually dispersed in soil, water, sediment and other matrices, and then reach into the living systems through various pathways. Of these compounds, especially in acceptable particulate forms such as $PM_{2.5}$ and PM_1, are mutagenic, carcinogenic, and toxic. Their health impacts are exacerbated by their ability to penetrate deeply into the respiratory system and potentially enter systemic circulation. Furthermore, cardiovascular system, digestive system are also affected by these compounds. Additionally, the neurotoxicity, immunotoxicity and reproduction toxicity are also the serious impacts of PAHs and their oxidised derivatives. Case studies included in present chapter have shown that, (a) nitro-PAHs are substantially more hazardous than benzo[a]pyrene, (b) 9-nitroanthracene contributes to lung genotoxicity by causing DNA damage in lung tissue, (c) zebrafish embryos exposed systemically to A-PAHs showed cardiovascular damage, (d) inhalation of PAHs is the major cause of more pancreatic cancer and diabetes, in pregnant women, (e) PAHs and B[a]P are neurotoxic, (f) 1-Nitropyrene, for example, forms hydrogen bonds and does π–π interactions with the aryl hydrocarbon receptor, reduces cytokine levels, activates CYP1A1 transcription, and can strongly inhibit phagocytosis, and (g) following two weeks of oral therapy with B[a]P and pyrene, C57BL/6 mice displayed damage to their testicular tissue. Studies found that highest concentration (314.55 ng/m^3) of $PM_{2.5}$ bound PAHs was observed in Taiyuan, China, and largest concentration (142.6 ng/m^3) of $PM_{2.5}$ bound O-PAHs was detected in Estarreja [Portugal].

This chapter also provides information about the detection methods. Enhanced monitoring and regulatory approaches, supported by sophisticated analytical techniques, will be key in reducing the risks associated with airborne particulate pollution and informing public health initiatives. It was found that HPLC technique is better than GC for analysis of nitro-PAHs, as these are more stable at room temperature.

References

Abbasi, S., Keshavarzi, B.: Source identification of total petroleum hydrocarbons and polycyclic aromatic hydrocarbons in PM10 and street dust of a hot spot for petrochemical production: Asaluyeh County, Iran. Sustain. Cities Soc. **45**, 214–230 (2019)

Albinet, A., Leoz-Garziandia, E., Budzinski, H., Villenave, E.: Simultaneous analysis of oxygenated and nitrated polycyclic aromatic hydrocarbons on standard reference material 1649a (urban dust) and on natural ambient air samples by gas chromatography–mass spectrometry with negative ion chemical ionisation. J. Chromatogr. A **1121**(1), 106–113 (2006)

Alhamdow, A., Zettergren, A., Kull, I., Hallberg, J., Andersson, N., Ekström, S., Berglund, M., Wheelock, C.E., Essig, Y.J., Krais, A.M., Georgelis, A.: Low-level exposure to polycyclic aromatic hydrocarbons is associated with reduced lung function among Swedish young adults. Environ. Res. **197**, 111169 (2021)

Alves, C., Evtyugina, M., Vicente, E., Vicente, A., Rienda, I.C., de la Campa, A.S., Tomé, M., Duarte, I.: $PM_{2.5}$ chemical composition and health risks by inhalation near a chemical complex. J. Environ. Sci. **124**, 860–874 (2023)

Anderson, K.E., Kadlubar, F.F., Kulldorff, M., Harnack, L., Gross, M., Lang, N.P., Barber, C., Rothman, N., Sinha, R.: Dietary intake of heterocyclic amines and benzo (a) pyrene: associations with pancreatic cancer. Cancer Epidemiol. Biomark. Prev. **14**(9), 2261–2265 (2005)

Andreou, G., Rapsomanikis, S.: Polycyclic aromatic hydrocarbons and their oxygenated derivatives in the urbanatmosphere of Athens. J. Hazard. Mater. **172**, 363–373 (2009)

Bamford, H.A., Bezabeh, D.Z., Schantz, M.M., Wise, S.A., Baker, J.E.: Determination and comparison of nitrated-polycyclic aromatic hydrocarbons measured in air and diesel particulate reference materials. Chemosphere **50**(5), 575–587 (2003)

Bandowe, B.A.M., Meusel, H.: Nitrated polycyclic aromatic hydrocarbons (nitro-PAHs) in the environment—a review. Sci. Total. Environ. **581**, 237–257 (2017)

Bandowe, B.A.M., Meusel, H., Huang, R.J., Ho, K., Cao, J., Hoffmann, T., Wilcke, W.: $PM_{2.5}$-bound oxygenated PAHs, nitro-PAHs, and parent-PAHs from the atmosphere of a Chinese megacity: seasonal variation, sources, and cancer risk assessment. Sci. Total. Environ. **473**, 77–87 (2014)

Bandowe, B.A.M., Wei, C., Han, Y., Cao, J., Zhan, C., Wilcke, W.: Polycyclic aromatic compounds (PAHs, oxygenated PAHs, nitrated PAHs, and azaarenes) in soils from China and their relationship with geographic location, land use, and soil carbon fractions. Sci. Total. Environ. **690**, 1268–1276 (2019)

Bangia, K.S., Symanski, E., Strom, S.S., Bondy, M.A.: Cross-sectional analysis of polycyclic aromatic hydrocarbons and diesel particulate matter exposures and hypertension among individuals of Mexican origin. Environ. Health **14**, 51 (2015)

Barbas, J.T., Sigman, M.E., Dabestani, R.: Photochemical oxidation of phenanthrenesorbed on silica gel. Environ. Sci. Technol. **30**(5), 1776–1780 (1996)

Barrado, A.I., García, S., Barrado, E., Pérez, R.M.: PM2.5-bound PAHs and hydroxy-PAHs in atmospheric aerosol samples: correlations with season and with physical and chemical factors. Atmos. Environ. **49**, 224–232 (2012)

Besis, A., Gallou, D., Avgenikou, A., Serafeim, E., Samara, C.: Size-dependent in vitro inhalation bioaccessibility of PAHs and O/N PAHs—implications to inhalation risk assessment. Environ. Pollut. **301**, 119045 (2022)

Boldo, E., Medina, S., Le Tertre, A., Hurley, F., Mücke, H.G., Ballester, F., Aguilera, I., Daniel Eilstein on Behalf of the Apheis Group: Apheis: health impact assessment of long-term exposure to PM2.5 in 23 European cities. Eur. J. Epidemiol. **21**(6), 449–458 (2006)

Bortey-Sam, N., Ikenaka, Y., Akoto, O., Nakayama, S.M., Asante, K.A., Baidoo, E., Obirikorang, C., Saengtienchai, A., Isoda, N., Nimako, C., Mizukawa, H.: Oxidative stress and respiratory symptoms due to human exposure to polycyclic aromatic hydrocarbons (PAHs) in Kumasi, Ghana. Environ. Pollut. **228**, 311–320 (2017)

Brown, J.S., Gordon, T., Price, O., Asgharian, B.: Thoracic and respirable particle definitions for human health risk assessment. Part. Fibre Toxicol. **10**(1), 12 (2013)

Brucker, N., Moro, A.M., Charao, M.F., Durgante, J., Freitas, F., Baierle, M., Nascimento, S., Gauer, B., Bulcao, R.P., Bubols, G.B., Ferrari, P.D., Thiesen, F.V., Gioda, A., Duarte, M.M.M.F., de Castro, I., Saldiva, P.H., Garcia, S.C.: Biomarkers of occupational exposure to air pollution, inflammation, and oxidative damage in taxi drivers. Sci. Total. Environ. **463–464**, 884–893 (2013)

Cao, Z., Wang, M., Shi, S., Zhao, Y., Chen, X., Li, C., Li, Y., Wang, H., Bao, L., Cui, X.: Size-distribution-based assessment of human inhalation and dermal exposure to airborne parent, oxygenated, and chlorinated PAHs during a regional heavy haze episode. Environ. Pollut. **263**, 114661 (2020)

Chen, H., Mao, X., He, J., Yu, B., Huang, Z., Yu, J., Zheng, P., Chen, D.: Dietary fibre affects intestinal mucosal barrier function and regulates intestinal bacteria in weaning piglets. Br. J. Nutr. **110**(10), 1837–1848 (2013)

Chuesaard, T., Chetiyanukornkul, T., Kameda, T., Hayakawa, K., Toriba, A.: Influence of biomass burning on the levels of atmospheric polycyclic aromatic hydrocarbons and their nitro derivatives in Chiang Mai, Thailand. Aerosol Air Qual. Res. **14**(4), 1247–1257 (2014)

Chung, J.Y., Kim, Y.J., Kim, J.Y., Lee, S.G., Park, J.E., Kim, W.R., Yoon, Y.D., Yoo, K.S., Yoo, Y.H., Kim, J.M.: Benzo [a] pyrene reduces testosterone production in rat Leydig cells via a direct disturbance of testicular steroidogenic machinery. Environ. Health Perspect. **119**(11), 1569–1574 (2011)

Cochran, R.E., Dongari, N., Jeong, H., Beránek, J., Haddadi, S., Shipp, J., Kubátová, A.: Determination of polycyclic aromatic hydrocarbons and their oxy-, nitro-, and hydroxy-oxidation products. Anal. Chim. Acta **740**, 93–103 (2012)

Crépeaux, G., Bouillaud-Kremarik, P., Sikhayeva, N., Rychen, G., Soulimani, R., Schroeder, H.: Late effects of a perinatal exposure to a 16 PAH mixture: increase of anxiety-related behaviours and decrease of regional brain metabolism in adult male rats. Toxicol. Lett. **211**(2), 105–113 (2012)

Crimmins, B.S., Baker, J.E.: Improved GC/MS methods for measuring hourly PAH and nitro-PAH concentrations in urban particulate matter. Atmos. Environ. **40**(35), 6764–6779 (2006)

Cunha, V., Vogs, C., Le Bihanic, F., Dreij, K.: Mixture effects of oxygenated PAHs and benzo [a] pyrene on cardiovascular development and function in zebrafish embryos. Environ. Int. **143**, 105913 (2020)

Danion, M., Le Floch, S., Kanan, R., Lamour, F., Quentel, C.: Effects of in vivo chronic hydrocarbons pollution on sanitary status and immune system in sea bass (Dicentrarchuslabrax L.). Aquat. Toxicol. **105**(3–4), 300–311 (2011)

De Guidi, G., Falciglia, P.P., Catalfo, A., Fagone, S., Vagliasindi, F.G.A.: Soil contaminated with PAHs and nitro-PAHs: contamination levels in an urban area of Catania (Sicily, Southern Italy) and experimental results from simulated decontamination treatment. Clean Technol. Environ. Policy. **19**(4), 1121−1132 (2017)

De Oliveira Galvao, M.F., de Oliveira Alves, N., Ferreira, P.A., Caumo, S., de Castro Vasconcellos, P., Artaxo, P., de Souza Hacon, S., Roubicek, D.A., de Medeiros, S.R.B.: Biomass burning particles in the Brazilian Amazon region: mutagenic effects of nitro and oxy-PAHs and assessment of health risks. Environ. Pollut. **233**, 960–970 (2018)

Deng, W.J., Louie, P.K.K., Liu, W.K., Bi, X.H., Fu, J.M., Wong, M.H.: Atmospheric levels and cytotoxicity of PAHs and heavy metals in TSP and $PM_{2.5}$ at an electronic waste recycling site in southeast China. Atmos. Environ. **40**(36), 6945–6955 (2006)

Dos Santos, R.R., Leal, L.D.V., de Lourdes Cardeal, Z., Menezes, H.C.: Determination of polycyclic aromatic hydrocarbons and their nitrated and oxygenated derivatives in coffee brews using an efficient cold fiber-solid phase microextraction and gas chromatography mass spectrometry method. J. Chromatogr. A **1584**, 64–71 (2019)

Dos Santos, R.R., de Lourdes Cardeal Z., Costa Menezes, H.: Phase distribution of polycyclic aromatic hydrocarbons and their oxygenated and nitrated derivatives in the ambient air of a Brazilian urban area. Chemosphere **250**, 126223 (2020)

Elhassaneen, Y.A.E.A., El-Badawy, A.M.: Influence of charcoal broiled meat consumption on the liver functions and non-enzymatic antioxidants in human blood. Food Nutr. Sci. **4**(1), 26785 (2013)

EU: Directive 2004/107/EC of the European Parliament and of the Council of 15 December 2004 relating to arsenic, cadmium, mercury, nickel and polycyclic aromatic hydrocarbons in ambient air (2005)

Fu, L., Zhao, H., Xiang, Y., Xiang, H.X., Hu, B., Tan, Z.X., Lu, X., Gao, L., Wang, B., Wang, H., Zhang, C.: Reactive oxygen species-evoked endoplasmic reticulum stress mediates 1-nitropyrene-induced epithelial-mesenchymal transition and pulmonary fibrosis. Environ. Pollut. **283**, 117134 (2021)

Gao Y, Yang L, Chen J, Li Y, Jiang P, Zhang J, Wang W.: Nitro and oxy-PAHs bounded in $PM_{2.5}$ and $PM_{1.0}$ under different weather conditions at Mount Tai in Eastern China: sources, long-distance transport, and cancer risk assessment. Sci. Total. Environ. **622–623**, 1400–1407 (2018)

García-Alonso, S., Barrado-Olmedo, A.I., Pérez-Pastor, R.M.: An analytical method to determine selected nitro-PAHs in soil samples by HPLC with fluorescence detection. Polycycl. Aromat. Compd. **32**(5), 669–682 (2012)

Gerde, P., Muggenburg, B.A., Lundborg, M., Dahl, A.R.: The rapid alveolar absorption of diesel soot-adsorbed benzo[a]pyrene: bioavailability, metabolism and dosimetry of an inhaled particle-borne carcinogen. Carcinogenesis **22**(5), 741–749 (2001)

Guo, H., Zhang, Z., Wang, H., Ma, H., Hu, F., Zhang, W., Liu, Y., Huang, Y., Zeng, Y., Li, C., Wang, J.: Oxidative stress and inflammatory effects in human lung epithelial A549 cells induced by phenanthrene, fluorene, and their binary mixture. Environ. Toxicol. **36**(1), 95–104 (2021)

Han, X., Zhou, N., Cui, Z., Ma, M., Li, L., Cai, M., Li, Y., Lin, H., Li, Y., Ao, L., Liu, J.: Association between urinary polycyclic aromatic hydrocarbon metabolites and sperm DNA damage: a population study in Chongqing, China. Environ. Health Perspect. **119**(5), 652–657 (2011)

Han, M., Kong, J., Yuan, J., He, H., Hu, J., Yang, S., Li, S., Zhang, L., Sun, C.: Method development for simultaneous analyses of polycyclic aromatic hydrocarbons and their nitro-, oxy-, hydroxy-derivatives in sediments. Talanta Open **205**, 120128 (2019)

Hayakawa, K.: Chemistry of polycyclic aromatic hydrocarbons (PAHs), nitropolycyclic aromatic hydrocarbons (NPAHs), and oxidative derivatives of PAHs. In: Hayakawa, K. (ed.) Polycyclic Aromatic Hydrocarbons: Environmental Behavior and Toxicity in East Asia, pp. 3–10 (2018). ISBN: 978–981–10-6774-7

Hu, B., Tong, B., Xiang, Y., Li, S. R., Tan, Z. X., Xiang, H. X., Fu, L., Wang, H., Zhao, H., Xu, D.X.: Acute 1-NP exposure induces inflammatory responses through activating various inflammatory signaling pathways in mouse lungs and human A549 cells. Ecotoxicol Environ. Saf. **189**, 109977 (2020)

Huang, B., Liu, M., Bi, X., Chaemfa, C., Ren, Z., Wang, X., Sheng, G., Fu, J.: Phase distribution, sources and risk assessment of PAHs, NPAHs and OPAHs in a rural site of Pearl River Delta region, China. Atmos. Pollut. Res. **5**(2), 210–218 (2014)

Jacobs, L., Buczynska, A., Walgraeve, C., Delcloo, A., Potgieter-Vermaak, S., Van Grieken, R., Demeestere, K., Dewulf, J., Van Langenhove, H., De Backer, H., Nemery, B.: Acute changes in pulse pressure in relation to constituents of particulate air pollution in elderly persons. Environ. Res. **117**, 60–67 (2012)

Jafarabadi, A.R., Bakhtiari, A.R., Yaghoobi, Z., Yap, C.K., Maisano, M., Cappello, T.: Distributions and compositional patterns of polycyclic aromatic hydrocarbons (PAHs) and their derivatives in three edible fishes from Kharg coral Island, Persian Gulf, Iran. Chemosphere **215**, 835–845 (2019)

Jakovljević, I., Pehnec, G., Vađić, V., Čačković, M., Tomašić, V., Jelinić, J.D.: Polycyclic aromatic hydrocarbons in PM10, $PM_{2.5}$ and PM_1 particle fractions in an urban area. Air Qual. Atmos. Health **11**(7), 843–854 (2018)

Jiping, M., Hou, L., Wu, G., Wang, L., Wang, X., Chen, L.: Multi-walled carbon nanotubes for magnetic solid-phase extraction of six heterocyclic pesticides in environmental water samples followed by HPLC-DAD determination. Materials **13**(24), 5729 (2020)

Jung, J.H., Hicken, C.E., Boyd, D., Anulacion, B.F., Carls, M.G., Shim, W.J., Incardona, J.P.: Geologically distinct crude oils cause a common cardiotoxicity syndrome in developing zebrafish. Chemosphere **91**(8), 1146–1155 (2013)

Jung, K.H., Perzanowski, M., Rundle, A., Moors, K., Yan, B., Chillrud, S.N., Whyatt, R., Camann, D., Perera, F.P., Miller, R.L.: Polycyclic aromatic hydrocarbon exposure, obesity, and childhood asthma in an urban cohort. Environ. Res. **128**, 35–41 (2014)

Karimi, P., Peters, K.O., Bidad, K., Strickland, P.T.: Polycyclic aromatic hydrocarbons and childhood asthma. Eur. J. Epidemiol. **30**(2), 91–101 (2015)

Keith, L.H.: The source of US EPA's sixteen PAH priority pollutants. Polycycl. Aromat. Compd. **35**(2–4), 147–160 (2015)

Kic, P.: Influence of housing technology on dust level in cowsheds. In: Proceedings of 17th International Scientific Conference 'Engineering for Rural Development', Jelgava, Latvia, pp. 24–27 (2018)

Kitanovski, Z., Shahpoury, P., Samara, C., Voliotis, A., Lammel, G.: Composition and mass size distribution of nitrated and oxygenated aromatic compounds in ambient particulate matter from Southern and Central Europe—implications for the origin. Atmos. Chem. Phys. **20**(4), 2471–2487 (2020)

Koike, E., Yanagisawa, R., Takano, H.: Toxicological effects of polycyclic aromatic hydrocarbons and their derivatives on respiratory cells. Atmos. Environ. **97**, 529–536 (2014)

Kuang, H., Zhou, W., Zeng, Y., Xu, D., Zhu, W., Lin, S., Fan, R.: Dose makes poison: insights into the neurotoxicity of perinatal and juvenile exposure to environmental doses of 16 priority-controlled PAHs. Chemosphere **298**, 134201 (2022)

Kurnia, A., Lim, L.W., Takeuchi, T.: Determination of polycyclic aromatic hydrocarbons (PAHs) using environmentally friendly liquid chromatography. Makara J. Sci. **22**(1), 6 (2018)

Lammel, G.: Polycyclic aromatic compounds in the atmosphere—a review identifying research needs. Polycycl. Aromat. Compd. **35**(2–4), 316–329 (2015)

Leachi, H.F.L., Marziale, M.H.P., Martins, J.T., Aroni, P., Galdino, M.J.Q., Ribeiro, R.P.: Polycyclic aromatic hydrocarbons and development of respiratory and cardiovascular diseases in workers. Rev. Enferm. **73**(3), 20180965 (2020)

Lee, H.H., Choi, N.R., Lim, H.B., Yi, S.M., Kim, Y.P., Lee, J.Y.: Characteristics of oxygenated PAHs in PM10 at Seoul, Korea. Atmos. Pollut. Res. **9**(1), 112–118 (2018)

Li, R., Zhao, L., Zhang, L., Chen, M., Shi, J., Dong, C., Cai, Z.: Effects of ambient $PM_{2.5}$ and 9-nitroanthracene on DNA damage and repair, oxidative stress and metabolic enzymes in the lungs of rats. Toxicol. Res. **6**(5), 654–663 (2017)

Li, R., Wang, X., Wang, B., Li, J., Song, Y., Luo, B., Chen, Y., Zhang, C., Wang, H., Xu, D.: Gestational 1-nitropyrene exposure causes fetal growth restriction through disturbing placental vascularity and proliferation. Chemosphere **213**, 252–258 (2018)

Li, D., Yun, Y., Gao, R.: Oxygenated polycyclic aromatic hydrocarbons (Oxy-PAHs) facilitate lung cancer metastasis by epigenetically regulating the epithelial-to-mesenchymal transition (EMT). Environ. Pollut. **255**(2), 113–261 (2019)

Li, D., Zhao, Y., Du, W., Zhang, Y., Chen, Y., Lei, Y., Wu, C., Wang, G.: Characterization of PM2. 5-bound parent and oxygenated PAHs in three cities under the implementation of clean air action in Northern China. Atmos. Res. **267**, 105932 (2022)

Lisowski, P., Zarzycki, P.K.: Microfluid paper-based analytical devices (μPADs) and micro total analysis systems (μTAS): development, applications, and future trends. Chromatographia **76**(19), 1201–1214 (2013)

Lui, K.H., Bandowe, B.A.M., Tian, L., Chan, C.S., Cao, J.J., Ning, Z., Lee, S.C., Ho, K.F.: Cancer risk from polycyclic aromatic compounds in fine particulate matter generated from household coal combustion in Xuanwei, China. Chemosphere **169**, 660–668 (2017)

Ma, C., Ye, S., Lin, T., Ding, X., Yuan, H., Guo, Z.: Source apportionment of polycyclic aromatic hydrocarbons in soils of wetlands in the Liao River Delta, Northeast China. Mar. Pollut. Bull. **80**(1–2), 160–167 (2014)

Ma, W.L., Zhu, F.J., Hu, P.T., Qiao, L.N., Li, Y.F.: Gas/particle partitioning of PAHs based on equilibrium-state model and steady-state model. Sci. Total. Environ. **706**, 136029 (2020)

Miet, K., Le Menach, K., Flaud, P.M., Budzinski, H., Villenave, E.: Heterogeneous reactions of ozone with pyrene, 1-hydroxypyrene, and 1-nitropyrene adsorbed on particles. Atmos. Environ. **43**, 3699–3707 (2009)

Mirivel, G., Riffault, V., Galloo, J.C.: Simultaneous determination by ultra-performance liquid chromatography–atmospheric pressure chemical ionisation time-of-flight mass spectrometry of nitrated and oxygenated PAHs found in air and soot particles. Anal. Bioanal. Chem. **397**(1), 243–256 (2010)

Mueller, A., Ulrich, N., Hollmann, J., Sanchez, C.E.Z., Rolle-Kampczyk, U.E., von Bergen, M.: Characterization of a multianalyte GC-MS/MS procedure for detecting and quantifying polycyclic aromatic hydrocarbons (PAHs) and PAH derivatives from air particulate matter for an improved risk assessment. Environ. Pollut. **255**, 112967 (2019)

Nassar, H.F.: Assessment of dissolved and particulate polycyclic aromatic hydrocarbons (PAHs) and their nitro derivatives (NPAHs) in the Mediterranean sea surface waters along Alexandria Coast. J. Environ. Sci. Technol. **10**(6), 290–298 (2017)

Niu, X., Ho, K.F., Chuang, H.C., Sun, J., Huang, Y., Hu, T., Xu, H., Duan, J., Lui, K.H., Cao, J.: Comparison of cytotoxicity induced by PM2. 5-bound polycyclic aromatic compounds from different environments in Xi'an, China. Atmos. Environ. **216**, 116929 (2019)

Nowakowski, M., Rykowska, I., Wolski, R., Andrzejewski, P.: Polycyclic aromatic hydrocarbons (PAHs) and their derivatives (O-PAHs, N-PAHs, OH-PAHs): determination in suspended particulate matter (SPM)—a review. Environ. Process. **9**(1), 2 (2022)

Nyiri, Z., Novák, M., Bodai, Z., Szabó, B.S., Eke, Z., Záray, G., Szigeti, T.: Determination of particulate phase polycyclic aromatic hydrocarbons and their nitrated and oxygenated derivatives using gas chromatography–mass spectrometry and liquid chromatography–tandem mass spectrometry. J. Chromatogr. A **1472**, 88–98 (2016)

O'Connell, S.G., Haigh, T., Wilson, G., Anderson, K.A.: An analytical investigation of 24 oxygenated-PAHs (OPAHs) using liquid and gas chromatography–mass spectrometry. Anal. Bioanal. Chem. **405**(27), 8885–8896 (2013)

Olasehinde, T.A., Olaniran, A.O.: Neurotoxicity of polycyclic aromatic hydrocarbons: a systematic mapping and review of neuropathological mechanisms. Toxics **10**(8), 417 (2022)

Oliveira, M., Slezakova, K., Delerue-Matos, C., Pereira, M.C., Morais, S.: Children environmental exposure to particulate matter and polycyclic aromatic hydrocarbons and biomonitoring in school environments: a review on indoor and outdoor exposure levels, major sources and health impacts. Environ. Int. **124**, 180–204 (2019)

Ou, K., Song, J., Zhang, S., Fang, L., Lin, L., Lan, M., Chen, M., Wang, C.: Prenatal exposure to a mixture of PAHs causes the dysfunction of islet cells in adult male mice: association with type 1 diabetes mellitus. Ecotoxicol. Environ. Saf. **239**, 113695 (2022)

Øvrevik, J., Arlt, V.M., Øya, E., Nagy, E., Mollerup, S., Phillips, D.H., Låg, M., Holme, J.A.: Differential effects of nitro-PAHs and amino-PAHs on cytokine and chemokine responses in human bronchial epithelial BEAS-2B cells. Toxicol. Appl. Pharmacol. **242**(3), 270–280 (2010)

Pedersen, D.U., Durant, J.L., Penman, B.W., Crespi, C.L., Hemond, H.F., Lafleur, A.L., Cass, G.R.: Human-cell mutagens in respirable airborne particles in the northeastern United States. 1. Mutagenicity of fractionated samples. Environ. Sci. Technol. **38**(3), 682–689 (2004)

Peiffer, J., Cosnier, F., Grova, N., Nunge, H., Salquèbre, G., Decret, M.J., Cossec, B., Rychen, G., Appenzeller, B.M., Schroeder, H.: Neurobehavioral toxicity of a repeated exposure (14 days) to the airborne polycyclic aromatic hydrocarbon fluorene in adult Wistar male rats. PLOS One **8**(8), 71413 (2013)

Peiffer, J., Grova, N., Hidalgo, S., Salquèbre, G., Rychen, G., Bisson, J.F., Appenzeller, B.M., Schroeder, H.: Behavioral toxicity and physiological changes from repeated exposure to fluorene administered orally or intraperitoneally to adult male Wistar rats: a dose–response study. Neurotoxicology **53**, 321–333 (2016)

Pham, C.T., Boongla, Y., Nghiem, T.D., Le, H.T., Tang, N., Toriba, A., Hayakawa, K.: Emission characteristic of polycyclic aromatic hydrocarbons and nitro-polycyclic aromatic hydrocarbons from open burning of rice straw in the North of Vietnam. Int. J. Environ. Res. Public Health **16**(13), 2343 (2019)

Qiao, M., Qi, W., Liu, H., Qu, J.: Oxygenated, nitrated, methyl and parent polycyclic aromatic hydrocarbons in rivers of Haihe River System, China: occurrence, possible formation, and source and fate in a water-shortage area. Sci. Total. Environ. **481**, 178–185 (2014)

Qiao, M., Bai, Y., Cao, W., Huo, Y., Zhao, X., Liu, D., Li, Z.: Impact of secondary effluent from wastewater treatment plants on urban Rivers: polycyclic aromatic hydrocarbons and derivatives. Chemosphere **211**, 185–191 (2018)

Qiao, M., Fu, L., Cao, W., Bai, Y., Huang, Q., Zhao, X.: Occurrence and removal of polycyclic aromatic hydrocarbons and their derivatives in an ecological wastewater treatment plant in South China and effluent impact to the receiving river. Environ. Sci. Pollut. Res. **26**(6), 5638–5644 (2019)

Qiao, M., Fu, L., Li, Z., Liu, D., Bai, Y., Zhao, X.: Distribution and ecological risk of substituted and parent polycyclic aromatic hydrocarbons in surface waters of the Bai, Chao, and Chaobai Rivers in Northern China. Environ. Pollut. **257**, 113600 (2020)

Reizer, M.: What is particulate matter? Fine dust in the atmosphere—a compendium of knowledge about air pollution with particulate matter in Poland. Environ. Monit. Libr. (in polish) (2016)

Ribière, C., Peyret, P., Parisot, N., Darcha, C., Déchelotte, P.J., Barnich, N., Peyretaillade, E., Boucher, D.: Oral exposure to environmental pollutant benzo [a] pyrene impacts the intestinal epithelium and induces gut microbial shifts in murine model. Sci. Rep. **6**(1), 31027 (2016)

Richter-Brockmann, S., Dettbarn, G., Jessel, S., John, A., Seidel, A., Achten, C.: Ultra-high sensitive analysis of 3-hydroxybenzo [a] pyrene in human urine using GC-APLI-MS. J. Chromatogr. B **1118**, 187–193 (2019)

Ringuet, J., Albinet, A., Leoz-Garziandia, E., Budzinski, H., Villenave, E.: Reactivity of polycyclic aromatic compounds (PAHs, NPAHs and OPAHs) adsorbed on natural aerosol particles exposed to atmospheric oxidants. Atmos. Environ. **61**, 15–22 (2012)

Rojas, G.A., Saavedra, N., Saavedra, K., Hevia, M., Morales, C., Lanas, F., Salazar, L.A.: Polycyclic aromatic hydrocarbons (PAHs) exposure triggers inflammation and endothelial dysfunction in BALB/c mice: a pilot study. Toxics **10**(9), 497 (2022)

Saldarriaga-Noreña, H., López-Márquez, R., Murillo-Tovar, M.A., Arias-Montoya, M.I., Guerrero-Álvarez, J.A., Vergara-Sánchez, J.: Recent advances for polycyclic aromatic analysis in airborne particulate matter. In: Hydrocarbon Pollution and Its Effect on the Environment (2018). https://doi.org/10.5772/intechopen.79714

Santos, L.O., dos Anjos, J.P., Ferreira, S.L., de Andrade, J.B.: Simultaneous determination of PAHS, nitro-PAHS, and quinones in surface and groundwater samples using SDME/GC-MS. Microchem. J. **133**, 431–440 (2017)

Schauer, C., Niessner, R., Pöschl, U.: Analysis of nitrated polycyclic aromatic hydrocarbons by liquid chromatography with fluorescence and mass spectrometry detection: air particulate matter, soot, and reaction product studies. Anal. Bioanal. Chem. **378**(3), 725–736 (2004)

Schwartz, C.R.: A Social History of American Technology. Oxford University Press (1997). ISBN: 9780195046052

Singh, A., Kamal, R., Ahamed, I., Wagh, M., Bihari, V., Sathian, B., Kesavachandran, C.N.: PAH exposure-associated lung cancer: an updated meta-analysis. Occup. Med. **68**(4), 255–261 (2018)

Song, Y., Zhang, Y., Li, R., Chen, W., Chung, C.K.A., Cai, Z.: The cellular effects of $PM_{2.5}$ collected in Chinese Taiyuan and Guangzhou and their associations with polycyclic aromatic hydrocarbons (PAHs), nitro-PAHs, and hydroxy-PAHs. Ecotoxicol. Environ. Saf. **191**, 110225 (2020)

Souza, K.F., Carvalho, L.R., Allen, A.G., Cardoso, A.A.: Diurnal and nocturnal measurements of PAH, nitro-PAH, and oxy-PAH compounds in atmospheric particulate matter of a sugar cane burning region. Atmos. Environ. **83**, 193–201 (2014)

Srogi, K.: Monitoring of environmental exposure to polycyclic aromatic hydrocarbons: a review. Environ. Chem. Lett. **5**(4), 169–195 (2007)

Sun, Z., Zhu, Y., Zhuo, S., Liu, W., Zeng, E.Y., Wang, X., Xing, B., Tao, S.: Occurrence of nitro-and oxy-PAHs in agricultural soils in Eastern China and excess lifetime cancer risks from human exposure through soil ingestion. Environ. Int. **108**, 261–270 (2017)

Sun, C., Qu, L., Wu, L., Wu, X., Sun, R., Li, Y.: Advances in analysis of nitrated polycyclic aromatic hydrocarbons in various matrices. TrAC Trends Anal. Chem. **127**, 115878 (2020)

Szulejko, J.E., Kim, K.H., Brown, R.J., Bae, M.S.: Review of progress in solvent-extraction techniques for the determination of polyaromatic hydrocarbons as airborne pollutants. TrAC Trends Anal. Chem. **61**, 40–48 (2014)

Tokiwa, H., Sera, N., Horikawa, K., Nakanishi, Y., Shigematu, N.: The presence of mutagens/carcinogens in the excised lung and analysis of lung cancer induction. Carcinogenesis **14**(9), 1933–1938 (1993)

Tsay, J.J., Tchou-Wong, K.M., Greenberg, A.K., Pass, H., Rom, W.N.: Aryl hydrocarbon receptor and lung cancer. Anticancer. Res. **33**(4), 1247–1256 (2013)

Umbuzeiro, G.A., Franco, A., Martins, M.H., Kummrow, F., Carvalho, L., Schmeiser, H.H., Leykauf, J., Stiborova, M., Claxton, L.D.: Mutagenicity and DNA adduct formation of PAH, nitro-PAH, and oxy-PAH fractions of atmospheric particulate matter from Sao Paulo, Brazil. Mutat. Res. Genet. Toxicol. Environ. Mutagen. **652**(1), 72–80 (2008)

Uno, S., Nebert, D.W., Makishima, M.: Cytochrome P450 1A1 (CYP1A1) protects against non-alcoholic fatty liver disease caused by Western diet containing benzo[a]pyrene in mice. Food Chem. Toxicol. **113**, 73–82 (2018)

Ütnü, Y.E., Okutan, H., Aydın, A.A.: Polycyclic aromatic hydrocarbon (PAH) content of Malkara lignite and its ex-situ underground coal gasification (UCG) char residues. Fuel **275**, 117949 (2020)

Vione, D., Barra, S., De Gennaro, G., De Rienzo, M., Gilardoni, S., Perrone, M.G., Pozzoli, L.: Polycyclic aromatic hydrocarbons in the atmosphere: monitoring, sources, sinks and fate. II: sinks and fate. Ann. Chim. J. Anal. Environ. Cult. Herit. Chem. **94**(4), 257–263 (2004)

Walgraeve, C., Demeestere, K., Dewulf, J., Zimmermann, R., Van Langenhove, H.: Oxygenated polycyclic aromatic hydrocarbons in atmospheric particulate matter: molecular characterization and occurrence. Atmos. Environ. **44**(15), 1831–1846 (2010)

Wang, C., Yang, J., Zhu, L., Yan, L., Lu, D., Zhang, Q., Zhao, M., Li, Z.: Never deem lightly the "less harmful" low-molecular-weight PAH, NPAH, and OPAH—disturbance of the immune response at real environmental levels. Chemosphere **168**, 568–577 (2017)

Wang, Y., Zhang, H., Zhang, X., Bai, P., Zhang, L., Huang, S.J., Pointing, S.B., Nagao, S., Chen, B., Toriba, A., Tang, N.: Abundance, source apportionment and health risk assessment of polycyclic aromatic hydrocarbons and nitro-polycyclic aromatic hydrocarbons in $PM_{2.5}$ in the urban atmosphere of Singapore. Atmosphere **13**(9), 1420 (2022)

Wei, C., Bandowe, B.A.M., Han, Y., Cao, J., Watson, J.G., Chow, J.C., Wilcke, W.: Polycyclic aromatic compounds (PAHs, oxygenated PAHs, nitrated PAHs, and azaarenes) in air from four climate zones of China: occurrence, gas/particle partitioning, and health risks. Sci. Total. Environ. **786**, 147234 (2021)

Wester, P.W., Muller, J.J.A., Slob, W., Mohn, G.R., Dortant, P.M., Kroese, E.D.: Carcinogenic activity of benzo [a] pyrene in a 2 year oral study in Wistar rats. Food Chem. Toxicol. **50**(3–4), 927–935 (2012)

Wu, H., Zhan, T., Cui, S., Chen, J., Jin, Q., Liu, W., Zhang, C., Zhuang, S.: Endothelial barrier dysfunction induced by anthracene and its nitrated or oxygenated derivatives at environmentally relevant levels. Sci. Total. Environ. **802**, 149793 (2022)

Xu, Q., Hu, L., Chen, S., Fu, X., Gong, P., Huang, Z., Miao, W., Jin, C., Jin, Y.: Parental exposure 3-methylcholanthrene disturbed the enterohepatic circulation in F1 generation of mice. Chemosphere **286**(1), 131681 (2022)

Yao, Y., Wang, D., Ma, H., Li, C., Chang, X., Low, P., Hammond, S.K., Turyk, M.E., Wang, J., Liu, S.: The impact on T-regulatory cell related immune responses in rural women exposed to polycyclic aromatic hydrocarbons (PAHs) in household air pollution in Gansu, China: a pilot investigation. Environ. Res. **173**, 306–317 (2019)

Zaciera, M.: Research on the Determination of Nitro Derivatives of Polycyclic Aromatic Hydrocarbons in the Air. PhD thesis. University of Silesia, Katowice (2007)

Zhang, Y., Yang, B., Gan, J., Liu, C., Shu, X., Shu, J.: Nitration of particle-associated PAHs and their derivatives (nitro-, oxy-, and hydroxy-PAHs) with NO_3 radicals. Atmos. Environ. **45**(15), 2515–2521 (2011)

Zhang, B., Sun, J., Jiang, N., Zeng, Y., Zhang, Y., He, K., Xu, H., Liu, S., Ho, S.S.H., Qu, L., Cao, J.: Emission factors, characteristics, and gas-particle partitioning of polycyclic aromatic hydrocarbons in $PM_{2.5}$ emitted for the typical solid fuel combustions in rural Guanzhong Plain, China. Environ. Pollut. **286**, 117573 (2021)

Zhang, L., Ji, X., Ding, F., Wu, X., Tang, N., Wu, Q.: Apoptosis and blood-testis barrier disruption during male reproductive dysfunction induced by PAHs of different molecular weights. Environ. Pollut. **300**, 118959 (2022)

Zhao, T., Yang, L., Huang, Q., Zhang, W., Duan, S., Gao, H., Wang, W.: Polycyclic aromatic hydrocarbons (PAHs) and nitrated-PAHs (NPAHs) emitted by gasoline vehicles: characterization and health risk assessment. Sci. Total. Environ. **727**, 138631 (2020)

Zhu, Y., Liang, B., Xia, W., Gao, M., Zheng, H., Chen, J., Chen, Y., Tian, M.: Assessing potential risks of aquatic polycyclic aromatic compounds via multiple approaches: a case study in Jialing and Yangtze Rivers in downtown Chongqing, China. Environ. Pollut. **294**, 118620 (2022)

Zielinska, B., Samy, S.: Analysis of nitrated polycyclic aromatic hydrocarbons. Anal. Bioanal. Chem. **386**(4), 883–890 (2006)

Polycyclic Aromatic Hydrocarbons in Industrial Wastewater

6

Amanpreet Kaur, Aayasha Negi, Neha Bhatt, Man Vir Singh, Jitendra Pal Singh, and Santosh Kumar Verma

Abstract

Polycyclic aromatic hydrocarbons (PAHs) belong to a group of organic pollutants known for their persistence in the environment and frequently detected in industrial wastewater, attributed to their extensive application in the petrochemical, pharmaceutical and manufacturing sectors. These compounds pose significant threats to environmental and human health due to their carcinogenic, mutagenic, and toxic properties. Industrial activities like fossil fuel combustion, coal processing and petroleum refining result in elevated PAH levels in wastewater, potentially causing aquatic pollution and bioaccumulation in living organisms. Traditional approaches to treating wastewater, including coagulation, sedimentation, and biological methods, are frequently ineffective in fully removing PAHs. Considering the negative impacts of PAHs, it is crucial to identify and remove these compounds from water through dependable methods. A range of physical, chemical and biological approaches for the treatment of PAH-contaminated water has been explored. This chapter examined the occurrence of

A. Kaur (✉) · A. Negi
Department of Chemistry, School of Sciences, IFTM University, Moradabad, Uttar Pradesh, India
e-mail: Amanpreet2225@gmail.com

N. Bhatt
Department of Chemistry, SDS University, Pt. L.M.S., Rishikesh Campus, Tehri Garhwal, Uttarakhand, India

M. V. Singh
Division of Research and Innovation, School of Applied and Life Sciences, Uttaranchal University, Dehradun, Uttarakhand, India

J. P. Singh
Department of Physics, School of Sciences, IFTM University, Moradabad, Uttar Pradesh, India

S. K. Verma
Department of Chemistry and Chemical Engineering, Yulin University, Yulin, Shaanxi, China

H. Gupta et al. (eds.), *Ubiquitous Polycyclic Aromatic Hydrocarbon Contamination*, Emerging Contaminants and Associated Treatment Technologies,
https://doi.org/10.1007/978-3-032-11764-9_6

PAHs in industrial wastewater, their detrimental impacts on organisms, and the treatment approaches employed to address them. Future investigations ought to concentrate on creating economical and environmentally sustainable methods to address PAH contamination, thereby safeguarding both environmental integrity and public health.

Keywords

Polycyclic aromatic hydrocarbons • PAH sources in industry • PAH removal techniques • Advanced oxidation processes • Bioremediation

Introduction

Global industries are increasingly recognizing the critical significance of water in productivity and profitability. A multitude of nations have begun to experience water scarcity. Manufacturing, energy generation, and global material transportation are significantly dependent on water. Due to escalating economic pressures, industrial water usage has surged significantly (Abdel-Shafy and Mansour 2016). The global population has doubled in recent decades, leading to a sixfold rise in water usage. By the year 2025, it is anticipated that Asian and African countries would experience water shortages. Consequently, if the industrial water allocation increases from the current 5% to 17% by 2025, it would significantly jeopardize the water allocated for agricultural use. Consequently, several countries have enacted measures to control PAH pollution (Adeniji et al. 2023). The U.S. EPA has classified 16 PAHs as major pollutants owing to their significant health and environmental hazards. The carcinogenic properties of PAHs, have generated significant concern. Benzo[a]pyrene is a carcinogen, demonstrating significant evidence of its carcinogenicity in humans and animal models (Gupta and Gupta 2015). Methylated PAHs (Me-PAHs), a subset of PAH derivatives, exist in the environment analogous to their unsubstituted equivalents. Although research has examined the occurrence of PAH in industrial effluent sources, such as coking and landfill effluents (Krishnan et al. 2017), the behavior and transformation processes of PAHs in paper-making wastewater (PMWW) remain inadequately known. PMWW is characterized by complex complexes of inorganic and organic materials, including high levels of organic carbon and chemical oxygen demand. Each paper mill consumes significant volumes of water, ranging from 5 to 300 m^3 every ton of pulp produced. Medium-sized mills produce roughly 2000 m^3 of effluent per day (Ren et al. 2019). PAHs were reported to be the part of effluent and sludge samples of a paper industry in 2018 (Gupta 2018). Water supplies are strained by industrial activity, climate change, pollution, and increasing energy demands. A substantial quantity of water used in companies culminates in industrial effluent, then discharged into the ecosystem, elevating considerable problems and generating numerous threats (Berardi et al. 2019). This is particularly applicable to the chemical and related industrial industries. Consequently, it is essential to endeavor to minimize water use and to treat wastewater to render it reusable or, at the very least, less detrimental to the environment. Water is extensively used in chemical and related process industries, rendering them water-intensive. Its usual uses include serving as a solvent and cleaning agent, coolant, and water from boiler. These requirements are fulfilled using the finite sources of freshwater. Industrial effluent is a significant contaminant of aquatic ecosystems. Industrial activities generate significant quantities of pollutants, so damaging air and water (Bokade and Bajaj 2023). The discharge of substantial quantities of industrial wastewater into aquatic environments necessitates the implementation and design of tailored treatment processes for each effluent procedure. The volume of wastewater generated varies by the technological sophistication of processes across industries, although it may be mitigated by advancements in industrial technology (Chen et al. 2019). Industries are swiftly advancing, using substantial quantities of freshwater as both a raw resource and cooling agent. During industrial processes, various raw and intermediate products, and waste are introduced into the water. Despite being an inevitable result, wastewater significantly contributes to the contamination of aquatic habitats (García et al. 2020). From all waste-producing businesses, pharmaceutical sector, wood treatment and preservation industries, Dye, paint and plastic manufacturing plant generate the most dangerous wastewater (Singh et al. 2018, 2020). These industries are where wastewater treatment is crucial. This chapter offers an extensive analysis of the presence and dynamics of PAHs in industrial effluent. It examines the origins and properties of PAHs, delineates their toxicological importance, and emphasizes contemporary analytical and remediation strategies (Gbeddy et al. 2020). Table 6.1 summarize different PAH generated from industrial waste water. This chapter seeks to synthesize existing information to provide effective monitoring and management techniques aimed at mitigating PAH pollution and safeguarding environmental and public health.

Different Kinds of Wastewaters Containing PAHs

PAHs are prevalent in many industrial and municipal effluent streams as a result of their extensive production during the combustion and chemical processing of organic substances. The kinds and amounts of PAHs fluctuate markedly based on industrial activities, raw materials used, and treatment techniques implemented (Al Farraj et al. 2019).

Petrochemical and Oil Refinery Wastewater

The water generated from operations like water injection and drilling while petroleum extraction is greasy and often includes both organic and inorganic substances. Produced water comprises pollutants such as dissolved and distributed petroleum compounds, dissolved sedimentary minerals, production chemical substances, industrial solids, and gaseous components (Ghosal et al. 2016). Oil

Table 6.1 Different PAH generated from industrial waste water

PAHs	Formula	Molecular weight	Structure	References
Anthracene	$C_{14}H_{10}$	178.2	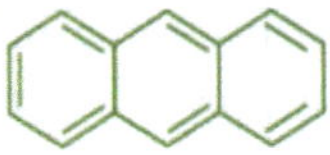	Ansari et al. (2023)
Acenaphthene	$C_{12}H_{10}$	154.2	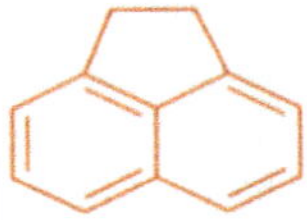	Ghasemi et al. (2017)
Acenaphthlene	$C_{12}H_{8}$	152.18		He et al. (2020)
Benzo (a) anthracene	$C_{18}H_{12}$	228.3	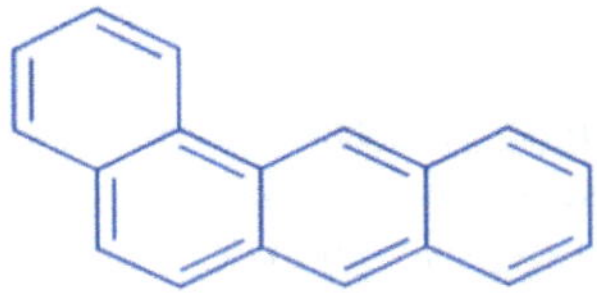	Kuppusamy et al. (2017)
Benzo (a) pyrene	$C_{20}H_{12}$	252.3	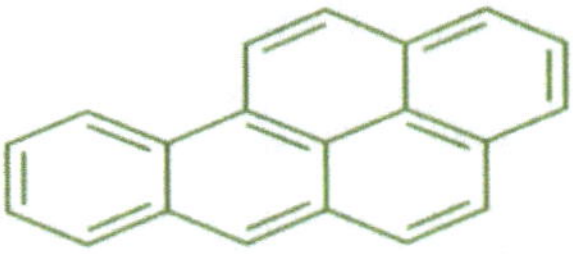	Ansari et al. (2023)
Benzo (b) fluoranthene	$C_{20}H_{12}$	252.3	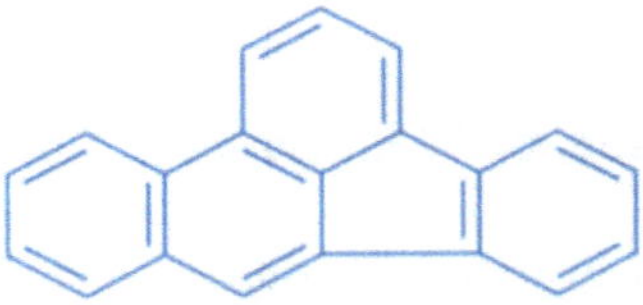	Dai et al. (2022)
Benzo (k) fluoranthene	$C_{20}H_{12}$	252.3	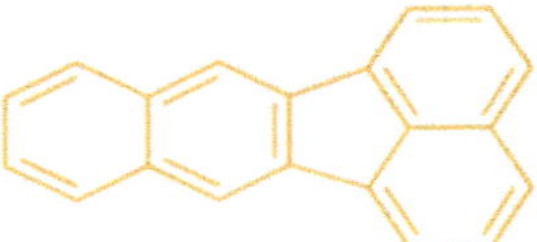	Ghasemi et al. (2017)

(continued)

Table 6.1 (continued)

PAHs	Formula	Molecular weight	Structure	References
Benzo (ghy) perylene	$C_{22}H_{12}$	276.3	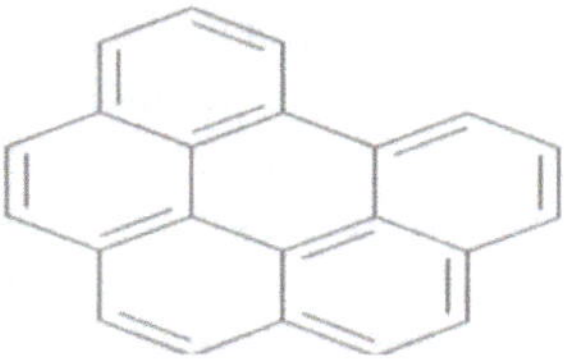	Ansari et al. (2023)
Chrysene	$C_{18}H_{12}$	228.2	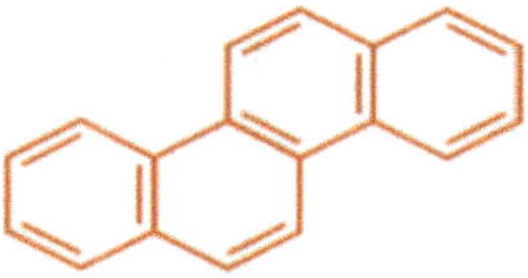	Gupta and Gupta (2016)
Dibenz (a,h) anthracene	$C_{22}H_{14}$	278.3	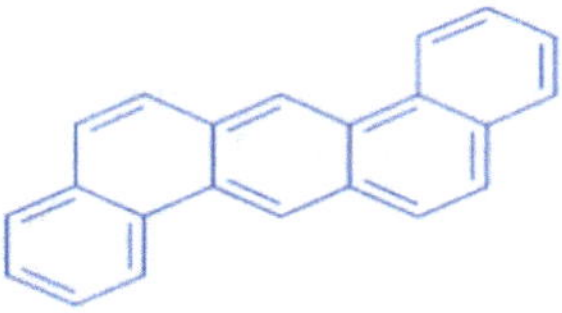	He et al. (2020)
Fluoranthene	$C_{16}H_{10}$	202.2	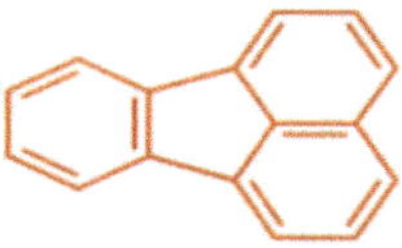	Abdullah et al. (2020)
Naphthalene	$C_{10}H_8$	128.91	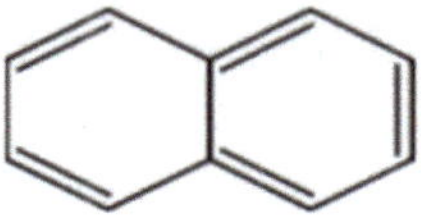	Sun et al. (2019)
Phenanthrene	$C_{14}H_{10}$	174.2	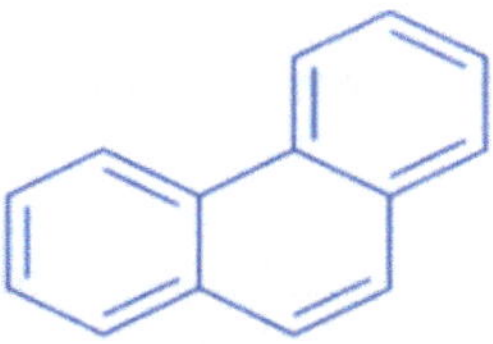	Kuppusamy et al. (2017)

components in solution are organic substances that are soluble in generated water, including benzenol, carboxylic acids, and low molecular weight aromatic compounds. PAHs and heavy alkyl phenols constitute organic components present in generated water and contribute to the formation of dispersed oil (Kong et al. 2023).

Coking and Steel Industry Wastewater

Worldwide, the industrial expansion is significantly rising, particularly in the critical industries of steel, coal gasification, textiles, and coke, which together generate millions of tons of industrial effluent. Industrial effluent is very hazardous and can cause cancer and genetic disorder, including PAHs like C_5H_5N and phenolic compounds. Over the previous ten years, considerable efforts have been devoted to research for the coking waste. The use of membrane bioreactors (MBR), moving bed biofilm reactors (MBBR) has been optimized to improve the extraction efficiency of conventional activated sludge methodology (Lamichhane et al. 2016). Despite an improved eradicate around 25%, significant costs and vary biofouling remain critical problems that must be addressed for the commercialization of these methods (Man et al. 2017).

Aluminum Smelting Wastewater

Numerous sectors release waste containing toxic metals into the atmosphere, comprising aquatic ecosystems. These substances may include acids and very poisonous Metallic compounds and mineral substances, such as aluminum. This contamination may render water inhospitable for aquatic creatures, unfit for residential use or irrigation, and may allow wastes to re-enter the food chain, adversely affecting people (Mortazavi et al. 2019). This pollution mostly results from atmospheric deposition from several sources, with industrial and vehicular emissions being the most significant contributors. Urbanized landscapes and industrial locations exhibited much more metal pollution than rural regions (Omolola et al. 2025). Although aluminum is one of the most prevalent metals, it is neither required nor biologically significant in living beings. Currently, there is no evidence demonstrating its biological significance in living organisms (Nasrollahzadeh et al. 2021).

Wood Treatment and Preservation Effluents

Solid wood is a naturally occurring polymer composite formed by the combination of cellulose, hemicelluloses, and lignin, which are structured into tubular structures that create a cylindrically layered composite (Mannina et al. 2020). Wood is regarded as a renewable resource since it may be organically replaced. Global wood preservative producers, such as KMG compounds, Koppers, Borax, Kop-Coat, BASF Wolman GmbH, and Arxada, are formulating new compounds in compliance with environmental agency standards (Pathak et al. 2022).

Pulp and Paper Industry Wastewater

The pulp and paper sector produces diverse waste during the paper manufacturing process at various operating phases. A significant amount of wastewater is produced during preparation of wood chips, pulping, bleaching, and paper manufacturing operations (Mahmoud 2020). Lignins were extracted from wood chips using Na_2S alkaline therapy circumstances throughout the wood chip preparation, washing, and pulping procedures. At this step, the produced lignin-rich effluent is referred to in the form of black liquor. The bleaching method involves the treatment of pulp with toxic chemicals, including H_2O_2 and CaO hence increasing the toxicity of the resultant effluents (Pulleyblank et al. 2019).

Dye, Paint, and Plastic Manufacturing Effluents

The fabric, printing, paper, processing of food, and leather industries mostly produce dye wastewater, with worldwide annual production of 800,000 tons of dyes, of which about 200,000 tons are textile dyes. Dye effluent contains many pollutants, including adhesives, salts, acids, toxic, and cancerous. These attributes contribute to ocular burns, dermal irritations, allergy, and asthma in the human body (Safo-Adu et al. 2023). The dye business produces many contaminants at different concentrations. Consequently, dye wastewater constitutes a threat to both human health and the ecosystem. Furthermore, it is exceedingly detrimental to use dye effluent in routine tasks such as showering, laundering and cooking. Despite being aware of the detrimental effects and existing national restrictions, after dyes have served their usefulness, they are discarded into the environment without treatment (Rayaroth et al. 2023).

Sources of PAHs in Industrial Wastewaters

Polycyclic aromatic hydrocarbons (PAHs) enter industrial wastewaters via many processes related to raw material use, combustion, high-temperature activities, and waste disposal. The composition and concentration of PAHs in industrial effluents are contingent upon the particular industrial activity and its related chemical or thermal processes (Sayara and Sánchez 2020). Comprehending these sources is essential for formulating targeted treatment and pollution mitigation methods. Table 6.2 represents different sources of PAHs in Industrial Wastewater.

The origins of PAHs in industrial wastewater are varied and mostly associated with combustion, chemical processing, and inadequate waste management. Comprehending these sources is essential for recognizing high-risk sectors, informing environmental policies, and developing efficient wastewater treatment systems. Figure 6.1 represents the different sources of PAHs in wastewater (Akinpelumi et al. 2023). Mitigating PAH emissions at their origin is a fundamental approach to

Table 6.2 Sources of PAHs in industrial wastewater

Source	Industries	Common PAH
Incomplete combustion	Steel, power plants, cement, incineration	Benzo[a]pyrene, Fluoranthene, Pyrene
PAH-containing chemicals	Wood treatment, asphalt, dyes, plastic manufacturing	Phenanthrene, Chrysene, Naphthalene
Waste and by-product disposal	Coke plants, smelters, incinerators, landfill sites	High-MW PAHs
Atmospheric deposition	Any high-temperature industry	Depends on local activity
Spills and leakages	Petroleum refining, chemical plants	Anthracene, Benzo[b]fluoranthene
Fuel/lubricant contamination	Heavy industries using diesel or furnace oil	Fluorene, Acenaphthene
Thermal decomposition and pyrolysis	Carbon black, aluminum, coke production	Pyrene, Benzo[a]anthracene
Runoff from contaminated surfaces	Mixed industrial zones	Depends on source

limiting environmental mitigating pollution and protecting environment and human well-being (Satouh et al. 2021).

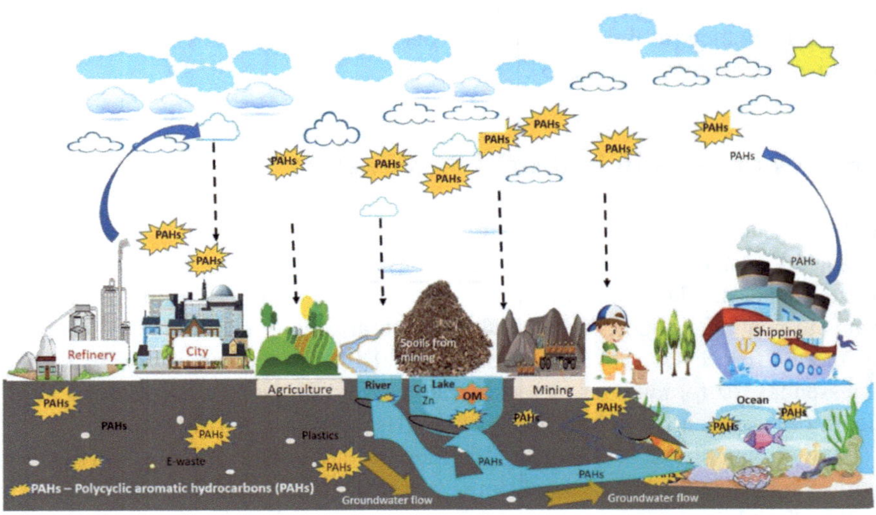

Fig. 6.1 Sources of PAHs in wastewater. Reproduced with permission from Akinpelumi et al., Journal of Hazardous Materials Advances, ©2023, Elsevier

Impact of Industrial Wastewater Containing PAH on Human Health and Environment

The general community is primarily exposed by inhalation of PAHs of indoor and outdoor air, consumption of polluted food, cigarette smoking, and contact with smoke from open fires. Common roots of PAHs encompass fossil fuels utilized in transportation, cooking, residential heating, and diverse industrial processes (Borji et al. 2020). Numerous studies on wild fish have established a connection between PAH exposure and the development of hepatic neoplasms or toxicopathic liver lesions linked to neoplasia. Such exposure is commonly identified through the presence of SPAHs in sediments, PAH metabolites or fluorescent aromatic compounds (FACs) in fish bile, PAH-DNA adducts in liver tissue, or through contaminated dietary intake (Zhao et al. 2021). Similar to mammals, fish possess immune defense systems that include both cell-mediated and humoral responses (Mukwevho et al. 2020). Research on innate immune responses indicates that fish macrophage activities are significantly influenced by PAH exposure, particularly with regard to the sensitivity of macrophage respiratory bursts. Lymphocyte proliferation is significantly affected by exposure to PAH in the context of acquired immune responses. Humans may be exposed to PAHs and their derivatives via many routes: inhalation of airborne PAHs, ingestion of PAHs in water, and dermal contact with PAHs in diverse surroundings. Moreover, eating represents another potential exposure pathway, since PAHs may readily infiltrate the food chain and bioaccumulate; moreover, food may get contaminated with PAHs during high-temperature cooking. The severity of the impacts of PAH exposure on human life is significantly influenced by the duration and method of exposure, the concentration and type of PAH, and the health status of the affected persons (Ma et al. 2017). However, outliers exist, such as naphthalene, which exhibits more carcinogenicity than other high molecular weight polycyclic aromatic hydrocarbons, such as benzo(ghi)perylene. Due to their varying carcinogenic potential, several global organizations have categorized PAHs into groups based on this criterion, including the International Agency for Research on Cancer (IARC), the US Environmental Protection Agency (EPA), and the European Union (EU) (Montano et al. 2025). Among the prioritized PAHs, benzo(a)pyrene (BaP) has attracted significant interest from the scientific community because to its elevated carcinogenic potential. Consequently, BaP is often used as an indicator of PAH carcinogenicity and is regarded as representative of PAHs. Chronic exposure to PAHs may result in long-term consequences such as cataracts, renal and hepatic damage, respiratory issues, and diminished immune system function. Reports indicate that PAHs may induce short-term effects after acute exposure to high concentrations of PAHs. The adverse effects including irritation and inflammation. Moreover, several PAHs, including anthracene, benzo(a)pyrene, and naphthalene, are recognized as skin irritants, with anthracene and benzo(a)pyrene also acting as skin sensitizers (Kouras et al. 2025). Figure 6.2 represents the effect of industrial waste water containing PAHs on human health.

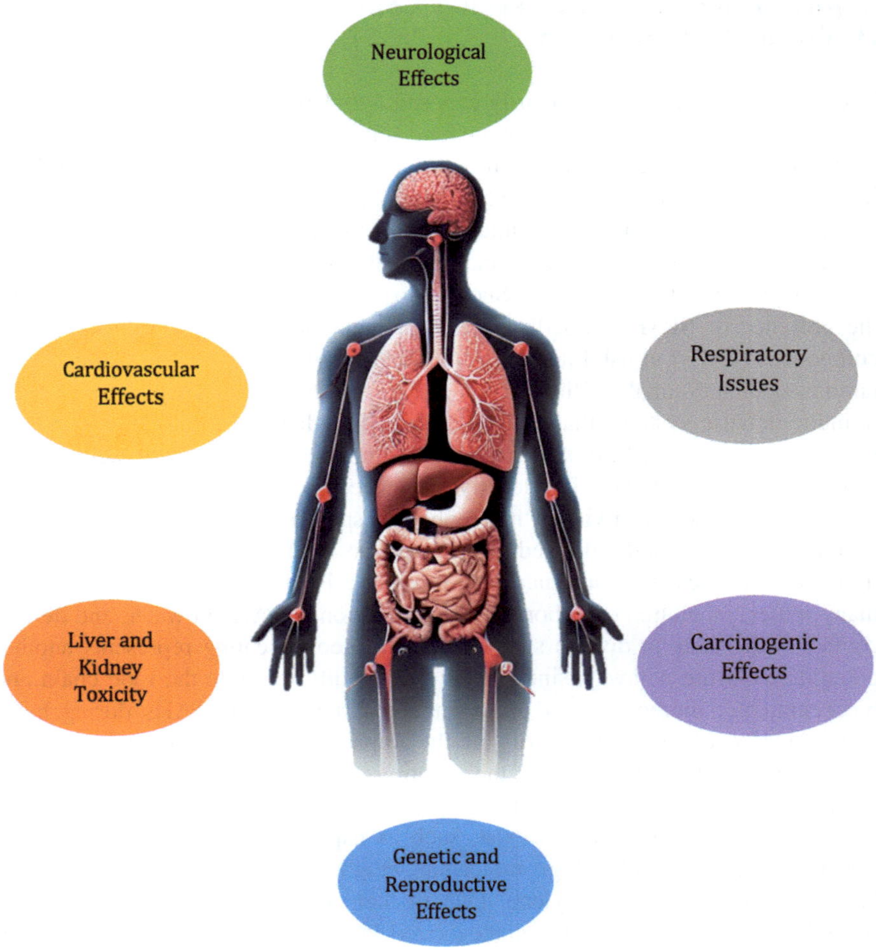

Fig. 6.2 Effect of industrial waste water containing PAHs on human health

Remediation of PAHs in Industrial Wastewaters

The isolation and accumulation of PAHs from industrial wastewater test material are critical for their precise determination. Remediation strategies seek to lower environmental pollutants to acceptable levels via Environmental degradation and transformation in atmospheric and aquatic media (Shen et al. 2020). Pollution from PAHs is a major worldwide issue, given their harmful impacts on human well-being and ecosystems. Multiple mitigation measures have been utilized to address PAH pollution involving physical and chemical components, and biological methods. PAHs, owing to their hydrophobic characteristics, demonstrate

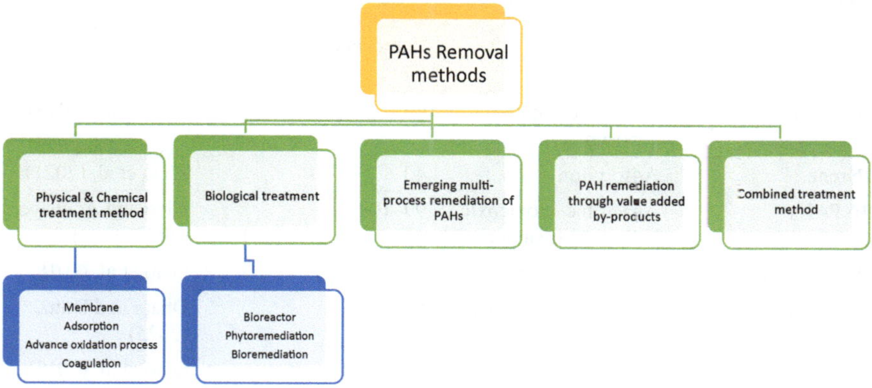

Fig. 6.3 Different methods of removal of PAHs from industrial waste water

significant solubility in organic solvents (Pandey et al. 2020). Figure 6.3 presents the different techniques employed for the elimination of PAHs.

Physical and Chemical Methods

Physical and chemical treatment methods are utilized to eliminate large particles along with pollutants from wastewater. Physical therapy is often utilized as a first phase prior to the implementation of advanced technologies (Fouda et al. 2021). Physical treatment methods include membrane filtration and adsorption. Most physical methods primarily transfer PAHs from water to another medium without modifying their chemical structure. Consequently, physical methods alone are insufficient for the complete removal of PAHs. Chemical procedures have garnered increased attention due to the inefficiency and time-consuming nature of physical methods for PAH removal. Chemical treatments encompass processes that can decompose or convert PAHs into less harmful or more manageable forms (Muthukumar et al. 2020). Table 6.3 provides an outline of the diverse methods employed for the recovery of PAHs.

Membrane

Membrane filtration treats water and wastewater by holding pollutants in a porous filter material. Sand is a filtered material and popular because to its accessibility, efficacy, and affordability (Bis et al. 2019). Advanced membrane-based pretreatment technologies remove PAHs from generated water. Ultrafiltration, microfiltration, nanofiltration, and reverse osmosis are popular. Integrating membrane-based techniques has also improved PAH removal efficiency. Nanofiltration technology has acquired interest due of its capability to remove organic micropollutants from water and wastewater (Han et al. 2022a). This method rejects monovalent salts little while maximizing organic micropollutant rejection. Ultrafiltration membrane

Table 6.3 Different method used in removal of PAHs

PAHs	Method	Removal effectiveness (%)	References
Naphthalene	Ultraviolet light exposure	62	Qiao et al. (2018)
Pyrene	Adsorption	40	Zhao et al. (2021)
16 PAHs	Anaerobic anoxicoxic biological treatment	99–100	Alao and Adebayo (2022)
Benzo (a) pyrene	Adsorption	48	Qiao et al. (2018)
16 PAHs	Photocatalyst ozonation	57	Dhara and Dutta (2025)
Naphthalene	Bioremediation	100	Qiao et al. (2018)
Anthracene	Bioremediation	73	Qiao et al. (2018)
Naphthalene	Green-derived sorbent material	76.20–105.60	Zhao et al. (2021)
16 PAHs	Bioremediation	42–77	Qiao et al. (2019)
Phenanthrene	Green-derived sorbent material	76.20–105.60	Zhao et al. (2021)
Acenaphthene	Green-derived sorbent material	76.20–105.60	Zhao et al. (2021)
Fluoranthene	Green-derived sorbent material	76.20–105.60	Zhao et al. (2021)
Pyrene	Green-derived sorbent material	76.20–105.60	Zhao et al. (2021)
Phenanthrene	Bioremediation	95	Alao and Adebayo (2022)
Pyrene	Bioremediation	54	Alao and Adebayo (2022)
Fluoranthene	Fenton method	62.95	Haneef et al. (2020)
Phenanthrene	Fenton method	63.16	Haneef et al. (2020)
Anthracene	Fenton method	85.47	Haneef et al. (2020)
Phenanthrene	Absorption and degradation	80	Gupta and Gupta (2016)
Pyrene	Absorption and degradation	65	Gupta and Gupta (2016)
Benzo (a) pyrene	Absorption and degradation	65	Gupta and Gupta (2016)
16 PAHs	Phytoremediation	89	Bhatti et al. (2024)
16 PAHs	Biodegradation	67.27	Ismail et al. (2022)
Pyrene,	Magnetic floatation	89.9	Mirzaee and Sartaj (2022)

(continued)

Table 6.3 (continued)

PAHs	Method	Removal effectiveness (%)	References
Benzo (a) pyrene	Magnetic floatation	66.9	Mirzaee and Sartaj (2022)
Indenopyrene	Magnetic floatation	78.2	Mirzaee and Sartaj (2022)
16 PAHs	Biodegradation	77.38	Ismail et al. (2022)
Phenanthrene	Oxidation	90.1	Wei et al. (2015)
Naphthalene	Oxidation	97.5	Wei et al. (2015)
Anthracene	Oxidation	55.4	Wei et al. (2015)
Benzo (a) pyrene	Oxidation	26.7	Wei et al. (2015)
Naphthalene	Air-assisted microextraction	82	Patel et al. (2020)
Phenanthrene	Adsorption	90	Gupta and Gupta (2016)
Naphthalene	Oxidation adsorption	92	Wei et al. (2015)
Fluorene	Oxidation adsorption	100	Wei et al. (2015)
Pyrene	Precipitation method	99	Sakshi et al. (2019)
Fluoranthene	Precipitation method	98	Sakshi et al. (2019)
Chrysene	Precipitation method	87	Sakshi et al. (2019)
Phenanthrene	Precipitation method	97	Sakshi et al. (2019)
Naphthalene	Oxidations	97	Wei et al. (2015)
Anthracene	Oxidations	95	Cao et al. (2018)
Fluorene	Oxidations	87	Jalili et al. (2020)
Benzo (b) fluoranthene	Adsorption	59–91	Ren et al. (2019)
15 PAHs	Dispersive liquid–liquid microextraction	90	Krishnan et al. (2017)
PAHs	Electrochemical advanced oxidation	99.9	He et al. (2020)
PAHs with low and high molecular weights	Biodegradation	86	Ghasemi et al. (2017)

bioreactors were used to test the efficacy of removal for three low-concentration polycyclic aromatic hydrocarbons phenanthrene, fluoranthene, and pyrene from wastewater. The removal efficiencies were 91%, 82%, and 92% (Yemele et al. 2024).

Adsorption

Adsorption processes are classified as physical (physisorption) or chemical (chemisorption) based on their interactions (Gupta and Gupta 2016). Adsorption is a proven method for removing PAHs from generated water. Petroleum, wastewater, and agricultural leftovers are used to make carbon-rich biochar (Mukwevho et al. 2020). Figure 6.4 represents different types of adsorbents used in removal of PAHs from Industrial waste water. Pyrolysis biochar absorbs PAHs and other organic contaminants due to its aromatic carbon structure, similar to graphene. The porous carbon substance activated carbon, made from agricultural waste is an excellent adsorbent (de Andrade et al. 2020). Graphene, a hexagonal honeycomb of carbon atoms, removes pollutants well. Adsorption systems operate in batch and continuous modes. Small-scale applications like lab pilot studies use batch adsorption. Combine a specific amount of adsorbent and adsorbate in a reactor until equilibrium is reached. To improve adsorption effectiveness, temperature, adsorbate concentration, adsorbent amount, agitation speed, and particle size are considered (Yang et al. 2019). Continuous flow adsorption systems suit large-scale industrial applications.

Fig. 6.4 Systematic presentation of different types of adsorbents used in removal of PAHs from Industrial waste water

Advance Oxidation Process (AOPs)

AOPs were created in the Twentieth century methods for the treatment of oil and gas wastewater, eliminating contaminants and reducing color and odor. Advanced oxidation processes (AOPs) often include O_3, Cl_2, H_2O_2, and Fenton reagents, such as UV/H2O/Fe^{3+} systems. The efficacy of AOPs ranges from 32 to 99%. This diversity underpins its dependability, especially with catalysts. PAH degradation uses molecular ozone, hydroxyl radicals, and reactive species (Jackulin et al. 2024). Ozone interacts with PAH aromatic rings due to its electrophilicity, whereas hydroxyl radicals enable non-selective oxidation. Hydrogen peroxide and ferrous salt decompose organic wastewater pollutants efficiently in the Fenton reaction. This method is commonly used to decompose complex organic contaminants (Gaurav et al. 2021). Table 6.4 summarizes various oxidation methods utilized for the elimination of PAHs from industrial wastewater.

Coagulation

Coagulation is a typical way to reduce colloidal suspensions and organic contaminants like PAHs in aqueous solutions. Coagulation and chemical precipitation are efficient methods for water and wastewater treatment, often used in advanced purification technologies (Nowacka and Włodarczyk-Makuła 2015). Alum, aluminum chloride, ferric chloride, and ferric sulfate have been studied as coagulants. Various polymeric coagulants, such as polyaluminium chloride, polyferric chloride, and polyferric sulfate, as well as organic polyelectrolytes such polydiallyldimethyl ammonium chloride and anionic polyacrylamides, have shown efficient in treatment operations (Rosińska and Dąbrowska 2021). Composite inorganic-organic coagulants have been tested. Coagulation is effective yet has several downsides. The procedure produces a lot of putrescible sludge, making handling difficult. Chemical dosing operations are expensive and constraining. Nowacka and Włodarczyk-Makuła (2015) found that coagulation-precipitation procedures had varying PAH removal efficiency. Pyrene, fluoranthene, anthracene, and phenanthrene had elimination efficiencies of 75%, 57%, 40%, and 30%.

Biological Methods

Biological methods have emerged as environmentally friendly alternatives for the remediation of polycyclic aromatic hydrocarbons (PAHs), garnering significant interest due to the limitations of physical and chemical approaches. Traditional approaches frequently encounter issues associated with elevated costs, intricate procedures, strict regulatory standards, and insufficient degradation of pollutants. Biological techniques have been recognized for their potential to achieve efficient and sustainable remediation of PAHs (Sakshi et al. 2019). Table 6.5 presents a comparative overview of the benefits and drawbacks linked to different methods for the removal of PAH from industrial wastewater.

Table 6.4 Different methods of removal of PAHs from industrial waste water

PAHs	Water sample	Method of oxidation	Removal efficacy (%)	Industry	Source
Point source PAHs	Coking wastewater treatment plant	Ozone and ultraviolent	75	Coking industry wastewater	Sun et al. (2019)
8PAHs	Coagulant water	Hydrogen peroxide and ultraviolent	76	Steel industry wastewater	Martínez-Álvarez et al. (2021)
8PAHs	Electro coagulated water	Hydrogen peroxide and ultraviolent	70	Petrochemical and oil refining industries	Ramesh et al. (2022)
8PAHs	Groundwater	Hydrogen peroxide and ultraviolent	76	Textile and dyeing industries	Ijah et al. (2022)
Fluorene	Treated water produced water synthetic wastewater	Hydrogen peroxide and ultraviolent	98	Phosphate fertilizer industry	Ramezanzadeh et al. (2020)
Dibenzofuran	Treated water produced water synthetic wastewater	Hydrogen peroxide and ultraviolent	99	Pulp and paper industry	Han et al. (2022b)
Dibenzothiophene	Treated water produced water synthetic wastewater	Hydrogen peroxide and ultraviolent	99	Coal processing/ combustion	Shen et al. (2020)
15 PAHs hydrocarbons	Water treatment landfill leachate	Fenton reaction	100	Wood preservation plants	Jinadasa et al. (2020)
6 PAHs	Cooking wastewater treatment plant	Hydrogen peroxide and ultraviolent	70	Aluminum smelting and metal foundries	Ansari et al. (2023)

Table 6.5 Summary of advantages and disadvantages all methods of PAHs removal from industrial waste water

Treatment method	Advantages	Disadvantages	Source
Adsorption	Minimal initial expenditure, compact modules, environmentally sustainable, adaptable method, and recyclable and recoverable adsorbent	Regular regeneration is required, influenced by salinity, elevated temperature, maximum retention duration, costly adsorbent restoration, and detrimental excess adsorbent	Sher et al. (2023)
Advanced oxidation process	Effortless operation, significant degradation and solubilize oil minerals	Proficient labour necessary, optimisation, oversight, and preparation processes needed	Wang et al. (2018)
Bioremediation	Access to inexpensive microorganisms, a straightforward method, and complete mineralisation result in the generation of CO_2, H_2O, and biomass	Prolonged deterioration Time and task optimisation is an excessive undertaking	Sher et al. (2023)
Chemical oxidation	Brief treatment duration and environmentally sustainable	Maximum operational and maintenance costs	Wang et al. (2018)
Chemical precipitation	Energy-efficient procedure, user-friendly, economical, and optimal recovery	Demand for chemicals, production of sludge, and ancillary waste	Sher et al. (2023)
Electrochemical technologies	Advantageous byproduct, environmentally sustainable, and devoid of chemical requirements	Essential skilled labour and challenges in scaling	Riley et al. (2018)
Flotation	The operation is straightforward, the integration enhances process efficiency, it is sturdy and durable, and has no moving components	Retention period of 4–5 min, maximum air production, and skim volume	Wang et al. (2018)
Membrane filtration	Cost-effective, reduced likelihood of membrane fouling, compact modules, appropriate for saline water	Mineral scaling, membrane pore wetting, and membrane fouling	Sher et al. (2023)

PAHs undergo degradation via aerobic and anaerobic biological processes. Research indicates that mixed microbial cultures typically exhibit superior performance compared to pure cultures in the biological degradation of less water-soluble PAHs. Comprehensive assessments of wastewater treatment plants utilizing biological treatment methods have demonstrated their effectiveness in eliminating organic compounds. PAH removal in these systems occurs through several mechanisms, including volatilization during aeration, microbial degradation, and adsorption onto sludge. In conventional wastewater treatment plants, volatilization generally accounts for less than 2% of the overall contribution. Effective biodegradation typically necessitates the introduction of specialized microorganisms to improve PAH removal, whereas adsorption is the primary removal mechanism owing to the strong affinity of PAHs for particulate matter (Dai et al. 2022).

Bioreactor

Municipal wastewater treatment plants (MWTPs) treat industrial sewage using biological, physical, and chemical methods to effectively reduce contaminants such as solids, nutrients, and organic pollutants. The objective of these treatments is to reduce ecotoxicity and safeguard the quality of surface and groundwater (Bao et al. 2023). PAHs can undergo biodegradation in both aerobic and anaerobic environments. Baniasadi et al. (2018) demonstrate that lower molecular weight organic pollutants undergo more effective degradation during biological treatment due to their greater biodegradability and transformability relative to higher molecular weight compounds.

Research indicates differing removal efficiencies for PAHs in municipal biological wastewater treatment facilities. Zhao et al. (2021) found that the total removal efficiency of PAHs in summer varied between 63.22 and 63.58%. A separate study indicated that aerobic activated sludge treatment attained a phenanthrene (PHE) removal efficiency ranging from 83 to 97%. Sequencing batch reactors (SBRs) have shown the capacity to eliminate around 55% of PAHs. PAHs such as benzo(ghi)perylene (BghiP), naphthalene (NAP), and pyrene (PYR) showed varying removal rates (Kuyukina et al. 2020).

Phytoremediation

Phytoremediation and bioremediation are effective strategies that employ plants, microorganisms, and enzymes to detoxify contaminated environments, facilitating their restoration to a natural condition without leading to additional harm to nature (Dhara and Dutta 2025). Phytoremediation is an innovative environmental technique that makes use of growing plants in polluted soils, waters to facilitate the removal or degradation pollutants. Some plant species exhibit notable effectiveness, indicating greater efficiency and appropriateness for phytoremediation processes. Bioremediation utilizes biological mechanisms to degrade environmental contaminants via metabolic processes, leading to the production of innocuous byproducts including cell biomass, carbon dioxide, and water. Microbial-based bioremediation for the elimination of PAHs in contaminated with oil areas was initially documented in the mid-twentieth century (Bhatti et al. 2024).

Bioremediation

Microbial Remediation

Bacteria exhibit significant metabolic versatility, enabling them to efficiently degrade PAH pollutants. Anaerobic PAH degradation operates through distinct mechanisms alternative electron acceptors to decompose and cleave aromatic rings

(Sui et al. 2021). Degradation by aerobic mechanisms of PAHs by bacteria primarily entails oxygenase-mediated metabolic pathways, encompassing the functions of monooxygenase and dioxygenase enzymes. The degradation process initiates with dioxygenase enzymes facilitating the hydroxylation of aromatic rings, resulting in the production of cis-dihydrodiol intermediates. The intermediates undergo further oxidation to form diol compounds via the action of dehydrogenase enzymes (Ismail et al. 2022).

Bacteria

Bacteria degrade PAH contaminants via different metabolic pathways due to their metabolic flexibility. When PAHs degrade aerobically, bacteria use O_2 as the ultimate acceptor of electrons to hydroxylate and cleave aromatic rings. Anaerobic PAH degradation decomposes aromatic structures via reductive processes and other electron acceptors (Liu et al. 2017). PAHs degrade aerobically via oxygenase-mediated mechanisms, including monooxygenases and dioxygenases. Dioxygenase enzymes hydroxylate aromatic rings to create cis-dihydrodiol intermediates, starting the degradation process. These intermediates are metabolized to diols by dehydrogenases. Mixed bacterial cultures and consortia, which collaborate on catabolic activities and have many degradation routes, often degrade PAH completely. Recent research has focused on PAH breakdown using mixed bacterial cultures and consortia (Wang et al. 2017). PAH breakdown is successful with immobilized and genetically engineered bacteria. Dispersion of bacterial inocula is a major issue in soil or sediment degradation. In subterranean soils, microbial mobility is limited and cells adhere strongly to soil organic materials, making dispersion difficult (Ismail et al. 2022).

Archaea

Saline environments, especially those linked to oil industries, are particularly vulnerable to petroleum pollution resulting from regular discharges of pollutants, such as PAHs. Extreme conditions frequently require the use of extremophiles, particularly archaea, for efficient bioremediation, given that standard microorganisms may fail to endure or operate effectively in high salinity environments (Patel et al. 2020). Current research has emphasized the crucial function of archaea in the bioremediation of PAHs. Nonetheless, research on their degradation mechanisms and pathways is comparatively sparse when contrasted with bacterial systems (Gou et al. 2022). The archaeon exhibited the ability to break down PAH in environments with elevated salinity (Banerjee et al. 2024).

Emerging Multi-process Remediation Of PAHs

Bioremediation with nanoparticle-based eco-engineering techniques are developing to remove contaminants including PAHs from atmosphere. Refinement is needed to develop functionalized nanoparticles by changing surface characteristics.

Functionalized nanoparticles improve bioremediation effectiveness and applicability by performing many activities. Enzymes, proteins, DNA, humic acids, and biosurfactants are used as functionalizing agents in bionanoremediation to remove petroleum hydrocarbons, including PAHs (Fouda et al. 2021). Nanoscale nanoparticles (NPs) have high reactivity and surface area, which are their principal benefits. Biofunctionalized nanoparticles efficiently remove PAH from polluted settings by nano-adsorption and catalytic degradation (Sam et al. 2023). After treatment, FeHCF NPs converted all PAHs into light weighted, non-hazardous metabolites, proving their photocatalytic and adsorbent properties. Physical, chemical, and biological strategies for PAH removal are successful, but combining them may boost effectiveness, especially for HMW PAHs (Barathi et al. 2023). Integrated remediation technologies overcome the limits of standalone procedures by tackling dead-end product creation, enabling faster environmental cleaning. Physical–biological coupling, chemical–biological coupling, multi-biological remediation, and complete physical–chemical–biological remediation systems have been studied for PAH cleanup (Gou et al. 2022). In the first and second cycles, alkaline precipitation and *Sphingobium* sp. PHE9 inoculation removed of PAHs from washing solvents. This shows how integrated techniques degrade PAHs and restore the ecosystem (Sharma et al. 2024).

PAH Remediation Through Value Added By-products

PAH biodegradation byproducts such biogas, bioelectricity, biosurfactant, and EPS have been studied for practical uses (Sam et al. 2023). Gitipour et al. (2018) used biostimulation to anaerobically degrade PAHs in polluted marine sediments. The study employed biostimulants such fresh organic byproducts from Solid waste from urban areas, digestate, and nutrients to degrade PAH by 55% in 120 days. Biohydrogen production peaked at 80 ml gVS^{-1} on days 3–30 under acidogenic settings, whereas biomethane production peaked at 140 ml gVS^{-1} on days 50–120 under methanogenic conditions. Biomethane production increased beyond 120 days, suggesting biostimulation might increase yields. Biohydrogen is a greener, sustainable fuel cell energy source. Vehicles, home electronics, and portable batteries might utilize it. Biogas, mostly methane, is a sustainable source of energy that may be utilized for heating, electrical production, and engine and turbine fuel. PAH degradation and bioelectricity production have been studied in microbial bioelectrochemical systems with one or more chambers that promote microorganism-mediated anode- and cathode-mediated redox reactions (Kumar et al. 2022). Microorganisms transmit electrons from organic materials to the anode, generating an electric current during deterioration. At the cathode, electrons reduce biotic or abiotic substances to generate electricity. This approach destroys PAHs and generates power, which may address energy needs (Lai et al. 2020). With alternate electron acceptors, biocatalytic mechanisms degrade PAHs better. Scientists created aerobic and anaerobic sediment microbial fuel cells. Anaerobic conditions resulted in degradation rates of 77%, 53%, and 37%, with a power production reduction of 3.6 mW m^{-2}. Biosurfactants and EPSs are produced during PAH decomposition. Amphiphilic biosurfactants produced by various PAH-degrading microorganisms increase PAH bioavailability and degradation efficiency (Sher et al. 2023). Biosurfactants from PAH degraders include rhamnolipids, lipopeptides, glycoproteins, and surfactins. Environmentally friendly biosurfactants are used in food, agriculture, oil, cosmetics, and medicines. They are suitable dispersion agents, emulsifiers, and chemically produced surfactant alternatives due to their excellent foaming ability, low critical micelle concentration, and robust surface activity (Wang et al. 2018).

Combine Treatment Method

High-molecular-weight PAHs may be degraded, solubilized, and removed from water via physical, chemical, and biological techniques. Biodegradation and Fenton oxidation remove PAHs, especially naphthalene and phenanthrene from water. The biodegradation method uses a Bacillus fusiformis (BFN) Fenton oxidation process, whereas strain from activated sludge synthesizes ferrous nanoparticles from tea extract in oxygen, nitrogen, and air (Gitipour et al. 2018). Biologically active filtration using biological and physical processes to remediate oil and gas effluent appears promising. This approach uses river ecosystem microorganisms and nutrients to produce biofilms that can survive high total dissolved solids. Granular activated carbon (GAC) removes organic contaminants and suspended particles well in this method. Integrated physical, chemical, and biological treatments may improve PAH solubilization and degradation. Strong oxidants, adsorption, and membrane technologies remove high-molecular-weight PAHs well (Chen et al. 2016). PAH removal rates in membrane bioreactors for wastewater treatment range from 50 to 100%, with low-molecular-weight PAHs being more efficient (Dai et al. 2022).

Future Outlook

Surface runoff and atmospheric deposition may transport combustion-derived PAHs to wastewater treatment plants. The destiny and behavior of PAHs and their derivatives in wastewater treatment plants are contingent upon physicochemical and environmental factors (Ziyaei et al. 2024). Water has a higher concentration of low molecular weight PAHs (LMW PAHs) than silt. Research indicates that wastewater treatment plant influent mostly contains LMW PAHs with two or three aromatic rings, whereas the effluent primarily consists of compounds with two rings. High Molecular Weight PAHs (HMW PAHs) adhere to sediment more effectively than to water. Sewage discharges LMW PAHs into Rivers, whereas municipal silt used as fertilizer emits HMW PAHs. SPAHs with higher molecular weight accumulate more in sediment (Li et al. 2025). Wastewater treatment plants exhibit insufficient biodegradation of PAHs. Hydrophobic and persistent PAHs adsorb onto soil. Full-scale membrane bioreactor wastewater treatment plants using activated sludge detected PAHs and their derivatives in the influent, effluent, and sludge. These devices degrade PAHs with 90% efficacy, despite their design and operation lacking optimization. To improve the detection of PAH and its derivatives, researchers should investigate trace nitro-PAH detection in complex matrices (Xie et al. 2025). Microbial activity requires investigation since PAHs may impede wastewater treatment plant (WWTP) efficacy. Certain wastewater treatment plants contravene regulations. The European Commission restricts sediment PAHs to 6000 ng/g prior to cropping. Improving pulmonary arterial hypertension treatment systems is essential research. PAH remediation strategies are chosen based on economic viability and ecological sustainability (Smetanová et al. 2025). Biotherapies are advantageous since they exhibit prolonged efficacy compared to physical and chemical treatments. Temperature, nutrient availability, aeration, and organic input rates influence microbial activity, hence affecting these processes. Wastewater treatment plants use biological techniques to eliminate PAHs, including anaerobic and anoxic treatments, albeit their effectiveness is limited (He et al. 2025). Research indicates that these circumstances reduce LMW PAHs by 35–52.9%. Seasonal and deleterious compounds in wastewater treatment plants may impair biological treatment efficacy, resulting in consistency challenges. Organic constituents, fungal immobilization, and substrate preparation enhance the deterioration of PAHs in soil. Ligninolytic fungi enhance biodegradation and microbial activity. Soil pH, temperature, and moisture influence biodegradation. Microbial activities operate most effectively in a pH range of 6.5–7.5. Beer grains and organic soil have the ability to decompose PAHs (Wu et al. 2025). Unidentified variables generate these outcomes. In sorption studies, lignin has superior binding affinity for pyrene compared to cellulose or hemicellulose. Phenol increases lignin pyrene sorption while decreasing cellulose and hemicellulose levels. Immobilization of lignocellulosic and PAH-degrading enzymes enhances PAH degradation. It is necessary to test these therapies in intricate hydrocarbon-contaminated environments (Zhao et al. 2025). This study should expand and evaluate the use of these approaches. The use of mushroom compost, charcoal, and distillery grains may facilitate the degradation of PAH. Laboratory investigations are promising; yet, empirical field studies are requisite. The bioavailability and degradation of PAH are enhanced by the water solubility of surfactants and the lowering of interfacial tension. Prolonged contact between soil and PAHs reduces enzyme activity, suggesting that surfactants may maintain enzyme functionality. Investigate white-rot fungi that decompose polycyclic aromatic hydrocarbons and lignocellulose in contaminated soils (Sayed et al. 2025).

Conclusion

This chapter examines PAHs and their alterations in extensive treatment of wastewater facilities. Their behavior, destiny, methods of analysis, biological therapies, viability evaluations approaches are discussed. The conversion of PAHs in WWTP has been hardly researched. Prior studies have shown that PAHs in industrial effluents may vary in concentration and often exceed permissible environmental threshold values. In a case study, concentrations of naphthalene, phenanthrene, and pyrene are documented in effluent from coke oven facilities, where levels of 50–300 μg/L of these compounds have been achieved by production processes and treatment methods. Petroleum refinery effluents may have benzo[a]pyrene concentrations reaching 15 mg/L, far above the WHO value of regulations of 0.7 mg/L for drinking water use. Furthermore, the concentrations of cumulative PAHs in surface waters receiving untreated or partly treated wastewater often exceed 500 μg/L, resulting in environmental contamination and bioaccumulation of PAHs in aquatic organisms. Many developing countries, although having regulatory frameworks and environmental standards like to those of the US EPA and EU, do not implement rigorous procedures and lack adequate technology for the removal of PAH. Conventional techniques for treatment like sedimentation and filtering are not always enough; thus, sophisticated procedures like adsorption, advanced oxidation processes, and bioremediation must be used to progressively eradicate them. In summary, the existence of PAHs in industrial wastewater is a critical issue that needs both immediate and long-term remedies. Addressing the issue not only safeguards ecological integrity but also improves public health and aligns industrial growth with principles of effective environmental management.

References

Abdel-Shafy, H.I., Mansour, M.S.: A review on polycyclic aromatic hydrocarbons: source, environmental impact, effect on human health and remediation. Egypt. J. Pet. **25**(1), 107–123 (2016)

Abdullah, S.R.S., Al-Baldawi, I.A., Almansoory, A.F., Purwanti, I.F., Al-Sbani, N.H., Sharuddin, S.S.N.: Plant-assisted remediation of hydrocarbons in water and soil: application, mechanisms, challenges and opportunities. Chemosphere **247**, 125932 (2020)

Adeniji, A.O., Okaiyeto, K., George, M.J., Tanor, E.B., Semerjian, L., Okoh, A.I.: A systematic assessment of research trends on polycyclic aromatic hydrocarbons in different environmental compartments using bibliometric parameters. Environ. Geochem. Health **45**(5), 1289–1309 (2023)

Akinpelumi, V.K., Kumi, K.G., Onyena, A.P., Sam, K., Ezejiofor, A.N., Frazzoli, C., Ekhator, O.C., Udom, G.J., Orisakwe, O.E.: A comparative study of the impacts of polycyclic aromatic hydrocarbons in water and soils in Nigeria and Ghana: towards a framework for public health protection. J. Hazard. Mater. Adv. **11**, 100336 (2023)

Al Farraj, D.A., Hadibarata, T., Yuniarto, A., Syafiuddin, A., Surtikanti, H.K., Elshikh, M.S., Al Khulaifi, M.M., Al-Kufaidy, R.: Characterization of pyrene and chrysene degradation by halophilic Hortaea sp. B15. Bioprocess Biosyst. Eng. **42**, 963–969 (2019)

Alao, M.B., Adebayo, E.A.: Fungi as veritable tool in bioremediation of polycyclic aromatic hydrocarbons-polluted wastewater. J. Basic Microbiol. **62**(3–4), 223–244 (2022)

Ansari, F., Ahmad, A., Rafatullah, M.: Review on bioremediation technologies of polycyclic aromatic hydrocarbons (PAHs) from soil: mechanisms and future perspective. Int. Biodeterior. Biodegrad. **179**, 105582 (2023)

Banerjee, S., Gupta, N., Pramanik, K., Gope, M., GhoshThakur, R., Karmakar, A., Gogoi, N., Hoque, R.R., Mandal, N.C., Balachandran, S.: Microbes and microbial strategies in carcinogenic polycyclic aromatic hydrocarbons remediation: a systematic review. Environ. Sci. Pollut. Res. **31**(2), 1811–1840 (2024)

Baniasadi, M., Mousavi, S.M., Zilouei, H., Shojaosadati, S.A., Rastegar, S.O.: Statistical evaluation and process optimization of biotreatment of polycyclic aromatic hydrocarbons in a bioreactor. Environ. Eng. Manag. J. **17**(8), 1869–1877 (2018)

Bao, S., Zhao, L., Liu, Y., Yang, L., Liu, P., Chen, C., Mao, X.: Influencing factors of bioaugmentation treatment of PAH-contaminated soils in slurry bioreactors. J. Environ. Chem. Eng. **11**(3), 109893 (2023)

Barathi, S., Gitanjali, J., Rathinasamy, G., Sabapathi, N., Aruljothi, K.N., Lee, J., Kandasamy, S.: Recent trends in polycyclic aromatic hydrocarbons pollution distribution and counteracting bioremediation strategies. Chemosphere **337**, 139396 (2023)

Berardi, C., Fibbi, D., Coppini, E., Renai, L., Caprini, C., Scordo, C.V.A., Checchini, L., Orlandini, S., Bruzzoniti, M.C., Del Bubba, M.: Removal efficiency and mass balance of polycyclic aromatic hydrocarbons, phthalates, ethoxylated alkylphenols and alkylphenols in a mixed textile-domestic wastewater treatment plant. Sci. Total. Environ. **674**, 36–48 (2019)

Bhatti, S.S., Bhatia, A., Bhagat, G., Singh, S., Dhaliwal, S.S., Sharma, V., Verma, V., Yin, R., Singh, J.: PAHs in terrestrial environment and their phytoremediation. In: Bioremediation for Sustainable Environmental Cleanup, pp. 67–85. CRC Press (2024)

Bis, M., Montusiewicz, A., Piotrowicz, A., Łagód, G.: Modeling of wastewater treatment processes in membrane bioreactors compared to conventional activated sludge systems. Processes **7**(5), 285 (2019)

Bokade, P., Bajaj, A.: Molecular advances in mycoremediation of polycyclic aromatic hydrocarbons: exploring fungal bacterial interactions. J. Basic Microbiol. **63**(3–4), 239–256 (2023)

Borji, H., Ayoub, G.M., Al-Hindi, M., Malaeb, L., Hamdan, H.Z.: Nanotechnology to remove polychlorinated biphenyls and polycyclic aromatic hydrocarbons from water: a review. Environ. Chem. Lett. **18**, 729–746 (2020)

Cao, W., Qiao, M., Liu, B., Zhao, X.: Occurrence of parent and substituted polycyclic aromatic hydrocarbons in typical wastewater treatment plants and effluent receiving rivers of Beijing, and risk assessment. J. Environ. Sci. Health Part A **53**(11), 992–999 (2018)

Chen, F., Tan, M., Ma, J., Zhang, S., Li, G., Qu, J.: Efficient remediation of PAH-metal co-contaminated soil using microbial-plant combination: a greenhouse study. J. Hazard. Mater. **302**, 250–261 (2016)

Chen, C.F., Ju, Y.R., Lim, Y.C., Hsieh, S.L., Tsai, M.L., Sun, P.P., Katiyar, R., Chen, C.W., Dong, C.D.: Determination of polycyclic aromatic hydrocarbons in sludge from water and wastewater treatment plants by GC-MS. Int. J. Environ. Res. Public Health **16**(14), 2604 (2019)

Dai, C., Han, Y., Duan, Y., Lai, X., Fu, R., Liu, S., Leong, K.H., Tu, Y., Zhou, L.: Review on the contamination and remediation of polycyclic aromatic hydrocarbons (PAHs) in coastal soil and sediments. Environ. Res. **205**, 112423 (2022)

de Andrade, J.R., Oliveira, M.F., Canevesi, R.L.S., Landers, R., da Silva, M.G.C., Vieira, M.G.A.: Comparative adsorption of diclofenac sodium and losartan potassium in organophilic clay-packed fixed-bed: X-ray photoelectron spectroscopy characterization, experimental tests and theoretical study on DFT-based chemical descriptors. J. Mol. Liq. **312**, 113427 (2020)

Dhara, A., Dutta, R.: A review on sources and distribution of polycyclic aromatic hydrocarbons (PAHs) in wetland ecosystem: focusing on plant-biomonitoring and phytoremediation. Environ. Sci. Pollut. Res. 1–23 (2025)

Fouda, A., Hassan, S.E.D., Saied, E., Azab, M.S.: An eco-friendly approach to textile and tannery wastewater treatment using maghemite nanoparticles (γ-Fe2O3-NPs) fabricated by Penicillium expansum strain (Kw). J. Environ. Chem. Eng. **9**(1), 104693 (2021)

García, J., García-Galán, M.J., Day, J.W., Boopathy, R., White, J.R., Wallace, S., Hunter, R.G.: A review of emerging organic contaminants (EOCs), antibiotic resistant bacteria (ARB), and antibiotic resistance genes (ARGs) in the environment: Increasing removal with wetlands and reducing environmental impacts. Bioresour. Technol. **307**, 123228 (2020)

Gaurav, G.K., Mehmood, T., Kumar, M., Cheng, L., Sathishkumar, K., Kumar, A., Yadav, D.: Review on polycyclic aromatic hydrocarbons (PAHs) migration from wastewater. J. Contam. Hydrol. **236**, 103715 (2021)

Gbeddy, G., Goonetilleke, A., Ayoko, G.A., Egodawatta, P.: Transformation and degradation of polycyclic aromatic hydrocarbons (PAHs) in urban road surfaces: influential factors, implications and recommendations. Environ. Pollut. **257**, 113510 (2020)

Ghasemi, N., Zare, F., Davari, P., Vilathgamuwa, M., Ghosh, A., Langton, C., Weber, P.: Dissimilar trend of nonlinearity in ultrasound transducers and systems at resonance and non-resonance frequencies. Ultrasonics **74**, 21–29 (2017)

Ghosal, D., Ghosh, S., Dutta, T.K., Ahn, Y.: Current state of knowledge in microbial degradation of polycyclic aromatic hydrocarbons (PAHs): a review. Front. Microbiol. **7**, 1369 (2016)

Gitipour, S., Sorial, G.A., Ghasemi, S., Bazyari, M.: Treatment technologies for PAH-contaminated sites: a critical review. Environ. Monit. Assess. **190**, 1–17 (2018)

Gou, Y., Song, Y., Yang, S., Yang, Y., Cheng, Y., Li, J., Zhang, T., Cheng, Y., Wang, H.: Polycyclic aromatic hydrocarbon removal from subsurface soil mediated by bacteria and archaea under methanogenic conditions: performance and mechanisms. Environ. Pollut. **313**, 120023 (2022)

Gupta, H.: PAH determination in effluent and sludge samples of paper industry. Environ. Technol. Innov. **9**, 115–121 (2018)

Gupta, H., Gupta, B.: Photocatalytic degradation of polycyclic aromatic hydrocarbon benzo[a]pyrene by iron oxides and identification of degradation products. Chemosphere **138**, 924–931 (2015)

Gupta, H., Gupta, B.: Adsorption of polycyclic aromatic hydrocarbons on banana peel activated carbon. Desalination Water Treat. **57**(20), 9498–9509 (2016)

Han, N., Wang, S., Rana, A.K., Asif, S., Klemeš, J.J., Bokhari, A., Long, J., Thakur, V.K., Zhao, X.: Rational design of boron nitride with different dimensionalities for sustainable applications. Renew. Sustain. Energy Rev. **170**, 112910 (2022a)

Han, N., Feng, S., Guo, W., Mora, O.M., Zhao, X., Zhang, W., Xie, S., Zhou, Z., Liu, Z., Liu, Q., Wan, K.: Rational design of Ruddlesden–Popper perovskite electrocatalyst for oxygen reduction to hydrogen peroxide. SusMat **2**(4), 456–465 (2022b)

Haneef, T., Ul Mustafa, M.R., Rasool, K., Ho, Y.C., Mohamed Kutty, S.R.: Removal of polycyclic aromatic hydrocarbons in a heterogeneous Fenton like oxidation system using nanoscale zero-valent iron as a catalyst. Water **12**(9), 2430 (2020)

He, Y., Yang, C., He, W., Xu, F.: Nationwide health risk assessment of juvenile exposure to polycyclic aromatic hydrocarbons (PAHs) in the water body of Chinese lakes. Sci. Total. Environ. **723**, 138099 (2020)

He, H., Ren, R., Jin, Q., Gong, L., Zhang, L., Xue, M., Fan, J., Wang, S.: Occurrence of EU-priority polycyclic aromatic hydrocarbons in edible oils and associated with human health risks in Hangzhou city of China. J. Food Compos. Anal. 107834 (2025)

Ijah, U.J.J., Musa, I.O., Amao, F., Maude, A.M., Victor-Ekwebelem, M.O.: Wastewater quality and unhygienic practices in Minna abattoir, North Central Nigeria. Sci. World J. **17**(2), 332–337 (2022)

Ismail, N.A., Kasmuri, N., Hamzah, N.: Microbial bioremediation techniques for polycyclic aromatic hydrocarbon (PAHs)—a review. Water Air Soil Pollut. **233**(4), 124 (2022)

Jackulin, F., Chitra, B., Karthick, S., Rangasamy, G.: A review on recent advancements in the treatment of polyaromatic hydrocarbons (PAHs) using sulfate radicals based advanced oxidation process. Environ. Res. 119124 (2024)

Jalili, V., Barkhordari, A., Ghiasvand, A.: Solid-phase microextraction technique for sampling and preconcentration of polycyclic aromatic hydrocarbons: a review. Microchem. J. **157**, 104967 (2020)

Jinadasa, B.K.K.K., Monteau, F., Fowler, S.W.: Review of polycyclic aromatic hydrocarbons (PAHs) in fish and fisheries products; a Sri Lankan perspective. Environ. Sci. Pollut. Res. **27**, 20663–20674 (2020)

Kong, X., Zhang, J., Ji, Q., Li, C., Chen, X., Cao, X., Zhu, F., Yang, S., Li, S., He, H.: Insights into adsorption mechanisms of nitro polycyclic aromatic hydrocarbons on common microplastic particles: experimental studies and modeling. Chemosphere **320**, 138050 (2023)

Kouras, A., Manoli, E., Avgenikou, A., Paraschaki, A., Voutsa, D., Samara, C., Foska, E.: Inhalation health risk assessment in the port of Volos, Greece, based on multiannual evolution of PM10-bound polycyclic aromatic hydrocarbons and metals/metalloids: implications for EU-regulated PM10 pollutants. Atmos. Environ. 121348 (2025)

Krishnan, S., Rawindran, H., Sinnathambi, C.M., Lim, J.W.: Comparison of various advanced oxidation processes used in remediation of industrial wastewater laden with recalcitrant pollutants. In: IOP Conference Series: Materials Science and Engineering, vol. 206, no. 1, p. 012089. IOP Publishing (2017)

Kumar, V.V., Venkataraman, S., Kumar, P.S., George, J., Rajendran, D.S., Shaji, A., Lawrence, N., Saikia, K., Rathankumar, A.K.: Laccase production by Pleurotus ostreatus using cassava waste and its application in remediation of phenolic and polycyclic aromatic hydrocarbon-contaminated lignocellulosic biorefinery wastewater. Environ. Pollut. **309**, 119729 (2022)

Kuppusamy, S., Thavamani, P., Venkateswarlu, K., Lee, Y.B., Naidu, R., Megharaj, M.: Remediation approaches for polycyclic aromatic hydrocarbons (PAHs) contaminated soils: technological constraints, emerging trends and future directions. Chemosphere **168**, 944–968 (2017)

Kuyukina, M.S., Krivoruchko, A.V., Ivshina, I.B.: Advanced bioreactor treatments of hydrocarbon-containing wastewater. Appl. Sci. **10**(3), 831 (2020)

Lai, X., Ning, X.A., Chen, J., Li, Y., Zhang, Y., Yuan, Y.: Comparison of the Fe2+/H2O2 and Fe2+/PMS systems in simulated sludge: removal of PAHs, migration of elements and formation of chlorination by-products. J. Hazard. Mater. **398**, 122826 (2020)

Lamichhane, S., Krishna, K.B., Sarukkalige, R.: Polycyclic aromatic hydrocarbons (PAHs) removal by sorption: a review. Chemosphere **148**, 336–353 (2016)

Li, R., Shen, Y., Shao, Y., Gao, Y., Yao, Z., Liu, Q., Liu, X., Shi, G.: Measurement Report: Polycyclic Aromatic Hydrocarbons (PAHs) and Their Alkylated (RPAHs), Nitrated (NPAHs) and Oxygenated (OPAHs) Derivatives in the Global Marine Atmosphere: Occurrence, Spatial Variations, and Source Apportionment, pp. 1–22. EGUsphere (2025)

Liu, S.H., Zeng, G.M., Niu, Q.Y., Liu, Y., Zhou, L., Jiang, L.H., Tan, X.F., Xu, P., Zhang, C., Cheng, M.: Bioremediation mechanisms of combined pollution of PAHs and heavy metals by bacteria and fungi: a mini review. Bioresour. Technol. **224**, 25–33 (2017)

Ma, Y., Liu, A., Egodawatta, P., McGree, J., Goonetilleke, A.: Quantitative assessment of human health risk posed by polycyclic aromatic hydrocarbons in urban road dust. Sci. Total. Environ. **575**, 895–904 (2017)

Mahmoud, A.E.D.: Nanomaterials: green synthesis for water applications. In: Handbook of Nanomaterials and Nanocomposites for Energy and Environmental Applications, pp. 1–21 (2020)

Man, Y.B., Chow, K.L., Cheng, Z., Mo, W.Y., Chan, Y.H., Lam, J.C.W., Lau, F.T.K., Fung, W.C., Wong, M.H.: Profiles and removal efficiency of polycyclic aromatic hydrocarbons by two different types of sewage treatment plants in Hong Kong. J. Environ. Sci. **53**, 196–206 (2017)

Mannina, G., Cosenza, A., Rebouças, T.F.: A plant-wide modelling comparison between membrane bioreactors and conventional activated sludge. Bioresour. Technol. **297**, 122401 (2020)

Martínez-Álvarez, I., Le Menach, K., Devier, M.H., Barbarin, I., Tomovska, R., Cajaraville, M.P., Budzinski, H., Orbea, A.: Uptake and effects of graphene oxide nanomaterials alone and in combination with polycyclic aromatic hydrocarbons in zebrafish. Sci. Total. Environ. **775**, 145669 (2021)

Mirzaee, E., Sartaj, M.: Activated carbon-based magnetic composite as an adsorbent for removal of polycyclic aromatic hydrocarbons from aqueous phase: characterization, adsorption kinetics and isotherm studies. J. Hazard. Mater. Adv. **6**, 100083 (2022)

Montano, L., Baldini, G.M., Piscopo, M., Liguori, G., Lombardi, R., Ricciardi, M., Esposito, G., Pinto, G., Fontanarosa, C., Spinelli, M., Palmieri, I.: Polycyclic aromatic hydrocarbons (PAHs) in the environment: occupational exposure, health risks and fertility implications. Toxics **13**(3), 151 (2025)

Mortazavi, M., Baghdadi, M., Javadi, N.H.S., Torabian, A.: The black beads produced by simultaneous thermal reducing and chemical bonding of graphene oxide on the surface of amino-functionalized sand particles: application for PAHs removal from contaminated waters. J. Water Process. Eng. **31**, 100798 (2019)

Mukwevho, N., Gusain, R., Fosso-Kankeu, E., Kumar, N., Waanders, F., Ray, S.S.: Removal of naphthalene from simulated wastewater through adsorption-photodegradation by ZnO/Ag/GO nanocomposite. J. Ind. Eng. Chem. **81**, 393–404 (2020)

Muthukumar, H., Shanmugam, M.K., Gummadi, S.N.: Caffeine degradation in synthetic coffee wastewater using silverferrite nanoparticles fabricated via green route using Amaranthus blitum leaf aqueous extract. J. Water Process. Eng. **36**, 101382 (2020)

Nasrollahzadeh, M., Sajjadi, M., Iravani, S., Varma, R.S.: Green-synthesized nanocatalysts and nanomaterials for water treatment: current challenges and future perspectives. J. Hazard. Mater. **401**, 123401 (2021)

Nowacka, A., Włodarczyk-Makuła, M.: Effectiveness of priority PAH removal in a water coagulation process. Water Sci. Technol. Water Supply **15**(4), 683–692 (2015)

Omolola, A.H., Ogunmoroti, E.A., Ayoola, P.B., Adedosu, T.A.: Assessment of physicochemical properties and polycyclic aromatic hydrocarbons (PAHs) in wastewater from New Atenda Abattoir, Ogbomoso and its environmental impacts. Int. J. Sci. Res. Technol. 3(01) (2025)

Pandey, N., Gusain, R., Suthar, S.: Exploring the efficacy of powered guar gum (Cyamopsis tetragonoloba) seeds, duckweed (Spirodela polyrhiza), and Indian plum (Ziziphus mauritiana) leaves in urban wastewater treatment. J. Clean. Prod. **264**, 121680 (2020)

Patel, A.B., Shaikh, S., Jain, K.R., Desai, C., Madamwar, D.: Polycyclic aromatic hydrocarbons: sources, toxicity, and remediation approaches. Front. Microbiol. **11**, 562813 (2020)

Pathak, S., Sakhiya, A.K., Anand, A., Pant, K.K., Kaushal, P.: A state-of-the-art review of various adsorption media employed for the removal of toxic polycyclic aromatic hydrocarbons (PAHs): an approach towards a cleaner environment. J. Water Process. Eng. **47**, 102674 (2022)

Pulleyblank, C., Cipullo, S., Campo, P., Kelleher, B., Coulon, F.: Analytical progress and challenges for the detection of oxygenated polycyclic aromatic hydrocarbon transformation products in aqueous and soil environmental matrices: a review. Crit. Rev. Environ. Sci. Technol. **49**(5), 357–409 (2019)

Qiao, M., Bai, Y., Cao, W., Huo, Y., Zhao, X., Liu, D., Li, Z.: Impact of secondary effluent from wastewater treatment plants on urban rivers: polycyclic aromatic hydrocarbons and derivatives. Chemosphere **211**, 185–191 (2018)

Qiao, M., Fu, L., Cao, W., Bai, Y., Huang, Q., Zhao, X.: Occurrence and removal of polycyclic aromatic hydrocarbons and their derivatives in an ecological wastewater treatment plant in South China and effluent impact to the receiving river. Environ. Sci. Pollut. Res. **26**, 5638–5644 (2019)

Ramesh, A., Harris, K.J., Archibong, A.E.: Reproductive toxicity of polycyclic aromatic hydrocarbons. In: Reproductive and Developmental Toxicology, pp. 759–778. Academic Press (2022)

Ramezanzadeh, M., Bahlakeh, G., Ramezanzadeh, B.: Green synthesis of reduced graphene oxide nanosheets decorated with zinc-centered metal-organic film for epoxy-ester composite coating reinforcement: DFT-D modeling and experimental explorations. J. Taiwan Inst. Chem. Eng. **114**, 311–330 (2020)

Rayaroth, M.P., Marchel, M., Boczkaj, G.: Advanced oxidation processes for the removal of mono and polycyclic aromatic hydrocarbons—a review. Sci. Total. Environ. **857**, 159043 (2023)

Ren, J., Li, J., Li, J., Chen, Z., Cheng, F.: Tracking multiple aromatic compounds in a full-scale coking wastewater reclamation plant: interaction with biological and advanced treatments. Chemosphere **222**, 431–439 (2019)

Riley, S.M., Ahoor, D.C., Regnery, J., Cath, T.Y.: Tracking oil and gas wastewater-derived organic matter in a hybrid biofilter membrane treatment system: a multi-analytical approach. Sci. Total. Environ. **613**, 208–217 (2018)

Rosińska, A., Dąbrowska, L.: Influence of type and dose of coagulants on effectiveness of PAH removal in coagulation water treatment. Water Sci. Eng. **14**(3), 193–200 (2021)

Safo-Adu, G., Attiogbe, F., Emahi, I., Ofosu, F.G.: A review of the sources, distribution sequences, and health risks associated with exposure to atmospheric polycyclic aromatic hydrocarbons. Cogent Eng. **10**(1), 2199511 (2023)

Sakshi, Singh, S.K., Haritash, A.K.: Polycyclic aromatic hydrocarbons: soil pollution and remediation. Int. J. Environ. Sci. Technol. **16**, 6489–6512 (2019)

Sam, K., Onyena, A.P., Zabbey, N., Odoh, C.K., Nwipie, G.N., Nkeeh, D.K., Osuji, L.C., Little, D.I.: Prospects of emerging PAH sources and remediation technologies: insights from Africa. Environ. Sci. Pollut. Res. **30**(14), 39451–39473 (2023)

Satouh, S., Martín, J., Orta, M.D.M., Medina-Carrasco, S., Messikh, N., Bougdah, N., Santos, J.L., Aparicio, I., Alonso, E.: Adsorption of polycyclic aromatic hydrocarbons by natural, synthetic and modified clays. Environments **8**(11), 124 (2021)

Sayara, T., Sánchez, A.: Bioremediation of PAH-contaminated soils: process enhancement through composting/compost. Appl. Sci. **10**(11), 3684 (2020)

Sayed, F.A., Eid, M.H., El-Sherbeeny, A.M., Abdel-Gawad, G.I., Mohamed, E.A., Abukhadra, M.R.: Environmental and health risk assessment of polycyclic aromatic hydrocarbons and toxic elements in the Red Sea using Monte Carlo simulation. Sci. Rep. **15**(1), 4122 (2025)

Sharma, S., Shaikh, S., Mohana, S., Desai, C., Madamwar, D.: Current trends in bioremediation and bio-integrated treatment of petroleum hydrocarbons. Environ. Sci. Pollut. Res. **31**(46), 57397–57416 (2024)

Shen, X., Zhang, J., Xie, H., Hu, Z., Liang, S., Ngo, H.H., Guo, W., Chen, X., Fan, J., Zhao, C.: Intensive removal of PAHs in constructed wetland filled with copper biochar. Ecotoxicol. Environ. Saf. **205**, 111028 (2020)

Sher, S., Waseem, M., Leta, M.K.: Review of techniques for the removal of polycyclic aromatic hydrocarbons from produced water. Environments **10**(3), 40 (2023)

Singh, S., Parveen, N., Gupta, H.: Adsorptive decontamination of rhodamine-B from water using banana peel powder: a biosorbent. Environ. Technol. Innov. **12**, 189–195 (2018)

Singh, S., Kumar, A., Gupta, H.: Activated banana peel carbon: a potential adsorbent for Rhodamine B decontamination from aqueous system. Appl. Water Sci. **10**, 185 (2020)

Smetanová, S., Jbebli, A., Kohoutek, J., Puklová, V., Černá, M., Krsková, A., Zvonař, M., Reguli, Z., Andrýsková, L., Piler, P., Přibylová, P.: Changing pattern of exposure to polycyclic aromatic hydrocarbons over time in the Central European population. J. Expo. Sci. Environ. Epidemiol. 1–15 (2025)

Sui, X., Wang, X., Li, Y., Ji, H.: Remediation of petroleum-contaminated soils with microbial and microbial combined methods: advances, mechanisms, and challenges. Sustainability **13**(16), 9267 (2021)

Sun, S.J., Zhao, Z.B., Li, B., Ma, L.X., Fu, D.L., Sun, X.Z., Thapa, S., Shen, J.M., Qi, H., Wu, Y.N.: Occurrence, composition profiles and risk assessment of polycyclic aromatic hydrocarbons in municipal sewage sludge in China. Environ. Pollut. **245**, 764–770 (2019)

Wang, J., Liu, J., Ling, W., Huang, Q., Gao, Y.: Composite of PAH-degrading endophytic bacteria reduces contamination and health risks caused by PAHs in vegetables. Sci. Total. Environ. **598**, 471–478 (2017)

Wang, L., Chen, L., Tsang, D.C., Li, J.S., Yeung, T.L., Ding, S., Poon, C.S.: Green remediation of contaminated sediment by stabilization/solidification with industrial by-products and CO2 utilization. Sci. Total. Environ. **631**, 1321–1327 (2018)

Wei, C., Bandowe, B.A.M., Han, Y., Cao, J., Zhan, C., Wilcke, W.: Polycyclic aromatic hydrocarbons (PAHs) and their derivatives (alkyl-PAHs, oxygenated-PAHs, nitrated-PAHs and azaarenes) in urban road dusts from Xi'an, Central China. Chemosphere **134**, 512–520 (2015)

Wu, D.D., Wei, N.N., Zhang, C.G., Yue, X.Z., Li, H., Zhang, W.Y., Jia, X.R., Zhu, J.K., Zhang, W.Q., Fan, Y.C., Yang, S.M.: Long-term atmosphere surveillance (2017–2022) of PM2. 5-bound polycyclic aromatic hydrocarbons and health risk assessment in a typical city in Northern China. BMC Public Health **25**(1), 1888 (2025)

Xie, J., Lan, R., Zhang, L., Yu, J., Liu, X., You, Z., Yang, F., Lin, T.: Global occurrence, food web transfer, and human health risks of polycyclic aromatic hydrocarbons in biota. Sci. Total. Environ. **958**, 177969 (2025)

Yang, X., Li, Z., Liu, Y., Xing, Y., Wei, J., Yang, B., Zhang, C., Yang, R.T., Tsai, C.J.: Research progress of gaseous polycyclic aromatic hydrocarbons purification by adsorption. Aerosol Air Qual. Res. **19**(4), 911–924 (2019)

Yemele, O.M., Zhao, Z., Nkoh, J.N., Ymele, E., Usman, M.: A systematic review of polycyclic aromatic hydrocarbon pollution: a combined bibliometric and mechanistic analysis of research trend toward an environmentally friendly solution. Sci. Total. Environ. **926**, 171577 (2024)

Zhao, C., Xu, J., Shang, D., Zhang, Y., Zhang, J., Xie, H., Kong, Q., Wang, Q.: Application of constructed wetlands in the PAH remediation of surface water: a review. Sci. Total. Environ. **780**, 146605 (2021)

Zhao, J.Y., Liu, Y., Shang, N.: Analysis of the potential disease burden of occupational polycyclic aromatic hydrocarbons exposure-associated lung cancer in China from 1990 to 2021. Chin. J. Ind. Hyg. Occup. Dis. **43**(6), 430–434 (2025)

Ziyaei, K., Mokhtari, M., Hashemi, M., Rezaei, K., Abdi, F.: Association between exposure to water sources contaminated with polycyclic aromatic hydrocarbons and cancer risk: a systematic review. Sci. Total. Environ. 171261 (2024)

Contamination of Polycyclic Aromatic Hydrocarbons in Foods and the Food Chain

7

Deepak Chandra Joshi, Pravesh Kumar Sharma, Somya Upadhyay, Lavkush Verma, and Keshav Raj Paudel

Abstract

Polycyclic Aromatic Hydrocarbons (PAHs) are persistent organic pollutants predominantly generated by the incomplete combustion of organic materials. Their presence in food and the food chain raises substantial public health concerns due to their carcinogenic, mutagenic, and endocrine-disrupting characteristics. This chapter examines PAH contamination, focussing on its occurrence and transit throughout food systems, and incorporates surveillance data, regulatory restrictions, and risk evaluations. Environmental deposition, absorption from polluted soil and water, bioaccumulation in aquatic and terrestrial species, and high-temperature cooking methods including smoking, grilling, and frying are all ways that PAHs get into food. The levels of benzo[a]pyrene (2 μg/kg) and PAH4 (30 μg/kg) in cereals, smoked fish, and grilled meats are often higher than the EU's standards. For example, in cereals, the levels vary from 2.5 to 276.7 μg/kg, in smoked fish, they may be as high as 222.7 μg/kg, and in grilled meats, they can be above 200 μg/kg. These pollutants go down the food chain, from soil and crops to cattle and fish, and then to people. This means that people

D. C. Joshi (✉) · P. K. Sharma
Department of Pharmacy, Faculty of Pharmaceutical sciences and Nursing, Vivekananda Global University, Jaipur, Rajasthan, India
e-mail: Deepakjoshi024@gmail.com

S. Upadhyay
Agra Public College of Higher Education and Research Center, Agra, Uttar Pradesh, India

L. Verma
School of Pharmacy, Chandigarh University, Unnao Campus, Unnao, Uttar Pradesh, India

K. R. Paudel
Centre for Inflammation, Centenary Institute and University of Technology Sydney, School of Life Sciences, Faculty of Science, Sydney, NSW 2007, Australia

H. Gupta et al. (eds.), *Ubiquitous Polycyclic Aromatic Hydrocarbon Contamination*, Emerging Contaminants and Associated Treatment Technologies,
https://doi.org/10.1007/978-3-032-11764-9_7

are constantly exposed to them through their diets, which can cause malignancies, reproductive and developmental problems, immune system problems, and neurotoxicity. Even as better methods for identifying problems and rules have improved monitoring, there are still issues with enforcement, maintaining uniform standards, and increasing consumer awareness. To protect food safety and human health worldwide, we need to employ a range of measures, including advanced food processing technology, microbial bioremediation, and innovative methods for detecting PAHs.

Keywords

Polycyclic aromatic hydrocarbons • Food safety • Carcinogens • Bioaccumulation • Food processing • Analytical methods • Bioremediation • Public health

Introduction

Polycyclic Aromatic Hydrocarbons (PAHs) are a big category of chemical pollutants that are particularly detrimental to health and food safety across the world. These chemicals, which have two or more aromatic rings that are stuck together, are typically formed when organic matter doesn't burn all the way (Montano et al. 2025). PAHs are defined as organic compounds containing multiple fused benzene rings, most commonly arranged in linear, angular, or cluster formations. They can have substituent groups like methyl, nitro, or amino, but their core classification hinges on the number of aromatic rings present. PAHs are broadly categorised into low-molecular-weight (LMW) types, which have two to three rings (such as naphthalene and anthracene), and high-molecular-weight (HMW) types, which have four to six rings (such as chrysene, fluoranthene, and benzo[a]pyrene). This classification is not merely academic; LMW PAHs are typically more volatile and prevalent in atmospheric vapor, while HMW PAHs tend to adhere to particulates and are more persistent in solid and aquatic environments (Patel et al. 2020). The United States Environmental Protection Agency (US-EPA) and the International Agency for Research on Cancer (IARC) have identified 16 priority PAHs, with benzo[a]pyrene classified as a Group 1 human carcinogen due to its genotoxicity and mutagenic potential. The carcinogenicity and toxicity across the PAH group, however, vary depending on their molecular structure, physical properties, and ability to undergo metabolic activation in biological systems (Halfadji et al. 2021). Global concerns about PAHs in food have intensified due to mounting evidence of their role in carcinogenesis, mutagenesis, and long-term health risks. Regulatory authorities in Europe, North America, Asia, and other regions have implemented guidelines and legislation to monitor and limit PAHs in foods, especially emphasizing the need for continuous surveillance of processed and raw food items (Montano et al. 2025). Notably, in recent years, research has highlighted alarming concentrations of PAHs in various foodstuffs from diverse geographical areas, reflecting differences in both environmental contamination and food processing practices.

Sources of PAHs in the Environment and Entry into Food Systems

The environmental presence of PAHs results from both natural and anthropogenic sources. Forest fires and volcanic emissions contribute naturally, but anthropogenic activities dominate the global PAH burden. Major sources include the combustion of fossil fuels, biomass, and waste, vehicular exhaust, industrial emissions, cigarette smoke, and certain manufacturing processes (Lawal 2017). High-temperature cooking practices—like grilling, roasting, frying, and smoking—are especially notorious for producing elevated PAH concentrations in food (Han et al. 2022). PAHs enter the food systems via multiple routes. Environmentally, they deposit onto crops from airborne particulates or are absorbed from contaminated soil and water. Bioaccumulation occurs in aquatic organisms such as fish and shellfish, where PAH concentrations can greatly exceed surrounding levels (Abdel-Shafy and Mansour 2016). When animals consume contaminated vegetation, PAHs are transferred up the food chain, intensifying dietary exposure in humans. In foods, PAHs are categorized as "endogenous" (produced during pyrolysis of carbohydrates, lipids, and proteins in high-temperature processes) and "exogenous" (derived from fuel combustion during cooking or smoking). Dietary sources of PAHs include cereals, vegetables, fruits, oils, meats, and processed foods. The contamination levels vary with the proximity to pollution sources, such as urban roads or industrial centers, and with agricultural practices. Grilled meats have been found to contain benzo[a]pyrene at levels up to 200 µg/kg, while smoked fish and meats can go even higher (Uiuiu et al. 2025). Even plant-based foods such as cereals and oils are susceptible, with drying processes and combustion fumes acting as additional contamination vectors (Paris et al. 2018). The schematic representation of occurrence, source, properties etc. of Polycyclic Aromatic Hydrocarbons mentioned in Fig. 7.1.

Relevance of Studying PAH Contamination in Food and Food Chains

Chronic exposure, primarily through diet, has been implicated in the rising incidence of several cancers—especially of the lung, skin, bladder, and gastrointestinal tract—as well as in cardiovascular, respiratory, reproductive, and neurodevelopmental disorders. The global relevance of PAHs in food systems arises from their ability to bioaccumulate and persist, not just in processed foods but in staple items from diverse cultures and climates (Sampaio et al. 2021). Their lipophilicity ensures they are readily absorbed into human tissues, and metabolic

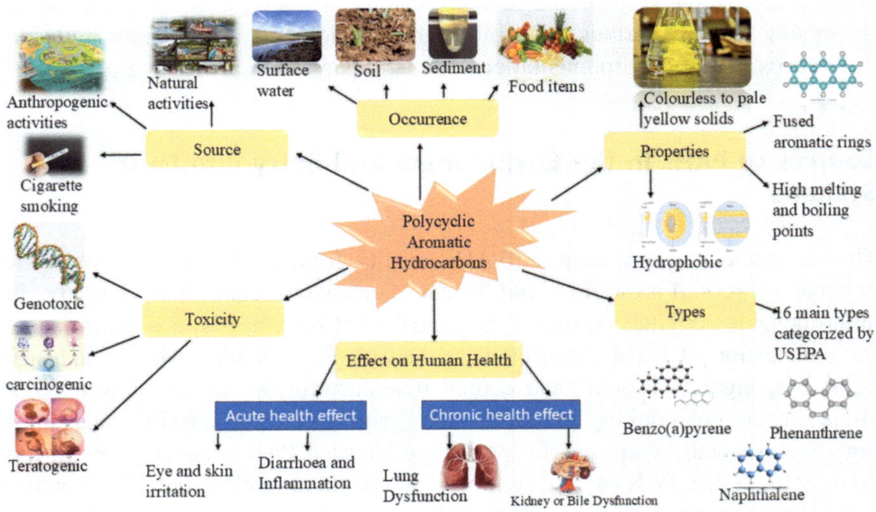

Fig. 7.1 Schematic representation of occurrence, source, properties etc. of polycyclic aromatic hydrocarbons

activation through enzymatic processes generates reactive intermediates that can interact covalently with DNA, proteins, and cell membranes, resulting in cellular and physiological damage. The cross-disciplinary nature of PAH research—spanning toxicology, analytical chemistry, food science, and public health—mirror the complex challenge posed by these compounds (D'Amore et al. 2025).

Chemistry and Characteristics of PAHs

Polycyclic Aromatic Hydrocarbons (PAHs) stand as one of the most studied classes of persistent organic pollutants due to their characteristic structural diversity, varied physicochemical properties, and complex toxicokinetics that drive their environmental behavior and health impacts. Over the past decades, a deepening scientific understanding has refined how we approach their analysis, fate, and risks (Berríos-Rolón et al. 2025). Central to this understanding are the detailed features of PAH chemistry, their distribution and persistence in environmental matrices, their classification based on molecular weight, and their toxicokinetics with bioaccumulation implications. The chemical structure of the various PAHs are mentioned in Fig. 7.2.

Structural Features of PAHs

PAHs are organic molecules composed solely of carbon and hydrogen, with their structures marked by multiple fused aromatic rings, typically benzene rings

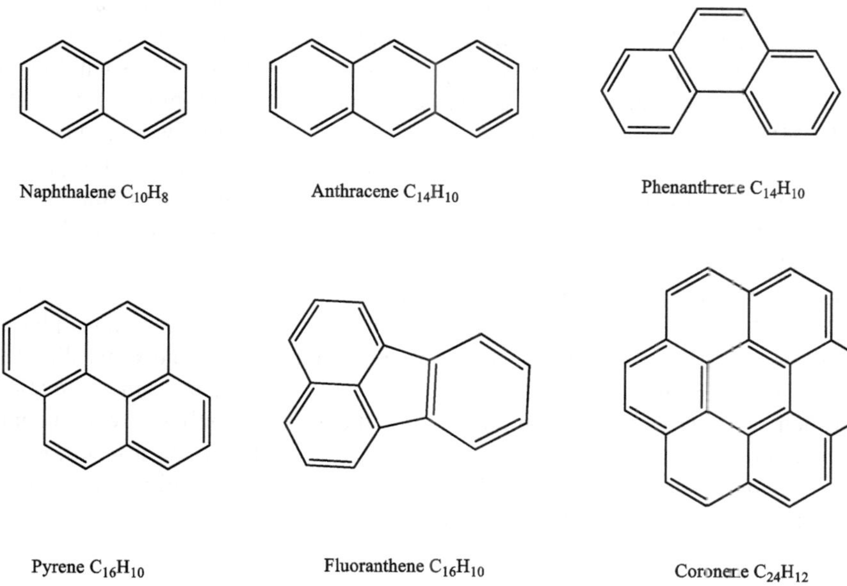

Fig. 7.2 The chemical structure of the various PAHs

arranged in linear, angular, or clustered fashion. The size and complexity of PAHs may vary from simple molecules like naphthalene (two rings) to large, multi-ringed compounds such as benzo[a]pyrene and coronene. These structural arrangements confer high thermodynamic stability and a substantial degree of chemical inertness, contributing to their persistence in the environment. Molecular geometry plays a crucial role in physical properties, reactivity, and environmental fate. For example, the dense π-electron clouds found in aromatic rings resist nucleophilic attacks, imparting biochemical recalcitrance and slowing biological degradation processes (Sahoo et al. 2020).

Recent advances have led to the synthesis and exploration of nitrogen-doped bowl-shaped PAHs, architectural innovations that further enhance electronic, chemical, and supramolecular properties for specialized applications. While not directly related to environmental toxicity, such molecular derivatives highlight the expanding relevance of PAHs beyond traditional pollution studies.

Physicochemical Properties: Molecular Weight, Volatility, Solubility

The physicochemical behavior of PAHs varies widely between individual compounds, but can be generally described according to molecular weight—one of the most influential factors. Low-molecular-weight PAHs (LMW-PAHs), comprised of two-three rings, such as naphthalene, fluorene, and anthracene, are generally more volatile and water-soluble than their heavier counterparts. Their vapor pressures and water solubility enable easier atmospheric transport and distribution in aquatic environments. For example, naphthalene possesses a water solubility of about 31 mg/L and relatively high volatility. Conversely, high-molecular-weight PAHs (HMW-PAHs), containing four or more rings (benzo[a]pyrene, chrysene, coronene), exhibit lower vapor pressure, extremely low water solubility, and heightened lipophilicity (LogKow often > 5), favoring strong adsorption to sediments, soils, and organic matter in biotic and abiotic environments (Gundlapalli et al. 2024).

Such disparities determine pathways by which PAHs partition within the environment. LMW-PAHs are released primarily into the atmospheric gaseous phase, while HMW-PAHs are found almost exclusively on particulates and soil/sediment-bound phases. High melting and boiling points in HMW compounds, coupled with their hydrophobic nature, further confine them to less mobile environmental compartments and slow degradation rates (Eldos et al. 2022). As a result, environmental processes impacting PAH fate (volatilization, dissolution, adsorption, partitioning) are driven strongly by molecular weight, solubility, and volatility.

Stability and Persistence in the Environment

PAHs display a remarkable ability to persist for extended durations in ecosystems. Their chemical stability arises from the fused aromatic system, which resists oxidation, reduction, and many forms of biological degradation, especially in HMW PAHs. Environmental half-lives for these compounds can stretch from months to years, with HMW PAHs exhibiting longer residence times due to stronger adsorption and reduced bioavailability (Sakshi and Haritash 2020). PAHs persist in multiple environmental matrices, including soil, water, sediments, and the atmosphere, forming reservoirs that facilitate complex cycles of deposition and re-distribution. Atmospheric PAHs may exist either as gases (LMW) or particulate-bound (HMW), with the former subject to long-range transport via wind and precipitation, while the latter deposit locally, contaminating soils and aquatic sediments. Photolysis and chemical transformations (by ozone, hydroxyl radicals) offer partial degradation routes, especially for LMW PAHs, but such mechanisms are limited for hydrophobic, particle-bound species. Biodegradation is further slowed in environments low in nutrients or oxygen (e.g., subaquatic sediments). Consequently, monitoring and remediation efforts must confront this persistence, especially in contaminated soils or dredged materials.

PAH Classification: Low Versus High Molecular Weight

Classifying PAHs by molecular weight illuminates their environmental fate and toxic potential. LMW PAHs, with two to four rings, are generally characterized by higher volatility, solubility, and tendency for atmospheric or water-bound distribution. HMW PAHs possess five or more rings and are less volatile, less soluble, and more prone to fixation in sediments and soil. This dichotomy profoundly influences both the analytical techniques used to detect and remove PAHs and the overall risk posed to ecosystems and human health. For example, LMW PAHs are more susceptible to biological and photochemical breakdown, making them less persistent but often more widely dispersed. HMW PAHs, due to their particle affinity and resistance to degradation, accumulate at possible "hot spots" and pose greater carcinogenic and mutagenic risk. The US Environmental Protection Agency has designated 16 PAHs as priority pollutants, including both LMW and HMW species, given their broad occurrence, recalcitrance, and significant health risks. Analytical techniques for environmental PAH detection rely upon the specific physicochemical properties of each class (Femi-Oloye et al. 2024). Conventional approaches such as gas chromatography-mass spectrometry (GC-MS), high-performance liquid chromatography (HPLC), and advanced microextraction methods each offer tailored solutions to address sample preparation and compound detection, with recent innovations leveraging nanomaterials, ionic liquids, and miniaturized sampling devices for improved sensitivity and selectivity.

Toxicokinetics and Bioaccumulation Potential

The toxicokinetics of PAHs encompass the absorption, distribution, metabolism, and excretion (ADME) processes that govern their fate in humans and wildlife. Due to their lipophilic character, PAHs are readily absorbed through dermal, inhalational, and gastrointestinal exposure, with efficient uptake into lipid-rich tissues, especially for HMW PAHs. Once inside the body, cytochrome P450 enzymes mediate metabolic activation, converting parent PAHs into reactive intermediates that can bind covalently to nucleic acids, proteins, and other biomolecules—initiating carcinogenic, mutagenic, and teratogenic responses.

Bioaccumulation potential is higher for HMW PAHs, as their low water solubility and high lipid solubility facilitate transfer and magnification up the food chain, particularly in aquatic organisms such as fish, mollusks, and crustaceans. Bioconcentration factors (BCFs) frequently exceed hundreds to thousands in certain species, especially in contaminated environments with high organic content and low degradation capacity. The cumulative effects of PAH bio-accumulation in upper trophic levels lead to significant human exposure through dietary intake, including consumption of contaminated seafood, meats, and crop products. However, metabolic and excretory pathways vary between organisms, resulting in complex risk profiles influenced by both compound-specific properties and organism-specific barriers. The Various PAHs contamination in environment and thermal processing mentioned in Fig. 7.3.

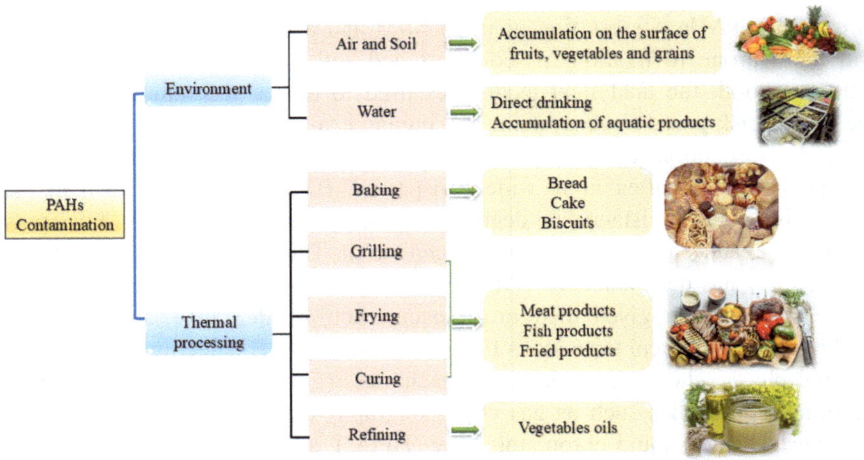

Fig. 7.3 Various PAHs contaminations in environment and thermal processing

Critical studies have shown that chronic exposure to PAHs, especially via ingestion, poses substantial incremental lifetime cancer risks, with pathways for toxicity also including immunotoxicity, endocrine disruption, and developmental impairment. Acute effects of LMW PAHs can include skin, eye, and respiratory irritation, while chronic impacts may involve multi-organ system damage and elevated risks for cancers, asthma, and reproductive dysfunction (Choi et al. 2023). The characteristics of Selected Polycyclic Aromatic Hydrocarbons (PAHs) are listed in Table 7.1.

Ecotoxicological and Regulatory Perspectives

The complex mixture of PAHs in environmental matrices yields not only direct toxicity but also multifaceted ecotoxicological effects, including disruption of aquatic food webs, impairment of microbial communities, and interference with vital biochemical processes in plants and animals. In freshwater systems, PAHs damage reproduction, growth, and survival of multiple taxa, with bioaccumulation and endocrine disruption presenting particular threats. Regulatory strategies and monitoring frameworks continue to evolve, informed by advances in analytical chemistry and risk assessment science (Honda and Suzuki 2020). Environmental standards remain especially stringent for food-relevant PAHs such as benzo[a]pyrene and benzo[ghi]perylene, with legal limits placed on allowable concentrations in foods and soils.

Table 7.1 Characteristics of selected polycyclic aromatic hydrocarbons (PAHs)

PAH	Molecular weight (g/mol)	Number of rings	Bioaccumulation potential	Comments/food contamination examples	References
Naphthalene	128	2	Low	More volatile, less bioaccumulative; found in air and water	Abdel-Shafy and Mansour (2016)
Phenanthrene	178	3	Moderate	Present in smoked/grilled foods; moderate lipophilicity	Patel et al. (2020)
Fluoranthene	202	4	High	Found in grilled meats and oils; higher lipophilicity	Sampaio et al. (2021)
Pyrene	202	4	High	Common in smoked foods; high persistence and bioaccumulation	Sampaio et al. (2021)
Benz[a]anthracene	228	4	High	Known carcinogen; accumulates in fatty tissues in food chains	Mohammed et al. (2025)
Chrysene	228	4	High	Detected in cooked meats and oils; toxic effects	Palade et al. (2023)
Benzo[a]pyrene	252	5	Very high	Potent carcinogen; highly bioaccumulative in fatty foods	Shimada and Fujii-Kuriyama (2004)
Dibenz[a,h]anthracene	278	5	Very high	Highly toxic, persistent; significant bioaccumulation in food	Kanaly and Harayama (2000)

Analytical and Remediation Approaches

Ongoing improvement in extraction and analysis techniques for PAHs have introduced microextraction, dispersive solid-phase extraction, and sorbents such as graphene or metal-organic frameworks for sample processing and detection. Hazardous solvent requirements in traditional methods are increasingly replaced by safer alternatives such as ionic liquids, and miniaturized methods enable improved throughput with lower sample and solvent volumes. Analytical platforms rely increasingly on combined chromatographic and spectrometric approaches to achieve enhanced selectivity, sensitivity, and rapid data output (Soursou et al. 2023; Alanazi 2025).

Remediation of PAH contaminated sites employs physical, chemical, and biological solutions. Techniques including soil washing, thermal treatment, advanced oxidation, and electrokinetic remediation complement a broad array of biological methods, such as microbial bioremediation, phytoremediation, and land farming. "In situ" and "ex situ" bioremediation strategies leverage bacteria, fungi, algae, and co-cultures, with the potential for multi-process combinatorial treatments. Emerging technologies have begun to generate value-added by-products during PAH treatment, further integrating remediation into circular economy principles (Gupte et al. 2016).

Occurrence and Levels of PAHs in Foods

Polycyclic Aromatic Hydrocarbons (PAHs) occur widely in foods of both plant and animal origin, as well as in a diverse array of processed products. Multiple factors, including environmental contamination, agricultural practices, food processing methods, and regional industrial activity influence their presence and concentrations (Sampaio et al. 2021). Over recent years, growing scientific scrutiny and strengthened surveillance by regulatory agencies—such as the FAO/WHO, EFSA, and FDA—have led to a deeper understanding of PAH distributions and potential health risks within global food systems (Santonicola et al. 2017; Khalili et al. 2023; Einolghozati et al. 2022). The reported levels of Polycyclic aromatic hydrocarbons (PAHs) in various food types from 2019 to 2025 are listed in Table 7.2.

PAHs in Cereal and Plant-Based Foods

Cereal grains, vegetables, and fruits represent key dietary staples, and monitoring PAH contamination in these foods remains a major public health priority. PAHs can enter plant-based foods primarily through deposition from contaminated air, irrigation with polluted water, or uptake from tainted soils in proximity to industrial zones or heavy traffic. Studies have shown that baby foods based on cereals can exhibit a wide range of PAH concentrations; individual sum PAH values have ranged from 2.5 μg/kg up to 276.7 μg/kg, with concerning exceedances of the WHO limit of 1.0 μg/kg for PAHs in nearly one-third of surveyed samples in some regions, such as the Greater Accra region in Ghana. These findings highlight the vulnerability of both raw crops and processed cereal products to contamination (Einolghozati et al. 2022).

Table 7.2 Reported levels of polycyclic aromatic hydrocarbons (PAHs) in various food types from 2019 to 2025

Year	Food type/ source	Typical PAHs monitored	Reported PAH levels (μg/kg)	Geographic or regulatory highlights	References
2019	Grilled and smoked meats	Benzo[a]pyrene, Benz[a]anthracene, Chrysene	Up to 450–700 μg/kg in some samples	Iran, Europe, high levels linked to charcoal grilling	Rigi et al. (2025)
2020	Cereal and baby foods	BaP, PAH4 composite	0.5–10 μg/kg	Europe monitoring infant formulas and cereals	Einolghozati et al. (2022)
2021	Smoked fish and seafood	BaP, BaA, BbF, Chr	3–15 μg/kg	European food safety reports, EFSA regulated PAH4	Santonicola et al. (2017)
2022	Oils and fats	PAH4	1–20 μg/kg	Contamination from processing; EU and global limits applied	Palade et al. (2023)
2023	Dried herbs, spices	BaP, PAH4, other PAHs	Up to 30–50 μg/kg	Higher maximum levels authorized for some dried, smoked products	Khalili et al. (2023)
2024	Meat products and ready foods	BaP, PAH4	1–12 μg/kg	EU Regulation No. 2023/915 sets max limits, ALARA principle followed	D'Amore et al. (2025)
2025	Multi-food monitoring	Broad range including BaP and PAH4	Ongoing monitoring; new limits enforced	Increased use of LC-MS/MS, GC-MS; stricter enforcement in EU	Alanazi (2025)

Vegetables and fruits also display variability in PAH levels depending on their cultivation environment and post-harvest handling. Foods grown near urban and industrial sources show higher contamination risks. Surveillance in mining-impacted agricultural zones has documented notable PAH levels in local crops, with health risk assessment studies regularly identifying contamination that can pose mutagenic or carcinogenic threats.

Animal-Based Foods: Meat, Milk, Eggs, Fish, and Seafood

Animal-derived foods are highly susceptible to bioaccumulation of PAHs, especially where feed, water, or air are contaminated. Meat and fish are particularly affected by culinary practices, such as grilling, smoking, and roasting, which induce pyrolytic formation of PAHs. Comprehensive profiling in Iranian markets found the maximum mean of 16 PAHs in smoked fish at 222.7 μg/kg, with lower concentrations in various grilled or fried meat products. Exposures in poultry and dairy (milk, eggs) are typically lower than in smoked meats, but risk remains, especially where feed sources are contaminated or processing involves direct heat. In most national and international studies, PAH levels in animal-based foods are generally below the current EU regulatory limits for the sum of key PAHs (e.g., 30 μg/kg for four principal PAHs, 5 μg/kg for benzo[a]pyrene), though outliers persist near industrial sources or among traditionally processed foods (Palade et al. 2023).

Fish and seafood are exposed to PAHs from aqueous sources and bioaccumulate these compounds in fat-rich tissues. Regional surveillance has often identified elevated levels of PAHs—including benzo[a]pyrene, chrysene, and fluoranthene—in both freshwater and marine species from areas with known pollution inputs. The risk in shellfish and oily fish is particularly pronounced due to their feeding and habitat preferences.

Processed Foods: Smoked, Fried, and Baked Products

The risk of PAH contamination is significantly influenced by cooking and processing techniques; smoked, fried, and baked foods typically exhibit higher PAH levels compared to their raw counterparts. Smoking processes cause the direct deposition of PAH-rich condensates, with concentrations linked to factors such as smoking duration, temperature, fat content, and wood type used. For example, smoked meats and cheeses have shown PAH4 concentrations ranging from 0.15 to 34.65 μg/kg. Fried snacks are another critical concern; deep frying at elevated temperatures using certain oils can generate additional PAHs, both in the oil and the final product. However, monitoring data, including in snack and frying oils, generally report benzo[a]pyrene concentrations below the EU regulatory threshold of 2 μg/kg in tested samples (Khalili et al. 2023).

Baked goods, especially those exposed to direct flame or high temperature for extended periods, have also demonstrated detectible PAH loads. Bread subjected to smoking, traditional baking (in open-flame or wood-fired ovens), or other similar processes often exhibits lower PAH concentrations than smoked meats or cheeses; however, it remains an important vector for human exposure in certain dietary contexts.

Occurrence and Levels of PAHs in Foods

Polycyclic Aromatic Hydrocarbons (PAHs) infiltrate food systems through multifaceted mechanisms, each underpinning their widespread occurrence and persistence in various foods. Deposition from air and soil onto crops plays a key role; airborne PAHs from urban, industrial, or traffic sources settle on the surfaces of fruits, vegetables, and grains, while root uptake and absorption from contaminated soils can further elevate PAH levels within edible portions. Bioaccumulation in fish and livestock is significant, especially in aquatic environments where PAHs may dissolve in water, absorb to particulates, and subsequently concentrate in lipid-rich tissues of aquatic organisms and terrestrial animals consuming contaminated feed (Wang et al. 2021). Endogenous formation of PAHs during high-temperature cooking through roasting, grilling, baking, and frying results from pyrolysis of carbohydrates, lipids, and proteins, generating both parent PAHs and hazardous derivatives like PAH4 (benzo[a]pyrene, chrysene, benzo[b]fluoranthene, benzo[a]anthracene); increased heat, exposure time, and lipid content all strongly enhance this process (Sampaio et al. 2021). Biomagnification through trophic levels occurs as PAHs, once bioaccumulated in lower organisms like algae or zooplankton, are transferred and concentrated within predators higher on the food chain manifesting as elevated PAH burdens in carnivorous fish, seafood, and livestock, thus amplifying the dietary exposure risk to humans (Wang et al. 2025). This interconnected suite of mechanisms demands vigilant monitoring of PAH contamination from soil, air, and water through all food production and processing stages.

Contamination in Food Chain and Food Web

Polycyclic aromatic hydrocarbons (PAHs) enter the food chain through several critical pathways, posing significant concerns for food safety and human health. One primary route is the environmental deposition and absorption by primary producers such as plants and phytoplankton. These organisms accumulate PAHs from contaminated soil, water, and atmospheric particulate matter, effectively introducing these pollutants at the base of both terrestrial and aquatic food webs (Barathan et al. 2024). Contaminated vegetation can further transfer PAHs to herbivorous animals, and through predation, these compounds move upward along trophic levels, accumulating in higher concentrations in predators due to biomagnification (Barathan et al. 2024). This gradual increase of PAHs through the food web is exacerbated by their hydrophobic nature and resistance to metabolism in many species, particularly those higher in the food chain such as certain fish, birds of prey, and marine mammals (Barathan et al. 2024).

In addition to environmental uptake, food contamination by PAHs is also widely influenced by human food processing methods. As mentioned in Fig. 7.4 the various sources of occurrence of PAHs in humans. Techniques involving high temperatures and open flames, like grilling, smoking, roasting, and frying, are known to generate PAHs through pyrolytic reactions, leading to direct contamination of food (Sampaio et al. 2021). These processes release volatile PAHs that can deposit

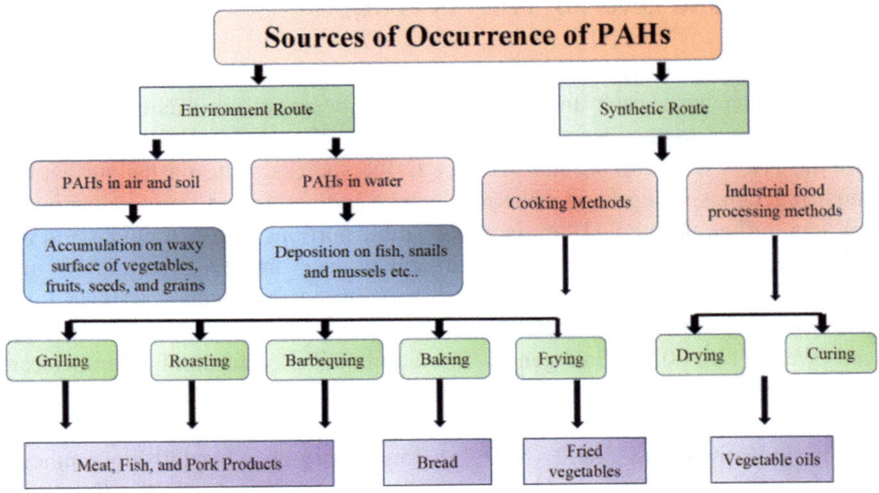

Fig. 7.4 Various sources of occurance of PAHs in humans

onto foods, especially those rich in fats, thereby increasing human dietary exposure (Sampaio et al. 2021). Furthermore, the contamination risk is compounded by the invisible nature of PAHs in food, as consumers cannot detect their presence sensorially, making them an insidious health threat (Sampaio et al. 2021).

Regional and National Surveillance: Case Studies

Regional and national surveillance programs have generated important data sets demonstrating geographic differences in PAH occurrence across food products. Annual national monitoring efforts in Europe (EFSA) compile extensive databases of PAHs detected in a broad spectrum of foods, with regular reporting on compliance with established legal limits and trends in contamination levels. The European Commission's Regulation 835/2011 represents a milestone in standardizing acceptable limits for both individual PAHs and the sum of four principal compounds (benzo[a]pyrene, chrysene, benzo[b]fluoranthene, benzo[a]anthracene) across key food groups. Summarized regional data repeatedly identify highest risk foods as smoked meats and fish, cocoa products, edible oils, and certain cereal-based processed foods (Santonicola et al. 2017).

International data confirm notable excesses in specific regions, often tied to industrial emissions, urbanization, and traditional processing methods. For example, systematic analysis of Ghanaian baby foods and Iranian grilled meats has uncovered routine exceedance of global PAH safety limits in locally produced items. Meanwhile, national health risk assessments in Europe, North America, and East Asia consistently demonstrate lower mean dietary exposures, but identify high-exposure clusters in certain subpopulations (e.g., indigenous groups relying on smoked fish or game, children consuming contaminated cereals).

FAO/WHO, EFSA, and FDA Reports on PAH Contamination

International agencies such as FAO/WHO, EFSA, and FDA have laid out recommendations, risk assessment frameworks, and legal thresholds intended to protect consumers from PAH-related harm. The Joint FAO/WHO Expert Committee on Food Additives, for instance, designates thirteen PAHs as genotoxic and carcinogenic, choosing benzo[a]pyrene as the benchmark for maximum allowable levels due to its strong carcinogenicity. Regulatory evolution, particularly in Europe, now requires reporting and maximum concentration controls for the "PAH4" group, reflecting advances in toxicological understanding. Codex Alimentarius and regional bodies align increasingly on setting PAH maximum residue levels for diverse food commodities, while emphasizing the need for continuous monitoring and corrective controls in food production and processing (Ingenbleek et al. 2019).

Data collected by agencies regularly underscores the close association between PAH levels and environmental pollution, agricultural practices, and regional diets. Annual calls for data submission (e.g., EFSA's continuous collection initiative) now feed into dynamic risk management and evidence-based updating of regulations. The FDA and other authorities in North America focus on dietary guidance and consumer advisories, particularly regarding risky processing methods and vulnerable populations (infants, young children, pregnant women).

Analytical Methods for Detection of PAHs in Food

Analytical determination of Polycyclic Aromatic Hydrocarbons (PAHs) in foods requires precise and robust methodologies due to their low concentrations, complex matrices, and potential co-contaminants. The evolution of analytical techniques encompasses classical extraction and purification methods, sophisticated chromatographic instrumentation, and the emergence of rapid, field-ready screening tools. Regulatory harmonization and standardization under AOAC, ISO, and EU frameworks ensure consistency and accuracy in PAH assessment in diverse food systems (Sereshti et al. 2023).

Sample Preparation Techniques

The initial phase of PAH analysis in foods revolves around efficient extraction and purification to isolate these hydrophobic compounds from the complex food matrix.

Soxhlet Extraction is a traditional approach utilizing continuous solvent reflux over an extended period (often 12–48 h), yielding high recovery for PAHs in solid foods. Its reliability makes it widely used in research, but time, solvent consumption, and labor intensity limit routine deployment (Bitwell et al. 2023).

QuEChERS (Quick, Easy, Cheap, Effective, Rugged, and Safe) is an increasingly popular sample preparation method for PAHs—particularly in plant-based matrices. Typically, food samples undergo acetonitrile extraction, followed by partitioning using salts and cleanup with dispersive solid-phase extraction. QuEChERS optimizes recoveries while reducing solvent and processing times, facilitating high-throughput multi-residue pesticide and PAH analysis (Musarurwa et al. 2019).

Solid-Phase Extraction (SPE) employs customized cartridges packed with sorbents (e.g., silica, C18, polymeric materials) that selectively retain PAHs while allowing unwanted matrix components to be washed away. Recent innovations include mesoporous silica and molecularly imprinted polymers, which greatly improve selectivity, capacity, and reproducibility for trace contaminants such as benzo[a]pyrene or chrysene. SPE can be integrated with other extraction protocols like gel permeation chromatography (GPC) or ultrasound-assisted solvent extraction, enabling even lower quantification limits and better accuracy (Młynarczyk et al. 2024).

Instrumental Chromatographic Techniques

PAH determination relies primarily on chromatographic separation coupled with sensitive detectors.

Gas Chromatography is a cornerstone technique, employing columns that separate PAHs with high resolution. Detection options include GC-MS (Gas Chromatography-Mass Spectrometry): Offers outstanding sensitivity, selectivity, and molecular specificity. GC-MS is preferred for quantifying and confirming identity of individual PAH congeners in complex matrices, with limits of detection often below 0.01 μg/kg. The mass spectrometer's ability to discriminate between isobaric compounds eliminates many false positives from matrix interferences (Wang et al. 2019). GC-FID (Flame Ionization Detection): Favored for routine, high-throughput work due to its broad linear response and simple operation. Although less selective than MS-based detection, GC-FID provides adequate detection for regulatory compliance samples, especially where matrix interferences are minimized through thorough extraction and clean-up (Zhou et al. 2020).

High-Performance Liquid Chromatography (HPLC) is another dominant method in PAH analysis. Key variants are HPLC-FLD (Fluorescence Detector): Capitalizes on the strong native fluorescence of PAHs. HPLC-FLD enables highly sensitive detection (down to ng/kg) but may encounter signal loss or false positives from co-eluting fluorescent compounds. It is often used for official confirmatory testing and compliance monitoring (Vistnes et al. 2022). LC-MS/MS (Liquid Chromatography-Tandem Mass Spectrometry): Expands analytical performance to poorly volatile or thermally labile compounds and enables multiple reaction monitoring for selectivity and multiplexing. LC-MS/MS is increasingly used for multi-class hazard assessment in foods but may require more intensive method validation (Alanazi 2025).

Advanced Methods: Biosensors, Portable Devices, Rapid Screening

Beyond laboratory-based approaches, a new generation of portable and biosensor-based tools is transforming field screening and rapid PAH detection. Biosensors leverage immunoassay platforms—with monoclonal antibodies or aptamers selective for PAH molecules—to produce electrical or optical signals upon target binding. Recent advances include portable, battery-operated biosensors capable of analyzing liquid samples in under 10 min, with results shown to correlate well with sediment and food toxicity levels. KinExA Inline Biosensor and microfluidic chips are examples of semi-automated platforms for near-real-time PAH quantitation. Immunosensors integrate immunoassay approaches into micro-devices, providing high-throughput, low-volume, in-situ analysis options while minimizing use of solvents and sample volumes. These devices are effective for initial screening and selection of samples for further laboratory-based confirmation. Portable screening devices—often microfluidic or chip-based—are now used for in-field environmental and food testing, integrating sample clean-up, detection, and digital data transmission for rapid regulatory decisions (Guo et al. 2025).

Standard Methods and Regulatory Frameworks

Standardization is essential for cross-lab comparability, risk management, and regulatory enforcement.

AOAC International publishes Official Methods of Analysis covering PAH detection in foods and feeds. Appendix F in AOAC guidelines lays out specific performance requirements including precision, accuracy, recovery, selectivity, and robustness—all vital for method validation and regulatory acceptance. ISO Standards (International Organization for Standardization) define performance characteristics for repeatability, reproducibility, and minimum limits of detection. ISO 5725 lays down criteria to quantify and manage measurement uncertainty for PAH analysis globally. EU Regulations, notably Commission Regulation (EU) 2023/915, provide harmonized maximum residue limits and specify analytical procedures for official PAH controls. Regulatory frameworks require regular review and reporting, with updates reflecting advances in toxicological knowledge and analytical practice. The most frequently referenced regulations are EC No 1881/2006 and EC No 333/2007, which detail matrices, limit values, extraction, and cleanup protocols for food safety surveillance.

Toxicological Impacts of PAHs

Polycyclic Aromatic Hydrocarbons (PAHs) are a diverse group of organic pollutants that exert complex toxicological impacts on humans and animals. These effects arise primarily due to mechanisms like mutagenicity, carcinogenicity, endocrine disruption, and extensive interactions with metabolic systems such as the cytochrome P450 (CYP450) enzyme family. Their health consequences span a wide range—cancer, reproductive and developmental toxicity, immunotoxicity, neurotoxicity—with strong support from both animal models and epidemiological studies.

Mechanisms of Toxicity: Mutagenicity, Carcinogenicity, Endocrine Disruption

PAHs are not inherently toxic in their native forms but become hazardous through metabolic activation in organisms. The key toxic mechanisms include the generation of reactive metabolites via enzymatic oxidation (notably by CYP450 enzymes), which leads to DNA adduct formation, chromosomal aberrations, and diverse genetic mutations. Among the most notorious PAHs, benzo[a]pyrene stands out due to its strong carcinogenicity; its epoxide-diol metabolites form covalent bonds with DNA bases, causing significant distortion and mutations that initiate carcinogenesis.

Carcinogenic effects are prominent when PAHs contain at least four fused aromatic rings—a structural prerequisite for metabolic transformation via CYP450 epoxidation. The resulting diol-epoxides, especially benzo[a]pyrene-7,8-diol-9,10-epoxide, are particularly genotoxic and induce frameshift mutations, double-strand breaks, and S-phase arrest in the cell cycle (Fig. 7.1). Beyond genotoxicity and carcinogenicity, PAHs also act as endocrine disruptors. They interfere with hormone synthesis and regulation, affecting reproductive hormones such as progesterone, and contribute to reduced birth weight, preterm birth, and congenital abnormalities (Shimada and Fujii-Kuriyama 2004). These endocrine effects are evident not only in experimental animals but also in human epidemiological studies linking maternal PAH exposure to adverse pregnancy outcomes. The Mechanisms of Toxicity: Mutagenicity, Carcinogenicity, Endocrine Disruption by PAH (Parent Compound's) are mentioned in Fig. 7.5.

Role of CYP450 Metabolism and Reactive Metabolites

The metabolism of PAHs is crucial to their toxicity, centering on CYP450 enzyme-mediated oxidations. After absorption, PAHs are transformed by enzymes such as CYP1A1, CYP1B1, CYP2C19, and CYP3A4, which catalyze the formation of

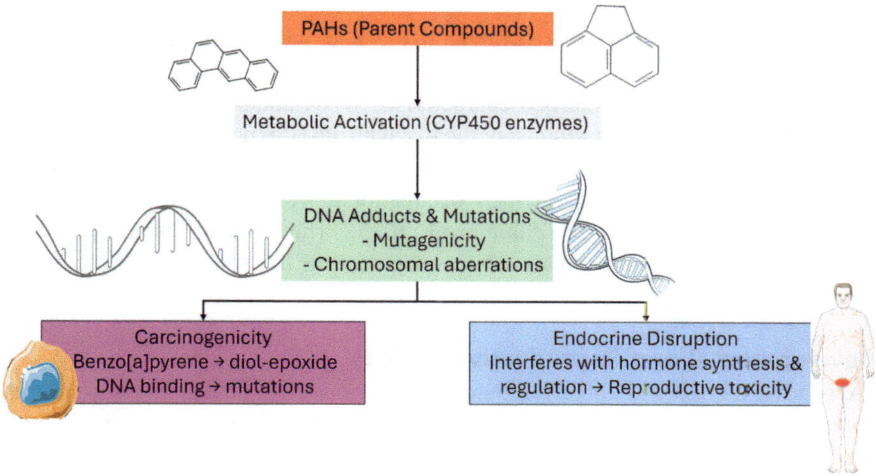

Fig. 7.5 Mechanisms of toxicity: mutagenicity, carcinogenicity, endocrine disruption by PAH (Parent compound)

hydroxylated metabolites, epoxides, and diol-epoxides. These reactive intermediates, notably BaP diol-epoxides, form DNA adducts that are resistant to repair, leading to persistent mutations and increased cancer risk.

The activation of the aryl hydrocarbon receptor (AHR), a ligand-dependent transcription factor, further enhances expression of CYP450 genes and modulates immune, stem cell, and apoptotic functions. This interplay not only accelerates the metabolic activation of PAHs but also impacts cell differentiation and immune regulation, linking genotoxic and immunotoxic responses.

Metabolic detoxification occurs through the formation of more water-soluble derivatives, mediated by glutathione transferases (GSTs) and uridine diphospho-glucuronosyltransferase (UGTs), allowing some elimination via urine or bile. However, bioactivation pathways often dominate, particularly with high-molecular-weight PAHs, resulting in significant genotoxic hazards (Chen et al. 2021).

Health Outcomes: Cancer, Developmental and Reproductive Toxicity, Immunotoxicity, Neurotoxicity

Animal and human biomonitoring data have confirmed a strong and consistent link between food-chain exposure to PAHs and various toxicological outcomes (Sun et al. 2020). PAH-DNA adducts have been detected in populations with high dietary intake of smoked or traditionally processed foods, and numerous biomarker studies reinforce dose-dependent relationships between exposure levels and the risk of disease (Ewa and Danuta 2017). The following Sub-sections detail the principal health effects observed from PAH exposure, including cancer, developmental and reproductive toxicity, immunotoxicity, and neurotoxicity.

Cancer

Chronic exposure to PAHs is strongly associated with multiple forms of cancer, including lung, skin, gastrointestinal, breast, bladder, and esophageal cancers. Molecular studies have identified characteristic DNA adducts and mutational signatures in tumor cells that align specifically with PAH exposure. Epidemiological investigations further support elevated cancer incidence in populations exposed to inhaled or dietary PAHs, such as occupational workers, urban dwellers, and consumers of smoked foods (Zahir et al. 2025). The evidence from epidemiology corroborates a causal relationship, underscoring the role of PAHs as potent environmental carcinogens.

Developmental and Reproductive Toxicity

PAHs adversely affect fetal development and reproductive health. Exposure during pregnancy is linked to reductions in birth weight, congenital malformations, spontaneous abortions, and intrauterine growth restriction. Clear associations emerge from both animal models and human cohort studies between PAH-DNA adduct formation and negative pregnancy outcomes. Furthermore, transplacental PAH exposure has been demonstrated to cause neurodevelopmental impairment, cognitive deficits, and increased behavioral disorders in offspring, highlighting the intergenerational risk (El-Sikaily et al. 2023).

Immunotoxicity

Exposure to PAHs results in immune system dysfunction and systemic inflammation, increasing susceptibility to infections and malignancies. Mechanisms include direct DNA damage to immune cells and altered gene expression mediated by the aryl hydrocarbon receptor (AHR). Experimental and clinical data reveal reductions in lymphocytes, disrupted cytokine profiles, and impaired pathogen defense in both exposed populations and laboratory animal models (Yu et al. 2022).

Neurotoxicity

Neurotoxic effects associated with PAHs are increasingly recognized, notably in models of developmental exposure. Benzo[a]pyrene induces prenatal and early postnatal neurobehavioral deficits in rodents and zebrafish, including cognitive dysfunction, memory impairment, and emotional hyperactivity. Similar deficits are reported in exposed children via epidemiological data. Chronic PAH exposure correlates with neurodevelopmental disorders, attention deficits, and behavioral changes, often manifesting a non-monotonic dose-response signature typical of endocrine disruptors (Das et al. 2016). Mechanistic pathways involve oxidative stress, neurotransmitter dysregulation, and direct neuronal DNA damage, culminating in impaired neurogenesis and synaptic function. These neurotoxic effects underscore the vulnerability of the developing brain to environmental PAHs.

Evidence from Animal Studies and Epidemiological Data

Animal models have played a pivotal role in clarifying the toxicological consequences of PAH exposure. Studies across rodents, fish, and other organisms demonstrate dose-dependent effects on cancer induction, immune suppression, and neurobehavioral abnormalities following dietary, inhalational, or dermal PAH exposures. Knockout models validate the importance of cytochrome P450-mediated metabolism and DNA adduct formation in toxicity pathways. Observations consistently show reproductive and developmental impairments tracking with prenatal or early-life exposure, with effects persisting into adulthood. Key toxicological outcomes related to dietary PAH exposure are summarized in Table 7.3.

Epidemiological investigations likewise suggest strong associations between environmental and occupational PAH exposure and adverse health outcomes—including increased rates of cancer, adverse reproductive events, asthma, and impaired respiratory function. Biomonitoring of PAH metabolites in urine, blood, or tissue confirms exposure in both general and high-risk populations (Wilson et al. 2023).

Emerging Technologies and Future Directions

Recent advances in mitigating, detecting, and managing Polycyclic Aromatic Hydrocarbon (PAH) contamination in food systems are being driven by the integration of nanotechnology, digital intelligence, bioremediation science, innovative food processing, and global policy reform. These approaches promise to enhance food safety, environmental restoration, and consumer awareness far beyond the capabilities of traditional methods.

Advanced Detection Technologies

Nanotechnology-Based Sensors

Nanotechnology-based sensors represent a breakthrough in food safety. PAH detection in oils, fish, and smoked meats rather than detecting microbial and chemical contaminants. Nano-sensors constructed from materials like gold nanoparticles, quantum dots, carbon nanotubes, and graphene exhibit exceptional sensitivity and specificity for detecting microbial and chemical contaminants including PAHs with rapid response times (Awlqadr et al. 2025). These sensors employ optical, electrochemical, and mass-sensitive mechanisms to identify toxins at very low concentrations. Some systems deliver colorimetric or fluorescent signals visible to the naked eye, dramatically reducing the need for complex laboratory analysis. Nanotechnology is also being integrated with biosensing, enabling real-time, multiplex analysis and deployment on smart packaging or at points of production (Nam et al. 2022). Deep learning and pattern recognition approaches help in correlating

Table 7.3 Toxicological outcomes of dietary PAH exposure

Aspect	Evidence from dietary studies	Epidemiological evidence	References
Cancer	Dose-response relationships between PAH exposure and tumors in lung, skin, liver, digestive tract; DNA adduct formation confirmed via CYP450 metabolism; knockout mice models show reduced carcinogenesis when metabolic pathways are disrupted	Increased incidence of lung, skin, breast, gastrointestinal cancers in populations with occupational/environmental PAH exposure; dose-dependent biomarker correlations with cancer risk in at-risk groups	Patel et al. (2020)
Reproductive and developmental toxicity	Maternal exposure causes fetal malformations, low birth weight, developmental neurotoxicity in rodents and fish; persistent effects into adulthood	Associations with adverse pregnancy outcomes including preterm birth, low birth weight, congenital defects; neurodevelopmental delays in exposed children documented	El-Sikaily et al. (2023)
Immunotoxicity	Suppression of B and T cell function, lymphoid tissue apoptosis, altered cytokine profiles; immune system oxidative stress mediated by PAH metabolites	Increased infections, immune dysfunction, hypersensitivity, and altered immune markers in exposed human populations	Olasehinde and Olaniran (2022)
Neurotoxicity	Behavioral and cognitive deficits in rodent and zebrafish models; oxidative stress and neurotransmitter disruption; prenatal exposure linked with neurodevelopmental abnormalities	Epidemiological data suggest associations with cognitive and behavioral disorders in children from PAH-exposed areas; prenatal and early-life exposure critical	Tartaglione et al. (2023)
Metabolism and mechanism	CYP450 enzymes metabolize PAHs into genotoxic intermediates; DNA adduct formation causes mutagenesis; AHR receptor activation modulates toxicity pathways	Biomonitoring in humans (urine, blood) confirms exposure; polymorphisms in metabolic enzymes correlate with differential susceptibility in populations	Smith et al. (2022)

contamination signatures, automating anomaly detection, and even in optimizing sensor efficacy through algorithmic calibration (Ding et al. 2023). These interventions allow more targeted surveillance, recall management, and resource allocation in food safety systems.

Bioremediation and Microbial Degradation of PAHs

Bioremediation especially microbial degradation offers eco-efficient, in situ solutions for managing PAH contamination. Specific bacterial strains (e.g., Pseudomonas aeruginosa, Mycobacterium, Sphingomonas) and fungal consortia break down high-molecular-weight PAHs into less hazardous substances. The processes are improved via biostimulation (optimizing conditions for indigenous microbes), bioaugmentation (adding specialized degraders), and utilization of biosurfactant-producing bacteria to increase solubility and bioavailability of persistent PAHs. Consortia approaches improve metabolic versatility, accelerating ring-hydroxylation and enhancing gene expressions (e.g., NidA, nahAC enzymes), resulting in higher degradation rates (Gupte et al. 2016). The advantages of bioremediation are its environmental friendliness, minimal disturbance to contaminated sites, and lower operational costs. However, degradation rates can vary due to abiotic factors temperature, pH, oxygen availability. Research priorities include optimizing microbial consortia, field-scale trials, and genetic engineering approaches for more consistent application. Aquaculture ponds, agricultural soils where PAHs bioaccumulate into crops and fish. Recent studies have also explored low-cost adsorbent-based removal of PAHs from aqueous systems to limit their entry into the food chain. For example, activated carbon synthesized from waste banana peel (surface area > 900 m^2/g) demonstrated efficient adsorption of PAHs, with equilibrium reached in 80 min and the process found to be spontaneous and endothermic (Gupta and Gupta 2016). Adsorption data fitted better to the Freundlich model, and regeneration of the adsorbent was successfully achieved using NaOH–ethanol mixtures, confirming its potential for reuse. Such approaches are inexpensive, sustainable, and scalable for mitigating PAH contamination in agricultural and aquaculture water sources, thereby indirectly protecting the food chain.

Novel Food Processing Techniques to Minimize PAHs

Recent innovations in food processing have demonstrated substantial reductions in PAH and other hazardous compound formation: Air Frying: Uses forced hot air rather than oil, significantly reducing the temperature and formation of PAHs during cooking. Air frying preserves quality while lowering carcinogenic risks. Sous-Vide: Involves vacuum-sealing food and cooking at low, tightly regulated temperatures, reducing contact with combustion gases and minimizing PAH formation. Studies confirm sous-vide yields dishes that are juicier and tenderer, with lower levels of process-induced contaminants. Vacuum Frying: Lower pressure frying allows for dehydration and cooking at temperatures below those that typically induce PAH genesis. Vacuum frying and microwave vacuum frying both dramatically decrease acrylamide and PAHs compared to conventional frying, while enhancing color, flavor, and nutritional retention (Yoon et al. 2024). These processing innovations are increasingly incorporated into commercial food production and culinary practices, providing industry and consumers with practical routes to safer foods.

Policy Gaps, Global Monitoring Needs, and Consumer Awareness Programs

Despite technological and scientific advances, major policy gaps remain in harmonizing international PAH control and mitigation efforts. Many national surveillance programs lack integration, consistent reporting standards, or broad-based biomonitoring frameworks for emerging contaminants. There is an urgent need for collaborative human biomonitoring programs, unified chemical priority lists, and international best practices among public health authorities like FAO, WHO, and the EU (de Paula Nunes et al. 2025) Consumer awareness also requires strengthening. Consumer education on safer cooking practices (air frying, avoiding charring) can reduce exposure. Surveys highlight insufficient knowledge about persistent pollutants, chemical safety, and risk minimisation. Tailored awareness initiatives, educational campaigns, and transparent labelling (enabled by smart sensors and digital traceability) are vital to empowering consumers and closing risk communication gaps (Rigi et al. 2025). Engaging farmers, producers, and consumers in participatory surveillance and hygiene practices can further reduce contamination risks from farm to table (Sharma et al. 2025).

Conclusion

The global food chain is at risk from persistent pollutants known as polycyclic aromatic hydrocarbons (PAHs), which are released by environmental pollution and high-temperature food processing. Compared to EU regulation limits of 2 and 30 μg/kg for PAH4, benzo[a]pyrene levels in grilled meats, smoked fish, and cereals routinely exceed international safety guidelines, with reported values as high as 200 μg/kg, 222.7 μg/kg, and 2.5–276.7 μg/kg, respectively. These excesses highlight the necessity of addressing PAH contamination in diets derived from plants and animals. DNA adduct formation, mutagenicity, carcinogenicity, and endocrine disruption are all brought on by metabolic activation. Lung, skin, gastrointestinal, reproductive, developmental, immunological, and neurological disorders are linked to long-term food exposure. People who eat grilled and smoked food run the risk of developing chronic health problems.

Potential technologies for reducing and detecting PAH contamination include microbial bioremediation, machine learning-assisted surveillance, sous-vide and vacuum frying, and biosensors based on nanotechnology. There is still a lack of worldwide harmonisation of regulatory limitations, thorough food monitoring, and consumer education regarding safer cooking practices. To lower the risks of PAHs, an integrated approach is required, which includes: (i) bolstering surveillance programs with cutting-edge analytical tools; (ii) encouraging safer food processing methods that lower the formation of PAHs; (iii) harmonising international standards and enhancing enforcement; and (iv) increasing consumer awareness through public health campaigns and transparent labelling. To safeguard the food supply and lower dietary PAH exposure worldwide, food scientists, toxicologists, legislators, and industry stakeholders must collaborate.

References

Abdel-Shafy, H.I., Mansour, M.S.M.: A review on polycyclic aromatic hydrocarbons: source, environmental impact, effect on human health and remediation. Egypt. J. Pet. **25**(1), 107–123 (2016). Available at: https://doi.org/10.1016/j.ejpe.2015.03.011

Alanazi, S.: Recent advances in liquid chromatography–mass spectrometry (LC–MS) applications in biological and applied sciences. Anal. Sci. Adv. **6**(1) (2025). Available at: https://doi.org/10.1002/ansa.70024

Awlqadr, F.H., et al.: Emerging trends in nano-sensors: a new frontier in food safety and quality assurance. Heliyon **11**(1), e41181 (2025). Available at: https://doi.org/10.1016/j.heliyon.2024.e41181

Berríos-Rolón, P.J., Cotto, M.C., Márquez, F.: Polycyclic aromatic hydrocarbons (PAHs) in freshwater systems: a comprehensive review of sources, distribution, and ecotoxicological impacts. Toxics **13**(4), 321 (2025). Available at: https://doi.org/10.3390/toxics13040321

Bharathan, N.K., Mattheyses, A.L., Kowalczyk, A.P.: The desmosome comes into focus. J. Cell Biol. **223**(9), e202404120 (2024). https://doi.org/10.1083/jcb.202404120. IF: 6.4 Q1. Epub 2024 Aug 9. PMID: 39120608; PMCID: PMC11317759

Bitwell, C., et al.: A review of modern and conventional extraction techniques and their applications for extracting phytochemicals from plants. Sci. Afr. **19**, e01585 (2023). Available at: https://doi.org/10.1016/j.sciaf.2023.e01585

Chen, C., et al.: Identification of structural properties influencing the metabolism of polycyclic aromatic hydrocarbons by cytochrome P450 1A1. Sci. Total. Environ. **758**, 143997 (2021). Available at: https://doi.org/10.1016/j.scitotenv.2020.143997

Choi, J.W., et al.: Toxicokinetic analyses of naphthalene, fluorene, phenanthrene, and pyrene in humans after single oral administration. Sci. Total. Environ. **870**, 161899 (2023). Available at: https://doi.org/10.1016/j.scitotenv.2023.161899

D'Amore, T., Smaoui, S., Varzakas, T.: Chemical food safety in europe under the spotlight: principles, regulatory framework and roadmap for future directions. Foods **14**(9), 1628 (2025). Available at: https://doi.org/10.3390/foods14091628

Das, S.K., Patel, B., Patri, M.: Neurotoxic effect of benzo[a]pyrene and its possible association with 6-hydroxydopamine induced neurobehavioral changes during early adolescence period in rats. J. Toxicol. 1–7 (2016). Available at: https://doi.org/10.1155/2016/8606410

de Paula Nunes, E., Abou Dehn Pestana, B., Pereira, B.B.: Human biomonitoring and environmental health: a critical review of global exposure patterns, methodological challenges and research gaps. J. Toxicol. Environ. Health Part B 1–19 (2025). Available at: https://doi.org/10.1080/10937404.2025.2529845

Ding, H., et al.: The application of artificial intelligence and big data in the food industry. Foods **12**(24), 4511 (2023). Available at: https://doi.org/10.3390/foods12244511

Einolghozati, M., et al.: Evaluation of polycyclic aromatic hydrocarbons (PAHs) in processed cereals: a meta-analysis study, systematic review, and health risk assessment. Heliyon **8**(12), e12168 (2022). Available at: https://doi.org/10.1016/j.heliyon.2022.e12168

Eldos, H.I., et al.: Recent advances in the treatment of PAHs in the environment: application of nanomaterial-based technologies. Arab. J. Chem. **15**(7), 103918 (2022). Available at: https://doi.org/10.1016/j.arabjc.2022.103918

El-Sikaily, A., et al.: Impacts of PAH accumulation on reproductive hormones, indices of oxidative stress and BPDE-albumin adduct in women with recurrent pregnancy loss. Toxicol. Res. **39**(3), 517–531 (2023). Available at: https://doi.org/10.1007/s43188-023-00181-5

Ewa, B., Danuta, M.-Š.: Polycyclic aromatic hydrocarbons and PAH-related DNA adducts. J. Appl. Genet. **58**(3), 321–330 (2017). Available at: https://doi.org/10.1007/s13353-016-0380-3

Femi-Oloye, O.P., et al.: An assessment of polycyclic aromatic hydrocarbons using estimation programs. Toxics **12**(8), 592 (2024). Available at: https://doi.org/10.3390/toxics12080592

Gundlapalli, M., et al.: Biodegradation of low molecular weight polycyclic aromatic hydrocarbons in soil: insights into bacterial activities and bioremediation techniques. Sustain. Chem. Environ. **7**, 100146 (2024). Available at: https://doi.org/10.1016/j.scenv.2024.100146

Guo, Z., et al.: Advances in research on novel technologies for the detection of exogenous contaminants in traditional Chinese medicine. Front. Pharmacol. 16 (2025). Available at: https://doi.org/10.3389/fphar.2025.1658241

Gupta, H., Gupta, B.: Adsorption of polycyclic aromatic hydrocarbons on banana peel activated carbon. Desalination Water Treat. **57**(20), 9498–9509 (2016). Available at: https://doi.org/10.1080/19443994.2015.1029007

Gupte, A., et al.: Bioremediation of polycyclic aromatic hydrocarbon (PAHs): a perspective. Open Biotechnol. J. **10**(1), 363–378 (2016). Available at: https://doi.org/10.2174/1874070701610010363

Halfadji, A., et al.: Human health assessment of sixteen priority polycyclic aromatic hydrocarbons in contaminated soils of Northwestern Algeria. J. Health Pollut. **11**(31) (2021). Available at: https://doi.org/10.5696/2156-9614-11.31.210914

Han, F., et al.: Modeling polycyclic aromatic hydrocarbons in India: seasonal variations, sources and associated health risks. Environ. Res. **212**, 113466 (2022). Available at: https://doi.org/10.1016/j.envres.2022.113466

Honda, M., Suzuki, N.: Toxicities of polycyclic aromatic hydrocarbons for aquatic animals. Int. J. Environ. Res. Public Health **17**(4), 1363 (2020). Available at: https://doi.org/10.3390/ijerph17041363

Ingenbleek, L., et al.: Polycyclic aromatic hydrocarbons in foods from the first regional total diet study in Sub-Saharan Africa: contamination profile and occurrence data. Food Control **103**, 133–144 (2019). Available at: https://doi.org/10.1016/j.foodcont.2019.04.006

Kanaly, R.A., Harayama, S.: Biodegradation of high-molecular-weight polycyclic aromatic hydrocarbons by bacteria. J. Bacteriol. **82**(8), 2059–67 (2000). https://doi.org/10.1128/JB.182.8.2059-2067. 2000IF: 3.0 Q3. PMID: 10735846; PMCID: PMC111252

Khalili, F., et al.: Polycyclic aromatic hydrocarbons (PAHs) in meat, poultry, fish and related product samples of Iran: a risk assessment study. J. Environ. Health Sci. Eng. **21**(1), 215–224 (2023). Available at: https://doi.org/10.1007/s40201-023-00854-1

Lawal, A.T.: Polycyclic aromatic hydrocarbons. A review. Cogent Environ. Sci. **3**(1), 1339841 (2017). Available at: https://doi.org/10.1080/23311843.2017.1339841

Młynarczyk, M., Belka, M., Hewelt-Belka, W.: Novel materials and approaches for solid-phase (micro) extraction in LC-MS-based metabolomics. TrAC Trends Anal. Chem. **180**, 117941 (2024). Available at: https://doi.org/10.1016/j.trac.2024.117941

Mohammed, H.A., Sulaiman, G.M., Al-Saffar, A.Z., Abomughaid, M.M., Hussein, N.N., Salih, Z.T., Hadi, N.A., Mohsin, M.H., Al-Kuraishy, H.M., Khan, R.A., Ismail, A.: Roles of foods and related components: an overview on cancer causatives, and plausible preventions. Appl. Food Res. **5**(2), 101132 (2025). https://doi.org/10.1016/j.afres.2025.101132

Montano, L., et al.: Polycyclic aromatic hydrocarbons (PAHs) in the environment: occupational exposure, health risks and fertility implications. Toxics **13**(3), 151 (2025). Available at: https://doi.org/10.3390/toxics13030151

Musarurwa, H., et al.: Recent developments and applications of QuEChERS based techniques on food samples during pesticide analysis. J. Food Compos. Anal. **84**, 103314 (2019). Available at: https://doi.org/10.1016/j.jfca.2019.103314

Nam, N.N., et al.: Recent progress in nanotechnology-based approaches for food monitoring. Nanomaterials **12**(23), 4116 (2022). Available at: https://doi.org/10.3390/nano12234116

Olasehinde, T.A., Olaniran, A.O.: Neurotoxicity of polycyclic aromatic hydrocarbons: a systematic mapping and review of neuropathological mechanisms. Toxics **10**(8), 417 (2022). Available at: https://doi.org/10.3390/toxics10080417

Palade, L.M., et al.: Polycyclic aromatic hydrocarbon occurrence and formation in processed meat, edible oils, and cereal-derived products: a review. Appl. Sci. **13**(13), 7877 (2023). Available at: https://doi.org/10.3390/app13137877

Paris, A., et al.: Polycyclic aromatic hydrocarbons in fruits and vegetables: origin, analysis, and occurrence. Environ. Pollut. **234**, 96–106 (2018). Available at: https://doi.org/10.1016/j.envpol.2017.11.028

Patel, A.B., et al.: Polycyclic aromatic hydrocarbons: sources, toxicity, and remediation approaches. Front. Microbiol. **11** (2020). Available at: https://doi.org/10.3389/fmicb.2020.562813

Rigi, P., et al.: Health risk assessment of polycyclic aromatic hydrocarbon compounds (PAHs) in grilled meats in Zahedan city of Iran. Sci. Rep. **15**(1), 24267 (2025). Available at: https://doi.org/10.1038/s41598-025-03807-w

Sahoo, B.M., et al.: Polyaromatic hydrocarbons (PAHs): structures, synthesis and their biological profile. Curr. Org. Synth. **17**(8), 625–640 (2020). Available at: https://doi.org/10.2174/1570179417666200713182441

Sakshi, Haritash, A.K.: A comprehensive review of metabolic and genomic aspects of PAH-degradation. Arch. Microbiol. **202**(8), 2033–2058 (2020). Available at: https://doi.org/10.1007/s00203-020-01929-5

Sampaio, G.R., et al.: Polycyclic aromatic hydrocarbons in foods: biological effects, legislation, occurrence, analytical methods, and strategies to reduce their formation. Int. J. Mol. Sci. **22**(11), 6010 (2021). Available at: https://doi.org/10.3390/ijms22116010

Santonicola, S., et al.: Study on the occurrence of polycyclic aromatic hydrocarbons in milk and meat/fish based baby food available in Italy. Chemosphere **184**, 467–472 (2017). Available at: https://doi.org/10.1016/j.chemosphere.2017.06.017

Sereshti, H., et al.: Analysis of polycyclic aromatic hydrocarbons using magnetic three-dimensional graphene solid-phase extraction coupled with gas chromatography–mass spectrometry. Separations **10**(11), 564 (2023). Available at: https://doi.org/10.3390/separations10110564

Sharma, B.M., et al.: Complementing global chemicals management through shaping consumer behavior. iScience **28**(6), 112700 (2025). Available at: https://doi.org/10.1016/j.isci.2025.112700

Shimada, T., Fujii-Kuriyama, Y.: Metabolic activation of polycyclic aromatic hydrocarbons to carcinogens by cytochromes P450 1A1 and 1B1. Cancer Sci. **95**(1), 1–6 (2004). Available at: https://doi.org/10.1111/j.1349-7006.2004.tb03162.x

Smith, J.N., Gaither, K.A., Pande, P.: Competitive metabolism of polycyclic aromatic hydrocarbons (PAHs): an assessment using in vitro metabolism and physiologically based pharmacokinetic (PBPK) modeling. Int. J. Environ. Res. Public Health **19**(14), 8266 (2022). Available at: https://doi.org/10.3390/ijerph19148266

Soursou, V., Campo, J., Picó, Y.: Revisiting the analytical determination of PAHs in environmental samples: an update on recent advances. Trends Environ. Anal. Chem. **37**, e00195 (2023). Available at: https://doi.org/10.1016/j.teac.2023.e00195

Sun, J., et al.: Biomonitoring of polycyclic aromatic hydrocarbons (PAHs) from Manila clam Ruditapes philippinarum in Laizhou, Rushan and Jiaozhou, bays of China, and investigation of its relationship with human carcinogenic risk. Mar. Pollut. Bull. **160**, 111556 (2020). Available at: https://doi.org/10.1016/j.marpolbul.2020.111556

Tartaglione, A.M., Racca, A., Ricceri, L.: Developmental exposure to polycyclic aromatic hydrocarbons (PAHs): focus on benzo[a]pyrene neurotoxicity. Reprod. Toxicol. **119**, 108394 (2023). Available at: https://doi.org/10.1016/j.reprotox.2023.108394

Uiuiu, P., et al.: Portable smoking ovens: what are the PAH levels in grilled and smoked rainbow trout? Fishes **10**(2), 82 (2025). Available at: https://doi.org/10.3390/fishes10020082

Vistnes, H., et al.: The determination of polycyclic aromatic hydrocarbons (PAHs) with HPLC-DAD-FLD and GC-MS techniques in the dissolved and particulate phase of road-tunnel wash water: a case study for cross-array comparisons and applications. Toxics **10**(7), 399 (2022). Available at: https://doi.org/10.3390/toxics10070399

Wang, S.-W., et al.: Determination of polycyclic aromatic hydrocarbons (PAHs) in cosmetic products by gas chromatography-tandem mass spectrometry. J. Food Drug Anal. **27**(3), 815–824 (2019). Available at: https://doi.org/10.1016/j.jfda.2019.01.003

Wang, H., et al.: Contribution of dietary uptake to PAH bioaccumulation in a simplified pelagic food chain: modeling the influences of continuous vs intermittent feeding in Zooplankton and fish. Environ. Sci. Technol. **55**(3), 1930–1940 (2021). Available at: https://doi.org/10.1021/acs.est.0c06970

Wang, H., et al.: Bioaccumulation and potential human health risks of PAHs in marine food webs: a trophic transfer perspective. J. Hazard. Mater. **485**, 136946 (2025). Available at: https://doi.org/10.1016/j.jhazmat.2024.136946

Wilson, L.B., et al.: Advances in PAH mixture toxicology enabled by zebrafish. Curr. Opin. Toxicol. **34**, 100392 (2023). Available at: https://doi.org/10.1016/j.cotox.2023.100392

Yoon, H., et al.: Effect of air-frying on formation and correlations of polycyclic aromatic hydrocarbons, acrylamide, and heterocyclic aromatic amines in foods and risk assessment. Food Sci. Biotechnol. **33**(10), 2399–2415 (2024). Available at: https://doi.org/10.1007/s10068-024-01639-4

Yu, Y., Jin, H., Lu, Q.: Effect of polycyclic aromatic hydrocarbons on immunity. J. Transl. Autoimmun. **5**, 100177 (2022). Available at: https://doi.org/10.1016/j.jtauto.2022.100177

Zahir, A., Ge, Z., Khan, I.A.: Public health risks associated with food process contaminants—a review. J. Food Prot. **88**(2), 100426 (2025). Available at: https://doi.org/10.1016/j.jfp.2024.100426

Zhou, J., et al.: A gas chromatography-flame ionization detection method for direct and rapid determination of small molecule volatile organic compounds in aqueous phase. 3 Biotech **10**(12), 520 (2020). Available at: https://doi.org/10.1007/s13205-020-02523-8

Toxicity of Polycyclic Aromatic Hydrocarbons in the Environment and Its Deteriorating Effects on Living Organisms

8

Rahul Maddheshiya, K. P. Singh, Soniya Dhiman, and Himanshu Gupta

Abstract

Polycyclic aromatic hydrocarbons (PAHs) are greatly concerned environmental pollutant. These get accumulated in the environment mainly due to incomplete combustion of oil, coal, wooden biomass, and petroleum products. Anthropogenic activities are also very responsible for PAH addition such as coal burning, extremely high heating from society towers, coal tar roads, asphalt production, liquefying plants, aluminum and coke production. PAH are known to cause carcinogenic and teratogenic effect in human. These chemicals have been also reported to effect immunity, neuronal integrity and activation of reactive oxygen species at large. PAHs also show similar effects in living organisms including pisces and invertebrates such as sea urchin and earthworms. Several biological remediation methods including algal, fungal, bacterial, and biostimulant has successfully employed for the eradication of higher PAH concentration in the environment or contaminated sites. The main aim of chapter is to highlight the impact of PAHs on environment and its magnitude of invertebrates, fishes as well as human beings.

R. Maddheshiya
IFTM University, Moradabad, Uttar Pradesh, India

K. P. Singh
Department of Zoology, University of Allahabad, Prayagraj, India

S. Dhiman
Helmholtz-Zentrum Dresden-Rossendorf, Helmholtz Institute Freiberg for Resource Technology, Dresden, Germany

H. Gupta (✉)
Department of Chemistry, School of Sciences, IFTM University, Moradabad, Uttar Pradesh, India
e-mail: hims.research@gmail.com

H. Gupta et al. (eds.), *Ubiquitous Polycyclic Aromatic Hydrocarbon Contamination*, Emerging Contaminants and Associated Treatment Technologies,
https://doi.org/10.1007/978-3-032-11764-9_8

Keywords

Polycyclic aromatic hydrocarbons (PAHs) • Carcinogenic • Environment • Impact • Human beings • Bioremediation

Introduction

Polycyclic aromatic hydrocarbons (PAHs) are burning issue because of their occurrence and negative impacts onto the environment and health of the animals including human beings (Sakshi and Haritash 2020; Gupta and Gupta 2016; Kumar and Gupta 2020). PAH is a group of hazardous chemicals made up of carbon and hydrogen atoms only and these atoms form two or more hydrocarbon rings fused together in linear or angular manner. These are light yellow or whitish amorphous powdery form, and this powdery state renders easy persistent of PAHs in the environment in particulate form (Faraji Ghasemi et al. 2020; Goudarzi et al. 2017a). PAHs also play vital role in our daily routine such as pyrene is employed for pigment manufacturing industries; fluorene is used as pesticides, pigments, dyes, and in pharmaceuticals; Phenanthrene is used for pesticidal and resin preparation; fluoranthene is widely taken into account for the manufacturing of agriculture related chemicals, dyes and pharmacy companies; acenaphthene is also used in agricultural as well as pharmaceuticals and plastics companies; anthracene acts as a diluent for wood preservatives (Honda and Suzuki 2020).

Generally, three types of processes are main victim of releasing PAHs in the environment i.e. biological, pyrogenic, and petrogenic. Heating the biomass in anaerobic condition is called pyrolysis. In the pyrolysis reaction pyrogenic PAHs are evolved. For example, formation of coal tar and coke occurs due to the high temperature burning of wooden biomass or thermal conversion of petroleum residues into lighter hydrocarbons. Pyrogenic PAHs are more concentrated in urban areas. Similarly, PAHs evolved at the period of crude oil processing and maturing are known as petrogenic. Oceans and freshwater oil spills by tankers, motor vehicle oils, gasoline are rich source of petrogenic PAHs. Biological PAHs accumulation in the environment may be attributed to the volcanic eruption, forest fires, algal synthesis (Abdel-Shafy and Mansour 2016a, b; Gupta and Kumar 2020).

These pollutants are less soluble in aqueous media and generally semi volatile compounds showing low vapor saturation pressure. With increasing molecular weight, PAHs show linearity in melting and boiling points but their solubility in water and vapor pressure gradually decreases. PAHs having tetra ringed structure are generally referred to as mid-ring PAHs and mostly exist in gaseous forms such as fluoranthene (FLa) and chrysene (CHr) (Kadri et al. 2017). Penta or hexa aromatic ringed PAHs are mostly particulate form such as benzo(a)pyrene (BaP), benzo(g,h,i) perylene(BghiP), and indeno(1,2,3-cd)pyrene(IcdP) etc. According to Sun et al. (2021), a total of 16 PAHs are in the dangerous pollutant list of USEPA.

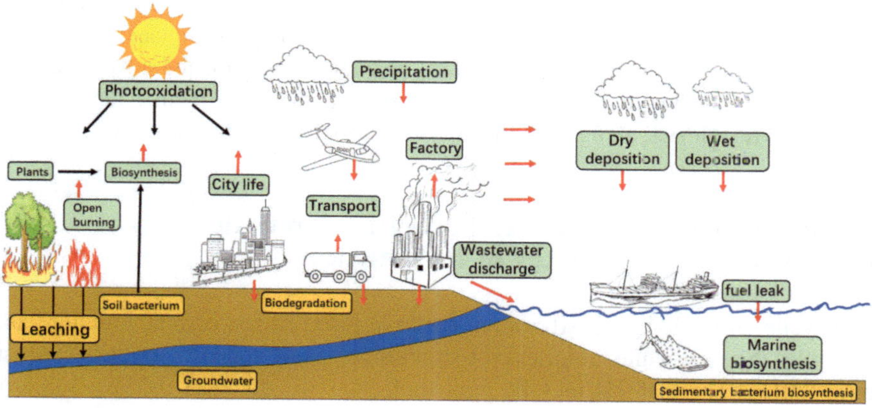

Fig. 8.1 Sources of PAHs accumulation in the environment and their transport means (Red arrow shows the direction of PAH). (Reprinted from original article by Sun et al. 2021, after taking permission from Elsevier)

Indulgence of PAHs in environment may be generally by two ways: biosynthesis of microbes and plants and secondly through burning of forests and grassland, eruption of volcanoes (Itoh et al. 2017; Shen et al. 2020). Biosynthetic pathway route of PAH formation is nearly negligible. Pyrolysis, petroleum products, black diamond, organic polymeric compounds, and wood and industries are severely increasing the PAH load on the environment (Abdel-Shafy and Mansour 2016a, b; Goudarzi et al. 2018). Due to the human interference the leakage of petroleum products in sea and land, discharge of coal tar and oil, carbon ink in the printers, asphalt, shale oil, municipal wastage, agricultural additives through chemical insecticides, pesticides and mining industries also add significant amounts of PAHs in the environment (Yang and Chen 2004; Hashemzadeh et al. 2019; Gupta and Dhiman 2023, 2024). They enter the environment through various receptors as large water bodies, rain, precipitation, and air (Fig. 8.1).

PAHs are known to be included in a special class of toxic substances (Jiang et al. 2014). The toxicity of PAHs shows positive response as the number of rings in it increases and benzo[a]pyrene (BaP) is considered highly toxic PAH (Simpson et al. 1996; Gupta and Gupta, 2015). Due to the complication in the ill effects caused by PAHs, a vast effort is pressured for understanding the health-related issues both in animals and human. Most of the PAHs are photosensitizers and therefore much of the emphasis has been drawn in studying the photon induced toxicity. Toxicity due to polycyclic aromatic hydrocarbons has been studied related to endocrinological disruption, reproductive biology and genotoxic effects (Barron et al. 2004). PAH has been also reported to cause toxicity related to liver, bone, embryological in pisces and invertebrates such as nibbler fishes, Chinese minnows, gold fishes, sea urchins, earthworms etc. (Cai et al. 2019; Nam et al. 2015; Ses et al. 2009; Zha et al. 2017).

Several measures have been implemented by researchers for the control and spread of PAH such as biological treatment with the help of archaea, algae, bacteria, fungi, and use of biostimulant. This chapter focuses on the adverse effects of PAHs on the health of both animals and human being. It also includes the concluded health risk along with the combined microplastics and PAHs.

Physico-chemical Properties Polycyclic Aromatic Hydrocarbons and Routes of Its Emission

Polycyclic aromatic hydrocarbons are multiple fused ringed organic structure and generally they are white, colorless, or slight dull yellow solid compounds (Suman et al. 2016). These PAHs may be light molecular weight PAH (LMW; 2–3 aromatic ringed; gaseous in nature); high molecular weight PAHs (HMW; contains 4 or more aromatic ringed structure; particulate form) (Lee and Vu 2010). Furthermore, Gupte et al. (2016) differentiated them into alternate (fused 6 carbon ring of benzene) and non-alternate PAHs (contains 6 carbon benzene rings along with ring with less than 6 carbon). PAHs are usually resistant towards nucleophilic reactions because of the presence π electrons arranged around aromatic carbon rings (Haritash and Kaushik 2009). These are very less soluble in water with usually very high boiling and melting points (Lee Vu 2010). As the molecular weight of the PAH increases, they become more water resistant and recalcitrant (Okere and Semple 2012).

Polycyclic aromatic hydrocarbons have been categorized into 3 types on the basis of their origin: (1) Pyrogenic: PAHs that are formed due to incomplete burning of organic substances at 350–1200 °C unintentionally. Sometimes intentional distillation of coal to coke and coal tar can also be a source of pyrogenic PAH emission. (2) Petrogenic: These PAHs are correlated to petroleum by-products or its widespread usage and leakage. (3) Biogenic: These PAHs can be accumulated in the environment by biological means such as phytoplanktons, microbes, algae, and during slow conversion rate of organic substances (Abdel-Shafy and Mansour 2016a, b; Mojiri et al. 2019). PAHs are added in the environment through two resources (Fig. 8.2): (1) Natural resource includes fire caused by lightening, forest fires, volcanic eruptions. (2) Human induced includes burning crops, industrial emissions etc. Industrial emission of PAH is remarkable because it adds highest amount of it to the environment such as by incomplete combustion during the waste incineration, cement-rubber tyre-and dye manufacturing, steel and iron production, power production, chemical insecticides, meat cooking, diesel engines, automobiles, coal gasification and from oil refineries (Srogi 2007; Ravindra et al. 2008; Abdel-Shafy and Mansour 2016a, b; Gupte et al. 2016; Mojiri et al. 2019; Gupta and Singh 2018; Gupta 2018). (3) Domestic emission may adhere to coal, kerosene oil, wood used for cooking, garbage burning etc. (Johnsen and Karlson 2007). Environment is most polluted with PAH in winters thereafter by spring,

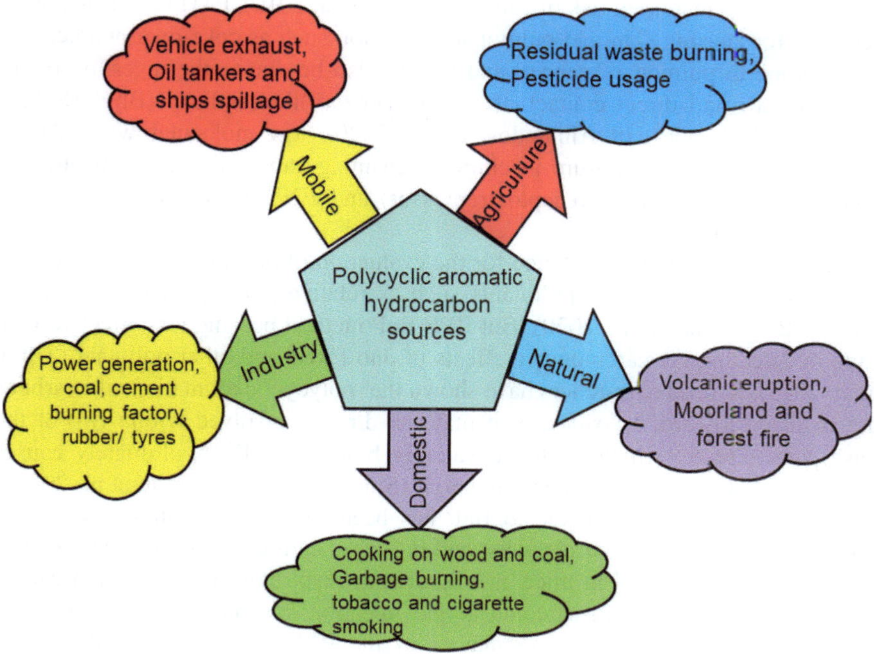

Fig. 8.2 Different modes of addition of polycyclic aromatic hydrocarbons to the environment

autumn, and summer. PAH is higher in winter, and spring is due to fossil fuel burning, use of heaters in residential places, reduced photo-degradation (Miura et al. 2019).

Adverse Effects of Polycyclic Aromatic Hydrocarbons on Human and Other Animals

Upon contamination with PAHs various adverse effects can be observed in both animals and human beings such as teratogenic effect, carcinogenic effects, reproductive and endocrinological derailing pathways, immuno- and neurological toxicity. Based on the complexity of the PAHs ill effects, we have summarized the health-related issues in both animals and the human beings.

Carcinogenic and Teratogenic Effects of PAHs

The common route of PAHs to human body may include smoking, occupational routine works such as workers in the cast iron factory and cold drink manufacturing hub, cooking food on wood, and gases from automobiles (St. Helen et al. 2012). PAHs are lipophilic in nature, and therefore it is easily accessible in animal body further leads to increased risk of cancer (Campo et al. 2010). PAH has been already reported for causing skin and pulmonary infection such as coke oven workers are more prone to pulmonary cancer, and it may also be due to the ingestion of the fumes (road) and direct contact of skin to benzo(b)fluoranthene (BbF) (Soltani et al. 2015). In view of Alhamdow et al. (2020), low molecular wt. PAHs as higher fluorine (FL) exposure in chimney cleaning causes lower methylation of AHRR (aryl hydrocarbon receptor repressor) and F2RL3 (coagulation factor II receptor-like 3).

Earlier a method was attributed for the evaluation of primary risk of cancer due to PAHs mixed with other pollutants such as relative potency factor (RPF) that include BaP (Chang et al. 2020). But RPF-BaP method became less effective with time in detecting the carcinogenic effects of duo PAHs and other pollutant (Tilton et al. 2015). Several researchers have shown that polycyclic aromatic hydrocarbons are oxidized to form epoxides, or it may get directly involved within two of the base pairs directly which further covalently binds with DNA ultimately causes mutation or aberrations in nuclear material (Sun et al. 2020b; Xiong et al. 2020).

The carcinogenic effect due to BaP has been reported in pulmonary system such as respiratory tract, nasal cavity, and in lungs (Latif et al. 2010). These results have been also observed in mice as a laboratory respiratory model. Colon cancer in experimental mice and sarcoma at the injection site has also been reported (Farombi et al. 2020). Some regulators of the PAHs (identified as the tumor inducers) that acts as new clinical target such as glycohydrolase, ADP-ribose and Interleukin 28B (Majumder et al. 2020; Zeng et al. 2020).

Genotoxicological Effects of PAHs

Polycyclic aromatic hydrocarbons are known to affect the genes in a long run. Increased level of BaP-DNA mainly epoxide has been observed in the placenta, foetus, and hepatic tissues of mother in a pregnancy periodic study (Agarwal et al. 2018). A significant level of lymphocyte hprt site and DNA aberration was also reported (WHO 1998). A PAH namely, BaP is known to actively block DNA replication by inappropriate insertion of nucleotidal bases at the point of original nucleotides that must be inserted (Wu et al. 2020). Carcinogenic genes have been reported to be activated by IcdP PAH (Boeckers et al. 2020). Gunter et al. (2007) has successfully shown the involvement of DNA adducts and PAH in cancerous patients.

Some PAHs such as Phenanthrene (Phe), Pyr, and Fla cause break in the single stranded DNA and results in hampered repair of it as observed in chicks (Nisha et al. 2017). Qin et al. (2020) has shown for the first time the facultative process of amino swap during Pyr to DNA binding along with inhibitory role it in hydroxyl substitution as well. PAHs genotoxicity and transmission levels studies are very important because it may lead to early assessment and prevention from the ill effects it in animals and human. Another study has shown that non-smoker females are more prone towards nuclear chromosome damage and oxidative stress due to

PAH as compared to males. This is because females have lower level of DNA repair capability and higher P 450 (CYP) cytochrome enzymatic activity that leads to higher DNA adducts as compared to males (Guo et al. 2014; Uppstad et al. 2011).

Reproductive and Endocrinological Pathways Derailing Effects of PAHs

Polycyclic aromatic hydrocarbons such as BaP, Pyr etc. are associated with the reproductive and endocrinological pathway disruption such as embryogenesis, foetal fate and sex related hormones (Kummer et al. 2013; Machado et al. 2014). Premature ovarian failure (POF) was significantly correlated with serum PAH concentration upon exposure to PAHs in women except Pyr (Ye et al. 2020). In an in vivo experiment, rats showed lowered hepatocyte metabolic rate upon exposure to BaP, and in adult rats this PAH exposure showed poor sperm quality in epididymis and testicular atrophy. This was mainly attributed to steroidogenic acute regulatory protein (StAR) expression inhibition by BaP resulted in lower testosterone level (Chung et al. 2011; Schraplau et al. 2015). In early gestational period of embryo implantation, the exposure of mammals with BaP at the time of pregnancy cause reduced formation of endometrial layer and blood vessels (Li et al. 2017). Increased risk of implantation failure due to lowered endometrial response upon exposure to BaP may be resulted. It is just because PAHs leads to derailing in the progesterone, estrogenic level, reduced gene expression and their receptor expression on the surface (Zhao et al. 2014).

Not only in lithospheric animals but aquatic animals are also affected by PAHs such as, in mussels lower gametic incidence and poor quality was noticed. Upon observance, tar used for waterproofing the pond was found the main culprit (Alonso et al. 2019). In another study in aquatic animal *Oryzias melastigma*, BaP retrogressively effected the genes related to embryonic development such as *ERα, cyp19a,* and *ivtg1* resulting in the adverse effect on embryonic sexual maturation (Sun et al. 2020a). In *Chlamys farreri* (female) BaP treatment showed inhibitory effect on sexual hormones such as 17β-estradiol (E2), testosterone (T) and progesterone (P) (Yang et al. 2020a). According to Vrinda et al. (2019), the Chr PAH was able to level up the genes related to immunity and neuroendocrine system in *Penaeus monodon* larvae. *Gadus morhua* when exposed to PAHs in a short-term tenure had slightly affected the physiological and gonadal development (Tollefsen et al. 2011). In *Danio rerio*, Aromatic hydrocarbon receptor pathway mediated reproductive toxicity was reported due to 120 h BF (Benzofluorene) (Hawliczek et al. 2012). Based on these experimental reports, it can be concluded that PAHs usually effect the development of an organism by the endocrinological disruption.

Neurotoxicological Effects of PAHs

This section addresses the effects of polycyclic aromatic hydrocarbons exposure on the neuronal system of the animals and human such as neural tissue damage, impairment of the learning and memory, behavioural issues etc. A disorder commonly known as attention deficit hyperactivity (ADHD) was observed in 9 year old child. Such behavioural issue was raised when pregnant mother was exposed to PAH (Perera et al. 2018). In another case, a group of 652 experimental coal miner where relationship between urinary PAHs markers and neuronal behaviour was carried out and result showed that the miners with high level of PAHs exposure during working hour led some irregularities such as less appropriate visual and auditory memory, less emotionally balanced, lower concentration of amino acids, amines (Mono-), and cholinergic transmitters (Niu et al. 2010; Wang et al. 2020. In another case, workers exposed to FL (Fluorene) below the limits had damaging impact upon the vestibular system of the factory workers (Bucolo et al. 2012).

In several researches worldwide, BaP role on modulating neuronal chemistry (memory and learning loss) has been exposed in animals such as zebra fish, mice etc. (Knecht et al. 2017; Mccallister et al. 2008). Mice upon oral PAH (Fluoranthene) exposure showed several abnormal neuronal behaviours such as spread foot upon landing, reduced grip, lowered aerial righting, comparatively more defecation and urinal, lower sensory stimuli, ataxia. These symptoms were highest up to 6 h of exposure and thereafter showed persistence up to 48 h (Saunders et al. 2003). Mice upon sub-acutely exposure to BaP showed lowered anxiety related behaviour and reduced expression of receptors for *N*-methyl-D-aspartate (NMDA) in brain (Grova et al. 2008). In another experiment, rats upon exposure to FL (Subcutaneously) showed an increased in anxiety against no impact learning capability (Peiffer et al. 2015).

Based on metabolic rates and estrogenic dependent metabolic-enzymes, males are more prone to PAH ill effects in comparison to female gender such as Fla and BaP show more severe neurological effect in males as compared to that of females (Ramesh et al. 2000).

In the urban areas, adult mice upon examination showed increased concentration of IFNδ and IL1β. This is attributed to the fact that in urban areas tested particulate matter were rich in PAHs, and as the mice inhaled the PAH showed their effect letting to the activation of TLR4-NFκB signalling pathway in the cerebral cortex and caused higher abnormal anxiety (Slotkin et al. 2017; Haghani et al. 2020). Some possible process by which these PAHs show sheerness in animals is the neuronal death induced by genotoxicity (Chepelev et al. 2015).

Activation of Reactive Oxygen Species (ROS) Due to PAHs

As the body gets intoxicated with pollutants that disturb the homeostatic redox balance of the body resulting in the accumulation of ROS, an oxidative stress condition internally. Yuan et al. (2017) have shown that BaP and Pyr exposure causes HL-7702 and HepG2 human hepatocyte mortality due to increase in ROS respectively (Ma et al. 2019).

In rats' hepatic tissue (WB-F344), BaP was found potent in inducing oxidative stress (Vondracek et al. 2009). In *Pinctada martensii*, Pyr is known to cause higher lipid peroxidation and increased hemolymph ROS (Xie et al. 2017). Similarly, Wang et al. (2008) also reported oxidative stress in *Phaeodactylum tricornutum* due to PAHs. The oxidative stress is the result of activation of cascade of immune cell response. Intracellular effect of the PAH is related to oxidative stress as shown by Wells et al. (2010), PAHs induced intracellular ROS system which in turn induced DNA molecule damage due to oxidative stress in correspondence to 8-oxoguanosine damage.

Immunotoxicological Effects of PAHs

Immunity related toxicity due to polycyclic aromatic hydrocarbons is well illustrated in both human beings and in other animals. Human when exposed to PAHs showed much stressed immune system such as in a case study, Bangladeshi adults males exposed to PAH showed higher concentration of activated CD19 + B cells (Burchiel et al. 2019). PAH exposure during pregnancy (early) alters umbilical cord lymphocytes concentration such as according to Herr et al. (2010), elevated percentage of $CD3^{+}$ and $CD4^{+}$ was observed along with lowered concentration of natural killer cell (NK cell) and $CD19^{+}$ cell. Similarly, hypermethylation of interferon-δ (INF- δ) in umbilical cord blood DNA was reported in a case study in Colombian naïve children (Total 53) resulting to the post exposure of the mother with PAHs. This provides a keen idea of the role of DNA methylation in immunity genes related to PAH contamination (Rychlik and Sille 2019; Tang et al. 2012). BaP a well-known PAH also acts as a deactivator of chemokine ligand 12-induced T-lymphocyte chemotaxis and activates IFN mediated signalling pathways (Liamin et al. 2018).

In laboratory exposed mice, BaP causes inactivation of concanavalin A-induced T –cell proliferation along with reduced concentration of interleukin-2 (IL-2), IL-4, and INF- δ and BbF causes damage to podocytes in rats by stopping the process of autophagy (Guan et al. 2017; Zhang et al. 2020). In the larvae of zebra fish, BaP exposure resulted in an elevated immune cells in correspondence to macrophages but lowered or suppressed immunity was observed in several model animals such as rainbow trout (*Oncorhynchus mykiss*), bluegill (*Lepomis macrochrius*), orange spotted grouper (*Epinephelus coioides*), and olive flounder (*Paralichthys olivaceus*) (Phelps et al. 2020; Connelly and Means 2010; Hur et al. 2013; Khaniyan et al. 2016; Phalen et al. 2014). According to Phalen et al. (2014), further injecting BaP in the abdominal tissues of *O. mykiss* caused increased level of cell line toxicity of B cells and T cells and a reduction in antibodies in the blood. Similarly, Hur et al. (2013) showed that BaP exposure results in the lowered activity of macrophages, lysozyme, and IgM in *P. olivaceus*. Another PAH named Chr has been also reported for its negative impact on immune system (total hemocyte count, pro-phenoloxidase (pro-PO) activity) in *Litopenaeus vannamei* (white shrimp) (Ren et al. 2014).

Aquatic Toxicity of Polycyclic Aromatic Hydrocarbons with Special Reference to Pisces

Synthetic chemical industries have forged much harm to aquatic animals with their increasing release of chemical pollutants directly (without treatment of pollutants especially PAHs) in the sea. These pollutants are responsible for different emerging diseases in quatic organisms especially to pisces such as cancer, neurological, derailing the embryonic pathways, and disruption in bone metabolism.

PAH Contamination During Early Developmental Phase of Pisces

In fishes such as *Platichthys stellatus* and *Parophrys vetulus* PAHs have been reported to cause carcinogenic effects. In juvenile sole BaP binding capability was higher (10 times) as compared to adult one. Similarly, BaP was able to bind with 90 times more power to sole juveniles to that of Sprague Dawley rats. In the same way, mutation (chemical change) of hepatic DNA of sole (juvenile) was 2–4 times higher compared to flounder juveniles and BaP 7,8-diol glucuronide concentration in the bile of flounder was much lower to that of former species (Honda and Suzuki 2020).

The sensitivity of the teleost fishes for PAHs have been reported at two developmental stages: Firstly, in the early cleavage stage PAH is known to distort the dorso-ventro axis formation by derailing the Wnt/β-catenin pathway. This disruption leads to distorted dead hyperpolarized embryo. Secondly, fishes are more susceptible to PAHs at the cardiac developmental period. Proper heart development causes proper movement and oxygen transport during metabolic demands. Embryonic heart is damaged by the presence of three ringed congeners such as phenanthrenes in two ways: (1) repolarization of the action potential by the inhibition of the potassium channels, (2) blockage of the ryanodine receptors in the cardiomyocytes (Cherr et al. 2017; Incardona and Scholz 2016).

PAH in the crude oil extracted is known to disrupt the normal development and functioning of the eye in fishes. According to Magnuson et al. (2018), *Sciaenops ocellatus* larvae (Red drum) and *Cyprinodon variegatus* larvae (Sheepshead) exposed to crude oil showed reduced retinal layer diameter and optometric response. The chorion of *Melanogrammus aeglefinus* (Atlantic haddock), was disrupted upon exposure to 0.7 μg/L tPAH of crude oil mainly because of the abnormal growth and cardiotoxicity (Sorensen et al. 2017). Chen et al. (2017), has also reported the toxic effect of monohydroxylated PAHs (OHPAHs) in the embryo of *Oryzias latipes* (Japanese medaka). In the laboratory experiment, injection of 3-OHBcP (1 nM) in the ova of medaka showed faster developmental period of the 1st, 4th, and 6th day. Upon injecting the 3-OHBcP (metabolite of benzo [c]phenanthrene) (1 nM concentration) at the 5th day of the post fertilization in the embryo of medaka resulted in the higher heart rate and several gene (nearly 780) expression was also found to be affected mainly related to the eye, muscle development, energy supply, and stress response proteins (Magnuson et al. 2018).

Effect of PAHs Exposure on the Bone Physiology in Pisces

Bone tissues are damaged due to PAHs such as 7, 12-dimethylbenz[a] anthracene and BaP, and cigarette etc. (Lee et al. 2002). Andreou et al. (2004) have shown that the cigarette smoke associated hydrocarbons when combined with lipopolysaccharide in *Porphyromonas gingivalis* (Gram negative anaerobic bacteria) causing periodontitis shows inhibitory action of PAHs on osteogenesis. In teleosts, body scales serve many functions out of which osteoclastic and osteoblastic are most important. In the regions of the crude oil export route fishes have shown deformed spinal sections (de Soysa et al. 2012). Goldfish were reared in the tank containing contaminated water with PAH and ALP (Alkaline phosphatase)-TRAP (Tartrate-resistant acid phosphatase) activity was measured. The result showed a remarkable suppression of ALP activity but no disturbance in TRAP activity. The result showed that osteoblast related mRNA expression (Osteocalcin, ALP, RANKL) was negatively regulated by PAH (naphthalene and acenaphthene) contaminated water (Fig. 8.3).

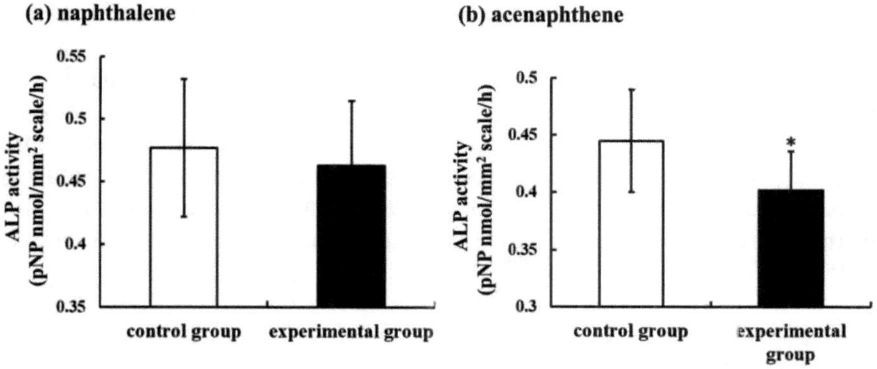

Fig. 8.3 Effects of PAHs on alkaline phosphatase (ALP) activity in goldfish cultured scales incubated for a period of 6 h.: **a** Naphthalene; **b** Acenaphthene (each 6 ng/L). (Reprinted from original open access article by Honda and Suzuki 2020 under license CC BY 4.0, https://creativecommons.org/licenses/by/4.0/)

PAH Mediated Hepatological Toxicity in Fishes

From the infected media, PAH gets primarily accumulated in the hepatic tissues because liver is associated with the process of biotransformation (Honda and Suzuki 2020). It is reported that most of the bioaccumulation of the PAHs has been shown by the fishes of the benthic region where sediments are mainly contaminated with the PAHs. For example, a bottom feeder fish *Solea solea* collected from three contaminant sites (Po Delta, Chioggia fishing area, and Venetian Lagoon) showed correlation with the bioaccumulations in the hepatic tissue to that of contaminants in the collected sites (Frapiccini et al. 2018). In *Gobiocypris rarus* a Chinese minnow clearly showed the possible apoptosis was due to BaP contamination. This PAH usually causes hyperactivity in the apoptotic genes such as bax, bcl-2, p53, and other caspases (Zha et al. 2017).

The concentrations of alkaline phosphatase lactate dehydrogenase in hepatic cells, total cholesterol, high density lipoprotein, and free cholesterol also significantly decreased when *Girella punctate* (Nibbler fish) were intraperitoneally injected with BaA during the breeding period (first 10 days). Hepatocarcinogenesis results from this aberrant concentration of vital bodily and blood components (Cai et al. 2019; Souza et al. 2016).

Aquatic Toxicity of Polycyclic Aromatic Hydrocarbons with Special Reference to Invertebrates

Toxicity in *Caenorhabditis Elegans, Chironomus Tentans, Daphnia Magna, Eisenia Fetida,* and *Artemia Salina*

Several PAHs have been reported which toxicity upon invertebrates (*Caenorhabditis elegans, Chironomus tentans, Daphnia magna, Artemia salina*) such as anthracene, pyrene, fluoranthene, phenanthrene, BaP, and acenapthene. The researches have shown that *C. elegans* sensitivity for PAHs was far less as compared to *C. tentans, Artemia salina* (Honda and Suzuki 2020). BaP was most toxic to *C. elegans* and fluoranthene was found to be most toxic to *D. magna.* Similarly, *D. magna* and *A. salina* were sensitive towards pyrene. In the same manner earthworm (*Eisenia fetida)* showed toxic sensitivity towards Phenanthrene (Nam et al. 2015).

Toxicity of PAH in *Hemicentrotus Pulcherrimus*

Sea urchins are an established experimental model for invertebrates and hence much work has been done regarding the toxicological effects (Meador et al. 1995). Fertilized eggs of sea urchin (*H. pulcherrimus*) were treated with 10^{-8} and 10^{-7} M doses of BaA and 4-OHBaA. The embryo did not show any remarkable change at blastula stage and prism stage. The abrupt changes began to be noticed at pluteus

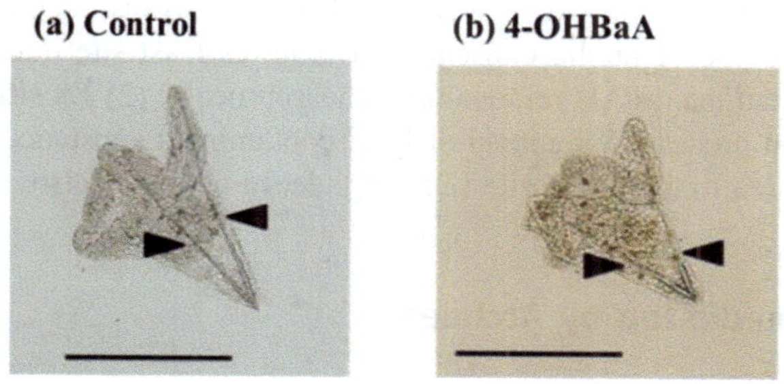

Fig. 8.4 Effect of PAH on the spicule formation during embryonic stage in *H. pulcherrimus*: **a** Control and **b** 4-hydroxybenz[a]anthracene (4-OHBaA)-treated embryos with 10^{-7} M. (Reprinted from original open access article by Honda and Suzuki 2020 under license CC BY 4.0, https://creativecommons.org/licenses/by/4.0/)

stage (Fig. 8.4) where spicule length was suppressed by the exposure of 4-OHBaA (at both doses) (Suzuki et al. 2015). It is clear from above that PAHs are involved in the inhibition of embryogenesis. Similarly, BaA also suppressed the growth of spicule but only at the dose concentration of 10^{-7} M. PAH (4-OHBaA) also showed effect at the mRNA level in *H. pulcherrimus,* where suppression of the mRNA related to spicule matrix protein was reported. Important transcription factors linked to spicule formation *H. pulcherrimus* E26 transformation-specific gene 1 and *H. pulcherrimus* Aristaless-like homeo-box gene 1 were markedly inhibited by 4-OHBaA (Honda and Suzuki 2020).

Bioremediation of Polycyclic Aromatic Hydrocarbon Using Biological Sources.

Besides using physical and chemical methods for the remediation of PAHs, employment of biological method has gained much attention in recent decade (Agnello et al. 2016; Ghosal et al. 2016; Gupta et al. 2025). Biological process of PAH remediation may use bioaugumentation, biostimulation (about 33%) followed by bioreactors (22%), compositing (13%), rhizo/phytoremediation (22%), enzyme mediated/vermin-remediation (4%) (Kuppusamy et al. 2017).

PAH bioremediation may involve two methods: (1) **In situ**: degradation of contaminants at the site of public consent as it requires least site interruption, more chances of contaminant PAH removal and cheap process. (2) **Ex situ**: it is a physical remediation process by removal of PAH contaminant substance removal from contaminated area to treatment site (Davletgildeeva and Kuznetsov 2024).

PAHs Bioremediation by Archaea

Petroleum industry is one of the key pollutants of PAH and is mainly associated with marine ecosystem contamination with it, therefore its bioremediation should involve much better extremophile instead of conventional microbes. *Halobellus rufus*, and *Halopenitus malekzadehii* have *nahE* gene that are known PAH-degrader (Al-Mailem et al. 2010). According to Bonfa et al. (2011), haloarchaea strains (Genus: *Haloferax)* can be used in the natural degradation of PAHs. Ten different strains of haloarchaea grown on a mixture of 5 PAHs namely pyrene, benzo(a)anthracene, anthracene, naphthalene, and phenanthrene. Without yeast in the mixture of PAHs, all the strains showed retarded growth except BO7, later on when yeast was added as a supplement in the media then a significant improvement in the growth of the colony BO7, MM17, PR13, and MM27 was observed. HPLC analysis showed that all these 5 strains: BO7, MM17, PR13, AA35, and CL47 strains gave positive. But BO7 was consistent for degrading all the 5 PAHs upto more than 45%. AA35 strain was able to reduce all PAHs except phenanthrene. CL47 strain efficiently reduced naphthalene concentration up to nearly 45%. Two strains of Archaea namely MM17 and PR13 was able to reduce only three ringed PAH (phenanthrene) only.

When yeast was added in the culture media the efficiency of all strains in degradation of PAHs increased. *H. volcanii,* PR13, AA35, CL47, and MM17 efficiently reduced all the PAHs in the media (Above 50% degradation rate). BO6 and MM27 were able to reduce only anthracene and phenanthrene.

PAH Bioremediation by Algae

Phylum algae contains miniature primary plants that acts as the important producer in estuarine as well as coast lines and may possess a very crucial role in in situ PAH bioremediation. These help in reduction of PAH from environment either by bioaccumulation in the body or by cellular degradation. (Ke et al. 2010). Algae during remediation of PAHs use two enzymatic pathways mainly mono- and dioxygenase resulting in the intermediate products formation such as hydroxylate and dihydroxylate (Chan et al. 2006). Cyanobacteria has been reported for aiding in bioremediation of HMW PAHs (De Llasera et al. 2016).

Fungi Mediated Bioremediation of PAHs

Mycological bioremediation is worldwide studied not only for PAHs but also in remediation of metallic pollutants, radiation absorbance etc. (Maddheshiya et al. 2024a, b). Usually, a fungus co-metabolizes the PAHs and causes the production of carbon dioxide with the help of mono-oxygenase enzymes (Gupta and Pathak 2020). For the bioremediation of LMW and HMW mainly two types of fungi are used:

(1) **Ligninolytic fungi**: They release lignin peroxidase, laccase, and manganese peroxidase enzymes for the breakdown of lignin in the wood. They simplify lignin into diphenol and later the intermediate gets converted into quinones by oxidation (Aydin et al. 2017). After catalytic breakdown of PAHs, ligninolytic enzyme generates polar products. These polar products are water soluble and become easily available for the fungi and other fauna (Gupta and Pathak 2020).
(2) **Non-ligninolytic fungi**: These fungi release cytochrome 450 mono-oxygenase like enzyme which easily degrade PAHs through the process of oxidation and results in the formation of water and arene oxide. This arene oxide further through non-enzymatic arrangement gets converted into gluconeric acid, xylose, and glucose (Ghosal et al. 2016). Some of the HMW polycyclic aromatic hydrocarbons show less solubility, therefore, specialized fungi release surfactants to counter the problem (Ojha et al. 2019).

Bacteria Mediated Biodegradation of PAHs

Bacteria is the most versatile PAH bioremediation agent used worldwide. It mainly uses aerobic pathway for the degradation of PHAs. In the aerobic pathway firstly hydroxylation of PAH rings occurs with the help of dioxygenase enzymes (Ferredoxin enzyme, reductase enzyme, and oxygenase subunit) resulting in the formation of cis-dihydrodiol and ultimately oxidizes to forms diol intermediates with the aid of dehydrogenase enzymes. Ortho-and meta- cleavage pathway helps dioxygenase enzyme to break the diol ringed structure and as a result some intermediates are formed such as gentisic acid, catechol, and procatechuic acid which ultimately assist tricarboxylic acid cycle (TCA cycle) (Mallick et al. 2011; Yang et al. 2020b). The final electron acceptor and co-substrate for the hydroxylation and oxygen-mediated cleavage of the aromatic ring during bacterial aerobic PAH degradation is oxygen (Chen et al. 2016).

The immobilization of delivering inoculum microbes can serve the PAH degradation better by increasing shelf life and activities of microbes in the soil system. Mrozik and Piotrowska- Seget (2010) noted higher percentage degradation of indeno[1,2,3-cd]pyrene (81%) and pyrene (71%) with the help of immobilized (on cinder beads) *Acinetobacter baumannii* INP1 and *Pseudomonas taiwanensis* PYR1. Immobilization renders the species of the microbes a comparatively more stable environmental condition, lower competition rate, and more stability.

Natural Attenuation and Usage of Stimulants in PAHs Remediation

In this process the degradation capacity of innate microbes is enhanced mainly by maintaining moisture and nutrient level and proper aeration. According to Chikere et al. (2017), degradation of 85% PHA and 98% total petroleum hydrocarbon is improved by attenuation procedure in crude oil polluted field and later PAH bioremediation. In the applied procedure a constant ratio of K:N:P was at 1:2:1, followed by tilling, proper periodic irrigation, mixing of the soil.

Biostimulation method is used for increasing the bioremediation by stimulating the overall capacity of microbes for the degradation of PAHs. The microbes are encouraged by nutritive addition to the soil such as potassium, nitrogen, sulphur, and phosphorus, addition of organic wastes, humic acid, and electron acceptors (Das and Das 2015). A remarkable study has been reported by Bianco et al. (2020), in which fresh organic solid waste, digestate, combination of macro and micronutrients (soil = 10:biostimulant = 1) was given to the mixture of pyrene, phenanthrene, anthracene and fluoranthene in marine sediment at 37 °C. the result showed remarkable enhancement from 12% in control to 55% in biostimulant treated media.

Conclusion

Many efforts have been made to know the possible adverse health risk due to PAHs on both animals and human. Human beings are exposed to PAHs through the occupational exposure that results in severe ill effects such as cancer, immunological and neurological problems etc. PAH also has significant effect not only to human but also fishes and invertebrates. In fishes, hepatic damage, neurotoxicity can be clearly noted. Along with it, invertebrate studies have also shown impact of PAHs in *C. elegans, E. fetida, C. tentans, A. salina, D. magna H. pulcherrimus.* To lower down the concentration of PAH in the environment some biological models such algae, fungi, bacteria can be involved. Adding to it the application of biostimulants ans enzymes can also degrade these toxicants. Thus, concerns should be raised for the prevention of PAH release and its biological degradation in the environment.

References

Abdel-Shafy, H.I., Mansour, M.S.: A review on polycyclic aromatic hydrocarbons: source, environmental impact, effect on human health and remediation. Egypt. J. Pet. **25**, 107–123 (2016a). https://doi.org/10.1016/j.ejpe.2015.03.011

Abdel-Shafy, H.I., Mansour, M.S.: A review on polycyclic aromatic hydrocarbons: source, environmental impact, effect on human health and remediation. Egypt. J. Pet. **25**, 107–123 (2016b). https://doi.org/10.1016/j.ejpe.2015.03.011

Agarwal, P., Singh, L., Anand, M., Taneja, A.: Association between placentalpolicyclic aromatic hydrocarbons (PAHS), oxidative stress, and preterm delivery: a case control study. Arch. Environ. Contam. Toxicol. **74**, 218–227 (2018). https://doi.org/10.1007/s00244-017-0455-0

Agnello, A.C., Bagard, M., van Hullebusch, E.D., Esposito, G., Huguenot, D.: Comparative bioremediation of heavy metals and petroleum hydrocarbons co-contaminated soil by natural attenuation, phytoremediation, bioaugmentation and bioaugmentation-assisted phytoremediation. Sci. Total. Environ. **563**, 693–703 (2016). https://doi.org/10.1016/j.scitotenv.2015.10.061

Alhamdow, A., Essig, Y.J., Krais, A.M., Gustavsson, P., Tinnerberg, H., Lindh, C.H., Hagberg, J., Graff, P., Albin, M., Broberg, K.: Fluorene exposure among PAH-exposed workers is associated with epigenetic markers related to lung cancer. Occup. Environ. Med. **77**, 488–495 (2020). https://doi.org/10.1136/oemed-2020-106413

Al-Mailem, D., Sorkhoh, N., Al-Awadhi, H., Eliyas, M., Radwan, S.: Biodegradation of crude oil and pure hydrocarbons by extreme halophilic archae from hypersaline coasts of the Arabian Gulf. Extremophiles. hypersaline coasts of the Arabian Gulf. Extremophiles **14**, 321–328 (2010)

Alonso, A., Suarez, P., Ruiz, Y., Dobal, V., San Juan, F.: Gonadal histopathological disorders in *Mytilus galloprovincialis* male exposed to tars used in mussel farms. Front. Mar. Sci. **6**, 577 (2019). https://doi.org/10.3389/fmars.2019.00577

Andreou, V., D'Addario, M., Zohar, R., Sukhu, B., Casper, R.F., Ellen, R.P., Tenenbaum, H.C.: Inhibition of osteogenesis in vitro by a cigarette smoke-associated hydrocarbon combined with *Porphyromonas gingivalis* lipopolysaccharide: reversal by resveratrol. J. Periodontol. **75**, 939–948 (2004)

Arias, A.H., Souissi, A., Glippa, O., Roussin, M., Dumoulin, D., Net, S., et al.: Removal and biodegradation of phenanthrene, fluoranthene and pyrene by the marine algae *Rhodomonas baltica* enriched from North Atlantic Coasts. B. Environ. Contam. Tox. **98**, 392–399 (2017). https://doi.org/10.1007/s00128-016-1967-4

Arulazhagan, P., Al-Shekri, K., Huda, Q., Godon, J.-J., Basahi, J.M., Jeyakumar, D.: Biodegradation of polycyclic aromatic hydrocarbons by an acidophilic *Stenotrophomonas maltophilia* strain AJH1 isolated from a mineral mining site in Saudi Arabia. Extremophiles **21**, 163–174 (2017). https://doi.org/10.1007/s00792-016-0892-0

Aydin, S., Karaçay, H.A., Shahi, A., Gökçe, S., Ince, B., Ince, O.: Aerobic and anaerobic fungal metabolism and Omics insights for increasing polycyclic aromatic hydrocarbons biodegradation. Fungal Biol. Rev. **31**, 61–72 (2017). https://doi.org/10.1016/j.fbr.2016.12.001

Baldantoni, D., Morelli, R., Bellino, A., Prati, M.V., Alfani, A., De Nicola, F.: Anthracene and benzo (a) pyrene degradation in soil is favoured by compost amendment: perspectives for a bioremediation approach. J. Hazard. Compost Amendment: Perspect. Bioremediat. Approach. J. Hazard. Mater. **339**, 395–400 (2017). https://doi.org/10.1016/j.jhazmat.2017.06.043

Barron, M.G., Heintz, R., Rice, S.D.: Relative potency of PAHs and heterocycles as aryl hydrocarbon receptor agonists in fish. Mar. Environ. Res. **58**, 95–100 (2004). https://doi.org/10.1016/j.marenvres.2004.03.001

Bianco, F., Race, M., Papirio, S., Esposito, G.: Removal of polycyclic aromatic hydrocarbons during anaerobic biostimulation of marine sediments. Sci. Tot. Environ. **709**, 136–141 (2020). https://doi.org/10.1016/j.scitotenv.2019.136141

Birolli, W.G., Santos, D.D.A., Alvarenga, N., Garcia, A.C., Romão, L.P., Porto, A.L.: Biodegradation of anthracene and several PAHs by the marine-derived fungus *Cladosporium* sp. CBMAI 1237. Mar. Pollut. Bull. **129**, 525–533 (2018). https://doi.org/10.1016/j.marpolbul.2017.10.023

Boeckers, M., Paul, N.W., Efferth, T.: Indeno[1,2,3-cd]pyrene and picenemediate actions via estrogen receptor alpha signalling pathway in in vitro cell systems, altering gene expression. Toxicol. Appl. Pharmacol. **396**, 114995 (2020). https://doi.org/10.1016/j.taap.2020.114995

Bonfá, M.R., Grossman, M.J., Mellado, E., Durrant, L.R.: Biodegradation of aromatic hydrocarbons by *Haloarchaea* and their use for the reduction of the chemical oxygen demand of hypersaline petroleum produced water. Chemosphere **84**, 1671–1676 (2011). https://doi.org/10.1016/j.chemosphere.2011.05.005

Bucolo, S., Palermo, A., Montemagno, A., Filidoro, M., Beatrice, F.: Proposal of ent3ering the mastoid vibration test in health surveillance of workers exposed to industrial solvents. G. Ital. Med. Lav. Ergon. **34**, 671–673 (2012)

Burchiel, S.W., Lauer, F.T., Factor-Litvak, P., Liu, X., Santella, R.M., Islam, T., Eunus, M., Alam, N., Islam, T., Rahman, M., Ahmed, A., Ahsan, H., Graziano, J., Parvez, F.: An increase in circulating B cells and B cell activation markers in peripheral blood is associated with cigarette smoking in a male cohort in Bangladesh. Toxicol. Appl. Pharmacol. **384**, 114783 (2019). https://doi.org/10.1016/j.taap.2019.114783

Cai, L., Li, J., Yu, L., Wei, Y., Miao, Z., Chen, M., Huang, R.: Characterization of transcriptional responses mediated by benzo[a]pyrene stress in a new marine fish model of goby *Mugilogobius Chulae*. Genes Genom. **41**, 113–123 (2019)

Campo, L., Rossella, F., Pavanello, S., Mielzynska, D., Siwinska, E., Kapka, L., Bertazzi, P.A., Fustinoni, S.: Urinary profiles to assess polycyclic aromatic hydrocarbons exposure in coke-oven workers. Toxicol. Lett. **192**, 72–78 (2010). https://doi.org/10.1016/j.toxlet.2008.12.018

Chan, S.M.N., Luan, T., Wong, M.H., TAM, N.F.Y.: Removal and biodegradation of polycyclic aromatic hydrocarbons by Selenastrum capricornutum. Environ. Toxicol. Chem. **25**(7),1772–17799 (2006). https://pubmed.ncbi.nlm.nih.gov/16833137/

Chang, Y., Huynh, C.T.T., Bastin, K.M., Rivera, B.N., Siddens, L.K., Tilton, S.C.: Classifying polycyclic aromatic hydrocarbons by carcinogenic potency using in vitro biosignatures. Toxicol. in Vitro **69**, 104991 (2020). https://doi.org/10.1016/j.tiv.2020.104991

Chen, B., Huang, J., Yuan, K., Lin, L., Wang, X., Yang, L., et al.: Direct evidences on bacterial growth pattern regulating pyrene degradation pathway and genotypic dioxygenase expression. Mar. Pollut. Bull. **105**, 73–80 (2016). https://doi.org/10.1016/j.marpolbul.2016.02.054

Chen, K., Tsutsumi, Y., Yoshitake, S., Qiu, X., Xu, H., Hashiguchi, Y., Honda, M., Tashiro, K., Nakayama, K., Hano, T., et al.: Alteration of development and gene expression induced by in ovo-nanoinjection of 3-hydroxybenzo[c]phenanthrene into Japanese medaka (*Oryzias latipes*) embryos. Aquatic. Toxicol. **182**, 194–204 (2017)

Chepelev, N.L., Moffat, I.D., Bowers, W.J., Yauk, C.L.: Neurotoxicity may be an overlooked consequence of benzo[a]pyrene exposure that is relevant to human health risk assessment. Mutat. Res. Rev. Mutat. Res. **764**, 64–89 (2015). https://doi.org/10.1016/j.mrrev.2015.03.001

Cherr, G.N., Fairbairn, E., Whitehead, A.: Impacts of petroleum-derived pollutants on fish development. Annu. Rev. Anim. Biosci. **5**, 185–203 (2017)

Chikere, C.B., Azubuike, C.C., Fubara, E.M.: Shift in microbial group during remediation by enhanced natural attenuation (RENA) of a crude oil-impacted soil: a case study of Ikarama Community. Bayelsa, Nigeria. 3 Biotech, **7**, 1–11 (2017)

Chung, J.-Y., Kim, Y.-J., Kim, J.Y., Lee, S.G., Park, J.-E., Kim, W.R., Yoon, Y.-D., Yoo, K.S., Yoo, Y.H., Kim, J.-M.: Benzo[a]pyrene reduces testosterone production in rat Leydig cells via a direct disturbance of testicular steroidogenic machinery. Environ. Health Perspect. **119**, 1569–1574 (2011). https://doi.org/10.1289/ehp.1003391

Connelly, H., Means, J.C.: Immunomodulatory effects of dietary exposure to selected polycyclic aromatic hydrocarbons in the bluegill (*Lepomis macrochirus)*. Int. J. Toxicol. **29**, 532–545 (2010). https://doi.org/10.1177/1091581810377518

Das, N., Das, D.: Strategies for remediation of polycyclic aromatic hydrocarbons from contaminated soil-an overview. J. Crit. Rev. **2**, 20–25 (2015)

Dastgheib, S.M.M., Amoozegar, M.A., Khajeh, K., Shavandi, M., Ventosa, A.: Biodegradation of polycyclic aromatic hydrocarbons by a halophilic microbial consortium. Appl. Microbiol. Biotechnol. **95**, 789–798 (2012). https://doi.org/10.1007/s00253-011-3706-4

Davletgildeeva, A.T., Kuznetsov, N.A.: Bioremediation of polycyclic aromatic hydrocarbons by means of bacteria and bacterial enzymes. Microorganisms **12**(9),1814 (2024). https://doi.org/10.3390/microorganisms12091814

De Llasera, M.P.G., de Jesús Olmos-Espejel, J., Díaz-Flores, G., Montaño Montiel, A.: Biodegradation of benzo (a) pyrene by two freshwater microalgae *Selenastrum capricornutum* and *Scenedesmus acutus*: a comparative study useful for bioremediation. Environ. Sci. Pollut. Res. **23**, 3365–3375 (2016). https://doi.org/10.1007/s11356-015-5576-2

de Soysa, T.Y., Ulrich, A., Friedrich, T., Pite, D., Compton, S.L., Ok, D., Bernardos, R.L., Downes, G.B., Hsieh, S., Stein, R., et al.: Macondo crude oil from the deepwater horizon oil spill disrupts specific developmental processes during *zebrafish* embryogenesis. BMC Biol. **10**, 40 (2012)

Faraji Ghasemi, F., Dobaradaran, S., Saeedi, R., Nabipour, I., Nazmara, S., Ranjbar Vakil Abadi, D., Arfaeinia, H., Ramavandi, B., Spitz, J., Mohammadi, M.J., Keshtkar, M.: Levels and ecological and health risk assessment of PM2.5-bound heavy metals in the northern part of the Persian Gulf. Environ. Sci. Pollut. Res. **27**, 5305–5313 (2020). https://doi.org/10.1007/s11356-019-07272-7

Farombi, E.O., Ajayi, B.O., Adedara, I.A.: 6-Gingerol delays tumorigenesis in benzo[a] pyrene and dextran sulphate sodium-induced colorectal cancer in mice. Food Chem. Toxicol. **142**, 111483 (2020). https://doi.org/10.1016/j.fri.2020.111483

Frapiccini, E., Annibaldi, A., Betti, M., Polidori, P., Truzzi, C., Marini, M.: Polycyclic aromatic hydrocarbon (PAH) accumulation in different common sole (*Solea solea*) tissues from the North Adriatic Sea peculiar impacted area. Mar. Pollut. Bull. **137**, 61–68 (2018)

Ghosal, D., Ghosh, S., Dutta, T.K., Ahn, Y.: Current state of knowledge in microbial degradation of polycyclic aromatic hydrocarbons (PAHs): a review. Front. Microbiol. **7**, 1369 (2016). https://doi.org/10.3389/fmicb.2016.01369

Goudarzi, G., Daryanoosh, S.M., Godini, H., Hopke, P.K., Sicard, P., De Marco A., Rad, H.D., Harbizadeh, A., Jahedi, F., Mohammadi, M.J., Savari, J., Sadeghi, S., Kaabi, Z., Khaniabadi, Y.O.: Health risk assessment of exposure to the Middle-Eastern Dust storms in the Iranian megacity of Kermanshah. Public Health **148**, 109–116 (2017). https://doi.org/10.1016/j.puhe.2017.03.009

Goudarzi, G., Geravandi, S., Alavi, N., Idani, E., Salmanzadeh, S., Yari, A.R., Jamshidi, F., Mohammadi, M.J., Ranjbarzadeh, A., Alamdari, F.A., Darabi, F., Rohban, A.: Association between cancer risk and polycyclic aromatic hydrocarbons' exposure in the ambient air of Ahvaz, southwest of Iran. Int. J. Biometeorol. **62**, 1461–1470 (2018). https://doi.org/10.1007/s00484-018-1543-1

Grova, N., Schroeder, H., Farinelle, S., Prodhomme, E., Valley, A., Muller, C.P.: Subacute administration of benzo[a]pyrene (B[a]P) reduces anxiety-related behaviour in adult mice and modulates regional expression of N-methyl-D-aspartate (NMDA) receptors genes in relevant brain regions. Chemosphere **73**, S295–S302 (2008). https://doi.org/10.1016/j.chemosphere.2007.12.037

Guan, S., Huang, Y., Feng, Z., Xu, L., Jin, Y., Lu, J.: The toxic effects of benzo[a]pyrene on activated mouse T cells in vitro. Immunopharmacol. Immunotoxicol. **39**, 117–123 (2017). https://doi.org/10.1080/08923973.2017.1299173

Gunter, M.J., Divi, R.L., Kulldorff, M., Vermeulen, R., Haverkos, K.J., Kuo, M.M., Strickland, P., Poirier, M.C., Rothman, N., Sinha, R.: Leukocyte polycyclic aromatic hydrocarbon-DNA adduct formation and colorectal adenoma. Carcinogenesis **28**, 1426–1429 (2007). https://doi.org/10.1093/carcin/bgm022

Guo, H., Huang, K., Zhang, X., Zhang, W., Guan, L., Kuang, D., Deng, Q., Deng, H., Xiaomin, Z., He, M., Christiani, D., Wu, T.: Women are more susceptible than men to oxidative stress and chromosome damage caused by polycyclic aromatic hydrocarbons exposure. Environ. Mol. Mutagen. **55**, 472–481 (2014). https://doi.org/10.1002/em.21866

Gupta, H.: PAH determination in effluent and sludge samples of paper industry. Environ. Technol. Innov. **9**, 115–121 (2018)

Gupta, H., Gupta, B.: Photocatalytic degradation of polycyclic aromatic hydrocarbon benzo [a] pyrene by iron oxides and identification of degradation products. Chemosphere **138**, 924–931 (2015)

Gupta, H., Gupta, B.: Adsorption of polycyclic aromatic hydrocarbons on banana peel activated carbon. Desalin. Water Treat. **57**(20), 9498–9509 (2016)

Gupta, H., Kumar, R.: Distribution of selected polycyclic aromatic hydrocarbons in urban soils of Delhi, India. Environ. Technol. Innov. **17**, 100500 (2020)

Gupta, H., Singh, S.: Kinetics and thermodynamics of phenanthrene adsorption from water on orange rind activated carbon. Environ. Technol. Innov. **10**, 208–214 (2018). https://doi.org/10.1016/j.eti.2018.03.001

Gupta, H., Dhiman, S., Kumar, A., Paliwal, S.K., Khan, G., Singh, H., Mishra, A., Mishra, A.K.: Photo- and bio-degradation of selected persistent organic pollutants. Sep. Purif. Rev. **54**(1), 3–16 (2025). https://doi.org/10.1080/15422119.2024.2332953

Gupta, H., Dhiman, S.: Synthetic nanoparticle-based remediation of soils contaminated with polycyclic aromatic hydrocarbons. In: Shanker, U., Hussain, C.M., Rani, M. (eds.) Handbook of Green and Sustainable Nanotechnology. Springer, Cham (2023). https://doi.org/10.1007/978-3-030-69023-6_102-1

Gupta, H., Dhiman, S.: Emerging contaminants in the environment introduction. In: Shanker, U., Rani, M. (eds.) Occurrence, Distribution and Toxic Effects of Emerging Contaminants, pp. 1–16. CRC Press (2024). https://doi.org/10.1201/9781003335757-1

Gupta, S., Pathak, B.: Mycoremediation of polycyclic aromatic hydrocarbons. In: Borthakur, A., Singh, P., Kumar, A. (eds.) Abatement of Environmental Pollutants, pp. 127–149. Amsterdam, Elsevier (2020). https://doi.org/10.1016/b978-0-12-818095-2.00006-0

Gupte, A., Tripathi, A., Patel, H., Rudakiya, D., Gupte, S.: Bioremediation of polycyclic aromatic hydrocarbon (PAHs): a perspective. Open Biotechnol. J. **10**, 363–378 (2016). https://doi.org/10.2174/1874070701610010363

Haghani, A., Johnson, R., Safi, N., Zhang, H., Thorwald, M., Mousavi, A., Woodward, N.C., Shirmohammadi, F., Coussa, V., Wise, J.P., Forman, H.J., Sioutas, C., Allayee, H., Morgan, T.E., Finch, C.E.: Toxicity of urban air pollution particulate matter in developing and adult mouse brain: comparison of total and filter-eluted nanoparticles. Environ. Int. **136**, 105510 (2020). https://doi.org/10.1016/j.envint.2020.105510

Haritash, A., Kaushik, C.: Biodegradation aspects of polycyclic aromatic hydrocarbons (PAHs): a review. J. Hazard. Mater. **169**, 1–15 (2009). https://doi.org/10.1016/j.jhazmat.2009.03.137

Hashemzadeh, B., Idani, E., Goudarzi, G., Ankali, K.A., Sakhvidi, M.J.Z., Babaei, A.A., Hashemzadeh, H., Vosoughi, M., Mohammadi, M.J., Neisi, A.: Effects of PM2.5 and NO2 on the 8-isoprostane and lung function indices of FVC and FEV1 in students of Ahvaz city. Iran. Saudi J. Biol. Sci. **26**, 473–480 (2019). https://doi.org/10.1016/j.sjbs.2016.11.008

Hawliczek, A., Nota, B., Cenijn, P., Kamstra, J., Pieterse, B., Winter, R., Winkens, K., Hollert, H., Segner, H., Legler, J.: Developmental toxicity and endocrine disrupting potency of 4-azapyrene, benzo[b]fluorene and retene in the zebrafish *Danio rerio* Reprod. Toxicol. **33**, 213–223 (2012). https://doi.org/10.1016/j.reprotox.2011.11.001

Herr, C.E., Dostal, M., Ghosh, R., Ashwood, P., Lipsett, M., Pinkerton, K.E., Sram, R., Hertz Picciotto, I.: Air pollution exposure during critical time periods in gestation and alterations in cord blood lymphocyte distribution: a cohort of live births. Environ. Health **9**, 46 (2010). https://doi.org/10.1186/1476-069X-9-46

Honda, M., Suzuki, N.: Toxicities of polycyclic aromatic hydrocarbons for aquatic animals. Int. J. Environ. Res. Public Health **17**, 1363 (2020). https://doi.org/10.3390/ijerph17041363

Hur, D., Jeon, J.-K., Hong, S.: Analysis of immune gene expression modulated by benzo[a]pyrene in head kidney of olive flounder (*Paralichthys olivaceus*). Comp. Biochem. Physiol. B-Biochem. Mol. Biol. **165**, 49–57 (2013). https://doi.org/10.1016/j.cbpb.2013.03.001

Incardona, J.P., Scholz, N.L.: The influence of heart developmental anatomy on cardiotoxicity-based adverse outcome pathways in fish. Aquat. Toxicol. **177**, 515–525 (2016)

Itoh, N., Naya, T., Kanai, Y., Kumon, F., Amano, K.: Historical changes in the aquatic environment and input of polycyclic aromatic hydrocarbons over 1000 years in Lake Kitaura, Japan. Limnology **18**, 51–62 (2017). https://doi.org/10.1007/s10201-0160488-5

Jiang, R.L., Xiao, B.C., Na, Y.U., Chen, L.Q.: Research advance in toxic effects of PAHs on aquatic animals. Mar. Fish. **36**, 372–383 (2014)

Johnsen, A.R., Karlson, U.: Diffuse PAH contamination of surface soils: environmental occurrence, bioavailability, and microbial degradation. Appl. Microbiol. Biotechnol. **76**, 533–543 (2007)

Kadri, T., Rouissi, T., Kaur Brar, S., Cledon, M., Sarma, S., Verma, M.: Biodegradation of polycyclic aromatic hydrocarbons (PAHs) by fungal enzymes: a review. J. Environ. Sci. (China) **51**, 52–74 (2017). https://doi.org/10.1016/j.jes.2016.08.023

Ke, L., Luo, L., Wang, P., Luan, T., Tam, N.F.-Y.: Effects of metals on biosorption and biodegradation of mixed polycyclic aromatic hydrocarbons by a freshwater green alga *Selenastrum capricornutum*. Bioresour. Technol. **101**, 6950–6961 (2010). https://doi.org/10.1016/j.biortech.2010.04.011

Khaniyan, M., Salamat, N., Safahieh, A., Movahedinia, A.: Detection of benzo[a] pyrene-induced immunotoxicity in orange spotted grouper (*Epinephelus coioides*). Environ. Toxicol. **31**, 329–338 (2016). https://doi.org/10.1002/tox.22047

Knecht, A.L., Truong, L., Simonich, M.T., Tanguay, R.L.: Developmental benzo[a] pyrene (B[a]P) exposure impacts larval behavior and impairs adult learning in zebrafish. Neurotoxicol. Teratol. **59**, 27–34 (2017). https://doi.org/10.1016/j.ntt.2016.10.006

Kumar, A., Gupta, H.: Activated carbon from sawdust for naphthalene removal from contaminated water. Environ. Technol. Innov. **20**, 101080 (2020). https://doi.org/10.1016/j.eti.2020.101080

Kummer, V., Maskova, J., Zraly, Z., Faldyna, M.: Ovarian disorders in immature rats after postnatal exposure to environmental polycyclic aromatic hydrocarbons. J. Appl. Toxicol. **33**, 90–99 (2013). https://doi.org/10.1002/jat.1714

Kuppusamy, S., Thavamani, P., Venkateswarlu, K., Lee, Y.B., Naidu, R., Megharaj, M.: Remediation approaches for polycyclic aromatic hydrocarbons (PAHs) contaminated soils: technological constraints, emerging trends and future directions. Chemosphere **168**, 944–968 (2017). https://doi.org/10.1016/j.chemosphere.2016.10.115

Latif, I.K., Karim, A.J., Zuki, A.B.Z., Zamri-Saad, M., Niu, J.P., Noordin, M.M.: Pulmonary modulation of benzo[a]pyrene-induced hemato- and hepatotoxicity in broilers. Poult. Sci. **89**, 1379–1388 (2010). https://doi.org/10.3382/ps.2009-00622

Lee, L.L., Lee, J.S.C., Waldman, S.D., Casper, R.F., Grynpas, M.D.: Polycyclic aromatic hydrocarbons present in cigarette smoke cause bone loss in an ovariectomized rat model. Bone **30**, 917–923 (2002)

Lee, B.-K., Vu, V.T.: Sources, distribution and toxicity of polyaromatic hydrocarbons (PAHs) in particulate matter. In: Air Pollution, pp. 99–122. London, IntechOpen (2010)

Li, X., Shen, C., Liu, X., He, J., Ding, Y., Gao, R., Mu, X., Geng, Y., Wang, Y., Chen, X.: Exposure to benzo[a]pyrene impairs decidualization and decidual angiogenesis in mice during early pregnancy. Environ. Pollut. **222**, 523–531 (2017). https://doi.org/10.1016/j.envpol.2016.11.029

Liamin, M., Le Mentec, H., Evrard, B., Huc, L., Chalmel, F., Boutet-Robinet, E., Le Ferrec, E., Sparfel, L.: Genome-wide transcriptional and functional analysis of human T lymphocytes treated with benzo[a]pyrene. Int. J. Mol. Sci. **19**, 3626 (2018). https://doi.org/10.3390/ijms19113626

Lors, C., Damidot, D., Ponge, J.-F., Périé, F.: Comparison of a bioremediation process of PAHs in a PAH-contaminated soil at field and laboratory scales. Environ. Pollut. **165**, 11–17 (2012). https://doi.org/10.1016/j.envpol.2012.02.004

Ma, J.-K., Eldin, W.F.S., El-Ghareeb, W.R., Elhelaly, A.E., Khedr, M.H.E., Li, X., Huang, X.-C.: Effects of pyrene on human liver HepG2 cells: cytotoxicity, oxidative stress, and transcriptomic changes in xenobiotic metabolizing enzymes and inflammatory markers with protection trial using lycopene. Biomed. Res. Int. **2019**, 7604851 (2019). https://doi.org/10.1155/2019/7604851

Machado, J. de B., Chatkin, J.M., Zimmer, A.R., Goulart, A.P.S., Thiesen, F.V.: Cotinine and polycyclic aromatic hydrocarbons levels in the amniotic fluid and fetal cord at birth and in the urine from pregnant smokers. PLoS One **9**, e116293 (2014). https://doi.org/10.1371/journal.pone.0116293

Maddheshiya, R., Dhiman, S., Gupta, H.: Emerging contaminants in wastewater and their treatment plans: a major focus on radioactive components emitted from medical waste (2024a)

Maddheshiya, R., Dhiman, S., Sahu, D.M., Gupta, H.: Role of phytoremediation and metagenomics in healing the wounded (polluted) planet: Ecofriendly and sustainable approach. Ed.

Environmental metagenomics, water quality and possible remediation measures of polluted waters. pp. 201–216. (2024b) https://doi.org/10.1016/B978-0-443-13659-7.00009-6
Magnuson, J.T., Khursigara, A.J., Allmon, E.B., Esbaugh, A.J., Roberts, A.P.: Effects of deepwater horizon crude oil on ocular development in two estuarine fish species, red drum (*Sciaenops ocellatus*) and sheepshead minnow (*Cyprinodon variegatus*). Ecotoxicol. Environ. Saf. **166**, 186–191 (2018)
Majumder, D., Debnath, R., Maiti, D.: IL-27 along with IL-28B ameliorates the pulmonary redox impairment, inflammation and immunosuppression in benzo(a) pyrene induced lung cancer bearing mice. Life Sci. **260**, 118384 (2020). https://doi.org/10.1016/j.lfs.2020.118384
Mallick, S., Chakraborty, J., Dutta, T.K.: Role of oxygenases in guiding diverse metabolic pathways in the bacterial degradation of low-molecular weight polycyclic aromatic hydrocarbons: a review. Crit. Rev. Microbiol. **37**, 64–90 (2011). https://doi.org/10.3109/1040841x.2010.512268
Mccallister, M.M., Maguire, M., Ramesh, A., Qiao, A., Hood, D.B.: Prenatal exposure to benzo(a)pyrene impairs later-life cortical neuronal function. Neurotoxicology **29**, 846–854 (2008). https://doi.org/10.1016/j.neuro.2008.07.008
Meador, J.P., Stein, J.E., Reichert, W.L., Varanasi, U.: Bioaccumulation of polycyclic aromatic hydrocarbons by marine organisms. Rev. Environ. Contam. Toxicol. **143**, 79–165 (1995)
Mehetre, G.T., Dastager, S.G., Dharne, M.S.: Biodegradation of mixed polycyclic aromatic hydrocarbons by pure and mixed cultures of biosurfactant producing thermophilic and thermotolerant bacteria. Sci. Total. Environ. **679**, 52–60 (2019). https://doi.org/10.1016/j.scitotenv.2019.04.376
Miura, K., Shimada, K., Sugiyama, T., Sato, K., Takami, A., Chan, C.K., et al.: Seasonal and annual changes in PAH concentrations in a remote site in the Pacific Ocean. Sci. Rep. **9**, 1–10 (2019)
Mojiri, A., Zhou, J.L., Ohashi, A., Ozaki, N., Kindaichi, T.: Comprehensive review of polycyclic aromatic hydrocarbons in water sources, their effects and treatments. Sci. Total. Environ. **2019**, 133971 (2019). https://doi.org/10.1016/j.scitotenv.2019.133971
Mrozik, A., Piotrowska-Seget, Z.: Bioaugmentation as a strategy for cleaning up of soils contaminated with aromatic compounds. Microbiol. Res. **165**, 363–375 (2010). https://doi.org/10.1016/j.micres.2009.08.001
Nam, T.H., Jeon, H.J., Mo, H.H., Cho, K., Ok, Y.S., Lee, S.E.: Determination of biomarkers for polycyclic aromatic hydrocarbons (PAHs) toxicity to earthworm (*Eisenia fetida*). Environ. Geochem. Health **37**, 943–951 (2015)
Nisha, A.R., Hazilawati, H., Mohd Azmi, M.L., Noordin, M.M.: DNA damage and adduct formation in immune organs of developing chicks by polycyclic aromatic hydrocarbons. Toxicol. Mech. Methods **27**, 215–222 (2017). https://doi.org/10.1080/15376516.2016.1273432
Niu, Q., Zhang, H., Li, X., Li, M.: Benzo[a]pyrene-induced neurobehavioral function and neurotransmitter alterations in coke oven workers. Occup. Environ. Med. **67**, 444–448 (2010). https://doi.org/10.1136/oem.2009.047969
Ojha, N., Mandal, S.K., Das, N.: Enhanced degradation of indeno (1, 2, 3-cd) pyrene using Candida tropicalis NN4 in presence of iron nanoparticles and produced biosurfactant: a statistical approach. 3 Biotech **9** (2019)
Okere, U., Semple, K.: Biodegradation of PAHs in 'pristine'soils from different climatic regions. J. Bioremed. Biodegrad. S **1**, 006 (2012). https://doi.org/10.4172/21556199.S1-006
Peiffer, J., Grova, N., Hidalgo, S., Salquèbre, G., Schroeder, H.: Behavioral toxicity and physiological changes from repeated exposure to fluorene administered orally or intraperitoneally to adult male *Wistar* rats: a dose-response study. Neurotoxicology **2016**(53), 321–333 (2015). https://doi.org/10.1016/j.neuro.2015.11.006
Perera, F.P., Wheelock, K., Wang, Y., Tang, D., Margolis, A.E., Badia, G., Cowell, W., Miller, R.L., Rauh, V., Wang, S.: Combined effects of prenatal exposure to polycyclic aromatic hydrocarbons and material hardship on child ADHD behavior problems. Environ. Res. **160**, 506–513 (2018). https://doi.org/10.1016/j.envres.2017.09.002
Phalen, L.J., Koellner, B., Leclair, L.A., Hogan, N.S., van den Heuvel, M.R.: The effects of benzo[a]pyrene on leucocyte distribution and antibody response in rainbow trout

(*Oncorhynchus mykiss*). Aquat. Toxicol. **147**, 121–128 (2014). https://doi.org/10.1016/j.aquatox.2013.12.017

Phelps, D.W., Fletcher, A.A., Rodriguez-Nunez, I., Balik-Meisner, M.R., Tokarz, D.A., Reif, D.M., Germolec, D.R., Yoder, J.A.: In vivo assessment of respiratory burst inhibition by xenobiotic exposure using larval zebra fish. J. Immunotoxicol. **17**, 94–104 (2020). https://doi.org/10.1080/1547691X.2020.1748772

Qin, C., Hu, X., Waigi, M.G., Yang, B., Gao, Y.: Amino and hydroxy substitution in fluences pyrene-DNA binding. Sci. Total Environ. **725**, 138542 (2020). https://doi.org/10.1016/j.scitotenv.2020.138542

Ramesh, A., Inyang, F., Hood, D.B., Knuckles, M.E.: Aryl hydrocarbon hydroxylase activity in F-344 rats subchronically exposed to benzo(a)pyrene and fluoranthene through diet. J. Biochem. Mol. Toxicol. **14**, 155–161 (2000). https://doi.org/10.1002/(sici)1099-0461(2000)14:3%3c155::aid-jbt5%3e3.0.co;2-3

Ravindra, K., Sokhi, R., Van Grieken, R.: Atmospheric polycyclic aromatic hydrocarbons: source attribution, emission factors and regulation. Atmos. Environ. **42**, 2895–2921 (2008). https://doi.org/10.1016/j.atmosenv.2007.12.010

Ren, X., Pan, L., Wang, L.: Immunotoxic effect of benzo[alpha]pyrene and chrysene in juvenile white shrimp *Litopenaeus vannamei*. Cent. Eur. J. Biol. **9**, 1048–1057 (2014). https://doi.org/10.2478/s11535-014-0317-y

Rychlik, K.A., Sille, F.C.M.: Environmental exposures during pregnancy: mechanistic effects on immunity. Birth Defects Res. **111**, 178–196 (2019). https://doi.org/10.1002/bdr2.1469

Sakshi, Haritash, A.K.: A comprehensive review of metabolic and genomic aspects of PAH-degradation. Arch. Microbiol. **202**, 2033–2058 (2020). https://doi.org/10.1007/s00203020-01929-5

Saunders, C.R., Shockley, D.C., Knuckles, M.E.: Fluoranthene-induced neurobehavioral toxicity in F-344 rats. Int. J. Toxicol. **22**, 263–276 (2003). https://doi.org/10.1080/10915810305114

Schraplau, A., Schewe, B., Neuschäfer-Rube, F., Ringel, S., Neuber, C., Kleuser, B., Püschel, G.P.: Enhanced thyroid hormone breakdown in hepatocytes by mutual induction of the constitutive androstane receptor (CAR, NR1I3) and arylhydrocarbon receptor by benzo[a]pyrene and phenobarbital. Toxicology **328**, 21–28 (2015). https://doi.org/10.1016/j.tox.2014.12.004

Sharma, A., Singh, S.B., Sharma, R., Chaudhary, P., Pandey, A.K., Ansari, R., et al.: Enhanced biodegradation of PAHs by microbial consortium with different amendment and their fate in in-situ condition. J. Environ. Manage. **181**, 728–736 (2016). https://doi.org/10.1016/j.jenvman.2016.08.024

Shen, H., Grist, S., Nugegoda, D.: The PAH body burdens and biomarkers of wild mussels in Port Phillip Bay, Australia and their food safety implications. Environ. Res. **188**, 109827 (2020). https://doi.org/10.1016/j.envres.2020.109827

Simpson, C.D., Mosi, A.A., Cullen, W.R., Reimer, K.J.: Composition and distribution of polycyclic aromatic hydrocarbon contamination in surficial marine sediments from Kitimat Harbor, Canada. Sci. Total. Environ. **181**, 265–278 (1996). https://doi.org/10.1016/0048-9697(95)05026-4

Slotkin, T.A., Skavicus, S., Card, J., Di Giulio, R.T., Seidler, F.J.: In vitro models reveal differences in the developmental neurotoxicity of an environmental polycylic aromatic hydrocarbon mixture compared to benzo[a]pyrene: neuronotypic PC12 cells and embryonic neural stem cells. Toxicology **377**, 49–56 (2017). https://doi.org/10.1016/jtox.2016.12.00

Soltani, N., Keshavarzi, B., Moore, F., Tavakol, T., Lahijanzadeh, A.R., Jaafarzadeh, N., Kermani, M.: Ecological and human health hazards of heavy metals and polycylic aromatic hydrocarbons (PAHs) in road dust of Isfahan metropolis. Iran. Sci. Total Environ. **505**, 712–723 (2015). https://doi.org/10.1016/j.scitotenv.2014.09.097

Sørensen, L., Sørhus, E., Nordtug, T., Incardona, J.P., Linbo, T.L., Giovanetti, L., Karlsen, Ø., Meier, S.: Oil droplet fouling and differential toxico-kinetics of polycyclic aromatic hydrocarbons in embryos of Atlantic haddock and cod. PLoS ONE **12**, e0180048 (2017)

Souza, T., Jennen, D., van Delft, J., van Herwijnen, M., Kyrtoupolos, S., Kleinjans, J.: New insights into BaP-induced toxicity: role of major metabolites in transcriptomics and contribution to hepatocarcinogenesis. Arch. Toxicol. **90**, 1449–1458 (2016)

Srogi, K.: Monitoring of environmental exposure to polycyclic aromatic hydrocarbons: a review. Environ. Chem. Lett. **5**, 169–195 (2007). https://doi.org/10.1007/s10311007-0095-0

St. Helen, G., Goniewicz, M.L., Dempsey, D., Wilson, M., Jacob, P., Benowitz, N.L.: Exposure and kinetics of polycyclic aromatic hydrocarbons (PAHs) in cigarette smokers. Chem. Res. Toxicol. **25**, 952–964 (2012). https://doi.org/10.1021/tx300043k

Subashchandrabose, S.R., Venkateswarlu, K., Venkidusamy, K., Palanisami, T., Naidu, R., Megharaj, M.: Bioremediation of soil long-term contaminated with PAHs by algal- bacterial synergy of *Chlorella* sp. MM3 and *Rhodococcus wratislaviensis* strain 9 in slurry phase. Sci. Total. Environ. **659**, 724–731 (2019). https://doi.org/10.1016/j.scitotenv.2018.12.453

Suman, S., Sinha, A., Tarafdar, A.: Polycyclic aromatic hydrocarbons (PAHs) concentration levels, pattern, source identification and soil toxicity assessment in urban traffic soil of Dhanbad. India. Sci. Total Environ. **545**, 353–360 (2016). https://doi.org/10.1016/j.scitotenv.2015.12.061

Sun, K., Song, Y., Zong, W., Tang, J., Liu, R.: Anthracene-induced DNA damage and oxidative stress: a combined study at molecular and cellular levels. Environ. Sci. Pollut. Res. **27**, 41458–41474 (2020b). https://doi.org/10.1007/s11356-020-10049-y

Sun, K., Song, Y., He, F., Jing, M., Tang, J., Liu, R.: A review of human and animals exposure to polycyclic aromatic hydrocarbons: health risk and adverse effects, photo-induced toxicity and regulating effect of microplastics. Sci. Total. Environ. **773**, 145403 (2021). https://doi.org/10.1016/j.scitotenv.2021.145403

Sun, D., Chen, Q., Zhu, B., Lan, Y., Duan, S.: Long-term exposure to benzo[a]pyrene affects sexual differentiation and embryos toxicity in three generations of marine medaka (*Oryzias melastigma*). Int. J. Environ. Res. Public Health **17**, 970 (2020a). https://doi.org/10.3390/ijerph17030970

Suzuki, N., Ogiso, S., Yachiguchi, K., Kawabe, K., Makino, F., Toriba, A., Kiyomoto, M., Sekiguchi, T., Tabuchi, Y., Kondo, T., et al.: Monohydroxylated polycyclic aromatic hydrocarbons influence spicule formation in the early development of sea urchins (*Hemicentrotus pulcherrimus*). Comp. Biochem. Physiol. Part C **171**, 55–60 (2015)

Tang, W., Levin, L., Talaska, G., Cheung, Y.Y., Herbstman, J., Tang, D., Miller, R.L., Perera, F., Ho, S.-M.: Maternal exposure to polycyclic aromatic hydrocarbons and 5′-CpG methylation of interferon-gamma in cord White blood cells. Environ. Health Perspect. **120**, 1195–1200 (2012). https://doi.org/10.1289/ehp.1103744

Tilton, S.C., Siddens, L.K., Krueger, S.K., Larkin, A.J., Loehr, C.V., Williams, D.E., Baird, W.M., Waters, K.M.: Mechanism-based classification of PAH mixtures to predict carcinogenic potential. Toxicol. Sci. **146**, 135–145 (2015). https://doi.org/10.1093/toxsci/kfv080

Tollefsen, K.E., Sundt, R.C., Beyer, J., Meier, S., Hylland, K.: Endocrine modulation in Atlantic cod (*Gadus morhua* L.) exposed to alkylphenols, polyaromatic hydrocarbons produced water, and dispersed oil. J. Toxicol. Environ. Health A **74**, 529–542 (2011). https://doi.org/10.1080/15287394.2011.550562

Uppstad, H., Osnes, G.H., Cole, K.J., Phillips, D.H., Haugen, A., Mollerup, S.: Sex differences in susceptibility to PAHs is an intrinsic property of human lung adenocarcinoma cells. Lung Cancer **71**, 264–270 (2011). https://doi.org/10.1016/j.lungcan.2010.09.006

Vondracek, J., Krcmar, P., Prochazkova, J., Trilecova, L., Gavelova, M., Skalova, L., Szotakova, B., Buncek, M., Radilova, H., Kozubik, A., Machala, M.: The role of aryl hydrocarbon receptor in regulation of enzymes involved in metabolic activation of polycyclic aromatic hydrocarbons in a model of rat liver progenitor cells. Chem. Biol. Interact. **180**, 226–237 (2009). https://doi.org/10.1016/j.cbi.2009.03.011

Vrinda, S., Abdulaziz, A., Abhilash, K.S., Jasmin, C., Kripa, V., Singh, I.S.B.: Neuroendocrine and immunotoxicity of polyaromatic hydrocarbon, chrysene in crustacean post larvae. Ecotoxicology **28**, 964–972 (2019). https://doi.org/10.1007/s10646-019-02094-2

Wang, F., Yang, L., Zhang, B., Liu, G., Wang, C., Zhang, Y., Wang, T.: Neurobehavioral performance of PAH exposure in male coal miners in Shanxi, China: a cross-sectional study.

Int. Arch. Occup. Environ. Health **93**, 707–714 (2020). https://doi.org/10.1007/s00420020-01521-w
Wang, L., Zheng, B., Meng, W.: Photo-induced toxicity of four polycyclic aromatic hydrocarbons, singly and in combination, to the marine diatom *Phaeodactylum tricornutum*. Ecotox. Environ. Safe. **71**, 465–472 (2008). https://doi.org/10.1016/j.ecoenv.2007.12.019
Wells, P.G., McCallum, G.P., Lam, K.C.H., Henderson, J.T., Ondovcik, S.L.: Oxidative DNA damage and repair in teratogenesis and neurodevelopmental deficits. Birth Defects. Res. C-Embryo Today **90**, 103–109 (2010). https://doi.org/10.1002/bdrc.20177
World Health Organization (WHO).: Selected heterocyclic Polycyclic Aromatic Hydrocarbon Environmental Health Criteria, vol. 202, World Health Organization, Geneva, Switzerland (1998)
Wu, Z., Peng, L., Hu, Y., Xie, T., Yan, H., Wan, H., Liu, W., Liang, L., Xie, Z., Liu, K., Shi, H., Zhao, G., Hu, J.: BP[dG]-induced distortions to DNA polymerase and DNA duplex; a detailed mechanism of BP adducts blocking replication. Food Chem. Toxicol. **140**, 111325 (2020). https://doi.org/10.1016/j.fct.2020.111325
Xie, J., Zhao, C., Han, Q., Zhou, H., Li, Q., Diao, X.: Effects of pyrene exposure on immune response and oxidative stress in the pearl oyster, *Pinctada Martensii*. Fish Shell Fish Immunol. **63**, 237–244 (2017). https://doi.org/10.1016/j.fsi.2017.02.032
Xiong, Y., Li, J., Huang, G., Yan, L., Ma, J.: Interacting mechanism of benzo(a)pyrene with free DNA in vitro. Int. J. Biol. Macromol. **167**, 854–861 (2020). https://doi.org/10.1016/jijbiomac.2020.11.042
Yang, X., Li, E., Liu, F., Xu, M.: Interactions of PAH-degradation and nitrate-/sulfate-reducing assemblages in anaerobic sediment microbial community. J. Hazard. Mater. **2020**, 122068 (2020a). https://doi.org/10.1016/j.jhazmat.2020.122068
Yang, H.H., Chen, C.M.: Emission inventory and sources of polycyclic aromatic hydrocarbons in the atmosphere at a suburban area in Taiwan. Chemosphere **56**, 0–887 (2004) https://doi.org/10.1016/j.chemosphere.2004.05.031
Yang, Y., Zhou, Y., Pan, L., Xu, R., Li, D.: Benzo[a]pyrene exposure induced reproductive endocrine-disrupting effects via the steroidogenic pathway and estrogen signalling pathway in female scallop *Chlamys farreri*. Sci. Total Environ. **726**, 138585 (2020a). https://doi.org/10.1016/j.scitotenv.2020.138585
Ye, X., Pan, W., Li, C., Ma, X., Yin, S., Zhou, J., Liu, J.: Exposure to polycyclic aromatic hydrocarbons and risk for premature ovarian failure and reproductive hormones imbalance. J. Environ. Sci. **91**, 1–9 (2020). https://doi.org/10.1016/j.jes.2019.12.015
Yuan, L., Liu, J., Deng, H., Gao, C.: Benzo[a]pyrene induces autophagic and pyroptotic death simultaneously in HL-7702 human normal liver cells. J. Agric. Food Chem. **65**, 9763–9773 (2017). https://doi.org/10.1021/acs.jafc.7b03248
Zeng, Z., Lu, J., Wu, D., Zuo, R., Li, Y., Huang, H., Yuan, J., Hu, Z.: Poly(ADP-ribose) glycohydrolase silencing-mediated H2B expression inhibits benzo(a)pyrene-induced carcinogenesis. Environ. Toxicol. (2020). https://doi.org/10.1002/tox.23034
Zha, J., Hong, X., Rao, H., Yuan, L., Wang, Z., Kumaran, S.S.: Benzo[a]pyrene-induced a mitochondria-independent apoptosis of liver in juvenile Chinese rare minnows (Gobiocypris rarus). Environ. Pollut. **231**, 191–199 (2017)
Zhang, Y., Liu, D., Liu, Z.: The benzo[b]fluoranthene in the atmospheric fine particulate matter induces mouse glomerular podocytes injury via inhibition of autophagy. Ecotox. Environ. Safe. **195**, 110403 (2020). https://doi.org/10.1016/j.ecoenv.2020.110403
Zhao, Y., Chen, X., Liu, X., Ding, Y., Gao, R., Qiu, Y., Wang, Y., He, J.: Exposure of mice to benzo(a)pyrene impairs endometrial receptivity and reduces the number of implantation sites during early pregnancy. Food Chem. Toxicol. **69**, 244–251 (2014). https://doi.org/10.1016/j.fct.2014.04.021

Detection and Measurement Techniques for the Polycyclic Aromatic Hydrocarbons

9

Naveen Kumar, Devesh Singh, Arvind Tiwari,
Pankaj Kumar Chaurasia, and Sunita Singh

Abstract

Polycyclic aromatic hydrocarbons (PAHs) are hazardous pollutants produced primarily from partial combustion of organic materials, including emissions from waste incinerators, fossil fuel burning, motor vehicle exhaust, petroleum refineries, and industrial processes. These are persistent in nature and posing serious environmental and health risks due to their carcinogenic and teratogenic properties. Therefore accurate PAH detection in environmental samples is essential for human health risk assessment and requires efficient and suitable pre-treatment and detection techniques. Various advanced pre-treatment methods, such as ultrasound-assisted extraction (UAE), supercritical fluid extraction (SFE), microwave-assisted extraction (MAE), counter-current chromatography (CCC), solid-Phase micro extraction (SPME) and accelerated solvent extraction (ASE) have been used for the extraction of PAHs which offer higher sensitivity and selectivity. Various chromatographic, spectroscopic immunological, electrochemical, and hyphenated detection methods are used to detect PAHs after sample pre-treatment. High-performance liquid chromatography (HPLC), Gas chromatography with flame ionization detection (GC-FID) and gas

N. Kumar (✉)
Department of Chemistry, Sanatan Dharm College Muzaffarnagar, Muzaffarnagar, Uttar Pradesh, India
e-mail: naveenkumariitr@gmail.com

D. Singh · A. Tiwari
Department of Chemistry, DAV College Muzaffarnagar, Muzaffarnagar, Uttar Pradesh, India

P. K. Chaurasia
P.G. Department of Chemistry, L.S. College, B.R. Ambedkar Bihar University, Muzaffarpur, India

S. Singh
Department of Chemistry, Navyug Kanya Mahavidyalaya Lucknow, University of Lucknow, Lucknow, Uttar Pradesh, India

H. Gupta et al. (eds.), *Ubiquitous Polycyclic Aromatic Hydrocarbon Contamination*, Emerging Contaminants and Associated Treatment Technologies,
https://doi.org/10.1007/978-3-032-11764-9_9

chromatography-mass spectrometry (GC–MS) are extensively used techniques due to their accuracy and convenience. This chapter discusses advancements in PAHs research, focusing on improved pre-treatment and detection techniques to enhance accuracy, efficiency, and environmental sustainability.

Keywords

Polycyclic aromatic hydrocarbons (PAHs) • GC–MS • LC–MS • HPLC • Pre-treatment methods • Detection techniques • Environmental and health risks

Introduction

Polycyclic aromatic hydrocarbons (PAHs) are pervasive environmental pollutants found in soil, air, and water. They are composed of two or more fused aromatic rings (Zhang et al. 2022a, b, c). Structurally, PAHs can be classified as alternant (which include only fused benzene rings, such as anthracene and chrysene) or non-alternant (which include rings with fewer than six carbon atoms, such as fluoranthene) (Famiyeh et al. 2021; Patel et al. 2020). PAHs typically appear as colourless or off white solids and they exhibit low vapour pressures, and minimal water solubility. These compounds are highly lipophilic, making them readily soluble in organic solvents but sparingly soluble in water, with solubility decreasing as the number of rings increases (Kumar et al. 2025a, b; Abdel-Shafy et al. 2016). Their stable chemical nature contributes to their persistence nature and difficulty in degradation across environmental media and can remain in the atmosphere for decades (Zhang et al. 2022a, b, c). Thus because of their migrate ability to the environment and accumulate over time PAHs create major ecological and health risks. In this context the USEPA has recognized 16 main concern PAHs due to their environmental prevalence and toxicity (Patel et al. 2020; Moazzen et al. 2022). Additionally, the WHO has expanded the list to include 33 PAHs under international regulation, acknowledging their global environmental and health impact (Pandey et al. 2011). Table 9.1 presents the physico-chemical properties of the foremost PAHs.

In general PAHs emanate from both natural and human-made sources. Natural sources contribute far less and include volcanic eruptions, forest fires, erosion of petroleum-containing rocks, and microbial or algal activity. The major contributors to PAH emissions include the partial combustion of wood, coal, petroleum, and other organic materials, commonly occurring during industrial processes, vehicle emissions, waste incineration, residential heating, and power generation (Kumar et al. 2025a, b; Famiyeh et al. 2021; Ravindra et al. 2008). Furthermore, Based on their formation, PAHs are categorized as pyrogenic (formed by high-temperature combustion with little oxygen), petrogenic (originating from crude oil and its derivatives), and biogenic (produced by microorganisms, plants, or the slow decay of organic matter) (Abdel-Shafy et al. 2016; Rocha et al. 2019). The common people are exposed to PAHs via ambient and indoor air, cigarette smoke, industrial

Table 9.1 Physico-chemical properties of EPA-listed 16 PAHs (Patel et al. 2020; Pandey et al. 2011)

Name of PAH	Abbreviation	Chemical formula	Molecular weight (g/mole)	Water solubility (ng/L)	Vapour pressure (mmHg)	Toxicity as per IARC
Acenaphthene,	ACE	$C_{12}H_{10}$	154.2	3.9×10^6	4.47×10^{-3}	3
Acenaphthylene,	ACY	$C_{12}H_8$	152.2	3.93×10^6	0.029	3
Anthracene,	ANT	$C_{14}H_{10}$	178.2	0.73×10^5	1.75×10^{-6}	3
Benz[a]anthracene	BaA	$C_{18}H_{12}$	228.2	0.01×10^6	2.5×10^{-6}	2B
Benzo[a]pyrene	BaP	$C_{20}H_{12}$	252.3	0.038×10^5	5.6×10^{-9}	1
Benzo[b]fluoranthene	BbF	$C_{20}H_{12}$	252.3	0.015×10^5	5.0×10^{-7}	2B
Benzo[ghi]perylene	BghiP	$C_{22}H_{12}$	276.3	0.026×10^4	1.03×10^{-10}	3
Benzo[k]fluoranthene	BkF	$C_{20}H_{12}$	252.3	0.008×10^5	9.59×10^{-11}	2B
Chrysene	CHY	$C_{18}H_{12}$	228.2	0.015×10^5	6.4×10^{-9}	2B
Dibenz[a,h]anthracene	DahA	$C_{22}H_{14}$	278.3	0.005×10^5	1.0×10^{-10}	2A
Fluoranthene	FLU	$C_{16}H_{10}$	202.2	0.26×10^6	5.0×10^{-6}	3
Fluorene	FL	$C_{13}H_{10}$	166.2	1.98×10^6	3.2×10^{-4}	3
Indeno[1,2,3-cd]pyrene	IcdP	$C_{22}H_{12}$	276.3	0.062×10^6	10^{-10}-10^{-16}	2B
Naphthalene	NAP	$C_{10}H_8$	128.2	3.17×10^7	0.087	2B
Phenanthrene	PHE	$C_{14}H_{10}$	178.2	1.29×10^6	6.8×10^{-4}	3
Pyrene	PYR	$C_{16}H_{10}$	202.2	0.13×10^6	2.5×10^{-6}	3

emissions, and contaminated foods, particularly aquatic products (Kumar et al. 2025a, b). Food is considered the key source of PAHs exposure in human being, especially when it is cooked using high-heat methods like grilling, roasting, or frying (Moazzen et al. 2022). Additionally, crops grown near industrial areas can absorb PAHs from contaminated soil, water, or air (Moazzen et al. 2022). Fish can accumulate PAHs directly from polluted water, posing health risks to populations relying on seafood as a dietary staple. PAHs created several forms of cancer, including lung, skin, bladder, stomach, and liver cancers. Long-term exposure,

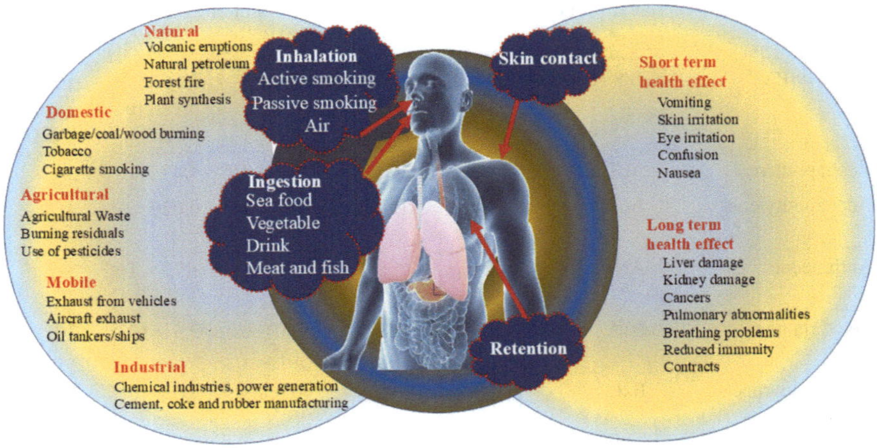

Fig. 9.1 Sources, exposure routes and health effects of PAHs

especially in occupational settings, increases the risk of cancer and may disrupt immune and endocrine systems (Kumar et al. 2025a, b; Lawal 2017; Kim et al. 2013; Rengarajan et al. 2015). Therefore, accurate monitoring and analysis of PAHs is essential for the defence of ecosystem. Figure 9.1 shows the sources, exposure routes and health effects of the PAHs.

Complex matrices of PAHs and their low environmental concentrations pose analytical challenges for their analysis. To address this, researchers have developed various sensitive and selective techniques for measuring PAHs in environmental samples for which the USEPA recommends standard procedures concerning extraction, cleanup, and analysis (Singh et al. 2022; Tumelo et al. 2024). Traditional PAH analysis involves pre-treatment of the sample followed by analysis by chromatographic techniques (Famiyeh et al. 2021; Zhang et al. 2020). More recently, advanced pretreatment methods SFE, UAE, MAE, ASE, and QuEChERS (quick, easy, cheap, effective, rugged, and safe) extraction method have gained attention (Famiyeh et al. 2021; Zhang et al. 2020; Cloutier et al. 2017). GC-FID or HPLC and GC–MS are widely adopted PAHs detection techniques due to their selectivity and sensitivity. However, they require large amounts of sample and consume more time (Poster et al. 2006; Tumelo et al. 2024; Mansouri et al. 2020; liu et al. 2019). To overcome these limitations, newer approaches such as field-portable GC–MS technique are being developed for onsite analysis (Okparanma et al. 2013a). These innovations, along with advanced methods SPME and ASE, have enhanced analytical precision and efficiency (Zhang et al. 2020). This chapter discusses advances in the measurement of PAHs with a special focus on hyphenated chromatographic and spectroscopic techniques.

Challenges in Detecting and Measuring PAHs

The analysis of PAHs in various samples is difficult because of the intricate molecular structure, trace concentrations, environmental persistence, and bioaccumulation nature of PAHs in various environmental matrices. With respect to structure many PAHs are isomeric and therefore not easy to distinguish by using conventional analytical techniques. To understand structural resemblance needs sophisticated instruments such as hyphenated or multi dimensional chromatographic techniques for separation and resolution of overlapping peaks (Chen et al. 2022; Poster et al. 2006). In addition, various derivatives of PAHs generated by environmental changes, such as alkyl-, oxy-, and nitro-, are usually not detected by routine screening instruments, although they are more toxic (O'Connell et al. 2013; Han et al. 2019; Lung et al. 2015). PAHs have a propensity to store in lipid-rich tissues, so their bioaccumulation further complicates accurate identification and risk assessment (Kumar et al. 2025a, b). Trace-level determination of PAHs is more challenging due to their presence in complex environmental matrices with very low concentrations (Tumelo et al. 2024). PAHs exhibited strong sorption affinity with sediments. Some PAHs accumulate in high amounts in sediments and therefore their bioavailability is reduced due to their migration through the sediment profile, complicating the extraction and detection of PAHs (Queb-Suarez et al. 2022). Matrix interference is another major challenge in PAH detection in biological samples, as chromatographic inhibition occurs due to co-extracted substances such as humic acids or lipids in soil, which can interfere with the PAH signal and reduce sensitivity (Williams et al. 2023; Patel et al. 2020). Therefore, it is the need of the hour to develop high resolution extraction methods and analytical techniques for PAHs analysis. Typically PAHs analysis begins with sample pretreatment methods such as SFE, UAE, MAE, and ASE, which are chosen based on the matrix type and analyte properties. As shown in Fig. 9.2, for further identification and quantification these extraction steps are typically followed by chromatographic separation—often GC or HPLC—coupled with detectors such as UV, fluorescence or MS, depending on sensitivity and selectivity requirements.

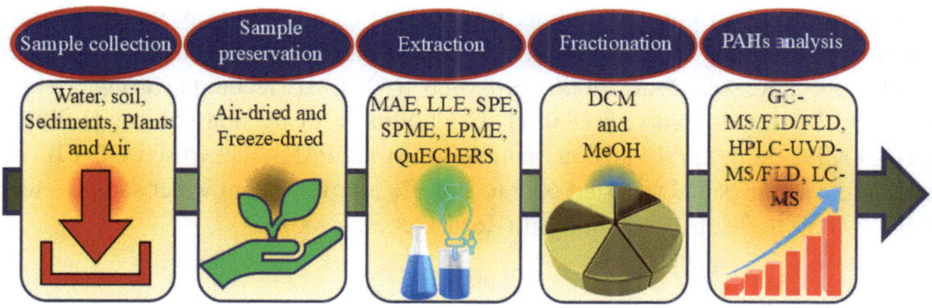

Fig. 9.2 General procedure for extraction and detection of PAHs

Pretreatment Methods/Extraction Techniques

The analysis of PAHs often relies on pre-concentration using collection media, making it essential to understand how effectively these compounds can be extracted during sampling. Due to the extensive range of physicochemical properties of PAHs such as volatility and reactivity, achieving efficient extraction of all target compounds is challenging (Pandey et al. 2011). Therefore, to decide proper pretreatment methods, solvents, and reliable recognition techniques is essential for accurate monitoring. Solvent choice is particularly important, as it directly impacts the recovery of PAHs. Commonly used solvents are methanol, acetone, dichloromethane, and mixtures like methanol/acetone, acetone/dichloromethane, toluene/methanol, and hexane/acetone (Tumelo et al. 2024; Rezaee et al. 2006). Typically, particulate matter is collected on quartz filters and extracted using high-efficiency techniques that are cost-effective, less labor-intensive, and environmentally friendly.

Liquid–Liquid Extraction (LLE)

This extraction technique generally separates organic pollutants from aqueous solutions. In this method, PAHs are extracted using an organic solvent, typically consisting of an aqueous phase mixed with an immiscible organic phase. Samples containing PAHs are placed in organic solvents, shaken vigorously, and extracted, repeating the process again and again to obtain the maximum amount of the desired compounds. Non-polar solvents such as dichloromethane, hexane, or acetone-hexane mixtures are best suited for high extraction efficiency, as PAHs are hydrophobic compounds (Tumelo et al. 2024; Rezaee et al. 2006). Wu et al. (2012) measured 16 PAHs in four edible oils using LLE followed by GC–MS technique. They used n-hexane and N,N-dimethylformamide solvents for extraction and reported LMW-PAHs conc. from 0.82 μg/kg (CHY) to 457.12 μg/kg (ACY) and HMW-PAHs conc. 0.68 μg/kg (BaP) to 16.35 μg/kg (BbF) (Wu et al. 2012). LLE requires large volume of solvent and to overcome this problem more efficient technique i.e. DLLME have been introduce in which a small amount of suitable solvent is dispersed into the sample, forming the cloudy solution, which improve extraction efficiency by enhancing the contact with the target compounds (Rezaee et al. 2006). Rezaei et al. (2006) proposed a DLLME technique using a small volume of extraction solvent (8.0 μL C_2Cl_4) and a dispersion solvent (1.00 mL CH_3COCH_3) in 5.00 mL aqueous sample to extract PAHs. They used GC-FID for testing the samples and reported 60.3 to 111.3% recovery from water samples with 0.007–0.030 μg/L of detection limit (Rezaee et al. 2006).

Soxhlet Extraction (SE)

It is a well-known process for extracting PAHs from complicated solid matrix, including solids, particulate matter and sediments. This technique allows for complete removal of PAHs by using organic solvents (often DCM, hexane, acetone or a combination of these) that repeatedly wash the sample (Wang et al. 2007). Even though Soxhlet extraction is highly rated for its effectiveness and reproducibility, it has disadvantages like long extraction time, more solvent consumption and labour-intensive operation (Lau et al. 2010). In recent time even though faster, greener and more reliable techniques like MAE and ASE are gradually taking place of Soxhlet in some labs, it is still a standard procedure, particularly for testing novel technologies (Wang et al. 2010).

Mechanical Agitation Method

In the approach, the sample is dispersed into suitable organic solvent or solvent blend (dichloromethane, acetone, or a mixture of the two, such as acetone/hexane, dichloromethane/acetone) in a flask or some vessel, and the mixture is continuously stirred by a mechanical shaker, an orbital mixer, or a magnetic stirrer. The stirring action allows an intimate contact between the solvent and the matrix favouring the dissolution and partition of PAHs in the liquid phase. Mechanical shaking uses less solvent, requires less energy, and can be used to handle multiple samples at one time. Efficiency of the procedure is affected by the type of solvent, duration of shaking, temperature and solvent/sample ratio (Lau et al. 2010; Giergielewicz et al. 2001). It has been shown that mechanical agitation can attain acceptable recoveries, especially for the lower-to-moderate molecular weight PAHs. For example, shear stirring showed good extraction efficiency for PAHs from contaminated soil and achieved extraction efficiency comparable to that of ultrasonic extraction (Song et al. 2002).

Solid-Phase Extraction (SPE)

This approach offers good selectivity, less solvent usage, good automation potential, and is an attractive and time-saving solution in greater frequency for routine environmental monitoring compared to liquid/liquid extraction. In this method, the analytes are trapped in the solid on the cartridge or disk and the unwanted material is then washed away. The retained PAHs are then eluted with an appropriate solvent for further determination (Marcé et al. 2000). The sorbent material is important for the effectiveness of the SPE. These are divided roughly on the basis of the type of material used, such as C18-bonded silica, polymeric sorbents, mixed-mode phases and so on, the latter being very efficient in retaining hydrophobic PAHs (Sajid et al. 2021). In a recent study priority PAHs were extracted from urban storm water using polymeric sorbent packed SPE cartridges with recoveries higher than 85% (Esfandiar et al. 2021). Recent developments have revolved around the miniaturization of SPE formats and the use of nanomaterials for improved extraction efficiencies and selectivity. In addition, the combination of SPE to on-line monitoring such as LC–MS/MS has significantly increased throughput and sensitivity. Given its adaptability and efficiency, SPE continues to be a cornerstone method for trace-level PAH analysis in environmental research (Ramirez et al. 2013).

Solid-Phase Microextraction (SPME)

SPME is especially well known for trace-level detection in soil, water and air because it provides a sharp, clean and sensitive substitute for conventional extraction methods. This solvent-free, combined sampling and pre-concentration method is being used more and more for the determination of PAHs in environmental matrices (Tumelo et al. 2024; Zhang et al. 2020). The targeted PAHs are directly absorbed or adsorbed from gaseous or aqueous sample using a fiber covered with an extracting phase. After the extraction phase, the fiber is thermally desorbed in a gas chromatograph's injection port for analysis following the extraction stage. Because it requires solvent free condition with no sample preparation and can be immediately interfaced with GC–MS or GC-FID systems, SPME is very beneficial. According to latest research, commercially available SPME fibers used to extract PAHs efficiently with an extensive range of molecular weights (Manousi et al. 2020).

Accelerated Solvent Extraction (ASE)

ASE is popular extraction methods, which uses high pressure and temperature to speed up PAHs extraction, and is known to be a quick, solvent saving, time efficient and environmentally friendly extraction technique (Wang et al. 2007). Samples are, in this method, inserted into an extraction cell and treated with hot solvents under high pressure. This heat and pressure mechanism significantly improves the extraction efficiency. It is applicable to a variety of solvents, whereas mixtures of organic solvent that include acetone, dichloromethane, and hexane are usually employed to maximize the recovery of PAHs (Tumelo et al. 2024; Zhang et al. 2020).

Microwave-Assisted Extraction (MAE)

In microwave assisted extraction, samples are kept to expose in microwave energy in the presence of desirable solvents, which enhance desorption and dissolution of target parts (analytes) into the solvent phase. It is the technique that offers many benefits over conventional methods such as Soxhlet extraction, including limited solvent usage with highest extraction efficiency (Wang et al. 2007). Smith et al.'s work from 2022 confirmed that, under ideal microwave circumstances, MAE allowed for more than 90% recovery of priority PAHs from polluted soils using an acetone: hexane ratio. With reduced matrix interference and increased sensitivity, researcher demonstrated its usefulness in coastal sediment analysis in another study (Chee et al. 1996).

Ultrasound-Assisted Extraction (UAE)

It is a widespread technique used for extraction of PAHs. It works through generating cavitation bubbles into the solvent using high-frequency ultrasonic waves. The bubbles crumple leading to the high mechanical agitation of the medium along the gas–liquid interface, which disturbs the matrix and the transfers of PAHs from sample to solvent, thus accelerating the extraction (Paris et al. 2019). UAE has several advantages comparing with the conventional extraction techniques. It reduces dramatically the extraction time and the amount of the extraction solvent, making it environmentally friendly. Furthermore, UAE can be used at room temperature and is applicable to a variety of PAHs including high molecular mass PAHs. It has been used extensively for the detection of PAHs in environmental samples, and it has promising performance in the reduction of solvent usage and throughput (Banjoo et al. 2005).

Supercritical Fluid Extraction

It is a new extraction method to extract PAH from environmental samples using supercritical fluids such as carbon dioxide (CO_2). It has been carried out by subjecting CO_2 to high pressure and temperature, thereby retaining CO_2 in supercritical state, at which its ability to dissolve hydrophobic compounds, such as PAHs, is significantly increased. CO_2 exhibits some peculiar properties in supercritical state, intermediate between gases and liquids, low viscosity, and high diffusivity (Tumelo et al. 2024; Zhang et al. 2020). So it can be used as an efficient PAHs extractant. The SFE extraction is especially favourable for high molecular weight PAHs, which cannot so easily be extracted by the use of the standard methods (Oliver et al. 2004). SFE works under elevated temperature and pressure also raises the solubility of PAHs in the supercritical fluid for efficient and fast extractions. Nonetheless, SFE is more expensive and requires specific apparatus compared with other methods like Soxhlet extraction. However, it is regarded as an efficient and green approach particularly for large-scale environmental monitoring (Anitescu et al. 2006).

QuEChERS Method

QuEChERS is a newly developed procedure for extraction of pesticides and other organic contaminants, from food, water and soil. The procedure typically includes two steps: the sample is initially homogenized with acetonitrile to extract PAHs and the subsequent adding of salts ($MgSO_4$ and NaCl) to promote the phase separation. The obtained extract can be further analysed by chromatography or mass spectrometry for PAHs quantification. Initially developed for pesticide-related analysis, QuEChERS has been increasingly used for the PAHs extraction due to its fast-processing time, inexpensive cost, and capability to tackle complex matrices. QuEChERS has been demonstrated to be a strong, reliable extraction technique for PAHs, providing a suitable sensitivity and precision for trace pollutants (Varela-Martínez et al. 2020; Cloutier et al. 2017).

Chromatographic Techniques for PAH Analysis

Because of complex environmental matrices and trace concentration of PAHS sensitive and reliable separation and detection techniques are required for exact measuring of PAHs. In this context following sections present the recent technological advances in analytical techniques for PAHs detection.

Gas Chromatography (GC)

GC-FID is the most favourite technique for the analysis of PAHs. This technique consists of a mobile and a stationary phase and works based on volatility and reaction of compounds with the stationary phase. In this technique sample passes across a capillary column which has a thin film of the stationary phase while a carrier gas is continuously moved through it (Poster et al. 2006). Thus the stationary phase selection is very important for the separation of PAHs by GC technique. Methyl- and phenyl-substituted polysiloxanes, as well as modified fluoro-substituted and diphenylfluoroxane-grafted polysiloxanes are the evergreen traditional materials used for stationary phases. Moreover, ionic liquids and their bonded polysiloxane provide excellent selectivity. Recently many advanced materials such as graphene, g-C_3N_4, CNTs, and MOFs have also been introduced for this. Moreover chiral mesoporous silica, pentiptycene quinone (PQ), hexagonal boron nitride (h-BN), Poly (3-hexylthiophene), oligothiophene and dichienyl benzothiazole derivatives are other emerging novel material have been used as stationary phase in GC which improve the PAHs separation efficiency (Zhang et al. 2020). Carvalho et al. (2019) used solid-phase extraction followed by GC-FID technique (equipped with a fused silica capillary column (OV-5) containing dimethylpolysiloxane with 5% phenyl as stationary phase) for identification of 16 PAHs in ground water. This procedure exhibited excellent linearity ($r^2 > 0.99$) and good precision (RSD < 5.0%) with 0.079–0.378 μg/L to 0.024–0.113 μg/L of LOQ (limit of quantification) and LOD (limit of detection) values, respectively (Carvalho et al. 2019). Emoyan et al. (2020) used GC-FID technique containing PHME siloxane column and a 30 cm/s linear velocity of helium as the mobile phase to measuring of 16 PAHs in soil from a particular vehicle-park in southern Nigeria. Results showed that the conc. of PAHs changes considerably ($P < 0.05$) with soil sampling depth and seasonal fluctuation (For wet season conc. vary from 365.8–1065.8 µg/kg to 142.9–5321.9 µg/kg and for dry season from 66.4–269.3 µg/kg to 212.1–1342.5 µg/kg for top and sub soils, respectively) (Emoyan et al. 2020). Kumar et al. (2011) examines the levels of PAHs concentration near road side soil in Jalandhar (India). They used silica gel micro column to eliminate impurities from extracts and then injected the samples into GC-FID having RH–5 capillary columns of 30 m length for analysis. The study conclude that the carcinogenic PAHs were found in higher conc. (6.41 µg/g) than to noncarcinogenic (2.17 µg/g) in all the samples with maximum conc. of 8.12–28.4 µg/g and minimum conc. of 2.6–28 µg/g of five foremost PAHs, respectively and 4.04 µg/g of total average PAHs concentration (Kumar et al. 2011). Likewise, Hishamuddin et al. (2023) studied the capability of GC-FID (polysiloxane HP-5 capillary column with 0.320 mm diameter and 30 m length) to measure the PAHs in road dust samples (Malaysia). The results reported PAH concentrations varied from 137.8 to 5813 ng/g, with the highest concentration being that of indeno[1,2,3-c,d]pyrene. In addition, benzo[a]pyrene equivalent concentrations (BaPeq) were also evaluated for assessing carcinogenic health risk and the maximum BaPeq was noticed for BaP (429.0 ng/g) (Hishamuddin et al. 2023). Additionally, hybrid techniques such as GC–MS combine the high resolution of chromatographic separations with the molecular specificity of mass spectrometry, further increasing the applicability of these techniques. These are well known for their excellent sensitivity, data acquisition and processing performance. These techniques involved comparing the mass spectra of compounds to the GC–MS spectrum database (Fig. 9.3) (Kumar et al. 2023b).

Colón et al. (2022) estimated the 16 PAHs in dairy products by using semi-automated SPE as pre-treatment methods and GC–MS as detection technique. It showed excellent performance with 1–200 ng/kg of LOD and 80–107% recoveries rates (Palacios Colón et al. 2022). Same method has been optimized by Liu et al. (2019) for simultaneously determination of 42 parent and halogenated PAHs in real drinking water samples which was achieved to ngL^{-1} level of LODs (0.34 to 3.37 ngL^{-1}) with 74.88–119.4% recoveries (Liu et al. 2019). Likewise, using UAE, semi-automated SPE and GC–MS techniques, Rascón et al. (2019) achieved 3–70 ng/kg of LODs with 85–105% recoveries of 16 PAHs in 32 meat samples (Rascón et al. 2019). Recently, to measure PAHs in infant food Ingegno et al. extracted the targeted PAHs (BaP, BaA, CHY, and BbF) by using QuEChERS method and then analyzed the extracts by gas chromatography–tandem mass spectrometry (GC–MS/MS). The proposed approach demonstrated 73.1–110.7% recoveries (R^2 value > 0.999) with LODs of 0.019–0.036 µg/kg and LOQs of 0.06–0.11 µg/kg, indicating the method's sensitivity and reliability (Mariateresa et al. 2024). Han et al. (2019) quantify the total of 44 target compounds (PAHs

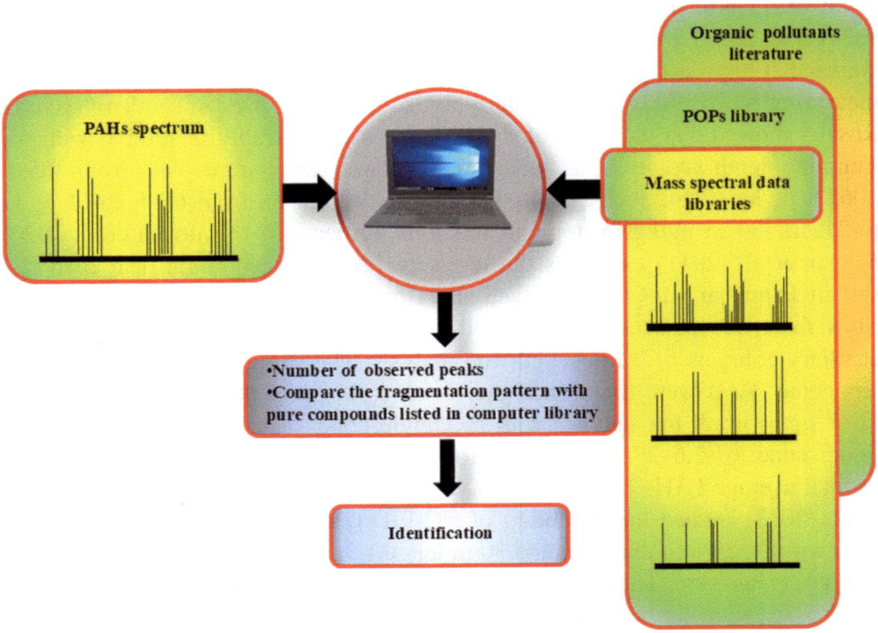

Fig. 9.3 Identification of PAHs by computer aided GC–MS technique

and derivatives) present in Jiuxiang river residue. First they extracted the desired compounds by ASE method using acetone and n-hexane (1:1, v/v) followed by SPE was applied for further cleaned up. After that the extracts was analysed by using GC and LC techniques with different modes of mass spectrometry. These techniques successfully detected the compounds with the good LODs i.e. 0.01–0.02 ng/g for PAHs, 0.002–0.067 ng/g for nitro derivatives of PAHs, 0.01–0.1 ng/g for oxy derivatives of PAHs and 0.003–0.006 ng/g for hydroxyl derivatives of PAHs. The recovery rates (52.8–114.1%) also was convincing with good repeatability (RSD < 10%) for the majority of analytes (Han et al. 2019). Incorporation of thermal techniques based quantification methods further enhanced the performance of these techniques. For example Biache et al. (2017) designed a Py-GC–MS/FID system founded on pyrolysis at 450 °C to quantify PAHs in soils. This approach was faster, environmentally friendly and less expensive as compare to classical methods with respect to PAH quantification in contaminated soils (Biache et al. 2017).

High-Performance Liquid Chromatography (HPLC)

It allows larger number of PAH isomer separation as compare to gas chromatography; however, it showed somewhat low efficiency due to low diffusion rates in both liquid and stationary phases. To separate PAHs C18, C8, cyano, phenyl and amino groups are the most common used materials as stationary phases used in HPLC technique. Several other material like ionic liquid, CNTs, MOFs and graphite carbon have been also used as more efficient stationary phases to increase the separation of PAHs (Zhang et al. 2020; Purcaro et al. 2012). For example an extraction protocol based on C18 coated fabric phase with HPLC-FLD detection technique for trace-level determination of 4 selected PAHS (ANT; PHE; PYR; FLU) in water samples has been introduced by Saini et al. (2017b). They reported that the proposed module showed good linearity (R^2 = 0.9983–0.9997) with 0.1 to 1 pg/mL of LODs and 0.3 to 3 pg/mL of LOQs for these PAHs (Saini et al. 2017b). A mesoporous TiO_2 sorbent based SPME–HPLC system showed good selectivity with low LODs values of 0.01–0.10 μg/L for eight PAHs estimation. With this sorbent the proposed approach exhibited 80.6–106.6 and 80.9–103.5% recovery rates for river water and coal ash water sample, respectively with good linearity of 0.03–30 μg/L (Wang et al. 2017). HPLC can detect various PAHs using FLD or UV detectors, or sometimes photo diode arrays combined with FLD detector. For example, Vistnes et al. (2022) used GC–MS and HPLC–DAD-FLD detection techniques for measuring 16 PAHs in road tunnel wastewater in Norway. They used ASE & SPE as pre-treatment methods and demonstrated that HPLC–DAD-FLD took less separation time (22.5 min) with higher LODs (In particulate phases from 0.020 to 2.0 μg/g and in dissolved phases from 0.0098 to 0.50 μg/L) for 13 out of 16 PAHs as compared to GC–MS method, showing the better performance of HPLC–DAD-FLD (Vistnes et al. 2022). Likewise, Yousefi et al. (2018) applied HPLC/FLD with combined deep eutectic solvent based micro extraction technique for analysis of PAHs which showed excellent LOD of 0.7–6.6 ng/L, recovery rates of 83–117% with coefficient of determination of 0.9983 to 0.9996 (Yousefi et al. 2018). Woźniak et al. (2023) found PAHs in herbal plants from Polish backyard gardens using HPLC-FL detection and the QEChERS purification technique. The study quantified eleven PAHs in mint, parsley and other herbs and reported 902.35 μg/g FW consisting mainly of ANT in mint collected from the mining area in Konin, making it promising for routine monitoring of PAHs (Woźniak et al. 2023). The development of ultra-high performance liquid chromatography (UHPLC) is further a promising approach, which increases the detection efficiency (Saini et al. 2017b; Mansouri et al. 2020). Recently, a very fast and sensitive method based on C18-functionalized ultrafine magnetic silica NPs extraction with UHPLC detection was introduced by Zhao et al. (2023) for trace level detection of 5 PAHs (ANT, BaA, BbF, FL, PYR) in tea leaves. In optimized conditions the method showed 84.8–105.4% recoveries with 1.69–9.97 ng/g of LODs and 5.12–30.21 ng/g of LOQs, respectively (Zhao et al. 2023). Purcaro et al. used SPE with UHPLC-FLD to quantify 16 EPA-listed PAHs in synthetic textile fibers. This method showed LODs and LOQs of 0.3 and 0.9 μg/kg, respectively, with > 0.99 of regression coefficients (Purcaro et al. 2012). A liquid–liquid phase microextraction prior to UHPLC-(+)APCI-MS/MS analysis enhanced the sensitivity of nitro- or oxy-PAHs analysis. The method achieved 0.02 to 0.85 ng/mL of LODs and 0.15 to 1.10 ng/mL of LOQs with up to 95–100% extraction recoveries (María Guiñez et al. 2018). The integration of UHPLC with APPI and

MS/MS has effective in simultaneous detecting of 20 PAHs and 9 nitro-PAHs in the environmental matrices (Aerosol samples) with LODs values < 10 pg and < 3 pg, respectively (Lung et al. 2015).

Capillary Electrophoresis (CE) or Capillary Electrochromatographic (CEC) Technique

This technique combines principles of electrophoresis and chromatography that uses a capillary tube and a high voltage electric field for liquid phase separation. It offers ease operation with high speed and separation efficiency. It is an effective method for detecting PAHs particularly PAHs having high boiling point (Zhang et al. 2022a, b, c; Ferey et al. 2014). Ferey et al. (2014) critically reviewed the applications of capillary electrophoresis in the analysis of PAHs with diverse approaches to capillary and microchip electrophoresis (Ferey et al. 2014). In a study the portable microchip capillary electrophoresis technique showed excellent LOD for PAHs in the low ppb range (6 ppb to 2000 ppb), representing a promising method for the analysis of PAHs (Stockton et al. 2008). Micellar Electrokinetic Chromatography, abbreviated as MEKC, is a mode of capillary electrophoresis was integrated with stir bar sorptive extraction and liquid desorption process for evaluating the PAHs in aqueous medium. The method achieved 2–11 mg/L of LODs with 0.99 of coefficient of determination (Pedro et al. 2006). Similarly Alzola et al. (2008) applied MEKC with a diode array detector (P/ACE) and HPLC with fluorescence detector for 16 PAHs detection in 89 samples of sewage sludge. The results showed by techniques was convincing and detection limit as low as 0.003, 0.17, 0.0055 and 0.0013 $mgkg^{-1}$ for NAP, ACY, FLU and PYR, respectively (Alzola et al. 2008). Ludivine Ferey et al. (2013) employed a novel method which was based on cyclodextrin (CD)-modified capillary zone electrophoresis (CZE) and laser-induced fluorescence (LIF) detector to analyze 19 PAHs in food samples. The method takes only 15 min separation time and showed LOQs < 100 μ/L for 16 out of 19 PAHs with efficiency $>1.5 \times 10^5$ (Ludivine Ferey et al. 2013). Wu et al. (2018) used C18 silica microspheres with CEC to analyze 16 PAHs in aqueous samples, achieved detection limits between 0.020 to 0.89 mg/L with good reproducibility and good recovery (87.6–112.4%) rates (Wu et al. 2018). Cheng et al. (2013) invented an online CEC–APCI-MS protocol using poly (stearyl methacrylate-divinyl benzene) monolithic column for the detection of 16 trace PAHs in seafood, which achieved 1.3 to 37.5 ng/g of LODs and 4.3 to 125.0 ng/g, of LOQs with good correlation coefficients (0.9950–0.9999) (Cheng et al. 2013).

Supercritical Fluid Chromatography (SFC)

It combines the benefits of both GC and UPLC and support diverse detectors. It can operate in normal or reverse phase mode making it a more versatile separation technique. It has several advantages including compatibility with thermally degradable and high boiling compounds, higher resolution than HPLC, and adjustable solvating power through temperature and pressure control. Initially CO_2 was used as the sole mobile phase; however modern SFC also uses subcritical conditions with polar organic additions (such as acetonitrile, methanol, isopropanol) to improve separation efficiency (Schulze et al. 2020; Klink et al. 2015). Lübeck et al. (2018) proposed a SFC method coupled with UV/QTOF-MS detectors to simultaneously investigate PAHs and oxy-PAHs in unconventional oils. The method showing good speed and efficiency with bare silica columns and incorporating water or formic acid as additives further enhanced oxy-PAHs signals significantly (Lübeck et al. 2018). Similarly a fast and sensitive SFC-MS technique (with QuEChERS) measuring oxygenated and parent PAHs within 15 min in three distinct kinds of milk samples. It achieved 67.66–118.46% recovery, 0.13–0.78 μg/kg LOQs and 0.04–0.24 μg/kg LODs with 1.45–14.68% precision (Zhang et al. 2022b). Yoshioka et al. (2018) used a method i.e. SFC/APCI-MS to identify 16 EU-listed PAHs in coffee and dark beer beverages, presenting a LOQ value of < 1 mg/kg only in 14 min of analysis time (Yoshioka et al. 2018). Likewise, SFC-APLI-TOF(MS) showed sub-μg/L of detection limits and give 0.5 μg/L LOD value for 1-hydroxypyrene in human urine sample (Klink et al. 2015). Wicker et al. (2018) designed an online SFE-SFC-MS method which reduced the sample extraction and preparation time along with minimizing the loss of samples. In context to PAHs analysis this technique showed 0.001 to 5 ng/g LODs and 5 to15 ng/g LOQs with $R^2 \geq 0.99$, representing a promising method for the analysis of contaminants environmental matrices (Wicker et al. 2018). Continuing advances in chromatographic methods are expected to further improve the monitoring and assessment of PAH pollution in the environment.

Spectroscopic Techniques For PAH Analysis

Ultraviolet–Visible (UV–Vis) Spectroscopy

Spectroscopy is broadly used detection techniques that measure the absorbance of a substance in a certain wavelength range. It is applied for qualitative as well as quantitative analysis of substances. Ultraviolet–visible spectrometers measure attenuation of chromophoric analytes based on molecular structure and obeys Beer's law, where the alteration in absorbance is proportional to the concentration of analyte. All the compounds absorb UV radiation at specific wavelength, and this leads to $\sigma \rightarrow \sigma^*$, $n \rightarrow \sigma^*$, $\pi \rightarrow \pi^*$ and $n \rightarrow \pi^*$ transitions (Mansouri et al. 2020). PAHs have characteristic UV absorbance due to their conjugated ring structures (Halasinski et al. 2005). Although UV spectrometer is easy to operate, cost-effective and fast to analyze, it is not a highly applicable measurement technique for measuring PAHs alone. However, chromatographic techniques based on UV detectors are widely adopted for the identification of PAHs due to their affordability and accessibility (Gupta et al. 2015; Gupta 2016). Researchers successfully used UV–Vis spectrophotometer to study the % absorption removal of some selected PAHs from aqueous samples (Gupta et al. 2016; Kumar et al. 2020; Balati et al. 2014). For example, to evaluate the % adsorptive removal of some selected PAHs (ACE, NAP, PHE) from wastewater, Balati et al. (2014) successfully used UV–Vis spectrophotometer (UV–Vis 2100 Shimadzu) model) at λ_{max} 225.4, 220.5, and 250.0 nm for ACE, NAP, and PHE respectively (Balati et al. 2014). Similarly, Fuat Topuz et al. (2017) also analyzed PAHs (ANT, FL, PHE) adsorption on cyclodextrin-functionalized mesostructured silica nanoparticles using UV–Vis spectrophotometer. The spectra were recorded by Cary 100 spectrophotometer between 200 and 800 nm using quartz cuvette and evaluated the conc. of ANT, FL, and PHE (Fuat Topuz et al. 2017). Jager et al. (2006) used various spectroscopic methods including UV–vis. and IR for the analysis of composition of the extract of Carbon soot. The UV–visible spectra of the extracts showed main bands at around 250 nm wavelength, confirming the abundant presence of three-ring PAHs such as phenanthrene and anthracene. The bands on the longer wavelength side (beyond 300 nm) are evidence of the existence of higher PAHs such as PYR, which was also confirmed by matrix isolation spectroscopy in solid argon (Jäger et al. 2006). Recent advances have notably improved the applicability of this method for example Fabiani et al. (2024) were fabricated a thin film of PAHs on to a transparent substrate and examined their solid phase UV–visible spectra in the far-UV and UV–vis. regions. Further, Time-Dependent Density Functional Theory simulated the spectra which showing good agreement with experimental data. This study provides valuable reference for thin films and interstellar medium technologies (Fabiani et al. 2024). Oña-Ruales et al. (2016) has been applied annellation theory with UV–Vis spectra to guess the regions of maximum absorbance (LMA) of 21 PAHs (dibenzo and naphtho derivatives of fluoranthene). This study offers a promising advancement in the prediction of LMA in the UV–Vis spectra for complex PAHs (Oña-Ruales et al. 2016).

Fluorescence Spectroscopy (FS)

Fluorescence or luminescence, are the emission of light by substances in electronically excited states. It is a rapid, transient process involving excitation of electrons usually caused by absorbance of ultraviolet (UV) light trough $\pi \rightarrow \pi^*$ transition (Tian et al. 2025). The fluorescence of a substance depends on molecular structure, functional groups, and also the background matrix. PAHs have excellent fluorescence qualities due to their rigid planar π-bonded conjugated structure. Therefore, Fluorescence spectroscopy is highly sensitive for detecting PAHs. However, structurally similar PAHs in complex samples can cause spectral overlap and interference, making them complicates to identify and quantify. To overcome these challenges synchronous fluorescence spectroscopy (SFS) have been evolved, which not only improves spectral resolution but also enhances the selective detection of PAHs in complex samples (Tian et al. 2025; Wei et al. 2023; Li et al. 2011). In synchronous fluorescence spectroscopy scanning of excitation and emission wavelength take place simultaneously at a constant difference of Stokes' shift ($\Delta\lambda = \lambda_{em} - \lambda_{ex}$) which enhances the sensitivity of SFS allowing reduced overlap during analysis PAHs containing complex samples. SFS is easy to analyze multiple fluorescent components in complex samples due to its low spectral overlap (Sunuwar et al. 2021). Recently, Tian et al. (2025) reviewed about the advancement in the SFS in order to detect the PAHs in petroleum and coal samples. Based on previous studies, they assigned SFS peak values for PAHs and their derivatives. It is evident from the SFL peaks that the fused aromatic nuclei with 1, 2, 3 and 4 rings showed signals at 300 nm, 300–340 nm, and 340–400 nm and above 400 nm respectively (Tian et al. 2025), which are presented in Table 9.2.

Wang et al. (2012) applied multi-peak fitting technique to resolve the overlapping SFS peaks and assigned the SFS peaks to PAHs in coal samples based on different ring numbers. They described that the PAHs show different fluorescence characteristic depending on the number of rings. PAHs which have more linearly condensed aromatic rings show fluorescence at longer wavelengths. They reported that the SFS peaks at 286, 325 and 359 nm might be due to the single-, double- and three-ring fused aromatic nucleuses, respectively (Wang et al. 2012). Similarly, in the analysis of a mixture of six standard compounds, Sunuwar et al. (2021) reported SFS peaks at 321.9, 329.6, 374 and 382 nm for NAP, ACE, PYR and ANT, respectively, using computer simulations with the SFS technique (Cary Eclipse Fluorescence Spectrophotometer) (Sunuwar et al. 2021). Hou et al. (2019) used GC–MS and FS together to analyze the PAHs in crude oil. The results showed that the amount of class II PAHs (NAP and ACE series) was more than 60% with high fluorescence intensity in the 320–350 nm range, while class IV PAHs (PYR and CHY) were very low in the 380–400 nm range (Hou et al. 2019). Recently Huang et al. (2024) examined the real-time migration of some selected PAHs (ANT, PYR, PHE) from plastic items to food materials using fluorescence spectroscopy with a TDR-ATLD algorithm based calibration method. The proposed method achieved excellent LOD (0.08, 0.06, and 0.21 ng/mL) and LOQ (0.24, 0.17, 0.65 ng/mL) with R^2 values of 0.9866, 0.9896, and 0.9840 for ANT, PYR, and PHE, respectively (Huang et al. 2024). Digambar Patra (2008) developed a fluorescence-based fiber optic sensor for the detection of PAHs in various motor oils and the study found that the kerosene and petrol oil have lower ring PAHs than mobil, diesel, and 2 T oil (Patra et al. 2008). Li et al. (2011) demonstrated the potential of SFS to detect trace concentrations of BaP, BkF, and ANT in complex green, black and oolong tea with MAE pre-treatment of samples, obtaining detection limits as low as 0.18–0.28, 0.55–0.89, and 0.64–3.58 µg/kg for BaP, BkF, and ANT, respectively with 77.1 to 116% recoveries (Li et al. 2011). Additionally, Laser-induced fluorescence (LIF) spectroscopy has excellent detection sensitivity for real-time field screening of PAHs (Taylor et al. 2021; Qazi et al. 2021). For example an intrinsic fluorescence based confocal laser scanning microscopy detect PHE, PYR and NAP distribution in contaminated soil without solvent extraction with rapid spatial and concentration data in a single scan (Qazi et al. 2021). A study demonstrated the capability of denuder technology coupled with thermal

Table 9.2 Synchronous fluorescence spectral peaks assignment of PAHs (Tian et al. 2025; Wang et al. 2012)

PAHs	Chemical structure	Number of fused ring	Wavelength (nm) at Δλ = 23 nm	Wavelength (nm) at Δλ = 17 nm	Wavelength (nm) at Δλ = 10 nm
Naphthalene	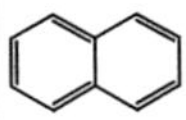	2	321	309	–
Acenaphthene	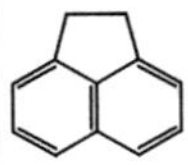	3	326	–	–
Acenaphthylene	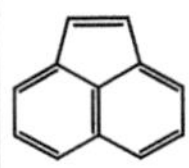	3	–	–	–
Fluorene	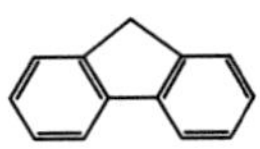	3	299	–	–
Anthracene	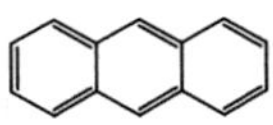	3	378	373	
Phenanthrene	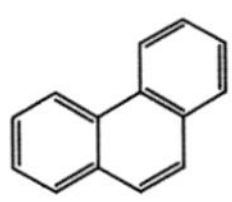	3	345	372	–
Pyrene	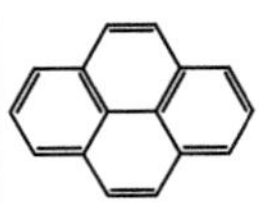	4	369	–	–
Fluoranthene	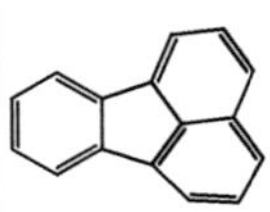	4	–	–	–
Benz[a]anthracene	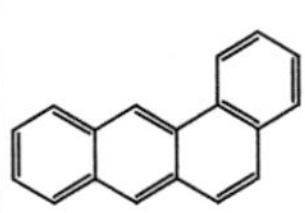	4	407	–	–
Chrysene	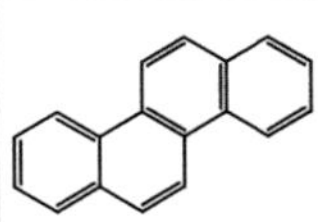	4	341	–	–
Benzo[b]fluoranthene	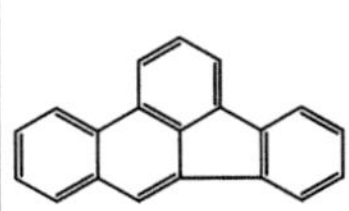	5	–	–	407

(continued)

Table 9.2 (continued)

PAHs	Chemical structure	Number of fused ring	Wavelength (nm) at $\Delta\lambda$ = 23 nm	Wavelength (nm) at $\Delta\lambda$ = 17 nm	Wavelength (nm) at $\Delta\lambda$ = 10 nm
Benzo[k]fluoranthene	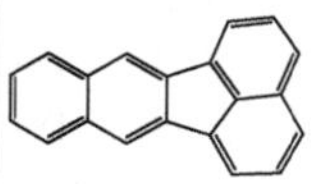	5	427		405
Benzo[a]pyrene	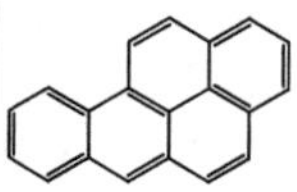	5	409		
Dibenz[a,h]anthracene	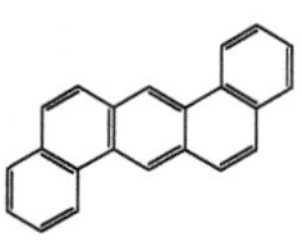	5	412	–	–
Benzo[ghi]perylene	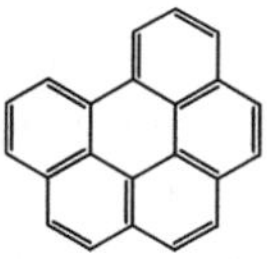	6	407	–	–
Indeno[1,2,3-cd]pyrene	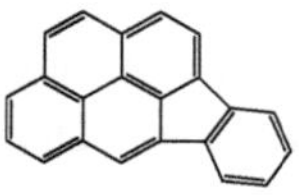	6	472		470

desorption GC–MS (TD-GC–MS) with LIF to monitor and quantify PAHs emissions from household fires, burning sugar cane, and industrial processes in South Africa. This method achieved detection limits for NAP varied from 1.8 μg/m^3 (diesel) to 26 μg/m^3 (industry) (Forbes et al. 2010). Spectral data analyses by using machine learning (ML) have simplified the multi-components analysis. For example to analyze binary mixtures of three PAHs (FL, ACY, and NAP) in water Wang et al. (2020a, b) invented a 3D FS with a bird swarm algorithm and a support vector machine model which achieved 100% classification accuracy with shortest running time (Wang et al. 2020a, b). Recently Wei et al. (2023) employed constant wavelength SFS with the BP algorithm to examine and quantify four PAHs (BaP, BbF, BaA, CHY) in edible oil. The proposed method showed strong predictive accuracy with more than 98.5% recoveries with LOD values of 0.014, 0.068, 0.026, and 0.013 μg/kg for BaP, BbF, BaA, and CHY, respectively (Wei et al. 2023). Finally, fluorescence spectroscopy-based methods, particularly SFS and LIF, are emerging techniques for rapid, sensitive, and selective investigation of PAHs in complex environmental matrices.

Fourier-Transform Infrared (FTIR) Spectroscopy

FTIR has the advantages include ease of sample processing without extraction, high sensitivity towards IR radiation and its ability to analyze complex molecules. Molecular structures have various functional groups that exhibit unique vibration modes such as stretching vibrations (symmetric or asymmetric) mode and bending vibrations (scissoring, rocking, wagging, and twisting) mode. Infrared spectroscopy or vibrational spectroscopy identified various types of vibrations between atoms in functional groups through absorption of infrared electromagnetic radiation of specific wavelengths (Zhang et al. 2024a, b). The presence of PAHs or their derivatives are confirmed by the FTIR signals corresponding to aromatic C–H stretching vibrations which are found within the wave number range of 3000 to 2900 cm^{-1} (Okparanma et al. 2013a; Kumar et al. 2023a, b). For example in a study the IR band within a range of 2960–650 cm^{-1} confirm the aromatic rings due to the presence of PAHs (Itodo et al. 2018). FTIR used universally in the identification of organic compounds including PAHs. For example, Jager et al. (2006) applied various spectroscopic techniques including FTIR for determination of composition of the extract of carbon soot. The FTIR spectra showed bands between 3300–2800 cm^{-1} due to C–H stretch, 3100–3000 cm^{-1} due to aromatic = C–H vibration, and a weak band at 3299 cm^{-1} due to triply bound C–H stretch revealing the existence of PAHs or substituted aromatics in the extract (Jäger et al. 2006). Elufisan et al. (2020) studied the PAH degradation efficacy of Stenotrophomonas species by using FT-IR spectrometer (Bruker Alpha) and some new peaks presented in FTIR spectra at 3200–2800 cm^{-1} (due to –OH group) at 1641 cm^{-1} (due to $-C = O$ group) and at 2911 cm^{-1} (due to $-CH2$) which confirms the degradation of PAHs (Elufisan et al. 2020). Similarly Qi et al. (2017) studied the PAHs degradation efficiency of Gordonia sp. nov. strain Q8 (isolated from oil-field produced water) using FTIR. They employed 32 scan by using FTIR model FTS-40 in the range of 4000–400 cm^{-1} and the peak at 1660.43 cm^{-1} showed that PAHs were degraded into the benzene molecules (Qi et al. 2017). IR spectra particularly, Vis–NIR spectra have broad and overlapping overtones and combinations band therefore difficulty in interpreting of these spectra can be overcome by using advanced data-processing techniques such as multivariate calibration methods (Eldos et al. 2020; Okparanma et al. 2013b). For example, Vis–NIR with PLS (Partial least-squares) regression analysis predicted PHE in diesel-contaminated soil samples very well with 0.75–0.83 value of coefficient of determination and 0.21–0.25 mg/kg of root-mean-square error (Okparanma et al. 2013b). Moreover, FTIR spectroscopy widely used in the field of asphalt research and predicting vital properties of asphalt materials (Zhang et al. 2024a, b;). For example, recently, Wetekam et al. (2024) modified FTIR to identify substances affecting the reusability of reclaimed asphalt (RA). The study used peak height or peak area analysis at specific absorption bands (band at 630–741 cm^{-1} for Antraphthylenes, at 660–749 cm^{-1} for Flourenes, at 640–749 cm^{-1} for Flouranthenes, at 650–732 cm^{-1} for Indenopyrene and at 640–742 cm^{-1} for Phenanthrenes) to detect and quantify

PAH contamination as well as binder and asphalt modifications, and ageing conditions. The results showed that extremely PAH-contaminated binders (8000 mg/kg PAHs) could be simply recognized visually, while less polluted binders had a spectrum similar to that of non-contaminated bitumen (Wetekam et al. 2024). Recently, an attenuated total reflection (ATR) integrated FTIR technique quantify the PAHs in 34 binders samples derived from reclaimed asphalt using multivariate analysis methods. It offered a promising protocol for the partitioning between PAH quantities below or above the 25 mg/kg limit recognized in Germany (Weigel et al. 2022).

Surface-Enhanced Raman Spectroscopy (SERS)

It is a non-destructive, rapid and ultra sensitive technique used for onsite and trace detection. It is an emerging method for detecting low concentrations of organic pollutants in complex environmental matrices, even for single molecule detection. However, PAHs showed poor SERS response due to lacking metal-affinity functional groups. Therefore researchers have focused on converting PAHs into a SERS active structure. For this purpose functional groups are modified onto the surface of metal NPs which can create trap or hydrophobic conditions to capture PAH molecules. These modified systems offer rapid and onsite analysis capabilities (Zhang et al. 2022a, b, c; Ni et al. 2025; Gu et al. 2016). For example, Zhang et al. (2022a, b, c) fabricated gold nanoparticles (AuNPs) modified with sulfhydryl-substituted cyclodextrin based-SERS for quantitative analysis of selected PAHs. The proposed method showed high sensitivity and high selectivity for PYR, ANT, PHE, and triketoneum with LODs of 100, 10, 100, and 1000 nM, respectively (Zhang et al. 2022a, b, c). Similarly, a SERS method using AuNPs decorated hydrophobic porous GMA-EDMA polymer is developed by Wang et al. (2014) in which performance of SERS significantly improved because porous GMA-EDMA polymer enhances PAHs adscrption via hydrophobic interactions and facilitates uniform AuNPs distribution. The LODs of ANT, PHE, and PYR by this method were reported to be 0.93×10^{-7}, 4.5×10^{-7}, and 1.1×10^{-7} M, respectively, with good reproducibility (RSD $\approx$ 16%) rates (Wang et al. 2014). In a study a bare gold nano particle-coupled-film based SERS was developed by Gu et al. (2016) for capturing PAHs from bulk solution, which showed excellent detection ability for PAHs as low as 1.2×10^{-8} M, 2.0×10^{-8} M, 5.5×10^{-8} M, 6.3×10^{-8} M for BbF, FLU, BaA, and PYR, respectively (Gu et al. 2016). Recently, Ni et al. (2025) designed and applied a glass fiber filter customized with AgNPs (AgNPs@GF filter)—based SERS technique for quick and on-site qualitative and quantitative measurment of PAHs (FLU, PHE, and PYR) in water and air. The developed strategy demonstrated excellent performance with 0.7, 1, and 0.1 ppb LODs for FLU, PHE, and PYR with good recovery (89.22 – 99.70%) and correlation coefficient (R^2 = 0.975/0.936/0.958; $P < 0.001$) in water. Similarly PAHs present in air were collected on AgNPs@GF filter and LODs of 9.11, 18.18, and 14.59 ppb were achieved for FLU, PHE, and PYR, respectively with

R^2 values of 0.967, 0.973 and 0.957, respectively (Ni et al. 2025). Tijunelyte et al. (2017) present a 4-dodecyl benzenediazonium-tetrafluoroborate based SERS nanosensor which achieved 0.026, 0.064, and 3.94 mg/L LODs for BaP, FLU, and NAP, respectively in water samples (Tijunelyte et al. 2017). Furthermore, various graphene or graphene oxide (GO) based material have been developed as SERS substrates which offers strong adsorption of PAHs through π-π interactions (Liu et al. 2022; Wang et al. 2020a, b). For example, an efficient SERS method using Fe_3O_4@PDA@Ag@GO nanocomposite is reported by Liu et al. (2022) which achieved 10^{-8} to 10^{-7} g/L of LODs for PHE, PYR, and benzanthene. In this method AgNPs provide strong electromagnetic enhancement, graphene oxide (GO) offers chemical enhancement through π-π interactions while polydopamine (PDA) prevent Ag from agglomeration and shedding (Liu et al. 2022). Likewise, Wang et al. (2020b) presented AgNPs/graphene based SERS for quick detection of PAHs in the environment, which achieved LOD as low as 0.73 ppb, 1.1 ppb and 0.57 ppb for PYR, ANT and PHE, respectively (Wang et al. 2020a, b). Zhou et al. (2020) confirmed the reliable performance of AgNPs-based SERS for the trace qualitative as well as quantitative detection of 16 PAHs in water samples at a conc. level as low as 0.1–100 μg/L. The integration of LLE with this method further enhances the detection sensitivity by 2–3 times and achieved LOD 100 ng/L for ANT, 50 ng/L for PYR and 5 ng/L for BaP (Zhou et al. 2020). Recently, molecularly imprinted polymer (MIP)-based sensors/substrates are emerging materials for SERS for highly selective ultra-trace detection of pollutants like pesticides, heavy metal, pharmaceuticals and PAHs etc. in complex environmental matrices (Lazarević-Pašti et al. 2023). MIPs are three dimensional smart materials that are formed in the presence of a template molecule (the target analyte such as atoms, ions or complex molecules). After polymerization reaction removal of template leaves behind cavities that selectively rebind the target analyte (Fig. 9.4). Thus MIPs have extraordinary molecular recognition properties and therefore, showed potential applications in catalysis, drug delivery, chemo/bio sensing, separation and purification technologies (Lazarević-Pašti et al. 2023; Castro-Grijalba et al. 2020).

Integration of MIPs with plasmonic nanostructures leads to synergistic hybrid platform where MIPs provide selective binding sites for the target pollutant and NPs enhance Raman signals through SERS (Castro-Grijalba et al. 2020). Castro-Grijalba et al. (2020) invented a hybrid Au@MIP system for SERS by layering gold nanoparticles (AuNPs) on glass followed by uniform spin coating with a MIP thin film. They used PYR or FLU as templates for the preparation of Au@MIP substrate. Therefore, the proposed system offered the detection of PYR and FLU in nM level concentrations with a 100-fold perfection compared to corresponding non-imprinted system, highlighting the potential applications of MIPs based hybrid SERS in the selective recognition of pollutants (Castro-Grijalba et al. 2020). The accuracy of detection of PAHs particularly in aqueous environment is low due to the complex water environment, and these problems can be overcome by the integration of chemometrics with SERS technology. In this context recently, Zhang et al. (2024b) developed a WT-SG-SiPLS-VIM-RF model by using chemometrics with SERS and random forest (RF) algorithm integrating as spectral preprocessing

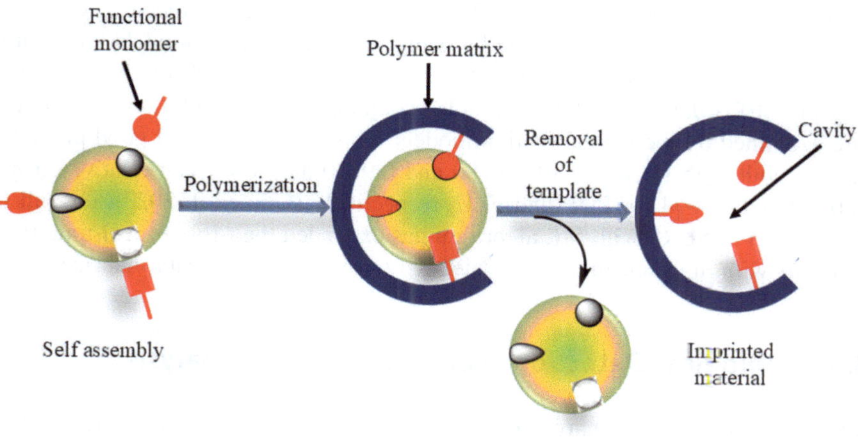

Fig. 9.4 Molecular imprinting polymer (MIP) process

for quantify and assess the carcinogenic threat of PAHs in water. The performance of the model studied on the SERS data of 36 water samples (mixed with AgNPs) collected from Yulan Lake, tap water and distilled water. The proposed model showed superb predictive ability for PHE and BaA, with the prediction determination coefficients of 0.9658 and 0.9537 and the average relative error of prophecy of 0.0646 and 0.0949, respectively (Zhang et al. 2024a, b). Furthermore, managing large volume of Raman spectral data with overlapping signals are further most challenges in SERS analysis. To address these challenges, the recent integration of machine learning (ML) or artificial intelligence (AI) has revolutionized the field by enhancing big data processing capabilities (Srivastava et al. 2024; Guo et al. 2023). For advanced data processing ML focuses on developing multiple multivariate data analysis algorithms and statistical models that automate data analysis and are capable to recognize vibrational fingerprints with precision and accuracy (Lussier et al. 2020). Presently, Raman spectroscopy analysis used several supervised learning algorithms include back propagation (BP), partial least square (PLS), convolutional neural network (CNN), and genetic algorithm (GA) which have advantages such as high prediction accuracy, easy of overfitting, and strong anti-noise ability (Srivastava et al. 2024; Guo et al. 2023). For example, Yan et al. (2021) was applied three types of calibration algorithms (PLS, GA and BP) with SERE for quantitative measurement of NAP, PHE, and PYR. In this approach first of all PLS helps to choose the characteristic variables and then GA is used for determining the initial weights due to its global search capability and BP neural network to avoid the local minima. The proposed model showed excellent coefficient of determination of 0.9975, 0.9710, 0.9643, and root mean square error of 13.10, 5.40, 5.10 nmol/L for NAP, PHE, and PYR, respectively (Yan et al. 2021). Recently Atta et al. (2023) used gold nanostars based portable SERS and CNN AI model to identify various PAHs in water samples. The module achieved the LODs for PAHs in nanomolar range with 1.07×10^{-1} (μM) $RMSE_{conc}$ and 90%

of prediction accuracy (Atta et al. 2023). In another research recently, Guo et al. (2023) developed a SERS module by using RF as calibration models for quantify PHE and FLU in water samples. The experiment begins with firstly the collecting the SERS data of 31 PAHs samples containg AgNPs, and after that the data are examined by the RF calibration models. This method showed good prediction ability with high R^2 values of 0.9780 and 0.9770 and low mean relative error of 0.0369 and 0.0694 for PHE and FLU, respectively (Guo et al. 2023). Overall the integration of SERS with ML algorithm has expended their role in real-time PAHs analysis with high sensitivity and selectivity across environmental matrices.

Immunological Detection Methods for PAHs Analysis

Immunoassays are bioanalytical procedures that are capable to analyses PAHs based on their physical characteristics. These methods utilize the selective binding affinity between antibodies and PAHs antigens and offers rapid screening of environmental samples. In recent years various Immunoassay techniques including chemiluminescence immunoassays (CLIA), fluoroimmunoassays (FIA), and particularly enzyme–linked immunosorbent assays (ELISA) is very popular for the detection of PAHs (Qiao et al. 2017; Sun et al. 2018). For example in water samples a monoclonal antibody (McAb) based ELISA achieved 99.08% and 98.45% of recovery rates for ACE and PYR, respectively with 12.17 $ngmL^{-1}$ of sensitivity (Qiao et al. 2017). Similar approach has been introduced by Li et al. (2016) for real time analysis of 3–5 ring PAHs in aqueous samples which measured a conc. minimum of 0.2 μg/L within 10 min of analysis time (Li et al. 2016). Furthermore, Fluorescence-linked immunoassays (FLISA) provide enhanced sensitivity due to their stronger optical signals. To detect PHE and its homolog in water samples, Sun et al. (2018) fabricated a FLISA in conjugation of McAb in which rhodamine B isothiocyanate acts as fluorescent dye. This system achieved > 100% of recovery rates with 1.05 ng/mL of LOD for PHE and its homolog (Sun et al. 2018). Moreover, for detecting PAHs real-time polymerase chain reaction based Immunoassay (RT-IPCR) have been introduce which showed 100-to-10000-fold of sensitivity as compare to the traditional immunoassays (Meng et al. 2015, 2016). For example, a DNA based RT-IPCR assay achieved 450 fmol/L of LOD for PYR with recoveries rates of 96.8%, 101.4% and 99.6% from lake, tap and commercial mineral water, respectively (Meng et al. 2015). A similar RT-IPCR assay based on pyrene-modified DNA system demonestrated ultrasensitive recognition performance and achieved LOD of 3.5 pmol/L for PYR with recoveries rates of 98.4%, 98.2% and 99.7% for tap, lake and mineral water, respectively. Moreover, the system also exhibited a strong correlation ($R^2 = 0.9906$) with GC–MS method (Meng et al. 2016). Immunosensors, which convert antigen–antibody interactions into measurable electrical and biomechanical signals via transducers, further, enhance the applicability of these techniques. They also allow in-situ analysis and fast screening for multiple environmental samples. For example, Recently

Wu et al. (2023) developed an immunosensor using a multi-walled carbon nanotubes (MWCNTs)-chitosan oligosaccharide (COS)- nanocomposite loaded with phenanthrene antibody for detecting PHE in seawater. The sensor's strong electron transfer ability and biocompatibility were used to amplify the detection signal and showed a sensitive and accurate real-time monitoring with LOD of 0.30 ng/mL (Wu et al. 2023). Likewise a same approach have been presented by Yan et al. (2022) for detection of BaP in sea water which achieved 0.27 ng/ml of LOD with 96.6 to 100% of recovery rate (Yan et al. 2022). The widely used techniques for the detection of PAHs in environmental matrices are summarized in Table 9.3.

Conclusion and Future Perspectives

The detection and quantification of PAHs remain a critical area in environmental analytical chemistry due to their carcinogenicity, environmental persistence, and complex behavior across various matrices. Traditional methods involving SE, LLE, and mechanical agitation combined with GC and HPLC have been foundational for PAH analysis, offering reproducibility and robust quantification. Spectroscopic methods—particularly UV–Vis, fluorescence, FTIR, and Raman—provide additional advantages for rapid screening and structural elucidation, although they often suffer from limited sensitivity and specificity in complex mixtures. Hybrid approaches such as GC–MS and LC–MS have significantly enhanced detection capabilities, enabling trace-level quantification and isomer resolution, especially in heavily contaminated or complex matrices. These techniques detect conc. of PAHs with detection limit as low as 0.34 ng/L. Similarly, advanced techniques like CEC-MS and SFC-MS have shown the capacity of achieving high resolution (with 0.001 ng/g of LOD) with minimal solvent usage and faster run times. Immunological techniques, including ELISA and biosensor-based assays, provide portable and cost-effective alternatives for rapid on-site screening, although they are currently limited by antibody specificity and susceptibility to cross-reactivity. Emerging technologies such as SERS and portable sensor arrays are addressing many limitations of traditional techniques, offering real-time monitoring and field-deployable systems with up to 5–100 ng/L of LODs. Looking ahead, future efforts should focus on integrating sample preparation and detection into automated, miniaturized platforms that can operate in real-time under field conditions. Additionally, expanding the library of reference materials for emerging and substituted PAHs, improving matrix deconvolution algorithms, and harmonizing global standard protocols are essential steps for broader application. Interdisciplinary approaches that combine analytical chemistry, nanotechnology, and data science will be pivotal in pushing the boundaries of PAH detection, ensuring both environmental safety and public health protection.

Table 9.3 Summary of various techniques for the detection of PAHs

Type of sample	PAHs/number of PAHs	Extraction method	Recoveries	Detection technique	LOD/LOQ/PAHs concentrations range	Refs.
Ground water	16 PAHs	SPE	–	GC-FID	LOD: 0.024–0.113 μg/L, LOQs: 0.079–0.378 μg/L	(Carvalho et al. 2019)
Soil	16 PAHs	–	–	GC-FID	Conc.: 66.4–1342.5 μg/kg	(Emoyan et al. 2020)
Soil near road side	16 PAHs	–	–	GC-FID	Total average conc.: 4.04 μg/g	(Kumar et al. 2011)
Road dust samples	16 PAHs	–	–	GC-FID	Conc.: 137.8 to 5813 ng/g	(Hishamuddin et al. 2023)
Dairy products	16 PAHs	Semi-automated SPE	80–107%	GC–MS	LODs: 1–200 ng/kg	(Palacios Colon et al. 2022)
Drinking water	42 parent and halogenated PAHs	SPE	74.88–119.4%	GC–MS	LODs: 0.34 to 3.37 ngL^{-1}	(Liu et al. 2019)
Meat and fish samples	16 PAHs	UAE with semi-automated SPE	85–105%	GC–MS	LODs: 3–70 ng/kg	(Rascón et al. 2019)
Infant foods	BaP, BaA, CHY, and BbF	QuEChERS	73.1–110.7%	GC–MS/MS	LODs:0.019–0.036 μg/kg; LOQs: 0.06–0.11 μg/kg	(Mariateresa et al. 2024)

(continued)

Table 9.3 (continued)

Type of sample	PAHs/number of PAHs	Extraction method	Recoveries	Detection technique	LOD/LOQ/PAHs concentrations range	Refs.
River sediment	44 PAHs	ASE, SPE	52.8–114.1%	GC-EI-SIM, GC-NICI-SRM, GC-EI-SRM, LC–ESI–MS/MS	LODs: 0.01–0.02 ng/g (PAHs); 0.002–0.067 ng/g (Nitro-PAHs); 0.01–0.1 ng/g (Oxy-PAHs); 0.003–0.006 ng/g (OH-PAHs)	(Han et al. 2019)
Water samples	ANT; PHE; PYR; FLU	–	–	HPLC-FLD	LODs: 0.1–1 pg/mL; LOQs: 0.3–3 pg/mL	(Saini et al. 2017b)
River water, Coal ash water	8 PAHs	SPME	River water: 80.6–106.6% Coal ash water: 80.9–103.5%	SPME–HPLC	LODs: 0.01–0.10 μg/L	(Wang et al. 2017)
Wastewater	16 PAHs	ASE & SPE	–	HPLC–DAD-FLD	LODs: 0.020 to 2.0 μg/g (Particulate phases); 0.0098 to 0.50 μg/L (Dissolved phases)	(Vistnes et al. 2022)
Environmental water samples	6 PAHs	Deep eutectic solvent based micro extraction	83–117%	HPLC/FLD	LODs: 0.7–6.6 ng/L	(Yousefi et al. 2018)
Herbal plants	16 PAHs	QEChERS	–	HPLC-FL	LODs: 902.35 μg/g	(Woźniak et al. 2023)
Tea leaves	ANT, BaA, BbF, FL, PYR	–	84.8–105.4%	UHPLC	LODs: 1.69–9.97 ng/g LOQs: 5.12–30.21 ng/g	(Zhao et al. 2023)

(continued)

Table 9.3 (continued)

Type of sample	PAHs/number of PAHs	Extraction method	Recoveries	Detection technique	LOD/LOQ/PAHs concentrations range	Refs.
Synthetic textile fibers	16 PAHs	SPE	–	UHPLC-FLD	LODs: 0.3 μg/kg, LOQs: 0.9 μg/kg	(Purcaro et al. 2012)
Natural and lake waters	Nitro- or oxy-PAHs	DLLME-SFO-SBE	95–100%	UHPLC-(+)APCI-MS/MS	LODs: 0.02–0.85 ng/mL LOQs: 0.15 –1.10 ng/mL	(María Guiñez et al. 2018)
Aerosol samples	20 PAHs and 9 nitro-PAHs	–	–	UHPLC-APPI-MS/MS)	LODs: <10 pg (PAHs) LOQs: <3 pg (Nitro-PAHs)	(Lung et al. 2015)
-	10 PAHs	–	–	Portable microchip CE	LODs: 6 ppb to 2000 ppb	(Stockton et al. 2008)
Aqueous samples	5 PAHs	Stir bar sorptive extraction	–	MEKC	LODs: 2–11 mg/L	(Pedro et al. 2006)
Sewage sludge	16 PAHs	–	–	MEKC and HPLC-FL	LODs: 0.003 (NAP), 0.17 (ACY), 0.0055 (FLU) and 0.0013 (PYR) $mgkg^{-1}$	(Alzola et al. 2008)
Food samples	19 PAHs	–	–	CZE/LIF	LOQs < 100 μ/L	(Ludivine Ferey et al. 2013)
Water samples	16 PAHs	–	87.6–112.4%)	CEC	LODs: 0.020 to 0.89 mg/L	(Wu et al. 2018)
Seafood samples	16 PAHs	–	–	CEC–APCI-MS	LODs: 1.3–37.5 ng/g LOQs: 4.3–125.0 ng/g	(Cheng et al. 2013)

(continued)

Table 9.3 (continued)

Type of sample	PAHs/number of PAHs	Extraction method	Recoveries	Detection technique	LOD/LOQ/PAHs concentrations range	Refs.
Unconventional oils	PAHs, oxy-PAHs	–	–	SFC- UV/QTOF-MS	Conc.: 4 μg/mL	(Lübeck et al. 2018)
Milk samples	PAHs, oxy-PAHs	QuEChERS	67.66–118.46%	SFC-MS	LODs: 0.04–0.24 μg/kg LOQs: 0.13–0.78 μg/kg	(Zhang et al. 2022a, b, c)
Coffee and dark beer	16 PAHs	–	–	SFC/APCI-MS	LOQs < 1 mg/kg	(Yoshioka et al. 2018)
Human urine sample	1-hydroxypyrene	–	–	SFC-APLI-TOF(MS)	LODs: 0.5 μg/L	(Klink et al. 2015)
Soil	16 PAHs	SFE	–	SFC-MS	LODs: 0.001–5 ng/g LOQs: 5–15 ng/g	(Wicker et al. 2018)
Wastewater	ACE, NAP, PHE	–	–	UV–Vis spectroscopy	–	(Balati et al. 2014)
Water	ANT, FL, PHE	–	–	UV–Vis spectroscopy	–	(Fuat Topuz et al. 2017)
Carbon soot	PAHs	–	–	UV–Vis spectroscopy	–	(Jäger et al. 2006)
Petroleum and coal samples	PAHs	–	–	SFS	–	(Tian et al. 2025)
Plastic items, food materials	NAP, ACE, PYR, ANT	–	–	EEM-FS	LODs: 0.08, 0.06, and 0.21 ng/mL LOQs: 0.24, 0.17, 0.65 ng/mL	(Huang et al. 2024)

(continued)

Table 9.3 (continued)

Type of sample	PAHs/number of PAHs	Extraction method	Recoveries	Detection technique	LOD/LOQ/PAHs concentrations range	Refs.
Green, black and oolong tea	BaP, BkF, ANT	MAE	77.1–116%	SFS	LODs: 0.18–0.28 μg/kg (BaP) 0.55–0.89 μg/kg (BkF) and 0.64–3.58 μg/kg (ANT)	(Li et al. 2011)
Household fires, Burning sugar cane, Industrial processes	NAP	–	–	TD-GC–MS/ LIF	LODs: 1.8 μg/m^3 (diesel); 26 μg/m^3 (industry)	(Forbes et al. 2010)
Edible oil	BaP, BbF, BaA, CHY	–	–	SFS with BP algorithm	LODs: 0.014 μg/kg (BaP), 0.068 μg/kg (BbF), 0.026 μg/kg (BaA), and 0.013 μg/kg (CHY)	(Wei et al. 2023)
Diesel-contaminated soil	PYR, ANT, PHE	–	–	SERS	LODs: 100 nM (PYR), 10 nM (ANT), 100 nM (PHE)	(Zhang et al. 2022a, b, c)
Water	ANT, PHE, PYR			SERS	LODs: 0.93×10^{-7} M (ANT), 4.5×10^{-7} M (PHE), 1.1×10^{-7} M (PYR)	(Wang et al. 2014)
Water, Air	FLU, PHE, and PYR	–	89.22 – 99.70%)	SERS	LODs: 0.7 ppb (FLU), 1 ppb (PHE), 0.1 ppb (PYR)	(Ni et al. 2025)

(continued)

Table 9.3 (continued)

Type of sample	PAHs/number of PAHs	Extraction method	Recoveries	Detection technique	LOD/LOQ/PAHs concentrations range	Refs.
Water	BaP, FLU, and NAP			SERS	LODs: 0.026 mg/L (BaP, 0.064 mg/L (FLU), 3.94 mg/L (NAP	(Tijunelyte et al. 2017)
–	PYR, ANT, PHE,	–	–	SERS	LODs: 0.73 ppb (PYR), 1.1 ppb (ANT), 0.57 ppb (PHE)	(Wang et al. 2020a, b)
–	16 PAHs	–	–	SERS	LODs: 0.1–100 µg/L	(Zhou et al. 2020)
Water samples	ACE, PYR	–	99.08% (ACE); 98.45% PYR	ELISA	Conc.: 12.17 $ngmL^{-1}$	(Qiao et al. 2017)
Aqueous environmental samples	3–5 ring PAHs	–	–	ELISA	Conc.:0.2 µg/L	(Li et al. 2016)
Water	PHE	–	> 100%	FLISA	LODs: 1.05 ng/mL	(Sun et al. 2018)
Lake water, tap water, commercial mineral water	PYR			RT-IPCR	LODs: 450 fmol/L	(Meng et al. 2015)
Tap water, lake water, mineral water	PYR			RT-IPCR	LODs: 3.5 pmol/L	(Meng et al. 2016)
Seawater	PHE	–	–	Immunosensors	LODs: 0.30 ng/mL	(Wu et al. 2023)
Seawater	BaP		96.6–100%	Immunosensors	LODs: 0.27 ng/ml	(Yan et al. 2022)

References

Abdel-Shafy, H.I., Mansour, M.S.M.: A review on polycyclic aromatic hydrocarbons: source, environmental impact, effect on human health and remediation. Egypt. J. Petroleum [online] **25**(1), 107–123 (2016). https://doi.org/10.1016/j.ejpe.2015.03.011

Alzola, R., Pons, B., Bravo, D., Arranz, A.: Determination of polynuclear aromatic hydrocarbons (pahs) in sewage sludge by micellar electrokinetic capillary chromatography and hplc-fluorescence detection: a comparative study. Environ. Technol. **29**(11), 1219–1228 (2008). https://doi.org/10.1080/09593330802270673

Anitescu, G., Tavlarides, L.L.: Supercritical extraction of contaminants from soils and sediments. J. Supercrit. Fluids **38**(2), 167–180 (2006). https://doi.org/10.1016/j.supflu.2006.03.024

Atta, S., Li, J.Q., Vo-Dinh, T.: Multiplex SERS detection of polycyclic aromatic hydrocarbon (PAH) pollutants in water samples using gold nanostars and machine learning analysis. Analyst **148**(20), 5105–5116 (2023). https://doi.org/10.1039/d3an00636k

Balati, A., Shahbazi, A., Amini, M.M., Hashemi, S.H.: Adsorption of polycyclic aromatic hydrocarbons from wastewater by using silica-based organic–inorganic nanohybrid material. J. Water Reuse Desalination **5**(1), 50–63 (2014). https://doi.org/10.2166/wrd.2014.013

Banjoo, D., Nelson, P.K.: Improved ultrasonic extraction procedure for the determination of polycyclic aromatic hydrocarbons in sediments. J. Chromatogr. A **1066**(1–2), 9–18 (2005). https://doi.org/10.1016/j.chroma.2005.01.033

Biache, C., Lorgeoux, C., Saada, A., Colombano, S., Faure, P.: Fast method to quantify PAHs in contaminated soils by direct thermodesorption using analytical pyrolysis. Talanta **166**, 241–248 (2017). https://doi.org/10.1016/j.talanta.2017.01.055

Carvalho, F.I.M., Dantas Filho, H.A., Dantas, K. das G.F.: Simultaneous determination of 16 polycyclic aromatic hydrocarbons in groundwater by GC-FID after solid-phase extraction. SN Appl. Sci. **1**(7) (2019). https://doi.org/10.1007/s42452-019-0839-z

Castro-Grijalba, A., Montes-García, V., Cordero-Ferradás, M.J., Coronado, E., Pérez-Juste, J., Pastoriza-Santos, I.: SERS-based molecularly imprinted plasmonic sensor for highly sensitive PAH detection. ACS Sensors **5**(3), 693–702 (2020). https://doi.org/10.1021/acssensors.9b01882

Chee, K.K., Wong, M.K., Lee, H.K.: Optimization of microwave-assisted solvent extraction of polycyclic aromatic hydrocarbons in marine sediments using a microwave extraction system with high-performance liquid chromatography-fluorescence detection and gas chromatography-mass spectrometry. J. Chromatogr. A **723**(2), 259–271 (1996). https://doi.org/10.1016/0021-9673(95)00882-9

Chen, T., Zheng, X., He, X., You, Y., Huang, G., Cao, Y., He, L., Wu, Y.: Comprehensive characterization of polycyclic aromatic hydrocarbon emissions from heavy-duty diesel vehicles utilizing GC × GC-ToF-MS. Sci. Total Environ. [online] **833**, 155127 (2022). https://doi.org/10.1016/j.scitotenv.2022.155127

Cheng, Y.-J., Huang, S.-H., Chiu, J.-Y., Liu, W.-L., Huang, H.-Y.: Analyses of polycyclic aromatic hydrocarbons in seafood by capillary electrochromatography–atmospheric pressure chemical ionization/mass spectrometry. J. Chromatogr. A **1313**, 132–138 (2013). https://doi.org/10.1016/j.chroma.2013.08.035

Cloutier, P.-L., Fortin, F., Groleau, P.E., Brousseau, P., Fournier, M., Desrosiers, M.: QuEChERS extraction for multi-residue analysis of PCBs, PAHs, PBDEs and PCDD/Fs in biological samples. Talanta **165**, 332–338 (2017). https://doi.org/10.1016/j.talanta.2016.12.080

Eldos, H.I., Ashfaq, M.Y., Al-Ghouti, M.A.: Rapid assessment of the impact of microwave heating coupled with UV-C radiation on the degradation of PAHs from contaminated soil using FTIR and multivariate analysis. Arab. J. Chem. **13**(11), 7609–7625 (2020). https://doi.org/10.1016/j.arabjc.2020.08.031

Elufisan, T.O., Rodríguez-Luna, I.C., Oyedara, O.O., Sánchez-Varela, A., Hernández-Mendoza, A., Dantán Gonzalez, E., Paz-González, A.D., Muhammad, K., Rivera, G., Villalobos-Lopez, M.A., Guo, X.: The polycyclic aromatic hydrocarbon (PAH) degradation activities and genome analysis of a novel strain Stenotrophomonas sp Pemsol Isolated from Mexico. PeerJ **8**, e8102 (2020). https://doi.org/10.7717/peerj.8102

Emoyan, O.O., Onocha, E.O., Tesi, G.O.: Concentration assessment and source evaluation of 16 priority polycyclic aromatic hydrocarbons in soils from selected vehicle-parks in southern Nigeria. Scient. African [online] 7, e00296. https://doi.org/10.1016/j.sciaf.2020.e00296

Esfandiar, N., Suri, R., McKenzie, E.R.: Simultaneous removal of multiple polycyclic aromatic hydrocarbons (PAHs) from urban storm water using low-cost agricultural/industrial byproducts as sorbents. Chemosphere **274**, 129812 (2021). https://doi.org/10.1016/j.chemosphere.2021.129812

Fabiani, L.M., Reiker, T., Zacharias, H.: Vuv/vis absorption spectroscopy of different PAHs. Appl. Phys. A **130**(11) (2024). https://doi.org/10.1007/s00339-024-07959-w

Famiyeh, L., Chen, K., Xu, J., Sun, Y., Guo, Q., Wang, C., Lv, J., Tang, Y.-T., Yu, H., Snape, C., He, J.: A review on analysis methods, source identification, and cancer risk evaluation of atmospheric polycyclic aromatic hydrocarbons. Sci. Total Environ. [online] **789**, 147741 (2021). https://doi.org/10.1016/j.scitotenv.2021.147741

Ferey, L., Delaunay, N.: Capillary and microchip electrophoretic analysis of polycyclic aromatic hydrocarbons. Anal. Bioanal. Chem. **407**(10), 2727–2747 (2014). https://doi.org/10.1007/s00216-014-8390-7

Ferey, L., Delaunay, N., Rutledge, D.N., Cordella, C.B.Y., This, H., Huertas, A., Raoul, Y., Gareil, P.: Optimizing separation conditions of 19 polycyclic aromatic hydrocarbons by cyclodextrin-modified capillary electrophoresis and applications to edible oils. Talanta **119**, 572–581 (2013). https://doi.org/10.1016/j.talanta.2013.11.062

Forbes, C., Rohwer, E.R.: Application of laser induced fluorescence to monitor atmospheric polycyclic aromatic hydrocarbons. WIT Trans. Ecol. Environ. [online] **1**, 367–377 (2010). https://doi.org/10.2495/air100331

Giergielewicz-Możajska, H., Dąbrowski, Ł, Namieśnik, J.: Accelerated solvent extraction (ASE) in the analysis of environmental solid samples—some aspects of theory and practice. Crit. Rev. Anal. Chem. **31**(3), 149–165 (2001). https://doi.org/10.1080/20014091076712

Gu, H.-X., Hu, K., Li, D.-W., Long, Y.-T.: SERS detection of polycyclic aromatic hydrocarbons using a bare gold nanoparticles coupled film system. Analyst **141**(14), 4359–4365 (2016). https://doi.org/10.1039/c6an00319b

Guiñez, M., Bazán, C., Martínez, L.D., Cerutti, S.: Determination of nitrated and oxygenated polycyclic aromatic hydrocarbons in water samples by a liquid–liquid phase microextraction procedure based on the solidification of a floating organic drop followed by solvent assisted back-extraction and liquid chromatography–tandem mass spectrometry. Microchem. J. **139**, 164–173 (2018). https://doi.org/10.1016/j.microc.2018.02.027

Guo, M., Li, M., Fu, H., Zhang, Y., Chen, T., Tang, H., Zhang, T., Li, H.: Quantitative analysis of polycyclic aromatic hydrocarbons (PAHs) in water by surface-enhanced Raman spectroscopy (SERS) combined with random forest. Spectrochim. Acta Part A Mol. Biomol. Spectrosc. **287**, 122057 (2023). https://doi.org/10.1016/j.saa.2022.122057

Gupta, H.: Photocatalytic degradation of phenanthrene in the presence of akaganeite nano-rods and the identification of degradation products. RSC Adv. **6**(114), 112721–112727 (2016). https://doi.org/10.1039/c6ra24602h

Gupta, H., Gupta, B.: Photocatalytic degradation of polycyclic aromatic hydrocarbon benzo[a]pyrene by iron oxides and identification of degradation products. Chemosphere **138**, 924–931 (2015). https://doi.org/10.1016/j.chemosphere.2014.12.028

Gupta, H., Gupta, B.: Vehicular tire as potential adsorbent for the removal of polycyclic aromatic hydrocarbons. Polycyclic Aromat. Compd. **38**(4), 354–368 (2016). https://doi.org/10.1080/10406638.2016.1217890

Halasinski, T.M., Salama, F., Allamandola, L.J.: Investigation of the ultraviolet, visible, and near-infrared absorption spectra of hydrogenated polycyclic aromatic hydrocarbons and their cations. Astrophys. J. **628**(1), 555–566 (2005). https://doi.org/10.1086/430631

Han, M., Kong, J., Yuan, J., He, H., Hu, J., Yang, S., Li, S., Zhang, L., Sun, C.: Method development for simultaneous analyses of polycyclic aromatic hydrocarbons and their nitro-, oxy-, hydroxy- derivatives in sediments. Talanta **205**, 120128–120128 (2019). https://doi.org/10.1016/j.talanta.2019.120128

Hishamuddin, N.H., Khan, M.F., Suradi, H., Siraj, B.M.Z., Islam, M.T., Sairi, N.A., Tajuddin, H.A., Jamil, A.K.M., Akanda, M.J.H., Yusoff, S.: The sources of polycyclic aromatic hydrocarbons in road dust and their potential hazard. Sustainability [online] **15**(16), 12532 (2023). https://doi.org/10.3390/su151612532

Hou, Y., Li, Y., Liu, Y., Li, G., Zhang, Z.: Effects of polycyclic aromatic hydrocarbons on the UV-induced fluorescence spectra of crude oil films on the sea surface. Mar. Pollut. Bull. **146**, 977–984 (2019). https://doi.org/10.1016/j.marpolbul.2019.07.058

Huang, K., Wu, H.-L., Wang, T., Dong, M.-Y., Yan, X.-Q., Yu, R.-Q.: Chemometrics-assisted excitation-emission matrix fluorescence spectroscopy for real-time migration monitoring of multiple polycyclic aromatic hydrocarbons from plastic products to food simulants. Spectrochim. Acta Part A Mol. Biomol. Spectrosc. **304**, 123360 (2024). https://doi.org/10.1016/j.saa.2023.123360

Itodo, A.U., Sha'Ato, R., Arowojolu, M.I.: Distribution of polycyclic aromatic hydrocarbons (PAHs) in soil samples from regions around Loda-Irele Bitumen field, Nigeria. Pakistan J. Analyt. Environ. Chem. **19**(1), 71–78 (2018). https://doi.org/10.21743/pjaec/2018.06.07

Jäger, C., Krasnokutski, S., Staicu, A., Huisken, F., Mutschke, H., Henning, T., Poppitz, W., Voicu, I.: Identification and spectral properties of polycyclic aromatic hydrocarbons in carbonaceous soot produced by laser pyrolysis. Astrophys. J. Suppl. Ser. **166**(2), 557–566 (2006). https://doi.org/10.1086/507272

Kim, K.-H., Jahan, S.A., Kabir, E., Brown, R.J.C.: A review of airborne polycyclic aromatic hydrocarbons (PAHs) and their human health effects. Environ. Int. [online] **60**, 71–80 (2013). https://doi.org/10.1016/j.envint.2013.07.019

Klink, D., Schmitz, O.J.: SFC-APLI-(TOF)MS: hyphenation of supercritical fluid chromatography to atmospheric pressure laser ionization mass spectrometry. Anal. Chem. **88**(1), 1058–1064 (2015). https://doi.org/10.1021/acs.analchem.5b04402

Kumar, A., Gupta, H.: Activated carbon from sawdust for naphthalene removal from contaminated water. Environ. Technol. Innov. **20**, 101080 (2020). https://doi.org/10.1016/j.eti.2020.101080

Kumar, V., Kothiyal, N.C.: Distribution behavior and carcinogenic level of some polycyclic aromatic hydrocarbons in roadside soil at major traffic intercepts within a developing city of India. Environ. Monit. Assess. **184**(10), 6239–6252 (2011). https://doi.org/10.1007/s10661-011-2416-9

Kumar, N., Kumar, B., Gupta, H., Kumar, A.: Development and evaluation of cellulose/graphene-oxide based composite for removing phenol from aqueous solutions. Polymers **15**(3), 572–572 (2023a). https://doi.org/10.3390/polym15030572

Kumar, N., Kumar, P., Sharma, P.: In: Technological Advancement for the Chemical Characterization of Essential Oils. pp. 23–40. CRC Press eBooks (2023b). https://doi.org/10.1201/9781003389774-2

Kumar, N., Ratn, A., Joshi, N.C.: Low molecular weight polycyclic aromatic hydrocarbons (LMW-PAHs) in water: sources, their effects and treatments by physical methods. Environ. Sci. Eng. 307–329 (2025). https://doi.org/10.1007/978-3-031-82579-8_13

Kumar, N., Dhaka, S., Dhaka, R.K., Singh, S.: Sensing and efficient removal of polycyclic aromatic hydrocarbons (PAHs) in (Waste)water using nanomaterials. In: Nanotechnology in the Life Sciences, pp. 125–156. (2025). https://doi.org/10.1007/978-3-031-82794-5_6

Lau, E.V., Gan, S., Ng, H.K.: Extraction techniques for polycyclic aromatic hydrocarbons in soils. Int. J. Analyt. Chem. [online] e398381 (2010). https://doi.org/10.1155/2010/398381

Lawal, A.T.: Polycyclic aromatic hydrocarbons. A review. Cogent Environ. Sci. [online] **3**(1) (2017). https://doi.org/10.1080/23311843.2017.1339841

Lazarević-Pašti, T., Tasić, T., Milanković, V., Potkonjak, N.: Molecularly imprinted plasmonic-based sensors for environmental contaminants—current state and future perspectives. Chemosensors **11**(1), 35–35 (2023). https://doi.org/10.3390/chemosensors11010035

Li, X., Kaattari, S.L., Vogelbein, M.A., Vadas, G.G., Unger, M.A.: A highly sensitive monoclonal antibody based biosensor for quantifying 3–5 ring polycyclic aromatic hydrocarbons (PAHs) in aqueous environmental samples. Sensing Bio-Sens. Res. **7**, 115–120 (2016). https://doi.org/10.1016/j.sbsr.2016.02.003

Li, X.-Y., Li, N., Luo, H.-D., Lin, L.-R., Zou, Z., Jia, Y., Li, Y.-Q.: A novel synchronous fluorescence spectroscopic approach for the rapid determination of three polycyclic aromatic hydrocarbons in tea with simple microwave-assisted pretreatment of sample. **59**(11), 5899–5905 (2011).https://doi.org/10.1021/jf104873g

Liu, Q., Xu, X., Wang, L., Lin, L., Wang, D.: Simultaneous determination of forty-two parent and halogenated polycyclic aromatic hydrocarbons using solid-phase extraction combined with gas chromatography-mass spectrometry in drinking water. Ecotoxicol. Environ. Saf. **181**, 241–247 (2019). https://doi.org/10.1016/j.ecoenv.2019.06.011

Liu, J., Cui, W., Sang, S., Guan, L., Gu, K., Wang, Y., Wang, J.: Surface-enhanced Raman spectroscopic investigation of PAHs at a Fe3O4@GO@Ag@PDA composite substrates. Micromachines **13**(8), 1253–1253 (2022). https://doi.org/10.3390/mi13081253

Lübeck, J.S., Malmquist, L.M.V., Christensen, J.H.: Supercritical fluid chromatography for the analysis of oxygenated polycyclic aromatic compounds in unconventional oils. J. Chromatogr. A **1589**, 162–172 (2018). https://doi.org/10.1016/j.chroma.2018.12.056

Lung, S.-C.C., Liu, C.-H.: Fast analysis of 29 polycyclic aromatic hydrocarbons (PAHs) and nitro-PAHs with ultra-high performance liquid chromatography-atmospheric pressure photoionization-tandem mass spectrometry. Scientific Reports **5**(1) (2015). https://doi.org/10.1038/srep12992

Lussier, F., Thibault, V., Charron, B., Wallace, G.Q., Masson, J.-F.: Deep learning and artificial intelligence methods for Raman and surface-enhanced Raman scattering. TrAC Trends Anal. Chem. **124**, 115796 (2020). https://doi.org/10.1016/j.trac.2019.115796

Manousi, N., Zachariadis, G.A.: Recent advances in the extraction of polycyclic aromatic hydrocarbons from environmental samples. Molecules **25**(9), 2182 (2020). https://doi.org/10.3390/molecules25092182

Mansouri, E., Yousefi, V., Ebrahimi, V., Eyvazi, S., Hejazi, M.S., Mahdavi, M., Mesbahi, A., Tarhriz, V.: Overview of ultraviolet-based methods used in polycyclic aromatic hydrocarbons analysis and measurement. Separat. Sci. plus **3**(4), 112–120 (2020). https://doi.org/10.1002/sscp.201900077

Marcé, R.M., Borrull, F.: Solid-phase extraction of polycyclic aromatic compounds. J. Chromatogr. A **885**(1–2), 273–290 (2000). https://doi.org/10.1016/s0021-9673(00)00428-3

Mariateresa, I., Zianni, R., Rovere, I.D., Chiappinelli, A., Nardelli, V., Casamassima, F., Calitri, A., Quinto, M., Nardiello, D., Iammarino, M.: Development of a highly sensitive method based on QuEChERS and GC–MS/MS for the determination of polycyclic aromatic hydrocarbons in infant foods. Front. Nutrit. 11 (2024). https://doi.org/10.3389/fnut.2024.1403541

Meng, X.Y., Li, Y.S., Zhou, Y., Zhang, Y.Y., Qiao, B., Sun, Y., Yang, L., Hu, P., Lu, S.Y., Ren, H.L., Zhang, J.H., Wang, X.R., Liu, Z.S.: Real-time immuno-PCR for ultrasensitive detection of pyrene and other homologous PAHs. Biosens. Bioelectron. **70**, 42–47 (2015). https://doi.org/10.1016/j.bios.2015.03.019

Meng, X.Y., Li, Y.S., Zhou, Y., Sun, Y., Qiao, B., Si, C.C., Hu, P., Lu, S.Y., Ren, H.L., Liu, Z.S., Qiu, H.J., Liu, J.Q.: An improved RT-IPCR for detection of pyrene and related polycyclic aromatic hydrocarbons. Biosens. Bioelectron. **78**, 194–199 (2016). https://doi.org/10.1016/j.bios.2015.11.037

Moazzen, M., Shariatifar, N., Arabameri, M., Hosseini, H., Ahmadloo, M.: Measurement of polycyclic aromatic hydrocarbons in baby food samples in Tehran, Iran with magnetic-solid-phase-extraction and gas-chromatography/mass-spectrometry method: a health risk assessment. Front. Nutrit. **9** (2022).https://doi.org/10.3389/fnut.2022.833158

Ni, X., Wang, Y., Zhang, M., Cui, G., Meng, X., Chen, W., Jin, M., Shao, H., Zhang, F., Wang, C.: Rapid and on-site approaches for determination of polycyclic aromatic hydrocarbons in water and air by surface-enhanced Raman spectroscopy. ACS Omega **10**(6), 6258–6266 (2025). https://doi.org/10.1021/acsomega.5c00168

O'Connell, S.G., Haigh, T., Wilson, G., Anderson, K.A.: An analytical investigation of 24 oxygenated-PAHs (OPAHs) using liquid and gas chromatography–mass spectrometry. Anal. Bioanal. Chem. **405**(27), 8885–8896 (2013). https://doi.org/10.1007/s00216-013-7319-x

Okparanma, R.N., Mouazen, A.M.: Determination of total petroleum hydrocarbon (TPH) and polycyclic aromatic hydrocarbon (PAH) in soils: a review of spectroscopic and nonspectroscopic techniques. Appl. Spectrosc. Rev. **48**(6), 458–486 (2013a). https://doi.org/10.1080/05704928.2012.736048

Okparanma, R.N., Mouazen, A.M.: Visible and near-infrared spectroscopy analysis of a polycyclic aromatic hydrocarbon in soils. Scient. World J. **2013**, 1–9 (2013b). https://doi.org/10.1155/2013/160360

Oliver, Rost, H., Hirmann, D., Hasinger, M., Braun, R., Loibner, A.P.: Sequential supercritical fluid extraction (SSFE) for estimating the availability of high molecular weight polycyclic aromatic hydrocarbons in historically polluted soils. J. Environ. Quality **33**(1), 80–88 (2004). https://doi.org/10.2134/jeq2004.8000

Oña-Ruales, J.O., Ruiz-Morales, Y.: Prediction of the ultraviolet-visible absorption spectra of polycyclic aromatic hydrocarbons (Dibenzo and Naphtho) derivatives of fluoranthene. Appl. Spectrosc. **71**(6), 1134–1147 (2016). https://doi.org/10.1177/0003702816667517

Pandey, S.K., Kim, K.-H., Richard.: A review of techniques for the determination of polycyclic aromatic hydrocarbons in air. Trends Analyt. Chemis. **30**(11), 1716–1739 (2011). https://doi.org/10.1016/j.trac.2011.06.017

Paris, A., Gaillard, J.-L., Ledauphin, J.: Rapid extraction of polycyclic aromatic hydrocarbons in apple: ultrasound-assisted solvent extraction followed by Microextraction by packed sorbent. Food Anal. Methods **12**(10), 2194–2204 (2019). https://doi.org/10.1007/s12161-019-01568-7

Patel, A.B., Shaikh, S., Jain, K.R., Desai, C., Madamwar, D.: Polycyclic aromatic hydrocarbons: sources, toxicity, and remediation approaches. Front. Microbiol. [online] **11**(11) (2020). https://doi.org/10.3389/fmicb.2020.562813

Patra, D.: Distinguishing motor oils at higher concentration range by evaluating total fluorescence quantum yield as a novel sensing tool. Sens. Actuators, B Chem. **129**(2), 632–638 (2008). https://doi.org/10.1016/j.snb.2007.09.009

Pedro, Nogueira, F.: Combining stir bar sorptive extraction and MEKC for the determination of polynuclear aromatic hydrocarbons in environmental and biological matrices. Electrophoresis **27**(23), 4694–4702 (2006). https://doi.org/10.1002/elps.200600210

Purcaro, G., Moret, S., Bučar-Miklavčič, M., Conte, L.S.: Ultra-high performance liquid chromatographic method for the determination of polycyclic aromatic hydrocarbons in a passive environmental sampler. J. Sep. Sci. **35**(8), 922–928 (2012). https://doi.org/10.1002/jssc.201100922

Qazi, F., Shahsavari, E., Prawer, S., Ball, A.S., Tomljenovic-Hanic, S.: Detection and identification of polyaromatic hydrocarbons (PAHs) contamination in soil using intrinsic fluorescence. Environ. Pollut. **272**, 116010 (2021). https://doi.org/10.1016/j.envpol.2020.116010

Qi, Y.-B., Wang, C., ChengyuanLv, Lun, Z., Chang, Z.: Removal capacities of polycyclic aromatic hydrocarbons (PAHs) by a newly isolated strain from oilfield produced water. Int. J. Environ. Res. Public Health **14**(2), 215–215 (2017). https://doi.org/10.3390/ijerph14020215

Qiao, B., Li, Y., Meng, X., Sun, Y., Hu, P., Lu, S., Ren, H., Liu, Z., Zhou, Y.: Development of an indirect competitive ELISA for the detection of acenaphthene and pyrene. Food Hydrocolloids **28**(5), 789–800 (2017). https://doi.org/10.1080/09540105.2017.1313201

Palacios Colón, L., Rascón, A.J., Ballesteros, E.: Trace-level determination of polycyclic aromatic hydrocarbons in dairy products available in Spanish supermarkets by semi-automated solid-phase extraction and gas chromatography–mass spectrometry detection. Foods, **11**(5), 713 (2022). https://doi.org/10.3390/foods11050713

Poster, D.L., Schantz, M.M., Sander, L.C., Wise, S.A.. Analysis of polycyclic aromatic hydrocarbons (PAHs) in environmental samples: a critical review of gas chromatographic (GC) methods. Anal. Bioanal. Chem. **386**(4), 859–881 (2006). https://doi.org/10.1007/s00216-006-0771-0

Queb-Suarez, J.E., Ruiz-Marin, A., Canedo-Lopez, Y., Aguilar-Ucan, C.A., Montalvo-Romero, C., Flores-Trujillo, J.G., Perez-Morga, N.: Source identification, toxicity, and persistence of Pahs

in sediment core from a natural protected area in Mexico. Energies **15**(19), 7116–7116 (2022). https://doi.org/10.3390/en15197116
Ramirez, C.E., Wang, C., Gardinali, P.R.: Fully automated trace level determination of parent and alkylated PAHs in environmental waters by online SPE-LC-APPI-MS/MS. Anal. Bioanal. Chem. **406**(1), 329–344 (2013). https://doi.org/10.1007/s00216-013-7436-6
Rascón, A.J., Azzouz, A., Ballesteros, E.: Trace level determination of polycyclic aromatic hydrocarbons in raw and processed meat and fish products from European markets by GC-MS. Food Control **101**, 198–208 (2019). https://doi.org/10.1016/j.foodcont.2019.02.037
Ravindra, K., Sokhi, R., Vangrieken, R.: Atmospheric polycyclic aromatic hydrocarbons: source attribution, emission factors and regulation. Atmos. Environ. **42**(13), 2895–2921 (2008). https://doi.org/10.1016/j.atmosenv.2007.12.010
Rengarajan, T., Rajendran, P., Nandakumar, N., Lokeshkumar, B., Rajendran, P., Nishigaki, I.: Exposure to polycyclic aromatic hydrocarbons with special focus on cancer. Asian Pac. J. Trop. Biomed. **5**(3), 182–189 (2015). https://doi.org/10.1016/s2221-1691(15)30003-4
Rezaee, M., Assadi, Y., Milani Hosseini, M.-R., Aghaee, E., Ahmadi, F., Berijani, S.: Determination of organic compounds in water using dispersive liquid–liquid microextraction. J. Chromatogr. A **1116**(1–2), 1–9 (2006). https://doi.org/10.1016/j.chroma.2006.03.007
Rocha, A.C., Palma, C.: Source identification of polycyclic aromatic hydrocarbons in soil sediments: application of different methods. Sci. Total. Environ. **652**, 1077–1089 (2019). https://doi.org/10.1016/j.scitotenv.2018.10.014
Saini, S., Kabir, A., Rao, A., Malik, A., Furton, K.: A novel protocol to monitor trace levels of selected polycyclic aromatic hydrocarbons in environmental water using fabric phase Sorptive extraction followed by high performance liquid chromatography-fluorescence detection. Separations **4**(2), 22 (2017). https://doi.org/10.3390/separations4020022
Sajid, M., Nazal, M.K., Ihsanullah, I.: Novel materials for dispersive (micro) solid-phase extraction of polycyclic aromatic hydrocarbons in environmental water samples: a review. Anal. Chim. Acta **1141**, 246–262 (2021). https://doi.org/10.1016/j.aca.2020.07.064
Schulze, S., Paschke, H., Meier, T., Muschket, M., Reemtsma, T., Berger, U.: A rapid method for quantification of persistent and mobile organic substances in water using supercritical fluid chromatography coupled to high-resolution mass spectrometry. Anal. Bioanal. Chem. **412**(20), 4941–4952 (2020). https://doi.org/10.1007/s00216-020-02722-5
Singh, S., Ashesh, A., Linthoingambi Devi, N., Yadav, I.C.: A comprehensive review on occurrence, source, effect, and measurement techniques of polycyclic aromatic hydrocarbons in India. Microchem. J. **183**, 108005 (2022). https://doi.org/10.1016/j.microc.2022.108005
Song, Y.F., Jing, X., Fleischmann, S., Wilke, B.-M.: Comparative study of extraction methods for the determination of PAHs from contaminated soils and sediments. Chemosphere **48**(9), 993–1001 (2002). https://doi.org/10.1016/s0045-6535(02)00180-7
Srivastava, S., Wang, W., Zhou, W., Jin, M., Vikesland, P.J.: Machine learning-assisted surface-enhanced Raman spectroscopy detection for environmental applications: a review. Environ. Sci. Technol. **58**(47), 20830–20848 (2024). https://doi.org/10.1021/acs.est.4c06737
Stockton, A.M., Chiesl, T.N., Scherer, J.R., Mathies, R.A.: Polycyclic aromatic hydrocarbon analysis with the mars organic analyzer microchip capillary electrophoresis system. Anal. Chem. **81**(2), 790–796 (2008). https://doi.org/10.1021/ac802033u
Sun, Y., Li, Y., Meng, X., Qiao, B., Hu, P., Meng, X., Lu, S., Ren, H., Liu, Z., Zhou, Y.: Fluorescence-linked immunosorbent assay for detection of phenanthrene and its homolog. Anal. Biochem. **547**, 45–51 (2018). https://doi.org/10.1016/j.ab.2018.02.016
Sunuwar, S., Manzanares, C.E.: Excitation, emission, and synchronous fluorescence for astrochemical applications: experiments and computer simulations of synchronous spectra of polycyclic aromatic hydrocarbons and their mixtures. Icarus **370**, 114689 (2021). https://doi.org/10.1016/j.icarus.2021.114689
Taylor, A.T., Lai, E.P.C.: Current state of laser-induced fluorescence spectroscopy for designing biochemical sensors. Chemosensors **9**(10), 275 (2021). https://doi.org/10.3390/chemosensors9100275

Tian, M., Muhammad, T., Li, Y., Fan, X., Wang, L., Mi, F., Zheng, Y., Gao, J.: Application of synchronous fluorescence spectroscopy in the analysis of polycyclic aromatic hydrocarbons in petroleum and coal. Appl. Spectroscopy Rev. 1–27 (2025). https://doi.org/10.1080/05704928.2025.2476420

Tijunelyte, I., Betelu, S., Moreau, J., Ignatiadis, I., Berho, C., Lidgi-Guigui, N., Guénin, E., David, C., Vergnole, S., Rinnert, E., Lamy, M.: Diazonium salt-based surface-enhanced Raman spectroscopy nanosensor: detection and quantitation of aromatic hydrocarbons in water samples. Sensors **17**(6), 1198–1198 (2017). https://doi.org/10.3390/s17061198

Topuz, F., Uyar, T.: Cyclodextrin-functionalized mesostructured silica nanoparticles for removal of polycyclic aromatic hydrocarbons. J. Colloid Interface Sci. **497**, 233–241 (2017). https://doi.org/10.1016/j.jcis.2017.03.015

Tumelo Monty, M., Mokoena, L., Tshilongo, J.: A review on recent developments in the extraction and identification of polycyclic aromatic hydrocarbons from environmental samples. Water [online] 16(17), 2520–2520 (2024). https://doi.org/10.3390/w16172520

Varela-Martínez, D.A., González-Sálamo, J., Miguel Ángel González-Curbelo and Hernández-Borges, J.: Quick, easy, cheap, effective, rugged, and safe (QuEChERS) extraction. Elsevier eBooks 399–437 (2020). https://doi.org/10.1016/b978-0-12-816911-7.00014-1

Vistnes, H., Sossalla, N.A., Røsvik, A., Gonzalez, S.V., Zhang, J., Meyn, T., Asimakopoulos, A.G.: The determination of polycyclic aromatic hydrocarbons (PAHs) with HPLC-DAD-FLD and GC-MS techniques in the dissolved and particulate phase of road-tunnel wash water: a case study for cross-array comparisons and applications. Toxics [online] **10**(7), 399 (2022). https://doi.org/10.3390/toxics10070399

Wang, W., Meng, B., Lu, X., Liu, Y., Tao, S.: Extraction of polycyclic aromatic hydrocarbons and organochlorine pesticides from soils: a comparison between Soxhlet extraction, microwave-assisted extraction and accelerated solvent extraction techniques. Anal. Chim. Acta **602**(2), 211–222 (2007). https://doi.org/10.1016/j.aca.2007.09.023

Wang, P., Zhang, Q., Wang, Y., Wang, T., Li, X., Ding, L., Jiang, G.: Evaluation of Soxhlet extraction, accelerated solvent extraction and microwave-assisted extraction for the determination of polychlorinated biphenyls and polybrominated diphenyl ethers in soil and fish samples. Anal. Chim. Acta **663**(1), 43–48 (2010). https://doi.org/10.1016/j.aca.2010.01.035

Wang, Z., Wei, C., Shui, H., Ren, S., Pan, C., Wang, Z., Li, H., Lei, Z.: Synchronous fluorimetric characterization of heavy intermediates of coal direct liquefaction. Fuel **98**, 67–72 (2012). https://doi.org/10.1016/j.fuel.2012.03.046

Wang, X., Hao, W., Zhang, H., Pan, Y., Kang, Y., Zhang, X., Zou, M., Tong, P., Du, Y.: Analysis of polycyclic aromatic hydrocarbons in water with gold nanoparticles decorated hydrophobic porous polymer as surface-enhanced Raman spectroscopy substrate. Spectrochimica Acta Part AMolecular and Biomolecular Spectroscopy **139**, 214–221 (2014). https://doi.org/10.1016/j.saa.2014.11.104

Wang, X., Feng, J., Bu, Y., Tian, Y., Luo, C., Sun, M.: Mesoporous titanium oxide with high-specific surface area as a coating for in-tube solid-phase microextraction combined with high-performance liquid chromatography for the analysis of polycyclic aromatic hydrocarbons. J. Sep. Sci. **40**(11), 2474–2481 (2017). https://doi.org/10.1002/jssc.201700214

Wang, X., Xu, Q., Hu, X., Han, F., Zhu, C.: Silver-nanoparticles/graphene hybrids for effective enrichment and sensitive SERS detection of polycyclic aromatic hydrocarbons. Spectrochim. Acta Part A Mol. Biomol. Spectrosc. **228**, 117783–117783 (2020b). https://doi.org/10.1016/j.saa.2019.117783

Wang, S., Liu, S., Che, X., Wang, Z., Zhang, J., Kong, D.: Recognition of polycyclic aromatic hydrocarbons using fluorescence spectrometry combined with bird swarm algorithm optimization support vector machine. Spectrochimica Acta Part A: Molecular Biomolecular Spectroscopy [online] **224**, 117404 (2020). https://doi.org/10.1016/j.saa.2019.117404

Wei, J.-W., He, J.-R., Chen, S.-Y., Guo, Y.-H., Huo, X.-Z., Zheng, N., Cao, S.-H., Li, Y.-Q.: Synchronous fluorescence spectra-based machine learning algorithm with quick and easy accessibility for simultaneous quantification of polycyclic aromatic hydrocarbons in edible oils. Food Control **158**, 110205–110205 (2023). https://doi.org/10.1016/j.foodcont.2023.110205

Weigel, S., Wetekam, J., Mollenhauer, K.: Identification and classification of PAH in asphalt binders with FTIR spectroscopy and multivariate analysis methods. Fuel 126845 (2022). https://doi.org/10.1016/j.fuel.2022.126845

Wetekam, J., Mollenhauer, K.: FTIR spectroscopy analysis assessment of reclaimed asphalt at asphalt mixing plants to optimize the recycling. Transport. Eng. 100242–100242 (2024). https://doi.org/10.1016/j.treng.2024.100242

Wicker, A.P., Carlton, D.D., Tanaka, K., Nishimura, M., Chen, V., Ogura, T., Hedgepeth, W., Schug, K.A.: On-line supercritical fluid extraction—supercritical fluid chromatography-mass spectrometry of polycyclic aromatic hydrocarbons in soil. J. Chromatogr. B **1086**, 82–88 (2018). https://doi.org/10.1016/j.jchromb.2018.04.014

Williams, M., Olomukoro, A.A., Emmons, R.V., Goodage, N.H., Gionfriddo, E.: Matrix effects demystified: strategies for resolving challenges in analytical separations of complex samples. J. Separat. Sci. **46**(23) (2023). https://doi.org/10.1002/jssc.202300571

Woźniak, M., Hoppe, K., Drzewiecka, K.: Determination of polycyclic aromatic hydrocarbon content in garden herbal plants using liquid chromatographic analysis (HPLC-FL). Plants **12**(3), 551 (2023). https://doi.org/10.3390/plants12030551

Wu, S., Yu, W.: Liquid–liquid extraction of polycyclic aromatic hydrocarbons in four different edible oils from China. Food Chem. **134**(1), 597–601 (2012). https://doi.org/10.1016/j.foodchem.2012.02.155

Wu, J., Sun, J., Cheng, H., Liu, J., Wang, Y.: Analysis of polycyclic aromatic hydrocarbons by capillary electrochromatography by using capillary columns packed with polycyclic-aromatic-hydrocarbon-specific particles. Separat. Sci. plus **1**(6), 446–453 (2018). https://doi.org/10.1002/sscp.201800063

Wu, Y., Qu, W., Qiu, C., Chen, K., Zhuang, Y., Zeng, Z., Yan, Y., Gu, Y., Tao, W., Gao, J., Li, K.: The method and study of detecting Phenanthrene in seawater based on a carbon nanotube-chitosan oligosaccharide modified electrode immunosensor. Molecules **28**(15), 5701 (2023). https://doi.org/10.3390/molecules28155701

Yan, X., Shi, X., Ma, J.: Fast quantification of the mixture of polycyclic aromatic hydrocarbons using surface-enhanced raman spectroscopy combined with PLS-GA-BP network. J. Ocean Univer. China **20**(6), 1451–1458 (2021). https://doi.org/10.1007/s11802-021-4832-y

Yan, Y., Qiu, C., Qu, W., Zhuang, Y., Chen, K., Wang, C., Zhang, R., Wang, P., Wu, Y., Gao, J.: Detection of BaP in seawater based on multi-walled carbon nanotubes composites immunosenor. Front. Chem. **10** (2022). https://doi.org/10.3389/fchem.2022.950854

Yoshioka, T., Nagatomi, Y., Harayama, K., Bamba, T.: Development of an analytical method for polycyclic aromatic hydrocarbons in coffee beverages and dark beer using novel high-sensitivity technique of supercritical fluid chromatography/mass spectrometry. J. Biosci. Bioeng. **126**(1), 126–130 (2018). https://doi.org/10.1016/j.jbiosc.2018.01.014

Yousefi, S.M., Shemirani, F., Ghorbanian, S.A.: Hydrophobic deep eutectic solvents in developing microextraction methods based on solidification of floating drop: application to the trace HPLC/FLD determination of PAHs. Chromatographia **81**(8), 1201–1211 (2018). https://doi.org/10.1007/s10337-018-3548-7

Zhang, Q., Liu, P., Li, S., Zhang, X., Chen, M.: Progress in the analytical research methods of polycyclic aromatic hydrocarbons (PAHs). J. Liq. Chromatogr. Relat. Technol. **43**(13–14), 425–444 (2020). https://doi.org/10.1080/10826076.2020.1746668

Zhang, L., Li, W., Wu, S.: Rapid determination of oxygenated and parent polycyclic aromatic hydrocarbons in milk using supercritical fluid chromatography-mass spectrometry. Foods **11**(24), 3980–3980 (2022b). https://doi.org/10.3390/foods11243980

Zhang, Y., Yuan, L., He, S., Tao, H., Xie, W., Zhang, X., Ren, X., Jiang, T., Li, L., Zhu, Z.: Contemporary research progress on the detection of polycyclic aromatic hydrocarbons. Int. J. Environ. Res. Public Health **19**(5), 2790 (2022c). https://doi.org/10.3390/ijerph19052790

Zhang, H., Zheng, F., Zhang, J., Hu, D., Minelli, M., Pei, J., Sangiorgi, C.: Contributions of Fourier-transform infrared spectroscopy technologies to the research of asphalt materials: a comprehensive review. Fuel **371**, 132078–132078 (2024a). https://doi.org/10.1016/j.fuel.2024.132078

Zhang, Q., Chen, B., Wahid, F., Feng, W., Chen, X.: Intelligent detection and analysis of polycyclic aromatic hydrocarbons based on surface-enhanced raman scattering spectroscopy. Computat. Intell. Neurosci. [online] 1–7 (2022). https://doi.org/10.1155/2022/8330702

Zhang, R., Guo, M., Li, M., Tang, H., Zhang, T., Li, H.: Surface-enhanced Raman spectroscopy combined with chemometrics for quantitative analysis and carcinogenic risk estimation of polycyclic aromatic hydrocarbons in water with complex matrix. Chemometrics Intell. Laboratory Syst. 105293–105293 (2024b). https://doi.org/10.1016/j.chemolab.2024.105293

Zhao, X., Feng, X., Chen, J., Zhang, L., Zhai, L., SizheLv, Ye, Y., Chen, Y., Zhong, T., Yu, X., Xiao, Y.: Rapid and sensitive detection of polycyclic aromatic hydrocarbons in tea leaves using magnetic approach. Foods **12**(11), 2270–2270 (2023). https://doi.org/10.3390/foods12112270

Zhou, Z., Lu, J., Wang, J., Zou, Y., Liu, T., Zhang, Y., Liu, G., Tian, Z.: Trace detection of polycyclic aromatic hydrocarbons in environmental waters by SERS. Spectrochim. Acta Part A Mol. Biomol. Spectrosc. **234**, 118250–118250 (2020). https://doi.org/10.1016/j.saa.2020.118250

Sensors and Biosensors for Polycyclic Aromatic Hydrocarbons

10

Shaily and Danish Khan

Abstract

Among the organic pollutants, polyaromatic hydrocarbons are one of the utmost widely distributed in soil, water, and biological systems. The poly aromatic hydrocarbons comprise several aromatic fused rings and are made due to the partial burning of carbon-based matter such as fossil fuels, biomass, food processing, and industrial emissions. As they are easily accumulated in the environment leading to serious ecological and health risks. Long-term exposure to PAH be associated with various health issues like cardiovascular complications, carcinogenicity, and development disorders. Therefore, effective monitoring and mitigation strategies for their detection are very crucial mandates. The conventional techniques for the PAHs include GC, CE, HPLC, and SERS. Despite high sensitivity and specificity, these methods are expensive and need laborious sample preparations leading to the dependency on controlled laboratory conditions. In contrast, sensors and biosensors provide a promising alternative to detect PAHs selectively and quickly. The current chapter represents the recent developments and challenges of sensor and biosensor-based approaches for PAH detection.

Keywords

Analytical methods • Aromatic hydrocarbons • Biosensor • Electrochemical sensor • Surface-enhanced Raman spectroscopy

Shaily (✉)
Department of Chemistry, D. B. S. (P.G.) College, Dehradun, Uttarakhand, India
e-mail: shailydbs@gmail.com

D. Khan
Gene to Protein Pvt. Ltd. Hyderabad, Hyderabad, Telangana, India

H. Gupta et al. (eds.), *Ubiquitous Polycyclic Aromatic Hydrocarbon Contamination*,
Emerging Contaminants and Associated Treatment Technologies,
https://doi.org/10.1007/978-3-032-11764-9_10

Introduction

Polycyclic aromatic hydrocarbons (PAHs) are a special group of organic pollutants characterized by two or more attached benzene rings with carcinogenic effects on living organisms, particularly on human health. These chemicals are distributed worldwide with water, air, and soil (Hussain et al. 2018; Gupta and Gupta 2016; Kumar and Gupta 2020). PAHs are classified as the most popularly volatile or semi-volatile according to molecular weight contaminants having mutagenicity and carcinogenic properties (Abdel-Shafy and Mansour 2016). On the basis of these carcinogenic and mutagenic effects on human health, the existence of PAHs in the atmosphere and their monitoring by the government is a serious issue around the world (Jiang et al. 2023). Due to the toxic nature of these compounds, the analysis of PAHs in the environment by analysts pays significant consideration in the analytical scientific community. Various studies have been carried out by scientists all over the world for the recognition of PAHs in different environmental samples, likes air (Soury et al. 2019), soil (Ma et al. 2019; Gupta and Kumar 2020; Gupta and Dhiman 2023), food, (Farhadian et al. 2010; Gutierrez-Valencia and García de Llasera 2017), water (Reinert et al. 2018; Tijunelyte et al. 2017), effluents (Gupta 2018) and marine sediments (Nikolaou et al. 2009). The study of PAHs in these environmental samples shows a high level of these compounds. Among the wide class of PAHs, the list of sixteen PAHs, as shown in Fig. 10.1 (Boehm 1964), was selected by United States Environmental Protection Agency (US EPA), for reference for the sample analysis. The maximum acceptable concentration of these listed PAHs (Fig. 10.1) was well regulated at global level. The major difficulties tackled by the researchers is establishment of improved method for the detection of PAHs. The newly developed methods should be more sensitive and detect PAHs from the environmental samples more rapidly, quantitatively.

Classification and Properties of Polycyclic Aromatic Hydrocarbons (PAHs) Based on Their Structure

Chemically poly hydrocarbons are the class of chemicals in which two or extra benzene rings fused in linear, angular and clustered manor. According to the arrangement of benzene ring or molecular structure PAHs have been categories into, Low-molecular-weight (LMW) PAHs and high-molecular-weight (HMW) poly hydrocarbons.

Low-molecular-weight PAHs contains four and less benzene ring, whereas high-molecular-weight (HMW) poly hydrocarbons contains five or extra benzene ring in molecular structure. Many PAHs having the same number of benzene ring but their arrangement of rings is different, these different arrangement of benzene is responsible for different chemical and physical properties (Lundstedt 2003). PAHs are semi solid and solid compounds with high boiling and melting points. Moreover, they are chemically inactive but bound quickly with some particular matter. The existence of PAHs in the environment due to their physiological properties,

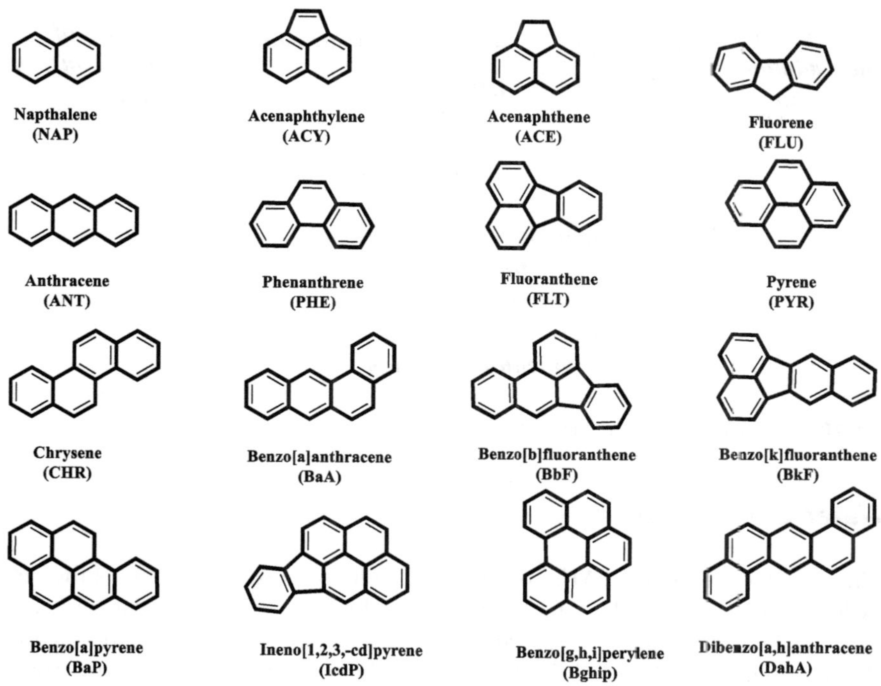

Fig. 10.1 Structure of PAHs generally present in the environment as pollutant. Reprinted from the original article by Boehm et al., 1964 "after taking permission from Elsevier"

surly because of hydrophobicity and non-polar nature (Srogi 2007). Therefore, the molecular structure of all the major PAHs is shown in Fig. 10.1. Although various PAHs are available in the environment, the studies done by the scientist and regulation by the government generally refers to the number of 14 up to 20 among the listed PAHs. The occurrence of PAHs in the atmosphere is natural, these are byproduct of the reaction mediated by micro-organisms. It is also formed by the partial burning of organic material, available as environmental pollutant (Bojakowska and Sokołowska 2001). Although, there are three key causes of the occurrences of PAHs in the environment as shown in Fig. 10.2. **Pyrolytic:** Partial incineration of fossil fuels such as petroleum and its component, coal tar, waste burning, and natural actions like's forest fires and volcanic explosions. **Petrogenic:** Availability of petroleum sources in water Such as gasoline, diesel fuel, kerosene, etc., **Diagenetic process:** PAHs content in the environment increase by the natural degradation of plants (Felemban et al. 2019).

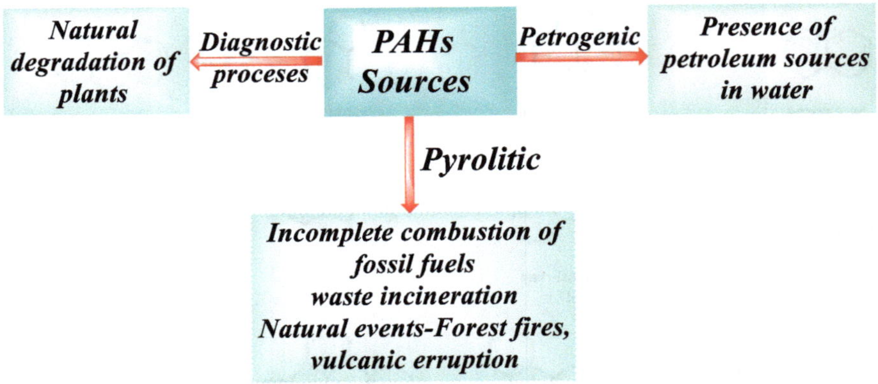

Fig. 10.2 Presence of PAHs in the environment

Water Contamination

The biodegradation of PAHs is very difficult because of its lipophilic and hydrophobic properties. Low molecular weight PAHs are easy degradable as compared to high molecular weight PAHs due to their high solubility and volatility (Behera et al. 2018; Ma et al. 2019). Whereas, higher concentration of low molecular weight PAHs is available in water because of its high soluble properties (Felemban et al. 2019). The analyzed, concentration of each and separate PAHs in the water exterior and coastal waters due to heavily polluted water comes from the industry is 50 ngl^{-1}. If the concentration reaches more than this measured value around 10 μgl^{-1}, the issue PAHs contamination in the area of industry is a major concern (Srogi 2007). According to the recent study of PAH and their global distribution in environment reveals that the concentration of PAHs in air is 0.11 ng/m^3, in water 0.06 ng l^{-1} and 21 ng g^{-1} in sediments (W.H.O 2018). Poly aromatic hydrocarbons (PAHs) are available in different sources of water with other organic pollutants (Ozaki et al. 2019). Instead of this, as the molecular weight of the PAHs rises, the water solubility of PAHs decreases gradually. The study done by Adeniji et al. group in 2018 reveals that PAHs like chrysene and benzo[a]pyrene are unevenly insoluble in water (Adeniji et al. 2017). The insoluble property of the former PAHs makes them their deposition easy on the surface of particular matters, and this matter work as courier for the deposit PAHs from land and air water. The uppermost suitable concentration of BaP in water is 0.7 μg/L given by WHO (Diener and Mudu 2021). The result concluded from the advanced study, 1% PAHs intake with water is acceptable from the total level (Mojiri et al. 2019).

Methods Used for the Recognition of PAHs

Analytical Methods

PAHs have been analyzed by various conventional methods like chromatography, including gas chromatography with mass spectrometry (Avino et al. 2017; Portoles et al. 2017; Li et al. 2017; Santos et al. 2017), high-performance liquid chromatography joined with UV–vis absorbance and fluorescence detector (Liu et al. 2017; Wang et al. 2017; Corps Ricardo et al. 2017), and electrophoresis techniques, i.e., capillary electrophoresis (CE) (Ferey et al. 2013; Ferey and Delaunay 2015). All the above-mentioned techniques show a large number of advantages, likes data accuracy, thoughtfulness, and consistency, with some limitations. However, the limitations of these chromatography techniques are given as follows. Loss of sample is one of the high risks during the sample separation. These techniques are very expensive compared to the other. Chromatography techniques are a time taking process and consume more than 30 min for one sample run. A large sample volume is required for these techniques. For the operation of the machine, a dedicated analytical chemist is required. For the withdrawal of PAHs from the sample organic solvent is required, as well as for the sample preparation.

Apart from GC and HPLC, other techniques such as spectroscopy are also available, which is acceptable only for research purposes, whereas GC and HPLC are worldwide accepted for the recognition of PAHs I environmental samples. The data generated using HPLC and GC are very trust full for both academic and industry purposes (Kumar et al. 2017). Though various methods have been established for the revealing of PAHs by using HPLC. Various other methods, likes biosensor, sensor, and immunoassay detection of PAHs and their metabolites, have been used as an alternative of HPLC and GC techniques. These methods show a larger number of advantages and attract the scientific community with low cost, fast detection, and higher sensitivity (Nsibande et al. 2019). Over the past decade, scientists have been trying to develop various alternative methods to develop a nanomaterial-based detector for PAHs detection. These nanomaterial-based devices are more sensitive and show rapid detection of PAHs from different samples collected from the environment. Various techniques based on surface enhanced Raman spectroscopy (SERS) (Du and Jing 2011; Du et al. 2016; Song et al. 2019), fluorescence spectrophotometry (Wang et al. 2015; Adegoke et al. 2017), and electrochemistry (Diaz-Gonzalez et al. 2016; Qiao et al. 2018; Mailu et al. 2019) are often used for the detection of PAHs.

Detection of PAHs via Electrochemical Sensors

In the current trends one of the versatile sensors used for the PAHs detection based on electrochemistry is called as electrochemical sensors (ECS). These electrochemical sensors have low making cost, easy to handle with resolving all the limitation of convention methods and make them these electrochemical sensors a portable device for PAHs detection (Nsibande et al. 2019). These sensors play a crucial role for the selective revealing of PAHs at very little concentration of analytes using different principle that includes potentiometry, amperometry/voltammetry, conductometry, electrochemical impedance spectroscopy and coulometry. Among the above-mentioned detection methods, one of the most rapidly method use for PAHs sensing is voltammetry, this method involves the scanning of sample over the latent over different variety and measuring the retort of current obtained from the redox reaction take place on the surface of working electrodes. The current obtained during the reaction and their variable response depending on the concentration of analyte and also depends on the voltammetry techniques which include CV, LSV, DPV and SWV (Comnea-Stancu et al. 2021). Beside this these electrochemical techniques have various limitations, including overlapping of signals due to coexisting compounds interference as well as fouling of electrode and problems of over-voltage caused by electron transfer kinetics (Zainal et al. 2020). However, the electrode used in this technique consequently need the modification time to time by using different material likes polymer, porous material and/or nanomaterial give them an electrode exceptional electrical and chemical properties for sensing of PAHs and also provide selectivity for numerous analytes in diverse matrices.

Detection of PAHs via Electrochemical Sensor Built on Self-assembled Monolayer Stage

The electrochemical sensor created on self-assembled monolayer (SAM) platforms involve the surface modification of electrode via organic molecules, that making them immobilization possible on the surface of electrode with increasing performance for application likes PAHs detection. The over-all structure of a SAM is exposed in Fig. 10.3 (Peng et al. 2020) and the chemical reaction for synthesis of SAM is presented by Munoz et al. (2017).

SAMs provide a simple and versatile method to modify electrode surfaces with specific functional groups (e.g., thiols, amines, silanes) (Munoz et al. 2017). The functionalized surface created by SAMs can be used to immobilize biomolecules like antibodies, proteins, and enzymes, enabling the establishment of biosensors. SAMs can improve the sensitivity, selectivity, and stability of electrochemical sensors. Application of SAM are as follows. Detection of specific biomolecules (e.g.,

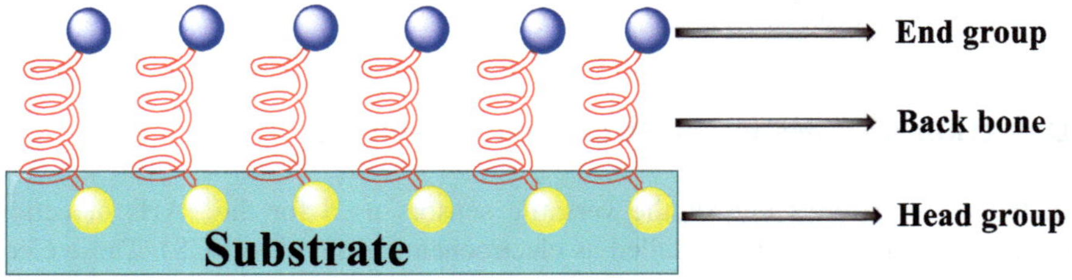

Fig. 10.3 Structure of self-assembled mono layer

antibodies, proteins, DNA). Detection of specific analytes (e.g., gases, pollutants). Detection of specific DNA sequences. There are some main advantages of using SAMs. SAM formation is a relatively simple process that does not require complex equipment. SAMs can be tailored to specific applications by selecting appropriate molecules and functional groups. SAMs can provide a stable and reproducible surface for electrochemical measurements (Comnea-Stancu et al. 2021).

Reprinted from the original article by Peng et al. 2020 "after taking permission from Elsevier".

Munoz et al. developed three different electrochemical sensors by the modification of indium tin oxide electrode for the recognition of PAHs. They used three different recognition elements.

In the primary work, the group synthesized or modified the electrode for PYR detection. The detection of PYR is due to π–π interactions.

The modification of synthesis of SAM initially involved several sequential steps, first the ITO electrode modified with an oxidizing agents NH_4OH, H_2O_2 and H_2O with the proportion of (1:1:5, v/v/v), then the treatment of electrode with solution of toluene containing 1% of 3-glycidoxy propyl trimethoxysilane and 1-aminopyrene (1 mM) for 36 h at 80 °C. For the detection of PYR, prepare KCl (0.1 M) solution that contains $[Fe(CN)_6]$ (10 mM) as redox marker, a platinum wire as counter electrode and Ag/AgCl as reference electrode, limit of detection of PYR is very selective even in the presence of another PAHs, by this method PYR detection of limit (LOD) is very low i.e. 1.75 ng/L (Munoz et al. 2017).

In the recent years, more and effective strategy was established by modifying electrochemical sensors created on SAM for the recognition of four PAHs, namely NAP, ANT, PYR and FLR. The interaction of these four PAHs with the fabricated ITO electrode involved a covalent interaction. This ITO electrode showed excellent selectivity for these four PAHs with excellent detection limit (LOD) of 0.79, 0.91, 1.70, and 0.79 ng/L, respectively. Additionally, the main advantage of this work was the remarkable selectivity, which was < 10% signal interference for all four PAHs and 95.7 to 108.0% recovery ranging. This information revealed that this kind of ITO electrodes was helpful to detect the PAHs using 4 ng/L samples (Munoz et al. 2020).

In one more study, where the electrode ITO was modified with -OH groups, then by the reaction with trimethoxysilanepropyl-3-amine for the formation of trimethoxysilanepropyl-3-amine based self-assembled monolayer, and carbon nanotubes (single-walled) were grafted by the synthesis of an amide bond. Although this type of CNT-rich element electrode provides a linear response for the detection of PAHs in the concentration range 1.75–7.00 ng/L and 1.75 ng/L limit of detection (LOD) (Munoz et al. 2018).

Electrochemical Sensors Based on Polymer and Molecularly-Imprinted Platform

Exploration of new and better-quality detection methods, the researchers developed one more advanced detection technique for the investigation of PAHs. This newly developed method is known as molecular imprinted polymers (MIPs). The MIPs tools are progressively used as sensor and gain much more attention in scientific community due to their easy production and their much higher stability. These sensors synthesis involve the production of polymer matrix via molecular imprinting procedure, with the similar molecular recognition site contains the same size and shape like the target PAHs. These MIPs sensors contain some chemical functional groups located inside the cavities of MIPs, that provide them a supplementary chemical selectivity via binding connections with the surface group available in micro-pollutant. MIPs are synthesized by the reaction functional monomer using the specific templet via copolymerization process (Bui and Haupt 2010).

The exclusion of template from the polymer system generally generates specific sites that shows connections with the template and its correspondents. Recently, MIPs paying the great attention in scientific community as an electrochemical sensor, because of their higher selectivity in the recognition of PAHs (Canfarotta et al. 2018).

Molecularly imprinted polymers (MIPs) are chemically produced by the copolymerization process by the reaction of monomer containing the appropriate functionalities in a porogenic solvent, using a template. The target substance and template should be similar, or the target substance itself that the selective reorganization of the analyte takes place. At the time of MIPs synthesis, the monomers used for the synthesis contain suitable functional groups, and the molecules of the template are pre-organized or self-assembled in solution to create a material by the formation of any non-covalent or covalent connections between their binding functionalities. During the cross-linked polymerization, these complexes or materials are conserved. Whereas the template removal creates a molecular cavity. These cavities selectively detect the PAHs by differentiating their size and molecular structures (Haupt 2003).

Electro-Synthesized Conducting Polymers in Molecular Imprinting

The more frequently used procedure for the preparation of conducting and non-conducting MIPs is electrochemical polymerization on the surface of transducer directly. The polymerization of this method is initiated via neither nor UV light or heat directly on transducer surface (Lakshmi et al. 2009; Öpik et al. 2009). However, the deposition of MIP film from the solution of monomer functional of porogenic solvent, sometime monomers polymerization take place by the use of cross-linking on the surface of electrode in the existence of template (Sharma et al. 2012). This deposited film creates roughness on the surface of electrode. The amount of charge transfer during the polymerization controls the factor of thickness. The morphology of material controlled by the supporting electrolytes and suitable solvent. Although the porosity and rigidity are due to the inclusion of ions from the electrolytes. The active monomers contain the electro polymerization moieties; these monomers are called as electro-active functional monomers. The most frequently used electro-polymerization moieties are Th, Py, EDOT, 3-aminophenylboronic acid (APBA), aniline (ANI), 2-aminothiophenol (2-AThPh), phenol (Ph), metalloporphyrin, 1,2-phenylenediamine (PD).

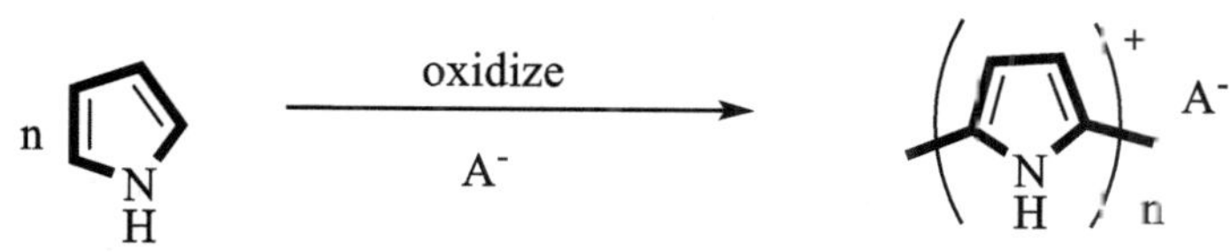

Fig. 10.4 Electro-polymerization of Polypyrrole. Reprinted from the original article by Herlem and Picaud 2021 "after taking permission from ACS

Notably, synthesis of molecularly imprinted ECPs (MIECPs) have been done in the existence of templates via polymerization of APBA, Th, ANI and Py, functional monomers. The interaction of atoms of recognition locations of the MIECP and particles of binding locations of the template via hydrogen bonding can be facilitate the template imprinting in MIECPs. Beside this one more interaction, that is π–π interactions between template and MIECP provide the selective cavity for the particular analysts. Additionally, the dispersive contact between the template and functional monomers increase stability due to π-electrons delocalization in the aromatic circle of functional monomers. The stabilization of MIECP increased if π-delocalized bond is delocalized in the π-conducting polymer chain (Pietrzyk et al. 2009a, b). Some MIP based sensor belonging to this category are described below.

(a) **Pyrrole-based functional MIP based sensor**

Among all the electro-polymerization functional monomers, PPy (poly pyrrole) is the maximum frequently used functional monomer for the synthesis of MIP systems. PPy have very good biocompatibility with excellent immobilization properties, PPy can be easily immobilized on he surface of polymers. The over oxidation of PPy at positive potentials, produced the undesired polymer and this polymer degrade easily with loss of conductivity and dedoping (Fig. 10.4) (Migdalshi et al. 1996; Wegner et al. 1987; Witkowski et al. 1991; Gao and Ivaska 1993; Gao et al. 1994; Mostany and Scharifer 1997; Vork et al., 1985; Krische and Zagorska 1989; Novak et al. 1991). Despite this over oxidation of PPy, one of the major disadvantages, PPy (OPPy) has been used widely in various electro-analytical techniques (Herlem and Picaud 2021).

The ring opening of PAH may be achieved via harsh chemical reagent with the full oxidation of PAH using chemical like fenton's agent due to its high oxidation potential. The PAH such as pyrene bounded to surface shows the electrochemical oxidation of pyrene into quinone at 1 V versus Ag/AgCl in pH 7, PBS at high area carbon electrodes. This electronic rearrangement with in the pyrene ring has been confirmed by the strident analytical peak at 1000 mV (versus Ag/AgCl). This peak at the 1000 m V also represent the complimentary effect between the analyst available in solution and tamplete cavity present in MIP. The aromatic ring of pyrene act as the π–π acceptor and NH groups of the pyrrole role as the π–π interaction donors. Due to these interactions, the signal obtained for pyne is highly specific and sensitive (Tiu et al. 2016).

(b) **Aminophenylboronic acid based functional MIP based sensor**

Understanding the philosophy for proper MIP design, it is necessary to understand the recognition of site of the analyst via molecular interaction between the binding site of the analyst. As a detecting component, polymer that contains APBA can bind in the reversible mode with hydroxyl groups of analysts using trans-esterification reaction method. This covalent approach for binding allows them to make APBA work as a chemosensor for PAHs detection. By using this technology, researchers have a high interest to use boronic acid-based MIP for the detection of PAHs that contain hydroxyl groups (Barathi and Kumar 2013).

One of the boronic acid-based MIPs has been synthesized via a coating of 4-mercaptophenylboronic functionalized ligand on the surface of modified Au nanoparticles. These Au nano-particles were modified through 4-mercaptoaniline, then with the electro-polymerization (Frasconi et al. 2010). The additional stabilization was achieved by Mercapto ethane sulfonic acid, this extra stabilization prevents the coagulation of MIPs. The 4-thioaniline, self-assembled monolayer MIP film was potentiodynamically grown up on an Au SPR. The removal of template was done by the acid hydrolysis that increase the detecting ability and selectivity of MIPs (Sharma et al. 2012; Frasconi et al. 2010).

(c) **Thiophene-based functional MIP based sensor**

The PTh is the second most useable ECPs (Teasdale and wallace 1993) due their high mechanical strength, process ability and high electrical conductivity (~ 106 Sm^{-1}) in both doeped and undoped forms (Schopf and Kossmehl 1995). PTh can be doped in the suitable solvent in either their oxidatively or reductively forms like PPy and PANI (Nalwa 1990). The possibility of reduction is due to the availability of sulfur in their ring structure and their p-doping. For the preparation of PTh films, adonic electro-polymerization techniques have been used widely and their applications in the sensing of PAH (Welzel et al. 1995; Hagen et al. 1990; Refaey and Schwitzgebel 2012; Atta et al. 1991).

The detection of analyst on the surface of transducer, the ECP layer was used to implant in the conventional manor in chronoamperometric sensor; Mazzotta et al. 2008). For more instance, templates like morphine in MIP surface, synthesized by the reaction of trimethylolpropane, trimethacrylate and methacrylic acid functional monomer cross-linker using free radical polymerization. The immobilization was done with the help of EDOT layer on an indium tin oxide electrode surface (Ho et al. 2005). The deposition of EDOT was done by electropolymerization potentiostatic at 1.2 V versus Ag/AgCl. However, methanol is used for extraction of morphine template from MIP particles. Moreover, for the imprinting of morphine, as a functional monomer EDOT was utilized (Yeh and Hoet 2005). For the selective imprinting the below mentioned following thiophene monomer have

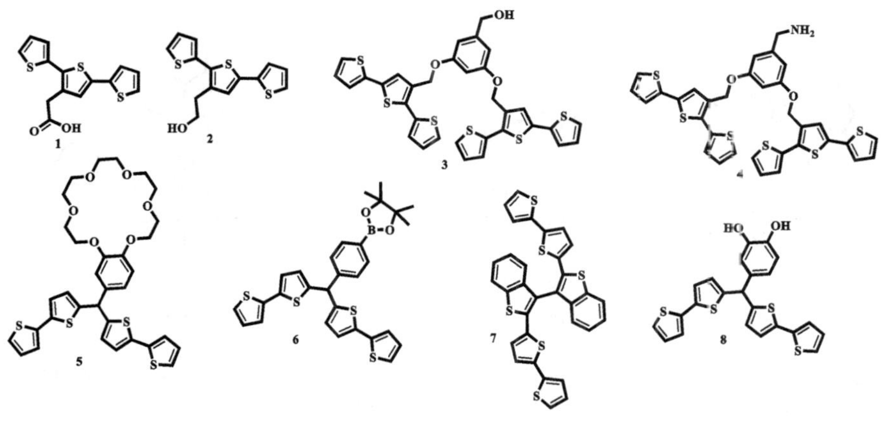

Fig. 10.5 Structure of thiophene Monomer used in MIP "Reprinted from original article by Pernites et al. 2011; Pietrzyk et al. 2009a, b; Pietrzyk et al. 2010;" "after taking permission from Elsevier and ACS"

been discovered for PAHs detection (Fig. 10.5) (Pernites et al. 2011; Pietrzyk et al. 2009a, b; Pietrzyk et al. 2010).

(d) **Porphyrin-based functional MIP sensor**

Preparation of MIP using Metello-porphyrins with stable and electrode-adherent properties was achieved by the use of various kinds of porphyrin monomers via electropolymerization. Whereas, two types of detection are offered by these monomers, i.e., one with the metal ion centers that offers the binding with the heteroatoms containing the lone pair of electrons or electron density and binds with the heteroatom of the template. The second is the outer part of the porphyrins derivatized with different groups, which binds with the electrostatic interaction or hydrogen bonds with the templates. By this technique, the chemosensor for nitrobenzene was obtained, and the chemosensor made by this technique is called as voltametric MIP chemosensor. For the preparation, the layering of MIP film of GCE was done by potentio-dynamic electro-polymerization in the existence of nitrobenzene pattern using nickel porphyrin (9) functional monomer (Fig. 10.6) (Panasyuk et al. 1998). Although the detection potential of nitrobenzene template is not good and selective, instead of this MIP sensor detect 3-nitroaniline and 2-nitrotoluene as well. The voltammetric MIP sensor was prepared of the sensing of 4-(2,4-dichlorophenoxy)butyric acid, an organo-halide by the use of cobalt porphyrin (**10**) as a functional monomer (Mazzotta and Cosimino 2008; Mazzotta and Cosimino 2010). For these chemosensor preparation, a platinum electrode was used, using cobalt porphyrin (10) monomers a MIP film was potentiodynamically implanted onto the exterior of Pt electrode. The extraction of template was done by washing with acetonitrile followed by washing with MeOH.

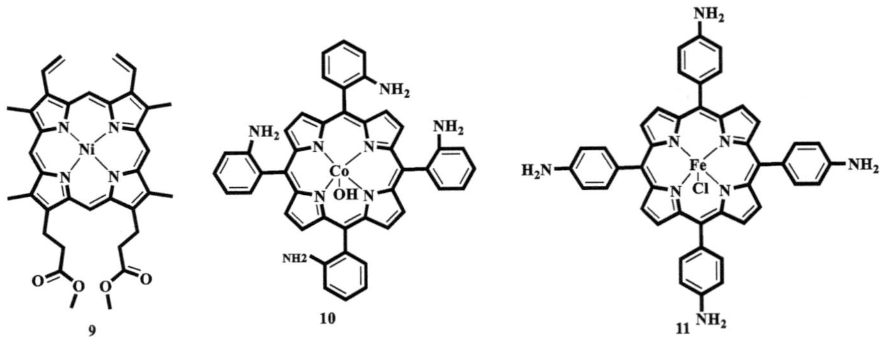

Fig. 10.6 Structure of porphyrins used in MIP "Reprinted from original article by Mazzotta and Cosimino 2010; Gómez-Caballero et al. 2010" "after taking permission from Elsevier"

The Oxidation iron porphyrin (11) contains amino groups, is favorable for the synthesis of porphyrin polymer in a very stable form, and contains phenazine like moieties with the porphyrin ring. Due to this ability, a novel sensor for the PAHs detection was obtained. These sensors are based on square wave voltammetry (SWV) sensor (Gómez-Caballero et al. 2010). For the preparation of these SWV sensors, electropolymerization of iron porphyrin (11) polymer was deposited on a carbon fiber microelectrode phosphate buffer (0.1 M, pH = 8.0) is used for the template removal.

The novel dendritic star co-polymer chemesensor for the detection PAHs was synthesis. This sensor was prepared by polymer of thiophenoimine derivative of 3-propylene and polymer of 3-hexythiophene on electrode based on gold metal (Au/G3PPT-co-P3HT) via in situ electrochemical co-polymerization. The chemosensor was based on ACV (alternating current voltammetry). These sensor response PHE from 0.50 to 6.71 mg/L concentration range with 0.25 mg/L, limit of detection. These electrochemical sensors based on MIP is stable up to 30 days at 4 °C and high recovery rate (93–107%) was observed. Reproducibility (RSD, 6.10%) and repeatability (RSD, 18.1%) is one of the reported advantages for the sensing of PAHs in wastewater (Makelane et al. 2016).

One more MIP based sensor, the same sensor described above was used for the detection of ANT, the technique using for sensing is stage choosy ACV. The detection in the linear range using this sensor is 0.62–10.05 mg/L with the LOD 0.47 mg/L (Makelane et al. 2017).

Some more examples describe the procedure of ECS-MIPs sensor for the detection of PAHs are given below.

For example, one more electrochemical MIP was synthesized by the copolymerization of ethylene glycol dimethacrylate (crosslinker) with the direct assimilate of ferrocenevinyl (tracer redox) into the binding cavities on carbon electrode. These sensors detect BaP by SWV with 0.24 mg/L LOD and the sensor detect BaP selectively in the occurrence of PHE, FLR and PYR with the linear range of 0–4.54 mg/L (Udomsap et al. 2018).

In the other example, one more ANT detecting sensor was developed by coating of MIP based polyurethane layer on the gold electrode (Latif et al. 2018). These conductometric sensor detect ANT selectively in the presence of PYR, PER and BaA with the detection limit 0.23 mg/L.

A glass carbon electrode covered with MIP polypyrrole was developed for detection of ANT by SWV (Mathieu-Scheers et al. 2019).

Electrochemical Biosensors

Detection of PAHs using biological units (like antibodies and enzymes) and electrode (as electrochemical transducer) is done by the sensor called as electrochemical biosensor. These sensor converts the binding of PAHs into electrical signals, which is measurable by using detector. The electrochemical biosensor serves as a rapid and selective detection of PAHs. Biosensors denote dominant analytical instrument for the recognition of PAHs using very small amount of analyst as shown in Fig. 10.7 (Munawar et al. 2020).

Mechanism of Bio-electrochemical Sensor

An electrochemical biosensor is instrument made from the biological detecting unit and the electrochemical transducer. The bio-sensor used to recognize the concentration and existence of PAHs in analytical samples. Various detection elements like nucleic acid, antibody, hormone, organelle, enzyme and whole cell, these biological elements interact with PAHs available in analytical samples and a biochemical signal produce. The biomedical signal convert into measurable electrical signal with the help of electrical transducer. The working principle of bio-electrochemical sensor is classified into two paths.

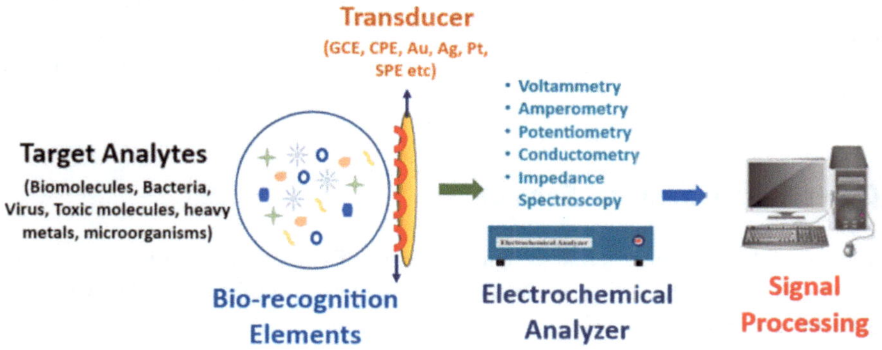

Fig. 10.7 Graphical representation of electrochemical biosensor. "Reprinted from original article by Shanbhag et al., 2023 "after taking permission from Elsevier"

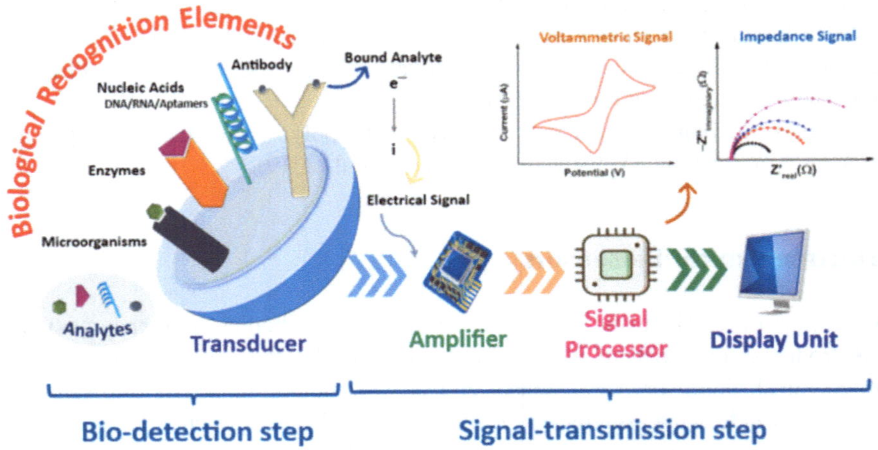

Fig. 10.8 Graphical representation of Bio-sensor mechanism. Reprinted from the original article by Shanbhag et al. 2023 "after taking permission from Elsevier

1. Bio-detection
2. Signal transmission.

The change in biochemical qualities such as antigen–antibody binding, hormone-receptor contact, DNA hybridization and enzyme activity due to bio-detection via analyte binding. This object the change in the electrical properties of the solution such as conductivity, redox potential and pH by generation and consumption of electrons. These change in electrical properties of the solution detect by the electrochemical transducer (electrode) and convert them into the measurable signal by calculating, how much concentration of analyte bind during the transduction stage. The electrode of bio-electrochemical sensor works on various principle, including conductometry, potentiometry, amperometry, and impedance spectroscopy (Fig. 10.8) (Shanbhag et al. 2023; Comnea-Stancu et al. 2021; Wu et al. 2023).

Other Electrochemical Sensors

Various other electrochemical sensor likes 3D-macrocyclic oligomers, 2D-structured nanomaterial clays (anionic) and reduced Graphene Oxide, were also used in recognition of PAHs.

A voltametric recognition of ANT using a 2D-nano structure anionic clay material was described these sensors made up of $Cd(OH)_2/Al(OH)_3$ via single-step approach based on green methodology on carbon glass electrode (GCE-Cd/Al-LDH) directly. The detection limit of this sensor is 0.09 pg/L, 0.02 to 17.82 ng/L linear range. These sensors are also helpful to detect PAHs in real sample like cloud/rain water with high recovery (Qiao et al. 2018).

In a later study, ERGO and 4-tertbutylcalix[4]arene (4TBC) are deposited on the carbon electrode to obtained screen printed carbon electrode (SPCE) for the selective recognition of ANT. An enhancement in the peak current was observed due non-covalent bond formation through π-π interaction and non-covalent bond formation. The LOD for the ANT is 2.85 ng/L with high linear response range from 0.36 to 1.43 mg/L (Zainal et al. 2019).

Detection of PAHs via SERS Based Optical Sensors

A most popular techniques for the detection of PAHs using different templates have been developed and used frequently. This technique is known as surface-enhanced Raman scattering (SERS). The SERS involves plasmonic nanostructures likes' silver and gold by the creation of plasmon surface resonance from localized electromagnetic fields. The techniques contain high enrichment of weak intrinsic Raman signals by plasmonic nanostructure (Caroleo et al. 2022). Due to this features SERS methods are suitable for the sensing of PAHs with high sensitivity. These SERS techniques involve the synthesis of nanostructure based on metal for specific reasons and also modified by the immobilization of ligand of functional group that used for the sensing for different analytes with enhanced capability (Kandi et al. 2022). Furthermore, the sample preparation for SERS is too simple no tedious methods is required even the sample does not affect by the water, this feature of SERS make them an important application for PAHs detection in from the sample collected with different sources.

PAHs are hydrophobic in nature; various challenges occur during the detection of PAHs from the SERS substrate. Consequently, cyclodextrin, calixarenes, thiols, humic acids, and dicarbamates, have been used for the functionalization of silver and gold nanoparticles (Wang et al. 2019). Instead of this several other SERS based sensor have been developed with high enhancement using gold with others material likes TiO_2 nanotube arrays, nickel 3D foam, coffee ring, alginate gel and porous polymers (Du et al. 2016). Regardless, these sensors have some disadvantages that include the poor solubility of PAHs in water, due to this sample requires a preconcentration step. These factors decrease the signal enhancement caused by the distortion of SERS signals.

Non-functionalized SERS Sensors

Many modest and non-functionalized SERS sensor have been described, for example, for the detection of PAHS likes PHE, NAP, and PYR by portable Raman spectrometer. Synthesized the unfunctionalized AuNPs colloid. By this sensor, there is an enhancement of 20-fold in SERS signals. Additionally, the LOD value for all the PAHs is 1.38 mg/L for NAP, 0.23 mg/L for PHE, and 0.45 mg/L for PYR (Gong et al. 2019).

Likewise, other non-functionalized SERS sensors have been developed; these sensors are capable of detect seven PAHs likes BaP, BbF, BkF, NAP, ACY, CHR and ANT (Shi et al. 2019). In the recent year, colloidal AuNPs have been synthesized by the reduction of citrate followed by enriched hydroxyl amine growth to grow the SERS device for the selective recognition of BaP (Su et al. 2021). Also, non-functional high-porous 3D SERS was established for the recognition of PYR (Capaccio et al. 2022). This sensor was prepared by the deposition of AgNPs film on Cr/Au bilayer trailed by the treatment includes inductively-coupled plasma and plasma action in argon environment.

SERS Sensors Based on Thiol and Amine Functionalities

Various SERS sensors were synthesized in which the metal nanoparticle, functionalized with amines and thiols. These amine and thiols-functionalized SERS sensors are used frequently in the sensing of PAHs. Raper Mason mechanism was applied for the synthesis of amine amine-functionalized sensor. Through this method, dopamine acts as a reducing reagent. These AuNPs sensor was prepared by the one pot synthesis, the synthesis id started with the oxidation of dopamine to dopamine quinone (DQ), 5,6-dihydroxyindole and polydopamine (PDA) (Du and Jing 2019). These AuNPs modified by PDA are capable of detecting PAHs close to AuNPs, whereas the AuNPs modified with DQ show the selective sensing of PAHs. In other studies, silver nanoparticles and gold nanoparticles were functionalized with 1-propanethiol and 5,50 dithiobis (succinimidyl-2-nitrobenzoate), respectively. An anti-BaP antibody anchoring and gold sulfur interaction make them these SERS sensors for the recognition of PYR from various sources of water (Zhang et al. 2017) and BaP in sea water (Dribek et al. 2017) simultaneously.

Cyclodextrins Based SERS Sensors

A supramolecule with a big cavity is also used for the functionalization of nanoparticles, especially for SERS sensor development. Whereas, for the functionalization of metal nanoparticles, Cyclodextrins (CDs) and their derivatives are in demand. For the synthesis of silver and gold nanoparticles, maximum of the time cyclodextrins and their derivatives are used directly as a stabilizing and reducing agent or bind on the surface of metal nanoparticles via linkers. Through these CDs containing sensor PAHs bind easily via host–guest chemistry by the formation of an inclusion complex. Moreover, the PAHs sensing is enhanced by the CDs molecules, which act as an internal standard for SERS. At the same time, two ratio-metric SERS sensors were synthesized. These sensors, the conjugation of β-CD with AuNPs with the help of 4-mercapto phenylboronic acid (4-MPBA) linker. The former SERS sensors were used for the sensing of PYR and ANT, whereas 3D mesoporous AuNPs were functionalized by mono-6-thio-β-CD (tb-CD) and used for the recognition of NAP and ANT detection (Yu et al. 2019; Zhang et al. 2020a, b). In the other example green synthetic route was used for the functionalization of AgNPs through β-CD. This β-CD acts as a stabilizing and reducing agent. These SERS sensors utilized for the recognition of ANT, PYR, and CHR/triphenylene (TPL) (Li et al. 2020).

SERS Sensors Based on Polymers and MIPs

The materials, such as 3D-supramolecular and polymers, are involved in the making of SERS sensor with extremely molecular detecting properties. The polymer-implanted SERS sensors are usually made in the presence of a template, such as ionic/molecular assembly, molecules, and ions (Castro-Grijalba et al. 2020). After the template removal, a vacant space was obtained that can be used for the detection of analytes. Although some of the limitations are also faced in MIPs as SERS sensor. These limitations include multiplex analysis and a quantifiable signal caused by conversion of binding events (Wang et al. 2019). A porous glycidyl methacrylate ethylene dimethacrylate (GMA-EDMA) composed 3D SERS sensor was developed in the syringe for the detection of four different PAHs, like PHE, BaP, PYR, and BkF (Shi et al. 2017). PYR OR FLU templates were used for the production of MIPs-based AuNPs SERS sensor (Castro-Grijalba et al. 2020). This SERS sensor is highly selective for the detection of PYR/FLU and PYR/FLU/BaP mixtures from the river water sample.

SERS Sensors Based on Metal Organic Frameworks

The metal organic framework (MOFs) paid the significant attraction in the scientific community as a porous material for the modification of metal nanoparticles to enhance the surface area, stability and impart the large quantity of SERS energetic metal nanoparticles inside cavity of MOFs. Due to the space inside the MOFs, they have good sensing ability generated by metal nanoparticles, by making MOFs-based chemosensors. These sensors are used without the elimination of preconcentration. MOFs-SESR based sensor for the detection of ANT, PYR and perylene (PER). These sensors were prepared by the deposition (MOFs, Cu_3(1,3,5-benzenetricarboxylic acid)$_2$, and AgNPs on a screen-printed carbon electrode by in situ electrodeposition (Li et al. 2019).

SERS Sensors Based on Graphene/graphene Oxide

The molecular structure of Graphene and graphene oxide material is recognized by the sp2-bonded carbon atoms and now are day graphene and graphene oxide are frequently used in various fields of sciences. In the molecular structure of graphene contains condense aromatic ring which make them graphene aromatic in nature. Due to this reason, graphene and graphene oxide behave as a good adsorbent via π–π stacking interaction (Migdalshi et al. 1996). Furthermore, SESR properties are enhanced by the formation of nanocomposite, nanocomposite obtained between metal nanoparticles and graphene/graphene oxide for detection of PAHs. A study was reported for the SERS-based sensor by decorated AgNPs with graphene (Wang et al. 2020a, b), for recognition of three PAHs, namely PYR, ANT, and PHE. In another report, gold nanoparticles anchored with graphene oxide were developed for PAHs detection via the SERS method.

SERS Sensors Based on Magnetic Nanoparticles

SERS sensors incorporated into magnetic nanoparticles have been used for the detection of PAHs and are generally known as MNPs in the scientific community.

These sensors offer improved sensitivity with synergistic single enhancement as plasmonic nanoparticles. MNPs are reusable and do not link with receptors, including aptamers and antibodies, for biosensing applications (Caroleo et al. 2022). Industrialized the technique for the recognition of PAHs by AuNPs nanoparticles grafted with Fe_3O_4 (Du et al. 2016). This immobilized SERS sensor grafted on magnetic Fe_3O_4 efficiently detects polyhydrocarbons from real samples like river water by the LOD reaching from 1.01 to 16.62 mg/L. The comparison of these SARS sensors with electrochemical sensors, A huge number of optical sensors based on SERS technology were developed. These optical SERS sensors displayed the capability of detecting 2–16 PAHs simultaneously. AuNPs deposited on graphene oxide has been used for the detection of 16 PAHs with lesser LOD (0.2–2 mg/L) (Wang et al. 2020a, b), while AuNPs deposited on Fe_3O_4 nanoparticles showed higher LOD (1.01–16.62 mg/L) (Dribek et al. 2017). Amongst, from the numerous SERS sensors with displayed the ability of sensing 2–7 PAHs simultaneously, with colloidal AuNPs/NaCl the deepest LOD was reported to 7 PAHs (0.01–0.14 mg/L), with copolymer-based porous colloid AuNPs, the LOD is 4 PAHs (0.04–0.15 mg/L) (Shi et al. 2017), sing the MOFs-based AgNPs the LOD was reported for 3 PAHs is 0.03–3.56 mg/L (Li et al. 2019) and β-cyclodextrin-based AuNPs used for the 2 PAHs detection with 0.08–0.78 mg/L (Yu et al. 2019). N-dodecanethiol-functionalized AuNPs (Zhang et al. 2020a, b) and hydroxylamine enriched colloidal AuNPs (Su et al. 2021) was used for the recognition of single polyhydrocarbons with low boundary of recognition. On the other hands, 1-propanethiol-functionalized AgNPs showed the low detection limit for PYR (Zhang et al. 2017). Despite of this, another factor likes recovery, reproducibility and stability considered for designing the sensor for PAHs detection.

Conclusion and Future Perspective

According to WHO, the uppermost satisfactory meditation of BaP in water at 0.7 μg/l. In an inclusive examination, it was projected that the mean ingestion of PAH with water is 1% of the overall tolerable level. In this chapter, we summarize the sensors and biosensors for the recognition of polycyclic aromatic hydrocarbons. This chapter contains in-depth information on the technology used for the recognition of PAHs. These technologies are based on analytical methods GC, HPLC, and SFC, and sensor-based, such as electrochemical and SERS-based optical sensing techniques. For the recovery of PAH from the real samples is a difficult task, so there is an urge for the substitution of conformist sample preparation techniques with progressive extraction/purification technologies. The flame ionization detector was used for the PAH recognition in GC techniques, in the various studies MS/MS detector was used for the detection, which is insufficient. Thoughtful optimization of mobile phase, stationary phase, length and type of column, temperature is the essential parameter for PAHs recognition with GC techniques within the shorter time. The other methods like HPLC required fluorescence detector, while the some more advanced methods are likes MS/MS recognition techniques is still absent. Whereas, SFC methods are limited to PAH detection in water samples, SFC techniques are not well suited for PAH analysis in food samples. Additionally, the grafted nanomaterial sensor paid more attention onwards optical and electrochemical sensors due to its cost effectiveness, shorter analysis time and capability of analyzing various PAHs simultaneously. The electrochemical sensors are important for the uncovering of PAHs because of the high sensitivity, the reason for the high sensitivity and selectivity is owing to the synergistic outcome of nanomaterials. Alongside, the other sensors based on SERS, these optical sensors showed an extraordinary improvement in Raman signals due to the incorporation of a diversity of nano-materials achieving high sensitivity. Though, the recognizing techniques and their applications in the single PAHs recognition is limited and the simultaneous analysis of multiple polyaromatic hydrocarbons and their method development is still in demands. A part from this most of the reported sensors are limited to water sample no such sensors are examination of PAHs in other samples like food etc.

References

Abdel-Shafy, H.I., Mansour, M.S.M.: A review on polycyclic aromatic hydrocarbons: source, environmental impact, effect on human health and remediation. Egypt. J. Pet. **25**, 107–123 (2016)

Adegoke, O., Montaseri, H., Nsibande, S.A., Forbes, P.B.C.: Alloyed quaternary/binary core/shell quantum dot-graphene oxide nanocomposite: preparation, characterization and application as a fluorescence "switch ON" probe for environmental pollutants. J. Alloy. Compd. **720**, 70–78 (2017)

Adeniji, A.O., Okoh, O.O., Okoh, A.I.: Analytical methods for the determination of the distribution of total petroleum hydrocarbons in the water and sediment of aquatic systems: a review. J. Chem. 5178937 (2017). https://doi.org/10.1155/2017/5178937

Avino, P., Notardonato, I., Perugini, L., Russo, M.V.: New protocol based on high-volume sampling followed by DLLME-GC-IT/MS for determining PAHs at ultra-trace levels in surface water samples. Microchem. J. **133**, 251–257 (2017)

Atta, N.F., Galal, A., Karagözler, E., Russell, G.C., Zimmer, H., Mark, H.B., Jr.: Electrochemistry and detection of some organic and biological molecules at conducting poly(3-methylthiophene) electrodes. Biosens. Bioelectron. **6**, 333–341 (1991)

Barathi, P., Kumar, S.T.: Electrochemical conversion of unreactive pyrene to highly redox active 1,2-quinone derivatives on a carbon nanotube-modified gold electrode surface and its selective hydrogen peroxide sensing. Langmuir **29**, 10617–10623 (2013)

Behera, B.K., Das, A., Sarkar, D.J., Weerathunge, P., Parida, P.K., Das, B.K., Thavamani, P., Ramanathan, R., Bansal, V.: Polycyclic Aromatic Hydrocarbons (PAHs) in inland aquatic ecosystems: perils and remedies through biosensors and bioremediation. Environ. Pollut. **241**, 212–233 (2018)

Boehm, P.D.: Polycyclic aromatic hydrocarbons (PAHs). Environ Forensics 313–337 (1964). https://doi.org/10.1016/B978-012507751-4/50037-9

Bojakowska, I., Sokołowska, G.: Polycyclic aromatic hydrocarbons (PAHs) in recent lake and river sediments in Poland. In: Weber, J., Jamroz, E., Drozd, J., Karczewska, A. (ed.) Biogeochemical Processes and Cycling of Elements in the Environment, 1st ed. Polish Society of Humic Substances (PTSH), Wroclaw, p. 187 (2001)

Bui, B.T.S., Haupt, K.: Molecularly imprinted polymers: synthetic receptors in bioanalysis. Anal. Bioanal. Chem. **398**, 2481–2492 (2010)

Canfarotta, F., Rapini, R., Piletsky, S.: Recent advances in electrochemical sensors based on chiral and nano-sized imprinted polymers. Curr. Opin. Electrochem. **7**, 146–152 (2018)

Capaccio, A., Sasso, A., Rusciano, G.: Feasibility of SERS-active porous Ag substrates for the effective detection of pyrene in water. Sensors **22**, 2764–2775 (2022)

Caroleo, F., Magna, G., Naitana, M. L., Di Zazzo, L., Martini, R., Pizzoli, F., Muduganti, M., Lvova., L., Mandoj, F., Nardis, S., Stefanelli, M., Di Natale, C., Paolesse., R.: Advances in optical Sensors for persistent organic pollutant environmental monitoring. Sensors **22**, 2649–2691 (2022)

Castro-Grijalba, A., Montes-García, V., Cordero-Ferradás, M.J., Coronado, E., Pérez-Juste, J., Pastoriza-Santos, I.: SERS-based molecularly imprinted plasmonic sensor for highly sensitive PAH detection. ACS Sensers **5**, 693–702 (2020)

Comnea-Stancu, I.R., van Staden, J.F., Stefan-van Staden, R.-I.: Review e trends in recent developments in electrochemical sensors for the determination of polycyclic aromatic hydrocarbons from water resources and catchment areas. J. Electrochem. Soc. **168**, 047504 (2021)

Ricardo, A.I.C., Guzmán Bernardo, F.J., Zougagh, M., Martín-Doimeadios, R.C.R., Ríos, A.: Magnetic nanoparticles-carbon nanotubes hybrid composites for selective solid-phase extraction of polycyclic aromatic hydrocarbons and determination by ultra-high performance liquid chromatography. Anal. Bioanal. Chem. **409**, 5125–5132 (2017)

Diaz-Gonzalez, M., Gutiérrez-Capitán, M., Niu., P., Baldi, A., Jiménez-Jorquera, C., Fernández-Sánchez, C.: Electrochemical devices for the detection of priority pollutants listed in the EU water framework directive. TrAC Trends Anal. Chem. **77**, 186–202 (2016)

Diener, A., Mudu, P.: How can vegetation protect us from air pollution? A critical review on green spaces' mitigation abilities for air-borne particles from a public health perspective—with implications for urban planning. Sci. Total. Environ. **796**, 148605 (2021)

Dribek, M., Rinnert, E., Colas, F., Crassous, M.-P., Thioune, N., David, C., de la Chapelle, M., Compère, C.: Organometallic nanoprobe to enhance optical response on the polycyclic aromatic hydrocarbon benzo[a]pyrene immunoassay using SERS technology. Environ. Sci. Pollut. Res. **24**, 27070–27076 (2017)

Du, J., Jing, C.: Preparation of thiol modified Fe_3O_4@Ag magnetic SERS probe for PAHs detection and identification. J. Phys. Chem. C **115**, 17829–17835 (2011)

Du, J., Xu, J., Sun, Z., Jing, C.: Au nanoparticles grafted on Fe_3O_4 as effective SERS substrates for label-free detection of the 16 EPA priority polycyclic aromatic hydrocarbons. Anal. Chim. Acta **915**, 81–89 (2016)

Du, J., Jing, C.: One-step fabrication of dopamine-inspired Au for SERS sensing of Cd^{2+} and polycyclic aromatic hydrocarbons. Anal. Chim. Acta **1062**, 131–139 (2019)

Farhadian, A., Jinap, S., Abas, F., Sakar, Z.I.: Determination of polycyclic aromatic hydrocarbons in grilled meat. Food Control **21**, 606–610 (2010)

Felemban, S., Vazquez, P., Moore, E.: Future trends for in situ monitoring of polycyclic aromatic hydrocarbons in water sources: the role of immunosensing techniques. Biosensors **9**, 142–158 (2019)

Ferey, L., Delaunay, N.: Capillary and microchip electrophoretic analysis of polycyclic aromatic hydrocarbons. Anal. Bioanal. Chem. **407**, 2727–2747 (2015)

Ferey, L., Delaunay, N., Rutledge, D.N., Huertas, A., Raoul, Y., Gareil, P., Vial, J.: Use of response surface methodology to optimize the simultaneous separation of eight polycyclic aromatic hydrocarbons by capillary zone electrophoresis with laser-induced fluorescence detection. J. Chromatogr. A **1302**, 181–190 (2013)

Frasconi, M., Tel-Vered, R., Riskin, M., Willner, I.: Surface plasmon resonance analysis of antibiotics using imprinted boronic acid-functionalized au nanoparticle composites. Anal. Chem. **82**, 2512–2519 (2010)

Gao, Z., Ivaska, A.: Electrochemical behaviour of dopamine and ascorbic acid at overoxidized polypyrrole(dodecyl sulphate) film-coated electrodes. Anal. Chim. Acta **284**, 393–404 (1993)

Gao, Z., Zi, M., Chen, B.: The influence of overoxidation treatment on the permeability of polypyrrole films. J. Electroanal. Chem. **373**, 141–148 (1994)

Gómez-Caballero, A., Ugarte, A., Sánchez-Ortega, A., Unceta, N., Goicolea, M.A., Barrio, R.J.: Molecularly imprinted poly[tetra(o-aminophenyl)porphyrin] as a stable and selective coating for the development of voltammetric sensors. J. Electroanal. Chem. **638**, 246–253 (2010)

Gong, X., Liao, X., Li, Y., Cao, H., Zhao, Y., Li, H., Cassidy, D.P.: Sensitive detection of polycyclic aromatic hydrocarbons with gold colloid coupled chloride ion SERS sensor. Analyst **144**, 6698–6705 (2019)

Gupta, H.: PAH determination in effluent and sludge samples of paper industry. Environ. Technol. Innov. **9**, 115–121 (2018)

Gupta, H., Dhiman, S.: Synthetic nanoparticle-based remediation of soils contaminated with polycyclic aromatic hydrocarbons. In: Shanker, U., Hussain, C.M., Rani, M. (eds.) Handbook of Green and Sustainable Nanotechnology. Springer, Cham (2023). https://doi.org/10.1007/978-3-030-69023-6_102-1

Gupta, H., Gupta, B.: Adsorption of polycyclic aromatic hydrocarbons on banana peel activated carbon. Desalin. Water Treat. **57**(20), 9498–9509 (2016)

Gupta, H., Kumar, R.: Distribution of selected polycyclic aromatic hydrocarbons in urban soils of Delhi. India. Environ. Technol. Innov. **17**, 100500 (2020)

Jiang, G., Song, X., Xie, J., Shi, T., Yang, Q.: Polycyclic aromatic hydrocarbons (PAHs) in ambient air of Guangzhou city: exposure levels, health effects and cytotoxicity. Ecotoxicol. Environ. Saf. **262**, 107–123 (2023)

Gutierrez-Valencia, T.M., García de Llasera, M.P.: On-line MSPD-SPE-HPLC/FLD analysis of polycyclic aromatic hydrocarbons in bovine tissues. Food Chem. **223**, 82–88 (2017)

Hagen, G., Thoresen, A.H., Sunde, S., Hesjevik, S.M., Ødegård, R.: Electrochemical response of poly-(3-Methyl-Thiophene) to various anions in aqueous solutions. Mol. Cryst. Liq. Cryst. **189**, 213–219 (1990)

Haupt, K.: Imprinted polymers—tailor-made mimics of antibodies and receptors. Chem. Commun. **2003**, 171–178 (2003)

Herlem, G., Picaud, F.: Breaking the controversy of the electropolymerization of pyrrole mechanisms by the effective screening medium quantum charged model interface. J. Phys. Chem. A **125**, 1860–1869 (2021)

Ho, K.-C., Yeh, W.-M., Tung, T.-S., Liao, J.-Y.: Amperometric detection of morphine based on poly(3,4-ethylenedioxythiophene) immobilized molecularly imprinted polymer particles prepared by precipitation polymerization. Anal. Chim. Acta **542**, 90–96 (2005)

Hussain, K., Hoque, R.R., Balachandran, S., Medhi, S., Idris, M.G., Rahman, M., Hussain, F.L.: Monitoring and risk analysis of PAHs in the environment. In: Hussain, C. (ed.) Handbook of Environmental Materials Management, pp. 973–1007. Springer, Cham (2018). https://doi.org/10.1007/978-3-319-58538-3_29-2

Kandi, S., Inbaraj, B.S., Chen, B.-H.: An improved surface enhanced Raman spectroscopic method using a paperbased grape skin-gold nanoparticles/graphene oxide substrate for detection of rhodamine 6G in water and food. Chemosphere **301**, 134702 (2022)

Kumar, A., Gupta, H.: Activated carbon from sawdust for naphthalene removal from contaminated water. Environ. Technol. Innov. **20**, 101080 (2020)

Kumar, S., Negi, S., Maiti, P.: Biological and analytical techniques used for detection of polyaromatic hydrocarbons. Environ. Sci. Pollut. Res. **24**, 25810–25827 (2017)

Krische, B., Zagorska, V.: Overoxidation in conducting polymers. Synth. Met. **28**, 257–262 (1989)

Lakshmi, D., Bossi, A., Whitcombe, M.J., Chianella, I., Fowler, S.A., Subrahmanyam, S., Piletska, E.V., Piletsky, S.A.: Electrochemical sensor for catechol and dopamine based on a catalytic molecularly imprinted polymer-conducting polymer hybrid recognition element. Anal. Chem. **81**, 3576–3584 (2009)

Latif, U., Ping, L., Dickert, F.L.: Conductometric sensor for PAH detection with molecular imprinted polymer as recognition layer. Sensors **18**, 767–776 (2018)

Li, D., Cao, X., Zhang, Q., Ren, X., Jiang, L., Li, D., Deng, W., Liuet, H.: Facile in situ synthesis of core-shell MOF@Ag nanoparticle composites on screen-printed electrodes for ultrasensitive SERS detection of polycyclic aromatic hydrocarbons. J. Mater. Chem. A **7**, 14108–14117 (2019)

Li, M., Yu, H., Cheng, Y., Guo, Y., Yao, W., Xie, Y.: Simultaneous and rapid determination of polycyclic aromatic hydrocarbons by facile and green synthesis of silver nanoparticles as effective SERS substrate. Ecotoxicol. Environ. Saf. **200**, 110780 (2020)

Liu, L.H., Yang, C.-X., Yan, X.-P.: Methacrylate-bonded covalent-organic framework monolithic columns for high performance liquid chromatography. J. Chromatogr. A **1479**, 137–144 (2017)

Lundstedt, S.: Analysis of PAHs and their transformation products in contaminated soil and remedial processes. PhD Dissertation, Department of Chemistry, Environmental Chemistry, (Umeå University, Sweden) (2003)

Ma, X., Huang, P., Dang, X., Ai, Y., Zheng, D., Chen, H.: MWCNTs/MnO_2 nanocomposite-based polythiophene coating for solid-phase microextraction and determination of polycyclic aromatic hydrocarbons in soil. Microchem. J. **146**, 1026–1032 (2019)

Makelane, H.R., John, S.V., Waryo, T.T., Baleg, A., Mayedwa, N., Rassie, C., Wilson, L., Baker, P., Iwuoha, E.I.: AC voltammetric transductions and sensor application of a novel dendritic poly(propylene thiophenoimine)-co-poly(3-hexythiophene) star co-polymer. Sensor Actuator B Chem. **227**, 320–327 (2016)

Makelane, H., John, S.V., Yonkeu, A.L.D., Waryo, T., Tovide, O., Iwuohaet, E.: Phase selective alternating current voltammetric signaling protocol: application in dendritic co-polymer sensor for anthracene. Electroanalysis **29**, 1887–1893 (2017)

Mathieu-Scheers, E., Bouden, S., Grillot, C., Nicolle, J., Warmont, F., Bertagna, V., Cagnon, B., Vautrin-Ul, C.: Trace anthracene electrochemical detection based on electropolymerized-molecularly imprinted polypyrrole modified glassy carbon electrode. J. Electroanal. Chem. **848**, 113253 (2019)

Mazzotta, E., Picca, R.A., Malitesta, C., Piletsky, S.A., Piletska, E.V.: Development of a sensor prepared by entrapment of MIP particles in electrosynthesised polymer films for electrochemical detection of ephedrine. Biosens. Bioelectron. **23**, 1152–1156 (2008)

Mazzotta, E., Cosimino, M.: Preliminary study on electrosynthesis of a Co-porphyrin based molecularly imprinted polymer for the selective detection of the herbicide 2,4-dichlorophenoxy carboxylic acid. Sens. Lett. **6**, 618–622 (2008)

Mazzotta, E., Cosimino, M.: Electrochemical detection of the toxic organohalide 2,4-DB using a Co-porphyrin based electrosynthesized molecularly imprinted polymer. Sens. Actuators B **148**, 186–194 (2010)

Mailu, S.N.: Simultaneous detection of anthracene and phenanthrene on Ag-Au alloy nanoparticles/overoxidized-polypyrrole composite modified carbon electrode. J. Appl. Chem. **12**, 65–80 (2019)

Migdalshi, J., Blaz, T., Lewenstam, A.: Conducting polymer-based ion-selective selectrodes. Anal. Chim. Acta **322**, 141–149 (1996)

Mojiri, A., Zhou, J.L., Ohashi, A., Ozaki, N., Kindaichi, T.: Comprehensive review of polycyclic aromatic hydrocarbons in water sources, their effects and treatments. Sci. Total. Environ. **696**, 133971 (2019)

Mostany, J., Scharifker, B.R.: Impedance spectroscopy of undoped, doped and overoxidized polypyrrole films. Synth. Met. **87**, 179–185 (1997)

Munawar, H., Mankar, J.S., Sharma, M.D., Garcia-Cruz, A., Fernandes, L.A.L., Peacock, M., Krupadam, R.J.: Highly selective electrochemical nanofilm sensor for detection of carcinogenic PAHs in environmental samples. Talanta **219**, 121273 (2020)

Munoz, J., Crivillers, N., Mas-Torrentet, M.: Carbon-rich mono layers on ITO as highly sensitive platforms for detecting polycyclic aromatic hydrocarbons in water: the case of pyrene. Chem. Eur. J. **23**, 15289–15293 (2017)

Munoz, J., Campos-Lendinez, Á., Crivillers, N., Mas-Torrent, M.: Selective discrimination of toxic polycyclic aromatic hydrocarbons in water by targeting p-stacking interactions. ACS Appl. Mater. Interfaces **12**, 26688–26693 (2020)

Munoz, J., Navarro-Senent, C., Crivillers, N., & Mas-Torrent, M.: Study of carbon nanotube-rich impedimetric recognition electrode for ultra-low determination of polycyclic aromatic hydrocarbons in water. Microchim Acta **185**, 255 (2018)

Nalwa, H.S.: Electrically conducting organometallic polymers. Appl. Organomet. Chem. **4**, 91–102 (1990)

Nikolaou, A., Kostopoulou, M., Lofrano, G., Mericet, S.: Determination of PAHs in marine sediments: analytical methods and environmental concerns. Glob. NEST J. **11**, 391 (2009)

Novak, P., Rasch, B., Vielstichet, W.: Overoxidation of polypyrrole in propylene carbonate: an in situ FTIR study. J. Electroanal. Soc. **138**, 3300 (1991)

Nsibande, S.N., Montaseri, H., Forbes, P.B.C.: Advances in the application of nanomaterial-based sensors for detection of polycyclic aromatic hydrocarbons in aquatic systems. Trends Analyt. Chem. **115**, 52–69 (2019)

Öpik, A., Menaker, A., Reut, J., Syritski, V.: Molecularly imprinted polymers: a new approach to the preparation of functional materials. Proc. Est. Acad. Sci. **58**, 3–11 (2009)

Ozaki, N., Yamauchi, T., Kindaichi, T., Ohashi, A.: Stormwater inflow loading of polycyclic aromatic hydrocarbons into urban domestic wastewater treatment plant for separate sewer system. Water Sci. Technol. **79**, 1426–1436 (2019)

Panasyuk, T., Dall'Orto, V.C., Marrazza, G., El'skaya, A., Piletsky, S., Rezzano, I., Mascini, M.: Molecular imprinted polymers prepared by electropolymerization of Ni-(Protoporphyrin IX). Anal. Lett. **31**, 1809–1824 (1998)

Peng, Y-Y., Srinivas, S., & Narain, R.: Modification of polymers. Polym. Sci. Nanotechnol. 95–104 (2020)

Pernites, R., Ponnapati, R., Felipe, M.J., Advincula, R.: Electropolymerization molecularly imprinted polymer (E-MIP) SPR sensing of drug molecules: pre-polymerization complexed terthiophene and carbazole electroactive monomers. Biosens. Bioelectron. **26**, 2766–2771 (2011)

Pietrzyk, A., Kutner, W., Chitta, R., Zandler, M.E., D'Souza, F., Sannicolò, F., Mussini, P.R.: Melamine acoustic chemosensor based on molecularly imprinted polymer film. Anal. Chem. **81**, 10061–10070 (2009a)

Pietrzyk, A., Suriyanarayanan, S., Kutner, W., Chitta, R., D'Souza, F.: Selective histamine piezoelectric chemosensor using a recognition film of the molecularly imprinted polymer of bis(bithiophene) derivatives. Anal. Chem. **81**, 2633–2643 (2009b)

Pietrzyk, A., Suriyanarayanan, S., Kutner, W., Maligaspe, E., Zandler, M.E., D'Souza, F.: Molecularly imprinted poly[bis(2,2′-bithienyl)methane] film with built-in molecular recognition sites for a piezoelectric microgravimetry chemosensor for selective determination of dopamine. Bioelectrochemistry **80**, 62–72 (2010)

Portoles, T., Garlito, B., Nácher-Mestre, J., Berntssen, M.H.G., Pérez-Sánchez, J.: Multi-class determination of undesirables in aquaculture samples by gas chromatography/tandem mass spectrometry with atmospheric pressure chemical ionization: a novel approach for polycyclic aromatic hydrocarbons. Talanta **172**, 109–119 (2017)

Qiao, X., Wei, M., Tian, D., Xia, F., Chen, P., Zhou, C.: One-step electrosynthesis of cadmium/aluminum layered double hydroxides composite as electrochemical probe for voltammetric detection of anthracene. J. Electroanal. Chem. **808**, 35–40 (2018)

Refaey, S.A.M., Schwitzgebel, G.: Effect of anion and cation exchange on the electrochemical properties of poly(3-methylthiophene) in aqueous solutions. Des. Monomers Polym. **3**, 389–398 (2012)

Reinert, N.P., Vieira, C.M.S., Da Silveira, C.B., Budziak, D., Carasek, E.: A low-cost approach using diatomaceous earth biosorbent as alternative SPME coating for the determination of PAHs in water samples by GC-MS. Separations **5**, 55–63 (2018)

Santos, L.O., dos Anjos, J.P., Ferreira, S.L.C., de Andrade, J.B.: Simultaneous determination of PAHS, nitro-PAHS and quinones in surface and groundwater samples using SDME/GC-MS. Microchem. J. **133**, 431–440 (2017)

Schopf, G., Kossmehl, G.: Polythiophenes: electrically conductive polymers. Springer 93–112 (1995). https://doi.org/10.1007/BFb0111624

Sharma, P.S., Pietrzyk-Le, A., D'Souza, F., Kutner, W.: Electrochemically synthesized polymers in molecular imprinting for chemical sensing. Anal. Bioanal. Chem. **402**, 3177–3204 (2012)

Shanbhag, M.M., Manasa, G., Mascarenhas, R.J., Mondal, K., Shetti, N.P.: Fundamentals of bioelectrochemical sensing. Chem. Eng. J. Adv. **16**, 100516 (2023)

Shi, X., Zhang, D., Jia, W., Yan, X., Yang, J., Wang, C., Ma, J.: Portable surface-enhanced Raman scattering sensor for the rapid detection of polycyclic aromatic hydrocarbons in coastal seawater. Mar. Technol. **53**, 46–55 (2019)

Song, D., Yang, R., Long, F., Zhu, A.: Applications of magnetic nanoparticles in surface-enhanced Raman scattering (SERS) detection of environmental pollutants. J. Environ. Sci. **80**, 14–34 (2019)

Soury, S., Bahrami, A., Alizadeh, S., Shahna, F.G., Nematollahi, D.: Development of a needle trap device packed with zinc based metal-organic framework sorbent for the sampling and analysis of polycyclic aromatic hydrocarbons in the air. Microchem. J. **148**, 346–354 (2019)

Srogi, K.: Monitoring of environmental exposure to polycyclic aromatic hydrocarbons: a review. Eviron. Chem. Lett. **5**, 169–195 (2007)

Su, M., Jiang, Q., Guo, J., Zhu, Y., Cheng, S., Yu, T., Du, S., Jiang, Y., Liu, H.: Qualityalert from direct discrimination of polycyclic aromatic hydrocarbons in edible oil by liquid-interfacial surfaceenhanced Raman spectroscopy. LWT **143**, 111143 (2021)

Shi, X., Yan, X., Zhang, X., Ma, L., Zhang, X., Wang, C., Ma, J.: Ultrasensitive detection of polycyclic aromatic hydrocarbons (PAHs) in water using three-dimensional SERS substrate based on porous material and pH 13 gold nanoparticles. J. Ocean Univ. China **18**, 1523–1531 (2017)

Teasdale, P.R., Wallace, G.G.: Molecular recognition using conducting polymers: basis of an electrochemical sensing technology-Plenary lecture. Analyst **118**, 329–334 (1993)

Tijunelyte, I., Betelu, S., Moreau, J., Ignatiadis, I., Berho, C., Lidgi-Guigui, N., Guénin, E., David, C., Vergnole, S., Rinnert, E., de la Chapelle, M.L.: Diazonium salt-based surface-enhanced Raman spectroscopy nanosensor: detection and quantitation of aromatic hydrocarbons in water samples. Sensors **17**, 1198–1215 (2017)

Tiu, B.D.B., Krupadam, R.J., Advincula, R.C.: Pyrene-imprinted polythiophene sensors for detection of polycyclic aromatic hydrocarbons. Sensors Actuators B Chem. **228**, 693–701 (2016)

Udomsap, D., Brisset, H., Culioli, G., Dollet, P., Laatikainen, K., Siren, H., Branger, C.: Electrochemical molecularly imprinted polymers as material for pollutant detection. Mater. Today Commun. **17**, 458–465 (2018)

Vork, F.T.A., Ubbink, M.T., Janssen, L.J.J., Barendrecht, E.: Degradation of polypyrrole-modified electrodes in a chloride containing medium. Recl. Trav. Chim. Pays-Bas **104**, 215–216 (1985)

Wang, L., Huang, Z., Gao, Q., Liu, Y., Kou, X., Xiao, D.: A novel pyrene fluorescent sensor based on the π–π interaction between pyrene and graphene of graphene-cadmium telluride quantum dot nanocomposites. Spectrosc. Lett. **48**, 748–756 (2015)

Wang, L.-L., Yang, C.-X., Yan, X.-P.: In-situ growth of covalent organic framework shells on silica microspheres for application in liquid chromatography. Chem. plus Chem. **82**, 933–938 (2017)

Wang, L., Pang, S., Zhou, G.: Recent advances in spectroscopy technology for trace analysis of persistent organic pollutants. Appl. Sci. **9**, 3439–3454 (2019)

Wang, S., Cheng, J., Han, C., Xie, J.: A versatile SERS sensor for multiple determinations of polycyclic aromatic hydrocarbons and its application potential in analysis of fried foods. Int. J. Anal. Chem. **2020**, 4248029 (2020a)

Wang, X., Xu, Q., Hu, X., Han, F., Zhu, C.: Silver-nanoparticles/graphene hybrids for effective enrichment and sensitive SERS detection of polycyclic aromatic hydrocarbons. Spectrochim. Acta Mol. Biomol. Spectrosc. **228**, 117783 (2020b)

Wegner, G., Wernet, W., Glatzhofer, D.T., Ulanski, J., Kröhnke, C.H., Mohammadi, M.: Chemistry and conductivity of some salts of polypyrrole. Synth. Met. **18**, 1–6 (1987)

Welzel, H.-P., Kossmehl, G., Schneider, J., Plieth, W.: Reactive groups on polymer-covered electrodes. 2. functionalized Thiophene polymers by electrochemical polymerization and their application as polymeric reagents. Macromolecules **28**, 5575–5580 (1995)

WHO (World Health Organization).: Guidelines for drinking-water quality: incorporating the first and second addenda. (Geneva) (World Health Organization) (2018). (https://apps.who.int/iris/bitstream/handle/10665/272345/9789241513760eng.pdf?ua=1)

Witkowski, A., Freund, M.S., Brajter-Toth, A.: Effect of electrode substrate on the morphology and selectivity of overoxidized polypyrrole films. Anal. Chem. **63**, 622–626 (1991)

Wu, J., Liu, H., Chen, W., Ma, B., Ju, H.: Device integration of electrochemical biosensors. Nat. Rev. Bioeng. **1**, 346–360 (2023)

Li, Y., Xian, Q., Li, L.: Development of a short path thermal desorption–gas chromatography/mass spectrometry method for the determination of polycyclic aromatic hydrocarbons in indoor air. J. Chromatogr. A **1497**, 127–134 (2017)

Yu, Z., Grasso, M.F., Sorensen, H.H., Zhang, P.: Ratiometric SERS detection of polycyclic aromatic hydrocarbons assisted by b-cyclodextrin-modified gold nanoparticles. Microchim. Acta **186**, 391 (2019)

Yeh, W.-M., Hoet, K.-C.: Amperometric morphine sensing using a molecularly imprinted polymer-modified electrode. Anal. Chim. Acta **542**, 76–82 (2005)

Zainal, P.N.S., Ahmad, S.A.A., Aziz, S.F.N.A., Rosly, N.Z.: Polycyclic aromatic hydrocarbons: occurrence, electroanalysis, challenges, and future outlooks. Crit. Rev. Anal. Chem. **52**, 878–896 (2020)

Zainal, P.N.S., Ahmad, S.A.A., Ngee, L.H.: Surface modification of screen-printed carbon electrode (SPCE) with calixarenefunctionalized electrochemically reduced graphene oxide (ERGO/C4) in the electrochemical detection of anthracene. J. Electochem. Soc. **166**, B110 (2019)

Zhang, M., Zhang, X., Qu, B., Zhan, J.: Portable kit for high through put analysis of polycyclic aromatic hydrocarbons using surface enhanced Raman scattering after dispersive liquid-liquid microextraction. Talanta **175**, 495–500 (2017)

Zhang, D., Hao, R., Zhang, L., You, H., Fang, J.: Ratiometric sensing of polycyclic aromatic hydrocarbons using capturing ligand functionalized mesoporous Au nanoparticles as a surface-enhanced Raman scattering substrate. Langmuir **36**, 11366–11373 (2020a)

Zhang, Q., Guo, W., He, L., He, L., Chen, Y., Shen, X., Wu, D.: A new SERS substrate of self-assembled monolayer film of gold nanoparticles on silicon wafer for the rapid detection of polycyclic aromatic hydrocarbons. Mater. Chem. Phys. **250**, 122994 (2020b)

Adsorption of Polycyclic Aromatic Hydrocarbons from Water Resources

11

Ambika Kumar, Anshu Kumar, Deepika Dimri, Neetu Yadav, Baliram Pd. Singh, Rajneesh Kumar, and Pramod Kumar

Abstract

Polycyclic aromatic hydrocarbons, also known as PAHs, are widely acknowledged to be the most common organic chemicals found in petroleum. The incomplete burning of organic materials or fossil fuels is the fundamental reason why PAHs are in the environment. Polycyclic aromatic hydrocarbons (PAHs) represent a worldwide problem because they may adversely impact human

A. Kumar (✉) · B. Pd. Singh
Department of Chemistry, Bhagalpur National College, Bhagalpur, (A Constituent Unit of T. M. B. U., Bhagalpur), Bhagalpur 812007, Bihar, India
e-mail: kumarambika.1115@gmail.com; ak.fche@bncollegebgp.ac.in

B. Pd. Singh
e-mail: baliram.pd.singh@gmail.com

A. Kumar
Department of Chemistry, T. N. B. College, Bhagalpur, Bihar, India
e-mail: dranshutnb@gmail.com

D. Dimri
HIMWAL (Himalayan Women Awareness and Livelihood) Society, Dehradun, Uttarakhand, India
e-mail: deepikadimri007@gmail.com

N. Yadav
Department of Chemistry, Indian Institute of Roorkee, Roorkee, Uttarakhand, India
e-mail: yadavneetu1986@gmail.com

R. Kumar
Department of Chemistry, Patna University, Patna, India
e-mail: rajneesh.scp@gmail.com

P. Kumar
Department of Chemistry & Chemical Sciences, School of Physical & Material Sciences, Central University of Himachal Pradesh, Dharamshala, India
e-mail: pramodgang03@hpcu.ac.in

H. Gupta et al. (eds.), *Ubiquitous Polycyclic Aromatic Hydrocarbon Contamination*, Emerging Contaminants and Associated Treatment Technologies,
https://doi.org/10.1007/978-3-032-11764-9_11

health and are poorly soluble in water. Researchers have tried various methods to remove PAHs in the environment such as photocatalytic degradation, electrokinetic remediation, advanced phytoremediation, extraction, bioremediation, chemical oxidation, adsorption and green nano remediation. Due to the remarkable adsorptive capability under lower aqueous solubility exhibited by PAH, the adsorption method has been found as one of the widely adopted methods for water decontamination. Various conventional and novel adsorbing materials function efficiently for removing PAHs from water sources. The materials used for PAH removal consist of biomass derivatives with zeolites and mesoporous silica, together with polymer nanocomposites and graphene-based materials. Environmental engineering research with material science forms an essential field for studying PAHs removal from water resources. Ongoing investigation of novel adsorbents and the optimization of conditions is essential to provide an effective solution to mitigate the detrimental effects of PAHs on ecosystems and human health. A summarise report on the adsorption with different inducing factors has been explored in this chapter, which can be optimised and implemented to remediate PAHs efficiently, particularly from the various water resources.

Keywords

Wastewater analysis • Polycyclic aromatic hydrocarbon (PAH) • Adsorption • Adsorption strategies • Adsorbent

Introduction

Polycyclic aromatic hydrocarbons (PAHs) are organic molecules made up of two or more fused benzene rings, characterised by their stable chemical structure and long-lasting presence in the environment (Sahoo et al. 2020). These chemicals exhibit limited solubility and are prone to adsorbing onto solid and carbonaceous objects in the environment. Although there are more than 100 types of PAHs, the United States Environmental Protection Agency (USEPA) designated 16 polycyclic aromatic hydrocarbons as priority pollutants in 1976 due to their harmful impacts on both the ecosystem and human health (Wise et al. 2015). These 16 PAHs are further divided into two groups based on their molecular structure and weight: low molecular weight (LMW) and high molecular weight (HMW) PAHs. LMW PAHs typically contain two or three fused benzene rings, while HMW PAHs consist of structures with four or more fused rings (Fig. 11.1).

Variation in structural properties of PAHs significantly alters the several characteristics of the molecules, including hydrophobicity, stability, toxicity, boiling and melting points, also increase, while studies show that low molecular weight (LMW) PAHs have higher water solubility and volatility compared to high molecular weight (HMW) PAHs (Vijayanand et al. 2023). In contrast, HMW PAHs are

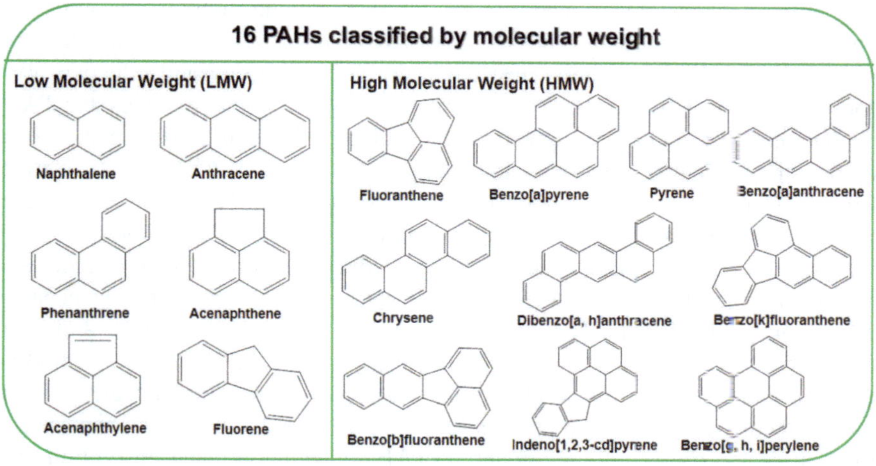

Fig. 11.1 The 16 PAHs priority pollutants, identified by the US EPA, are categorized based on their molecular weight (Reproduced with the permission from Elsevier (Queiroz et al. 2022); licence no. 6060610203285)

more hydrophobic than their LMW counterparts (Queiroz et al. 2022). The characteristics K_{OW} (octanol–water partition coefficient) value indicates the degree of preference for oil-like (octanol) material over a water-like environment. The physico-chemical properties significantly influence the environmental behaviour of PAHs (Gaurav et al. 2021). The semi volatile characteristics of LMW PAHs indicate their presence in the environment as vapours, rendering them especially vulnerable to atmospheric breakdown processes. In aqueous conditions, LMW PAHs are partially dissolved, rendering them readily accessible for diverse degradation processes. The HMW PAHs are predominantly linked to atmospheric and aquatic particles, rendering them less accessible for degradation. Furthermore, PAHs adsorbed onto particles can be carried across extensive distances in the atmosphere, rendering them pervasive throughout the ecosystem (Hussain et al. 2019). Both natural processes and anthropogenetic activities are responsible for the release of PAH contaminants (Abdel-Shafy and Mansour 2016). Natural process includes microbiological production and volcanic activity. But a significant number of PAHs are mostly emitted by anthropogenic activities, including vehicular emissions, military operations, agricultural production, fossil fuel combustion, residential waste incineration, petroleum industry leaks, carbon black manufacturing, internal combustion, carbon black manufacturing, and engine emissions also discharge substantial quantities of PAHs into the ecosystem (Manoli and Samara 1999; Ravindra et al. 2008). Polycyclic aromatic hydrocarbons are majorly contributors to pollutants in the environment because they are carcinogenic, mutagenic, and toxic, as well as being highly persistent in the ecosystem (Manoli and Samara 1999; Ravindra et al. 2008). Human exposure to PAHs, whose carcinogenic potential grows with molecular weight, is associated with a probability of cancer in many organs, and is also linked to immunosuppression and probable endocrine disruption (Ferreira Azevedo et al. 2025. Due to the harmful impacts of PAHs, researchers must begin developing speedy and effective ways for their removal, particularly from water. Major treatment methods for PAH-decontamination include Coagulation, Flocculation, Bioremediation, Solid-phase extraction, Solvent Extraction, Photocatalytic Degradation, Chemical Oxidation, Reverse Osmosis, and Adsorption (Adeola and Forbes 2021). Most of the treatment techniques are time-consuming and have high maintenance costs; some of the processes may generate secondary by-products, which are carcinogenic and mutagenic (e.g., trihalomethanes, halo-acetic acids), hence exacerbating negative effects on public health. The adsorption techniques are a sustainable and cost-effective treatment for PAHs. As a result, many studies have focused on finding efficient and cost-effective sorption media, along with investigating the environmental factors that affect the process for the PAH remediation. This chapter aims to compile the recent review articles on PAH remediation using adsorption as a tool, outlining the sources and adverse impact of PAHs, sorption media, influencing operational parameters, and various adsorbents used for the efficient and sustainable treatment of PAH molecules in aqueous media.

Various Sources of PAHs

The two main sources of polycyclic aromatic hydrocarbons (PAHs) in the environment are natural sources and human-made sources (Abdel-Shafy and Mansour 2016). Several human activities are associated with the incomplete combustion of fossil fuels and biomass. These activities include things like incomplete combustion of cooking at home, emissions from cars, industrial processes, smoking tobacco, and oil spills. In aquatic environment, the concentration of PAHs is commonly lower, and thus classified as a micropollutant (Fig. 11.2). Despite their ultra-trace levels, PAHs can have an important negative effect on the whole ecological system (Wang et al. 2024).

The occurrence of PAHs in the ecosystem is due to various combustion processes.

Potential sources of PAHs in water.

The primary sources of PAHs in aqueous medium are pyrogenic sources, urban and industrial runoff, shipping and port activities, wastewater and sewage discharges, atmospheric deposition (from industrial and vehicles emissions) and petrogenic sources (oil spills and petroleum discharges). Although the major contributors are human activities, natural sources like forest fires, volcanic activity and biogenic/diagenetic processes also contribute (Alegbeleye et al. 2017; Gupta 2018a, b). These hydrophobic chemicals are bound to organic matter and sediments, and are accumulated in the food via bioaccumulation in an organism (such as fish and shellfish). Figure 11.3 shows a global overview of sources of PAH in different water bodies based on diagnostic ratios for the discrimination of petroleum,

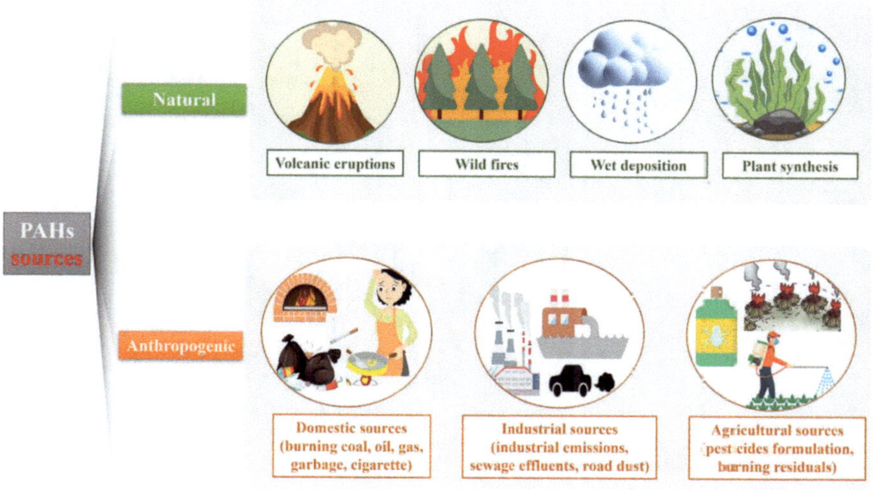

Fig. 11.2 Different sources and pathways of polycyclic aromatic hydrocarbons entering ecosystems (Reproduced with the permission from Elsevier (Tang et al. 2025); licence no. 6060630631225)

combustion, and traffic related emissions. Nigeria and China are the most studied countries with an average of 36.67% and 23.33% of the studies respectively, India has an average of 13.33% of the studies with focus on rivers and drinking water as shown in the mapped data. The map summarises the various sources of PAH with colour-coded bar charts showing the main sources of pollution in each region.

PAH compounds generally undergo transformations via mechanisms such as oxidation and photolysis, producing numerous PAH derivatives that are identified as even more dangerous than the original PAHs (Krzyszczak and Czech 2021). Based on their production sources, PAHs can hence be categorized into three distinct types: Petrogenic, Pyrogenic, and Biogenic (Stogiannidis and Laane 2014). The first category of PAHs, Pyrogenic, are accidentally produced through the partial combustion (350–1200 °C) in the absence of oxygen (Abdel-Shafy and Mansour 2016). The presence of petrogenic PAHs in petroleum and its derivatives is widely distributed in the environment as a result of various anthropogenic activities, including storage, transport, leakage, and utilisation of crude oil. The incorporation of microorganisms enables the synthesis of biogenic PAHs. (Mojiri et al. 2019).

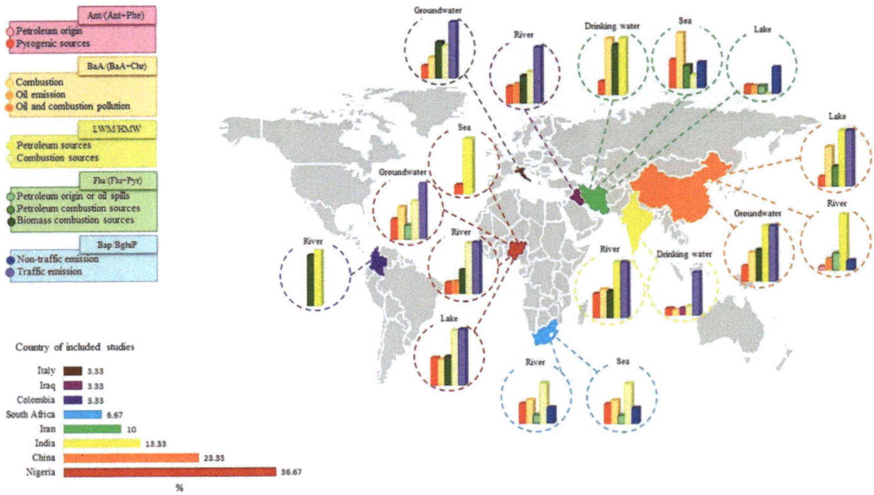

Fig. 11.3 Main sources of PAHs detected in water bodies across various global regions (Reproduced with the permission from Elsevier (Ziyaei et al. 2024); licence no. 6060621184006)

Toxicological and Ecological Risks of PAH Contamination in Water Bodies

PAHs are significant environmental and public health hazards in aquatic systems. They present a significant health hazard to both aquatic ecosystems and human health through many exposure pathways. Owing to lipophilic character, these organic compounds exhibit unusual environmental persistence by the resistance to biodegradation. Further, the hydrophobic character of these compounds facilitating the strong adsorption to suspended particles and organic matter in aquatic systems (Straif et al. 2005). DNA adducts are formed through the metabolic mechanism, thus resulting in mutagenic and carcinogenic effects, with studies showing that more than 70% of PAH contaminated water sources exceed the WHO and USEPA cancer risk thresholds (Xue and Warshawsky 2005). Beyond carcinogenicity, PAH exposure leads to acute health effects including respiratory symptoms and chronic exposure to PAH has more serious health effects. In aquatic ecosystems, PAHs have shown high bioconcentration factors with high accumulation potential in aquatic organisms, especially invertebrate and low trophic level organisms, which lack the efficiency of PAH metabolism pathways (Meador 2003).

Various Human Exposure Mechanisms to PAHs

The majority of human exposure arises from direct contact with pollutants, inhalation of contaminants, physical contact with toxins, or consumption of contaminated dietary substances. Inhalation is associated with both the route of exposure, i.e.,

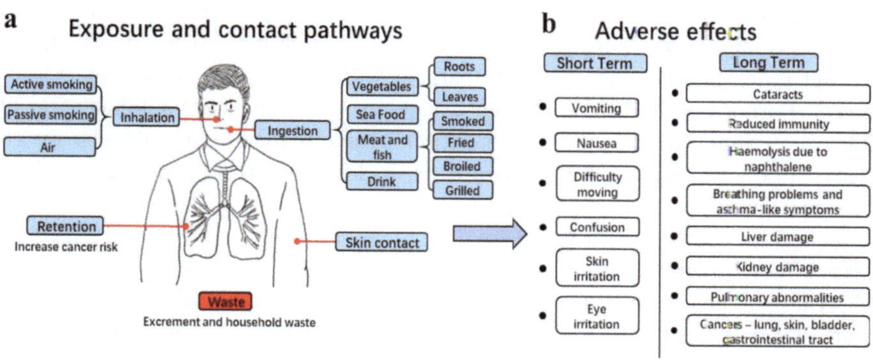

Fig. 11.4 **a** Exposure pathways and **b** adverse effects caused by of PAHs (Reproduced with the permission from Elsevier (Sun et al. 2021); licence no. 6060631144771)

through smoking and by breathing in polluted air (St. Helen et al. 2012). Research has shown that after the inhalation, benzo[a]pyrene (BaP) is capable to enter the pulmonary capillaries in humans (Chang et al. 2019). A significant pathway of exposure is direct contact with PAHs which occurs when the skin gets in contact with polluted water or surfaces allowing for the absorption of PAHs through the dermis. Mallah et al. (2022) and Kammer et al. (2011) determined that contact with sensitive body parts such as skin and eyes is central in exposure to PAHs. Kammer et al. (2011) performed an evaluation of skin absorption of pyrene (Pyr) and benzo[a]pyrene (BaP) during chimney sweeping using a tape stripping method. The results reveals that theses organic molecules were also potentially accumulating in the mid-layer of the stratum corneum in concentrations estimated to be nanograms per square centimetre. Many mechanisms of human exposure to PAHs often occur simultaneously. Outdoor workers such as road and roofing workers are prone to exposure to polycyclic aromatic hydrocarbon (PAHs) through direct exposure to contaminated substances or via inhalation of contaminated air (Hopf et al., 2018). Also, Chen et al. (2014) reported that another importance of human exposure to PAHs is dietary intake, especially smoked meat. A significant quantity of polycyclic aromatic hydrocarbons (PAHs) in the ecosystem accumulate in sediments and soil as a result of both dry and wet deposition. This leads to the uptake of these PAHs by crops, which ultimately leads to their assimilation into the food chain. (Kumar et al. 2021; Hussain et al. 2019). Several paths of exposure and contact are outlined in Fig. 11.4a.

Health Effects Associated with PAH Exposure

Human interaction with PAHs is typically carried through many pathways, and the detrimental consequences of such exposure are generally associated with the relative toxicity of chemicals, exposure length, health state, and other variables

(Geier et al. 2018). However, these compounds are recognized for their long-term hazardous impacts on living organisms. PAHs with high molecular weight (HMW) exhibit more carcinogenicity in comparison with low molecular weight (LMW) PAHs (Barbosa et al. 2023). Though exceptions exist, such as naphthalene (with 2 benzene rings), which exhibits more carcinogenicity than high molecular weight (HMW) polycyclic aromatic hydrocarbons, including benzo(ghi)perylene (with 6 benzene rings).

Short-Term Impacts

Short-term exposure to PAHs can impose an adverse impact on the function of the human lung, particularly among people with asthma, and also may raise the risk of blood clots in those with coronary artery disease (Kim et al. 2013). However, there is not much research on how short-term PAH exposure affects people with digestive, heart, blood, or muscle and bone disorders (Fig. 11.4b). In addition, being exposed to high levels of PAHs for a short time, especially at workstations, can often cause symptoms like confusion, vomiting, nausea and disorientation (Goudarzi et al. 2017).

Long-Term Impacts

Exposure to PAHs over an extended period of time has been linked to a wide variety of chronic health problems (Ravanbakhsh et al. 2023). It has been accompanying to jaundice, as well as liver and kidney dysfunction (Mumtaz and George 1995). In addition, studies have shown that coal furnace workers with long-term PAH exposure exhibit reduced serum immunoglobulin levels (Szczeklik et al. 1994). Naphthalene causes an erythrocyte haemolysis by inhalation (both orally and nasally) (Srogi 2007). Persistent and repeated eye irritation due to low-level exposure of PAHs might be involved in cataract formation (Mumtaz and George 1995). Furthermore, PAHs have been known to cause DNA damage (John et al. 2009). Figure 11.4b illustrates potential adverse effects associated with both acute and chronic PAH exposure.

The lipophilic properties of PAHs enable them to interfere with cellular functioning, leading to chronic diseases; therefore, the elimination of PAHs particularly from aquatic matrix is essential owing to their environmental persistence and potential for biomagnification, which presents long-term risks even at minimal concentrations. Effective remediation reduces direct exposure hazards and averts subsequent ecological and public health catastrophes, especially in areas dependent on aquatic food sources.

PAHs Removal Strategies

There are several traditional techniques in practice to mitigate the concentration of PAHs from wastewater, which include aerobic and anaerobic biodegradation, physical and chemical adsorption, biological methods, and their hyphenation (Ahmed et al. 2021). These strategies exhibit several constraints, including intricate procedures, substantial investment, and elevated maintenance expenses. Numerous processes generate byproducts that are carcinogenic and mutagenic (Pathak et al. 2022). Consequently, the pursuit of further alternative methodologies seems perpetual. Table 11.1 summarizes the merits and demerits of several PAH elimination strategies.

Among the several wastewater treatment technologies available for PAH removal, only a limited number have been commercially demonstrated. Since 1992, Condorchem Envitech, an environmental engineering firm headquartered in Barcelona, specializes in the treatment of industrial wastewater and air emissions, has employed different processes, including adsorption, nanofiltration, reverse osmosis, bioaugmentation, direct filtration, and coagulation-flocculation, to treat the contaminated water on a large scale (Gupta and Dhiman 2023). Due to elevated operating costs, several remediation techniques have not been commercialized or employed at commercial-scale water treatment. In this regard, adsorption is found as the predominant method for wastewater treatment, employing several materials including polymer nanocomposites (Lamichhane et al. 2016).

Adsorption-Based Remediation of PAHs in Water Systems

Adsorption facilitate various benefits, including environmental sustainability, energy efficiency, applicability at elevated temperature, and cost-effective adsorbents from agricultural byproducts, mineral resources and various solid waste (Lamichhane et al. 2016). Adsorbent and adsorbate are two important components that determine the technical and economic viability of the adsorption systems. Also, specific operating conditions used during adsorption play an important role. Most of the PAH removal techniques take advantage of the hydrophobic properties of the materials, allowing for the efficient adsorption of the PAH molecules at the interface. A wide range of adsorbing materials have been studied for the removal of PAHs, and each of them has different advantages in terms of efficiency and applicability (Table 11.2). For example, activated carbon has a removal efficiency during batch and column tests up to 100% and adsorption can be characterized by Freundlich and Langmuir isotherms (Lamichhane et al. 2016; Hu et al. 2024). Biochar is also similarly effective with efficiencies reported as high as 98.6% (Oleszczuk et al. 2012). Modified clay minerals and natural zeolites have also been reported to with removal capabilities over 99% and 85%, respectively (Lamichhane et al. 2016; Bhatnagar and Sillanpaa 2010). Compost and organic matter, and the use of synthetic polymers and iron oxide nanoparticles, have great adsorption capacity, ranging from 70 to 95% (Hale et al. 2011; Mahmoodi et al. 2014; Gupta 2015). The recent developments have focussed on the potential of nanomaterials with remarkable surface area, including carbon nanotubes (CNTs) and graphene revealing significant adsorption affinities (removal efficiencies ranging between 90–99%) (Yin et al. 2020; Liu et al. 2015; Wang et al. 2016). Granular

Table 11.1 Advantages and disadvantages of various remediation techniques for PAHs

Technique	Advantages	Disadvantages	References
Adsorption	Cost-effectiveness, environmental compatibility, operational flexibility, high removal efficiency, rapid process, applicability of recoverable and reusable adsorbents	Expensive adsorbents, secondary contamination	Lamichhane et al. 2016; Dowaidar et al. 2007; Rinawati et al. 2019; Gupta 2018a, b
Bioremediation	Low-cost, complete mineralization	Slow process, optimization challenges	Bisht et al. 2015; Gupte et al. 2016
Chemical oxidation	Short treatment time, effective	High chemical/ operation cost, toxic byproducts	Adeola and Forbes 2021; Haneef et al. 2020; Chen et al. 2015
Coagulation	Removes turbidity and particulates	Sludge formation, slow, less efficient for organics	Nowacka and Włodarczyk-Makuła 2015; Shabeer et al. 2014
Electrokinetic remediation	Effective for low-permeability soils	High energy use, possible secondary pollution	Saichek and Reddy 2005
Flocculation	High PAHs affinity	Secondary pollution requires microbial support	Nyström et al. 2020; Shabeer et al. 2014
Phytoremediation	Environmentally friendly, low-cost	Slow, limited to shallow contamination	Adeola and Forbes 2021; Gupta et al. 2025
Photocatalytic degradation	Effective at room temperature, eco-friendly	High-maintenance, costly catalysts	Silva et al. 2020; Karam et al. 2014
Reverse osmosis	Useful with large scale	Slow and cost-inefficient for small to medium-scale operation	Adeola and Forbes 2021; Smol et al. 2016
Solid-phase extraction	High extraction efficiency	Not robust, high solvent cost	Egli et al. 2015; Von Lau et al. 2014;
Soil vapor extraction	Removes volatile PAHs from soil	Ineffective for non-volatile PAHs, energy use	Cao et al. 2021
Solvent extraction	Fast, economical	Labor-intensive, solvent hazard	Von Lau et al. 2014; Lee et al. 1999

Table 11.2 Adsorption media for PAHs removal

Adsorption media	Methods employed	Removal efficiency (%)	Adsorption isotherm	Reference
Activated Carbon	Batch/ Column	Up to 100	Freundlich, Langmuir	Lamichhane et al. 2016Hu et al. 2024
Biochar	Batch	Up to 98.6	Freundlich, Langmuir	Lamichhane et al. 2016; Oleszczuk et al. 2012
Modified Clay Minerals	Batch	> 99	Freundlich	Lamichhane et al. 2016
Natural Zeolite	Batch	85–95	Langmuir	Bhatnagar and Sillanpää 2010
Compost/Organic Matter	Batch	70–90	Freundlich	Hale et al. 2011
Synthetic Polymers	Batch	80–95	Langmuir, Freundlich	Mahmoodi et al. 2014
Iron Oxide Nanoparticles	Batch	75–90	Langmuir	Gupta 2015
Graphene	Batch	90–99	Langmuir, Freundlich	Yan et al. 2020; Liu et al. 2015
Carbon Nanotubes (CNTs)	Batch	92–99	Langmuir, Freundlich	Wang et al. 2016; Lamichhane et al. 2016
Granular Activated Carbon (GAC)	Batch	> 90	Freundlich	Eeshwarasinghe et al. 2018

Activated Carbon (GAC) has been utilised to remove polycyclic aromatic hydrocarbons (PAHs) from contaminated water; however, this process required several hours to complete. (Eeshwarasinghe et al. 2018). The results emphasize the need to select suitable adsorbent media and optimize operational parameters for efficient PAH remediation.

Adsorption Efficiency and Mechanistic Insights

Adsorption methods exhibit the remarkable efficiency for the decontamination of water; hence, it is widely employed across various remediation fields (Rashid et al. 2021). Further, as a surface property, a relatively small increase in the adsorbent's surface area leads to a big increase in the amount of pollutants that the adsorbent can take in. However, substantial limitations persist regarding the broader application of adsorptive processes, especially in terms of cost management and the requirement for their regeneration. Adsorption processes are generally performed in sequential batch mode, either in continuous or discontinuous processes. Chemisorption and physisorption are two of the most essential features and ways to control the adsorption processes. (Lamichhane et al. 2016). The adsorption capacity (q) is mathematically defined through Eq. (11.1):

$$q = \frac{(C_o - C)V}{m_{ads}} \tag{11.1}$$

In Other words, the adsorbent's Removal Efficiency (RE) is defined as the percentage of contaminants removed from a system., which can be mathematically presented by Eq. (11.2):

$$RE = \frac{(Co - C)}{C} * 100 \tag{11.2}$$

where

q	adsorption capacity,
C_o	initial concentration of adsorbate
C	concentration of the adsorbate in the aqueous matrix post-adsorption,
V	volume of the solution,
m_{ads}	adsorbent mass
RE	Removal Efficiency.

Factors Affecting the Adsorption Phenomena

Time Dependent Adsorption Studies (Adsorption Kinetics)

This measure assesses the rate of transport of adsorbate molecules to the surface of an adsorbent under specific experimental conditions. Adsorption kinetics helps to evaluate the time consumed to attain the equilibrium, and during this specific time, the rate of molecules adsorbing onto the surface matches the rate of molecules desorbing (Walters and Luthy 1984). Several models have been established to explain these kinetics, which are well documented in published articles. In 1898, In 1898, Lagergren introduced the pseudo-first-order model, which is generally utilized when the concentration of the adsorbate stays relatively constant (Tien and Ramarao 2017), whereas Ho and McKay put forward the pseudo second order model, which suggests that a chemisorption process controls the rate of adsorption and is useful when the concentration of adsorbate is low (Ho and McKay 1999). Almost all the adsorption studies related to organic decontamination, can be characterized by both models. It has been observed that adsorption studies exhibited by nanoadsorbents are mostly characterized by the pseudo-second-order model (Ighalo et al. 2022). The pseudo-second-order model was originally developed to describe chemical adsorption, but it is now also utilized for physical adsorption processes. In addition, several other diffusion models are occasionally employed in kinetic studies of organic pollutant adsorption (Eeshwarasinghe et al. 2018).

Adsorption Isotherms

The primary aim of an adsorption isotherm study is to determine the relationship between the concentration of target molecules adsorbed onto the surface of an adsorbent and their residual concentration in a solution at a constant temperature (Majd et al. 2022). This makes the material qualities clearer and makes the adsorption processes better for real-world use. It also gives important information about the adsorption mechanisms, surface features, and adsorbent affinity. A number of different isotherm models are described in the literature, which are based on different assumptions and suited to different situations; the choice of models is very important for describing experimental data and mechanisms correctly.

(a). **Classical Two-Parametric Models**

The Langmuir and Freundlich models are the best and most useful ways to describe how organic pollutants, such polycyclic aromatic hydrocarbons (PAHs), are absorbed (Al-Ghouti and Da'ana 2020). The Langmuir model, originally developed for gas–solid adsorption studies, is commonly utilized to determine the adsorption of organic molecules from liquid solutions. This concept is based on the idea that a single layer of molecules can stick to a smooth surface, where all of the binding sites have the same energy levels and there are no interactions between molecules that are stuck to the surface. Linear representation of Langmuir equation is given by Eq. (11.3):

$$\frac{C_e}{q_e} = \frac{C_e}{q_{\max}} + \frac{1}{K_L q_{\max}} \tag{11.3}$$

in which q_{max} is the maximum adsorption capacity and K_L is the Langmuir constant. A critical parameter obtained from the Langmuir model is the dimensionless separation factor ($R_L = 1/(1 + K_L C_o)$), where values of R_L in the range of 0 to 1 indicate favourable adsorption, $R_L = 0$ represents irreversible adsorption, $R_L = 1$ indicates linear adsorption and $R_L > 1$ indicates unfavourable adsorption (Kheradmand et al. 2022). In contrast, the Freundlich model, developed by Herbert Freundlich in 1909, describes multilayer adsorption on heterogeneous surfaces in which the energy of adsorption decreases as the surface coverage increases (Vigdorowitsch et al. 2021). This empirical model indicates a preference for higher binding sites over those with lower binding energy, rendering it particularly effective for characterizing adsorption on surfaces exhibiting diverse energy levels. The Freundlich equation in linear form is written as (Eq. 11.4):

$$\ln q_e = \ln K_F + \frac{1}{n} \ln C_e \tag{11.4}$$

where K_F and n are called Freundlich constants, with n values greater than unity defining favourable adsorption conditions.

(b). **Three-Parametric Advanced Models**

Several three-parameter models have been constructed to overcome the limitations of the classical isotherms and offer better fitting capabilities for complex adsorption systems. The Sips (Langmuir–Freundlich) hybrid model is especially suitable for heterogeneous adsorption system, and it has demonstrated excellent fitting capability for PAH adsorption study. The Toth model of adsorption isotherm provides a three-parameter framework for heterogeneous adsorption systems, efficiently accommodating both low and high concentration limits (Yakout et al. 2013). For Toth exponent parameter approaching unity the isotherm becomes Langmuir like model, and for values less than unity heterogeneous adsorption with exponentially distributed site energies is indicated. The Redlich-Peterson isotherm combines aspects of the Langmuir and Freundlich equations, rendering it suitable for both heterogeneous and homogeneous systems over a broad concentration range (Walters and Luthy 1984). Studies indicate that the Redlich-Peterson model provides the most accurate match for PAH adsorption data compared to other isotherm models.

(c). **Specialized Models on Microporous Systems**

The Dubinin-Radushkevich (D-R) isotherm and its extended variant, Dubinin-Astakhov isotherm have been used for the prediction of adsorption of organic contaminants onto nano-adsorbents, to a lesser extent (Lamichhane et al. 2016). This model has been especially useful in finding the mean free energy of adsorption, which is useful in differentiating between physical ($E < 8$ kJ/mol) and chemical adsorption mechanism ($E > 8$ kJ/mol) (Puccia and Avena 2021). The D-R isotherm can describe three types of curves (S-, L-, and H-shaped isotherms) depending on the surface heterogeneity parameter and thus it is useful for different adsorbate-adsorbent systems.

(d). **Other Isotherm Models and Applications**

The Temkin isotherm makes the assumption of linear decrease in heat of adsorption and uniform distribution of binding energies and is especially useful for systems for which adsorbent-adsorbate interactions are important. The linear form is given by Eq. (11.5):

$$q_e = B \ln AT + B \ln C_e \tag{11.5}$$

B is related to heat of adsorption and AT is binding constant related to maximum binding energy, q_e is the adsorption efficiency (at equilibrium) and C_e is the equilibrium concentration. Alternative models like Henry's isotherm describe linear adsorption relationships that are appropriate for low concentration systems whereas the Jovanovich isotherm describes adsorption in terms of uncovered surface area and exponential relationships with equilibrium concentration (Ehiomogue et al. 2021; Musah et al. 2022).

The BET (Brunauer–Emmett–Teller) theory, which is an extension of Langmuir theory for multilayer adsorption, becomes relevant for systems associated with physical adsorption with the formation of multiple layers (Hamid and Chen 2024). However, it is very important to understand that isotherm fitting alone is not enough to definitively ascertain the mechanisms of adsorption, and analysis with other complementary techniques, such as FTIR, XPS and thermodynamic analysis, are of maximum importance. Hence, the temperature dependence of isotherm parameters gives further thermodynamic information, linear dependence of isotherm constants on temperature often being found in PAH adsorption studies. This effect of temperature is especially noteworthy for practical purposes, where the operational conditions may change and temperature-dependent isotherm correlations may be needed for accurate process design and optimization.

Release and Recovery: Desorption in Analytical Chemistry and Sorbent Regeneration

Desorption is the procedure through which adsorbed pollutants/molecules are released again to a liquid matrix. This procedure is followed by an adsorption process, which is crucial for the regeneration of the adsorbing materials, enabling their reuse (Bayuo et al. 2024). The reutilization of adsorbents is significant for economic considerations and holds environmental value as well. A number of different approaches, each of which requires optimization, can be utilized to accomplish the desorption of contaminants from the surface of the adsorbent. Desorption with organic solvents is a prevalent technique for removing organic contaminants. To achieve the proper desorption method, it is crucial to choose an appropriate solvent capable of effectively releasing pollutant molecules from the surface of the adsorbent; thus, the selected solvent should have a strong affinity for the targeted contaminants to ensure optimal desorption. For polycyclic aromatic hydrocarbons (PAHs), desorption solvents including acetone, methanol, dichloromethane, chloroform, hexane, acetonitrile, or various mixtures of these are being frequently used (Gachanja and Maritim 2019). To enhance the quantity of desorbed materials, several cycles of the desorption process are required to be executed. The desorption solvent should be cost-effective and environmentally sustainable, and must not compromise the characteristics of adsorbents. Several other approaches for desorbing contaminants and regenerating adsorbents include acid treatments (Baskar et al. 2022), ultrasound-assisted desorption (Hu et al. 2024), and supercritical fluid extraction (Librando et al. 2004); however, these practices are less commonly employed. Numerous investigations at the laboratory scale utilize organic solvents to regenerate or remove absorbed organic contaminants from adsorbents, achieving total or nearly total desorption under ideal conditions. Nonetheless, on a broader scale, particularly in wastewater treatment plants (WWTPs), the regeneration of adsorbents continues to pose significant issues, particularly due to the necessity of considering the environmental impact and associated costs of these processes.

Various Parameters Influencing the Extent of Adsorption

Understanding the various factors that influence adsorption is essential for optimizing this process in both industrial and laboratory settings. Below, we discuss the key parameters that significantly affect adsorption efficiency and capacity.

Particle Size and Surface Area

The surface area of an adsorbent has a big effect on its adsorption capacity, which in turn impacts how well it can remove PAHs from water. Studies indicate that reducing the particle size of the adsorbent typically improves its adsorption efficiency (Wang et al. 2015). Research indicates that chemicals like benzene, naphthalene, and anthracene adhere more readily to estuarine colloids than to larger soil or sediment particles (Wijayaratne and Means 1984; Hundal et al. 2001). This phenomenon arises from the fact that adsorption is predominantly a surface process (El-Sheikh et al. 2004). Research shows that adsorbents with smaller particles have a larger surface area for mass transfer, which makes them better at adsorbing (Lemic et al. 2007; Wang et al. 2015). As sediment particles get smaller, they have more surface area for compounds like phenanthrene to interact with, which makes them more likely to be absorbed (Cornelissen and Gustafsson 2004).

Solubility

The effectiveness of adsorbents regarding adsorption is significantly influenced by their solubility state (Zhang et al. 2019a, b). Studies have shown that there is a link between the higher solubility of PAHs in water and the lower adsorption of these chemicals. (Hu et al. 2008). In this context, the solubility of PAHs is inversely correlated with their molecular weight. A correlation has been established between an increase in the molecular mass of PAHs and a decrease in their solubility. The solubility of three analysed PAHs: naphthalene, acenaphthylene and phenanthrene decreases with the increasing molecular weight (Balati et al. 2015). The Eq. (11.6) describes the mathematical relationship of octanol–water partition coefficients with aqueous phase solubility about the partitioning of PAHs between organic matter and water.

$$K_{ow} = \frac{amount\ of\ organic\ chamical\ in\ octanol\ (in\ \text{mg per litre})}{amount\ of\ organic\ chemical\ in\ water (in\ \text{mg per litre})} \tag{11.6}$$

The octanol–water partition coefficient has been represented by the symbol log K_{ow} and is directly related to molecular mass of polycyclic aromatic hydrocarbons (PAHs); as the molecular mass increases, Kow value increases. Naphthalene and indeno(1,2,3-cd)pyrene have log K_{ow} values of 3.37 and 6.50, respectively, indicating that naphthalene is soluble more than indeno(1,2,3-cd)pyrene. This evaluation clarified the impact of chemical structure and molecular weight on solubility (Cho et al. 2002).

Salinity

The presence of salt ions in aqueous solutions has a notable impact on the adsorption capacity of polycyclic aromatic hydrocarbons (PAHs) (Pathak et al. 2022). Typically, a rise in the level of saline is associated with an increase in the adsorption capacity. Inorganic salt ions form strong interactions with water molecules, which lead to hydration shells that decrease the water solubility of polycyclic aromatic hydrocarbons (PAHs) (Yim 2003). This makes more room for organic solutes in the cavity volume, which lets more PAH molecules stick to the adsorbent's surface. With the addition of 0.34 M KCl, the adsorption coefficient went raised by a considerable 15% (Wang et al. 2015). The adsorption of phenanthrene was shown to increase in saline water, attributed to a reduction in solubility (1.12 mg/L) in pure water (0.97 mg/L) (Sun et al. 2009). Moreover, the adsorption efficiency of PAHs removal was observed to be elevated with rising salinity, with alterations in the physicochemical features of adsorbents (sediments here) (Hu et al. 2008).

Temperature (Thermodynamic Study)

Temperature is a crucial determinant affecting the adsorption of PAHs from aqueous solutions (Zhang et al. 2013) Van't Hoff's Eq. (11.7) describes the dependency on temperature, derived using the Gibbs free energy concept.

$$\ln K_{eq} = -\frac{\Delta H}{RT} + \frac{\Delta S}{R} \tag{11.7}$$

Here,

ΔH	change in enthalpy (kJ/mol),
T	absolute temperature (K),
ΔS	change in entropy (kJ/mol/K),
R	gas constant (8.314×10^3 kJ/K/mol),
Keq	equilibrium partition coefficient at a specified temperature (cm^3/g).

In the above Eq. (11.4), negative and positive values of ΔH/R represent endothermic, exothermic adsorption processes, respectively. HillEr et al. (2008) observed a decrease in adsorption with an increase in temperature from 4 to 27 °C, and the average reduction in adsorption for pyrene, naphthalene, and phenanthrene was evaluated as 27.4%, 27.3%, and 17.0%, respectively (HillEr et al. 2008). In summary, adsorption processes are typically exothermic, leading to enhanced adsorption as temperature decreases (Bekçi et al. 2006). Higher temperatures increased the adsorption of certain PAHs onto NH2-SBA-15, indicating an endothermic process (Balati et al. 2015). Reduced adsorption efficiency is typically anticipated at elevated temperatures for compounds whose solubility increases with a rise in temperature, but if the solubility decreases with a rise in temperature, then the adsorption capacities may be expected to be enhanced at elevated temperatures. (Kipling 2013). This shows how temperature affects how much adsorption happens by changing the balance between sorption and solubility.

pH

The extent of PAH absorption by a solution can be significantly influenced by the solution's pH level (Huang et al. 2019). For example, at low pH (acidic conditions), activated carbon and silica-based materials like NH_2-SBA-15 become more positively charged, attracting PAHs more strongly and increasing adsorption, such as over 90% removal for phenanthrene at pH 2 or over 79% for naphthalene on NH_2-SBA-15 at the same pH due to enhanced electrostatic attraction. In contrast, at high pH (alkaline conditions), the positive charge decreases, and hydroxide ions may block active sites, reducing adsorption; phenanthrene, for instance, drops below 40% at pH 12 (Gupta 2015). However, many studies show that pH does not influence much on PAH remediation for certain adsorbents, with rates remaining stable across a wide pH range. The impact varies by adsorbent and PAH, requiring further research to understand the mechanism.

Adsorption Media for PAHs Removal

The current section briefly reviews the adsorption studies of PAHs removal, employing a range of porous carbonaceous and non-carbonaceous materials.

Biochar

Biochar is a stable material that is rich in carbon, produced through the pyrolysis process of organic waste at temperatures between 300 and 700 °C, conducted in an oxygen-free environment (Sri Shalini et al. 2021). Biochar has demonstrated effectiveness as an adsorbent for polycyclic aromatic hydrocarbons (PAHs) due to its intrinsic properties, including hydrophobicity, enhanced surface area, and improved porosity (Zango et al. 2020). The polarity of PAHs and biochar properties influence the adsorption process. Electrostatic attraction is a key mechanism for the adsorption of PAH onto biochar surfaces. Solution pH and ionic strength impact biochar's PAH adsorption, while pyrolysis temperature affects its electron density and π-electron properties. Below 500 °C, the aromatic structures within biochar predominantly act as electron acceptors, whereas above 500 °C, these structures transition to function as electron donors. De Jesus et al. created coconut waste biochar to get rid of polycyclic aromatic hydrocarbons (PAHs). The biochar was able to adsorb 81.73% of benzo[a]anthracene, 78.51% of benzo[k]fluoranthene, and 81.97% of benzo[a]pyrene (De Jesus et al. 2019). Godlewska et al. also achieved remarkable adsorption of pyrene and PHE and by biochar synthesized from sludge (Godlewska et al. 2019). Liu et al. found that poplar catkin biochar absorbed 384 mg/g of naphthalene (NAP) (Liu et al. 2017). Higher temperatures may make substances more hydrophobic and aromatic, which could increase their ability to adsorb and allow π–π stacking and interactions with PAHs. Chen et al. studied pine needle biochar made at temperatures between 250 and 700 °C. They found that it worked better to remove naphthalene. The research indicated an adsorption capacity of 136.8 mg/g at the peak temperature of 700 °C (Chen et al. 2008). Studies suggest that adsorption by biochar doesn't always match up perfectly with temperature, partly because high temperatures can break down some groups and molecules in the biochar. Still, higher temperatures generally create more surface area resulting better adsorption, and Ding et al. confirmed that as the pyrolysis temperature increases, so does the surface area, leading to better adsorption (Ding et al. 2019).

Activated Carbon (AC)

Activated carbon is widely used to clean up polycyclic aromatic hydrocarbons (PAHs) in places that are wet. It does a good job of soaking up little amounts of contaminants. (Ania et al. 2007; Cabal et al. 2009; Valderrama et al. 2008). Carbonaceous materials can be used to make it by first carbonizing them and then activating them. The main interactions that happen during adsorption are between functional groups, π-π interactions, van der Waals forces, and hydrogen bonding at the surface of the adsorbent (Cheremisinoff 2019). Ge et al. improved activated carbon made from coal to make it better at adsorbing pyrene. The treatments made the adsorbents' surface area and pore volume bigger. The BET surface area grew from 764.96 m^2/g (CAC) to 1293.78 m^2/g (Ge et al. 2016). Danish and Ahmad also reviewed the activated carbon, made from wood biomass for removing PAHs, and found that the main factors affecting its adsorption ability are its surface area and the size of its pores (Danish and Ahmad 2018). The existence of functional groups that include oxygen, such as carboxyl, phenolic, carbonyl, lactonic, and quinonic, as well as sulfhydryl and amino groups, has a big effect on how pollutants interact with each other and how well they can be removed. Alves et al. studied sugarcane straw activated carbon (SCAC) for removing PAHs from wastewater, finding that SCAC showed higher adsorption capacity than commercial carbon (Alves et al. 2022). Their kinetic studies showed that biomass-derived sugarcane straw activated carbon (SCAC) performs as well as commercial activated carbon, highlighting the significance of surface properties and pore size during adsorption. Rinawati et al. introduced rice husk activated carbon (RHAC) for removing PAHs such as phenanthrene (PHE) and fluoranthene (Fl), achieving removal efficiencies of 98–99% for PHE and around 50% for Fl (Rinawati et al. 2019). Rasheed et al. reported a 99% removal efficiency for PAHs using Powdered Activated Carbon (PAC) (Rasheed et al. 2016). Similarly,

H. Kong et al. demonstrated that elevated activation temperatures augmented the PAH adsorption capacity of activated carbon derived from soybean stalks, which was ascribed to an increase in specific surface area (Kong et al. 2011). Raising the carbonization temperature from 300 °C to 600 °C enhanced the minimum and maximum NAP removal efficiencies for F-AC, going from 1.28–1.49 mg/g to 3.99–4.39 mg/g, and for MB-AC, going from 1.20–1.41 mg/g to 3.27–3.94 mg/g (Alade et al. 2012). Biochar has lower production costs and needs less activation energy than activated carbon (Li et al. 2014).

Carbon Nanotubes (CNTs)

CNTs are also widely used as adsorbents for removing PAHs. CNTs are highly stable both chemically and thermally, and their surface area is like that of biochar and activated carbon, making them well-suited for adsorbing polycyclic aromatic hydrocarbons. It is recognized for its better chemical and physical properties (Selvaraj et al. 2020). Carbon nanotubes (CNTs) are produced when two-dimensional (2D) hexagonally arranged carbon atom sheets are rolled into a tubular configuration. The physicochemical properties of CNTs are dictated by the various configurations of rolled graphene sheets. Carbon nanotubes (CNTs) have been developed as a prominent material in eliminating PAHs from wastewater. Carbon nanotubes (CNTs) and activated carbon (AC) exhibit analogous morphologies regarding their structure; furthermore, both are amenable to chemical modification (Abousalman-Rezvani et al. 2020). Carbon nanotubes may be able to absorb PAHs very well since they have a large surface area and many micropores. The chemical functional groups and specific surface area (SSA) of carbon nanotubes (CNTs) have a big effect on how well the adsorption process works (Glomstad et al. 2017). Zhang et al. created a magnetic composite out of single-walled carbon nanotubes (SWCNT), multi-walled carbon nanotubes (MWCNT), and iron oxide. This composite was able to adsorb phenanthrene (PHE) from water. When SWCNT and MWCNT were tested independently, the magnetic composite had a far higher adsorption capacity than either of them. The magnetization of CNTs enhanced the removal efficiency and facilitated regeneration of the adsorbents (Zhang et al. 2019a, b). PAH adsorption on CNTs mainly involves hydrophobic and electrostatic interactions (Engel and Chefetz 2019).

Graphene

Graphene is a flat, one-dimensional sheet comprised of carbon atoms that are organized in a honeycomb pattern. Each atom is strongly connected to three other atoms. Carbon nanotubes (CNTs) are like rolled-up graphene sheets, shaped into tiny tubes, so they interact with PAHs in similar ways, mainly through π–π stacking (where the flat surfaces of both materials attract each other), hydrogen bonding, and sometimes electrostatic forces (Yan et al. 2015). The strong π–π stacking interactions between graphene surfaces and the ring-shaped structures in polycyclic aromatic hydrocarbons (PAHs) improve the ability of graphene to capture these pollutants (Ersan et al. 2017). Graphene nanosheets (GNS) stand out because their huge surface area offers lots of places for contaminants to stick, and the π–π stacking between graphene and organic pollutants is very strong (Ersan et al. 2017). According to research, graphene nanosheets and graphene oxide (GO) are both effective at attracting polycyclic aromatic hydrocarbons. However, normal graphene often outperforms GO and reduced graphene oxide (rGO). Graphene oxide has more oxygen groups and less electron density than reduced graphene oxide, which changes how it adsorbs pollutants. Because PAHs are water-repelling (hydrophobic), they stick especially well to graphene's surface through π–π stacking. The oxygen groups on graphene oxide also help by allowing extra interactions like hydrogen bonding and electrostatic attraction with PAHs. For example, reduced graphene oxide can remove up to 95% of naphthalene, while GO removes about 60%; adding titanium dioxide to graphene boosts PAH removal to around 80% (Wang et al. 2014; Bai et al. 2017). Graphene wool has been found to adsorb 5 mg/g of phenanthrene and 20 mg/g of pyrene (Adeola and Forbes 2019). Other materials besides carbon, including silica, chitosan, imprinted polymers, molecularly imprinted, and metal–organic frameworks (MOF), are also used for PAH removal, sometimes working even better than traditional carbon-based adsorbents (Ahmed et al. 2020). These can be more porous, stable, and selective, with 90% removal of PAHs owing to their ability to form hydrogen bonds and π–π interactions. For example, chitosan beads with titanium dioxide or magnetite nanoparticles can adsorb more than 130 mg/g of PAHs, while molecularly imprinted polymers and mesoporous silica can remove between 79 and 97% of PAHs, respectively (Solano et al. 2021).

Challenges

Owing to the distinct characteristics of PAHs, several challenges are encountered in employing the adsorption procedure for the efficient remediation of PAHs. Lack of functional groups on PAHs enables the effective interconnection between the adsorbate-adsorbent system. They exhibit hydrophobic characteristics and are vulnerable to co-extraction, leading to significant matrix effects. Furthermore, PAHs have negligible charge, are typically found as complex environmental mixtures, and occur only in low concentration. Although mass spectrometry is generally considered the most effective detection technique, its sensitivity for PAHs is limited due to the absence of easily fragmentable sites for soft ionization. These unfavourable characteristics make adsorbents capable of effectively isolating PAHs from contaminated water. In this context, various adsorbents have been evaluated, including traditional ones like synthetic silica and alumina, as well as modern options such as molecular imprinting and nanoparticles (NPs).

Conclusion

PAHs are a worldwide problem in aquatic ecosystems because of their carcinogenic and poisonous characteristics. They have specific ability to bioaccumulate over long periods of time in ecosystems and humans. Their frequent presence in water bodies, by fossil fuel combustion, industrial effluents and urban runoff, are causing major challenges in maintaining water quality and protecting public health. Among others, adsorption is acknowledged as an environmentally sustainable, cost-effective, and highly efficient method for the removal of PAHs from aquatic environments. This is one of the numerous remedial strategies that are accessible. A diverse range of adsorption materials have been used, among them activated carbon, biochar, carbon nanotubes & graphene-based materials, synthetic polymers, and natural minerals have shown interesting removal efficiency (often more than 90–99% for advanced adsorption materials, such as activated carbon, CNTs and graphene). The adsorption behaviour of PAHs is complexly controlled by parameters, including surface area and particle size of adsorbent, pH of solution, temperature, salinity and molecular structure and solubility of PAH compounds. For example, a biochar made from poplar catkin showed a very high adsorption capacity of 384 mg/g for naphthalene, whereas coconut waste biochar showed the potential for removing more than 81% of several high molecular weight PAHs. Moreover, the adsorption process can most often be characterized by Langmuir and Freundlich isotherm models, where some systems exhibit favourable multilayer or heterogeneous adsorption. Despite these improvements, problems still exist with respect to adsorbent regeneration, removal of complex PAH mixtures at ultra-trace levels, effects of co-existing environmental contaminants and control of secondary pollution due to desorption processes. Recent innovations such as the strategies based on engineered nanomaterials, molecularly imprinted polymers and composite adsorbents are promising approaches to overcome these limitations. To improve removal efficiency and assure economic viability, it is important to optimize the operational circumstances, adsorbent design, and large-scale application protocols further. In summary, adsorption is also the predominant and most versatile method for efficient removal of PAHs from water resources because the removal efficiency is up to 100% as shown in batch and column studies with some adsorbents. Development of advanced materials and optimization of the process, as well as an enhanced mechanistic understanding of the adsorption phenomena, is critical to development of robust and sustainable solutions to minimise the environmental and health impacts of PAHs.

Acknowledgements Dr. Ambika Kumar is grateful to the Principal, Dr. Aniruddh Kumar, and faculty members of Bhagalpur National College, Bhagalpur, for their invaluable support and the conducive environment that facilitated the completion of this chapter. Dr. Kumar sincerely thanks his late father, Prof. Krishna Deo Roy, a devoted chemistry teacher at Dr. L. K. V. D. College, Tajpur, Samastipur, Bihar, whose passion and support inspired the academic journey of Dr. Kumar. Dr. Ambika Kumar is also grateful to Mrs. Rubi Roy, for her unwavering support and encouragement throughout this journey.

References

Abdel-Shafy, H.I., Mansour, M.S.: A review on polycyclic aromatic hydrocarbons: source, environmental impact, effect on human health and remediation. Egypt. J. Pet. **25**(1), 107–123 (2016)

Abousalman-Rezvani, Z., Eskandari, P., Roghani-Mamaqani, H., Salami-Kalajahi, M.: Functionalization of carbon nanotubes by combination of controlled radical polymerization and "grafting to" method. Adv. Coll. Interface. Sci. **278**, 102126 (2020)

Adeola, A.O., Forbes, P.B.: Optimization of the sorption of selected polycyclic aromatic hydrocarbons by regenerable graphene wool. Water Sci. Technol. **80**(10), 1931–1943 (2019)

Adeola, A.O., Forbes, P.B.: Advances in water treatment technologies for removal of polycyclic aromatic hydrocarbons: existing concepts, emerging trends, and prospects. Water Environ. Res. **93**(3), 343–359 (2021)

Ahmed, M.J., Hameed, B.H., Hummadi, E.H.: Review on recent progress in chitosan/chitin-carbonaceous material composites for the adsorption of water pollutants. Carbohyd. Polym. **247**, 116690 (2020)

Ahmed, S.F., Mofijur, M., Nuzhat, S., Chowdhury, A.T., Rafa, N., Uddin, M.A., Inayat, A., Mahlia, T.M.I., Ong, H.C., Chia, W.Y., Show, P.L.: Recent developments in physical, biological, chemical, and hybrid treatment techniques for removing emerging contaminants from wastewater. J. Hazard. Mater. **416**, 125912 (2021)

Alade, A.O., Amuda, O.S., Afolabi, T.J., Okoya, A.A.: Adsorption of naphthalene onto activated carbons derived from milk bush kernel shell and flamboyant pod. J. Environ. Chem. Ecotoxicol. **4**(7), 124–132 (2012)

Alegbeleye, O.O., Opeolu, B.O., Jackson, V.A.: Polycyclic aromatic hydrocarbons: a critical review of environmental occurrence and bioremediation. Environ. Manage. **60**, 758–783 (2017)

Al-Ghouti, M.A., Da'ana, D.A.: Guidelines for the use and interpretation of adsorption isotherm models: a review. J. Hazard. Mater. **393**, 122383 (2020)

Alves, L.S., Ferreira Neto, V.J., Costa, T.S., Gaspar, A.B., Mendes, F.M., Luna, A.S., Henriques, C.A.: Use of activated carbon obtained from sugarcane straw for PAH adsorption-a comparative study with commercial materials. Environ. Technol. **43**(6), 861–875 (2022)

Ania, C.O., Cabal, B., Pevida, C., Arenillas, A., Parra, J.B., Rubiera, F., Pis, J.J.: Effects of activated carbon properties on the adsorption of naphthalene from aqueous solutions. Appl. Surf. Sci. **253**(13), 5741–5746 (2007)

Bai, H., Zhou, J., Zhang, H., Tang, G.: Enhanced adsorbability and photocatalytic activity of TiO_2-graphene composite for polycyclic aromatic hydrocarbons removal in aqueous phase. Colloids Surf. B **150**, 68–77 (2017)

Balati, A., Shahbazi, A., Amini, M.M., Hashemi, S.H.: Adsorption of polycyclic aromatic hydrocarbons from wastewater by using silica-based organic–inorganic nanohybrid material. J. Water Reuse Desalination **5**(1), 50–63 (2015)

Barbosa, F., Jr., Rocha, B.A., Souza, M.C., Bocato, M.Z., Azevedo, L.F., Adeyemi, J.A., Santana, A., Campiglia, A.D.: Polycyclic aromatic hydrocarbons (PAHs): updated aspects of their determination, kinetics in the human body, and toxicity. J. Toxicol. Environ. Health Part B **26**(1), 28–65 (2023)

Baskar, A.V., Bolan, N., Hoang, S.A., Sooriyakumar, P., Kumar, M., Singh, L., Jasemizad, T., Padhye, L.P., Singh, G., Vinu, A., Sarkar, B.: Recovery, regeneration and sustainable management of spent adsorbents from wastewater treatment streams: a review. Sci. Total. Environ. **822**, 153555 (2022)

Bayuo, J., Rwiza, M.J., Choi, J.W., Mtei, K.M., Hosseini-Bandegharaei, A., Sillanpää, M.: Adsorption and desorption processes of toxic heavy metals, regeneration and reusability of spent adsorbents: economic and environmental sustainability approach. Adv. Colloid Interface Sci. 103196 (2024)

Bekçi, Z., Seki, Y., Yurdakoç, M.K.: Equilibrium studies for trimethoprim adsorption on montmorillonite KSF. J. Hazard. Mater. **133**(1–3), 233–242 (2006)

Bhatnagar, A., Sillanpää, M.: Utilization of agro-industrial and municipal waste materials as potential adsorbents for water treatment—a review. Chem. Eng. J. **157**(2–3), 277–296 (2010)

Bisht, S., Pandey, P., Bhargava, B., Sharma, S., Kumar, V., Sharma, K.D.: Bioremediation of polyaromatic hydrocarbons (PAHs) using rhizosphere technology. Braz. J. Microbiol. **46**, 7–21 (2015)

Cabal, B., Budinova, T., Ania, C.O., Tsyntsarski, B., Parra, J.B., Petrova, B.: Adsorption of naphthalene from aqueous solution on activated carbons obtained from bean pods. J. Hazard. Mater. **161**(2–3), 1150–1156 (2009)

Cao, W., Zhang, L., Miao, Y., Qiu, L.: Research progress in the enhancement technology of soil vapor extraction of volatile petroleum hydrocarbon pollutants. Environ Sci Process Impacts **23**(11), 1650–1662 (2021)

Chang, Y., Siddens, L.K., Heine, L.K., Sampson, D.A., Yu, Z., Fischer, K.A., Löhr, C.V., Tilton, S.C.: Comparative mechanisms of PAH toxicity by benzo [a] pyrene and dibenzo [def, p] chrysene in primary human bronchial epithelial cells cultured at air-liquid interface. Toxicol. Appl. Pharmacol. **379**, 114644 (2019)

Chen, F., Hong, M., You, W., Li, C., Yu, Y.: Simultaneous efficient adsorption of Pb^{2+} and MnO_4^- ions by MCM-41 functionalized with amine and nitrilotriacetic acid anhydride. Appl. Surf. Sci. **357**, 856–865 (2015)

Chen, J., Wong, M.H., Wong, Y.S., Tam, N.F.: Multi-factors on biodegradation kinetics of polycyclic aromatic hydrocarbons (PAHs) by Sphingomonas sp. a bacterial strain isolated from mangrove sediment. Mar. Pollut. Bull. **57**(6–12), 695–702 (2008)

Chen, Y., Shen, G., Su, S., Shen, H., Huang, Y., Li, T., Li, W., Zhang, Y., Lu, Y., Chen, H., Yang, C.: Contamination and distribution of parent, nitrated, and oxygenated polycyclic aromatic hydrocarbons in smoked meat. Environ. Sci. Pollut. Res. **21**, 11521–11530 (2014)

Cheremisinoff, P.N.: Handbook of Water and Wastewater Treatment Technology. Routledge (2019)

Cho, H.H., Park, J.W., Liu, C.C.: Effect of molecular structures on the solubility enhancement of hydrophobic organic compounds by environmental amphiphiles. Environ. Toxicol. Chem. **21**(5), 999–1003 (2002)

Cornelissen, G., Gustafsson, Ö.: Sorption of phenanthrene to environmental black carbon in sediment with and without organic matter and native sorbates. Environ. Sci. Technol. **38**(1), 148–155 (2004)

Danish, M., Ahmad, T.: A review on utilization of wood biomass as a sustainable precursor for activated carbon production and application. Renew. Sustain. Energy Rev. **87**, 1–21 (2018)

De Jesus, J.H.F., da S. Matos, T.T., da C. Cunha, G., Mangrich, A.S., Romão, L.P.C.: Adsorption of aromatic compounds by biochar: influence of the type of tropical biomass precursor. Cellulose **26**, 4291–4299 (2019)

Ding, Z., Zhang, F., Sun, N., Chi, J.: Sorption of polycyclic aromatic hydrocarbons by biochars of wheat straw with different pyrolysis temperatures. In: IOP Conference Series: Earth and Environmental Science, vol. 237, p. 052008. IOP Publishing (2019)

Dowaidar, A.M., El-Shahawi, M.S., Ashour, I.: Adsorption of polycyclic aromatic hydrocarbons onto activated carbon from non-aqueous media: 1. The influence of the organic solvent polarity. Sep. Sci. Technol. **42**(16), 3609–3622 (2007)

Eeshwarasinghe, D., Loganathan, P., Kalaruban, M., Sounthararajah, D.P., Kandasamy, J., Vigneswaran, S.: Removing polycyclic aromatic hydrocarbons from water using granular activated carbon: kinetic and equilibrium adsorption studies. Environ. Sci. Pollut. Res. **25**(14), 13511–13524 (2018)

Egli, S.N., Butler, E.D., Bottaro, C.S.: Selective extraction of light polycyclic aromatic hydrocarbons in environmental water samples with pseudo-template thin-film molecularly imprinted polymers. Anal. Methods **7**(5), 2028–2035 (2015)

Ehiomogue, P., Ahuchaogu, I.I., Ahaneku, I.E.: Review of adsorption isotherms models. Acta Technica Corviniensis-Bull. Eng. **14**(4) (2021)

El-Sheikh, A.H., Newman, A.P., Al-Daffaee, H., Phull, S., Cresswell, N., York, S.: Deposition of anatase on the surface of activated carbon. Surf. Coat. Technol. **187**(2–3), 284–292 (2004)

Engel, M., Chefetz, B.: The missing link between carbon nanotubes, dissolved organic matter and organic pollutants. Adv. Coll. Interface. Sci. **271**, 101993 (2019)

Ersan, G., Apul, O.G., Perreault, F., Karanfil, T.: Adsorption of organic contaminants by graphene nanosheets: a review. Water Res. **126**, 385–398 (2017)

Ferreira Azevedo, L., de Souza Rocha, C.C., Souza, M.C.O., Machado, A.R.T., Devóz, P.P., Rocha, B.A., Antunes, L.M.G., Uribe-Romo, F.J., Campiglia, A.D., Barbosa, F., Jr.: High molecular weight polycyclic aromatic hydrocarbon (HMW-PAH) isomers: unveiling distinct toxic effects from cytotoxicity to oxidative stress-induced DNA damage. Arch. Toxicol. **99**(2), 679–687 (2025)

Gachanja, A.N., Maritim, P.K.: Polycyclic aromatic hydrocarbons determination. Encycl. Anal. Sci. 328–340 (2019)

Gaurav, G.K., Mehmood, T., Kumar, M., Cheng, L., Sathishkumar, K., Kumar, A., Yadav, D.: Review on polycyclic aromatic hydrocarbons (PAHs) migration from wastewater. J. Contam. Hydrol. **236**, 103715 (2021)

Ge, X., Wu, Z., Wu, Z., Yan, Y., Cravotto, G., Ye, B.C.: Microwave-assisted modification of activated carbon with ammonia for efficient pyrene adsorption. J. Ind. Eng. Chem. **39**, 27–36 (2016)

Geier, M.C., Chlebowski, A.C., Truong, L., Massey Simonich, S.L., Anderson, K.A., Tanguay, R.L.: Comparative developmental toxicity of a comprehensive suite of polycyclic aromatic hydrocarbons. Arch. Toxicol. **92**, 571–586 (2018)

Glomstad, B., Sørensen, L., Liu, J., Shen, M., Zindler, F., Jenssen, B.M., Booth, A.M.: Evaluation of methods to determine adsorption of polycyclic aromatic hydrocarbons to dispersed carbon nanotubes. Environ. Sci. Pollut. Res. **24**, 23015–23025 (2017)

Godlewska, P., Siatecka, A., Kończak, M., Oleszczuk, P.: Adsorption capacity of phenanthrene and pyrene to engineered carbon-based adsorbents produced from sewage sludge or sewage sludge-biomass mixture in various gaseous conditions. Biores. Technol. **280**, 421–429 (2019)

Goudarzi, G., Idani, E., Alavi, N., Salmanzadeh, S., Babaei, A.A., Geravandi, S., Mohammadi, M.J., Mahboubi, M., Moradi, M.: Association of polycyclic aromatic hydrocarbons of the outdoor air in Ahvaz, southwest Iran during warm-cold season. Toxin Rev. **36**(4), 282–289 (2017)

Gupta, H.: Removal of phenanthrene from water using activated carbon developed from orange rind. Int. J. Sci. Res. Environ. Sci. **3**(7), 248 (2015)

Gupta, H.: Anthracene removal from water onto activated carbon derived from vehicular tyre. Sep. Sci. Technol. **53**(4), 613–625 (2018a)

Gupta, H.: PAH determination in effluent and sludge samples of the paper industry. Environ. Technol. Innov. **9**, 115–121 (2018b)

Gupta, H., Dhiman, S.: Synthetic nanoparticle-based remediation of soils contaminated with polycyclic aromatic hydrocarbons. In: Shanker, U., Hussain, C.M., Rani, M. (eds.) Handbook of Green and Sustainable Nanotechnology. Springer, Cham (2023)

Gupta, H., Dhiman, S., Kumar, A., Paliwal, S.K., Khan, G., Singh, H., Mishra, A., Mishra, A.K.: Photo-and bio-degradation of selected persistent organic pollutants. Sep. Purif. Rev. **54**(1), 3–16 (2025)

Gupte, A., Tripathi, A., Patel, H., Rudakiya, D., Gupte, S.: Bioremediation of polycyclic aromatic hydrocarbon (PAHs): a perspective. Open Biotechnol. J. **10**(1) (2016)

Hale, S., Hanley, K., Lehmann, J., Zimmerman, A., Cornelissen, G.: Effects of chemical, biological, and physical aging as well as soil addition on the sorption of pyrene to activated carbon and biochar. Environ. Sci. Technol. **45**(24), 10445–10453 (2011)

Hamid, U., Chen, C.C.: Generalized Brunauer–Emmett–Teller isotherm for mixed-gas multilayer adsorption equilibria. Ind. Eng. Chem. Res. **63**(31), 13853–13863 (2024)

Haneef, T., Mustafa, M.R.U., Wan Yusof, K., Isa, M.H., Bashir, M.J., Ahmad, M., Zafar, M.: Removal of polycyclic aromatic hydrocarbons (PAHs) from produced water by ferrate (VI) oxidation. Water **12**(11), 3132 (2020)

HillEr, E., Jurkovič, L., Bartal, M.V.U.V.: Effect of temperature on the distribution of polycyclic aromatic hydrocarbons in soil and sediment (2008)

Ho, Y.S., McKay, G.: Pseudo-second order model for sorption processes. Process Biochem. **34**(5), 451–465 (1999)

Hopf, N.B., Spring, P., Hirt-Burri, N., Jimenez, S., Sutter, B., Vernez, D., Berthet, A.: Polycyclic aromatic hydrocarbons (PAHs) skin permeation rates change with simultaneous exposures to solar ultraviolet radiation (UV-S). Toxicol. Lett. **287**, 122–130 (2018)

Hu, D., Liu, S., Qi, L., Liang, J., Zhang, G.: A critical review on ultrasound-assisted adsorption and desorption technology: mechanisms, influencing factors, applications, and prospects. J. Environ. Chem. Eng. 114307 (2024)

Hu, X., Liu, J., Mayer, P., Jiang, G.: Impacts of some environmentally relevant parameters on the sorption of polycyclic aromatic hydrocarbons to aqueous suspensions of fullerene. Environ. Toxicol. Chem. **27**(9), 1868–1874 (2008)

Huang, D., Xu, B., Wu, J., Brookes, P.C., Xu, J.: Adsorption and desorption of phenanthrene by magnetic graphene nanomaterials from water: roles of pH, heavy metal ions and natural organic matter. Chem. Eng. J. **368**, 390–399 (2019)

Hundal, L.S., Thompson, M.L., Laird, D.A., Carmo, A.M.: Sorption of phenanthrene by reference smectites. Environ. Sci. Technol. **35**(17), 3456–3461 (2001)

Hussain, K., Hoque, R.R., Balachandran, S., Medhi, S., Idris, M.G., Rahman, M., Hussain, F.L.: Monitoring and risk analysis of PAHs in the environment. In: Handbook of Environmental Materials Management, pp. 973–1007 (2019)

Ighalo, J.O., Yap, P.S., Iwuozor, K.O., Aniagor, C.O., Liu, T., Dulta, K., Iwuchukwu, F.U., Rangabhashiyam, S.: Adsorption of persistent organic pollutants (POPs) from the aqueous environment by nano-adsorbents: a review. Environ. Res. **212**, 113123 (2022)

John, K., Ragavan, N., Pratt, M.M., Singh, P.B., Al-Buheissi, S., Matanhelia, S.S., Phillips, D.H., Poirier, M.C., Martin, F.L.: Quantification of phase I/II metabolizing enzyme gene expression and polycyclic aromatic hydrocarbon–DNA adduct levels in human prostate. Prostate **69**(5), 505–519 (2009)

Kammer, R., Tinnerberg, H., Eriksson, K.: Evaluation of a tape-stripping technique for measuring dermal exposure to pyrene and benzo (a) pyrene. J. Environ. Monit. **13**(8), 2165–2171 (2011)

Karam, F.F., Hussein, F.H., Baqir, S.J., Halbus, A.F., Dillert, R., Bahnemann, D.: Photocatalytic degradation of anthracene in closed system reactor. Int. J. Photoenergy **2014**(1), 503825 (2014)

Kheradmand, A., Negarestani, M., Kazemi, S., Shayesteh, H., Javanshir, S., Ghiasinejad, H.: Adsorption behavior of rhamnolipid modified magnetic Co/Al layered double hydroxide for the removal of cationic and anionic dyes. Sci. Rep. **12**(1), 14623 (2022)

Kim, K.H., Jahan, S.A., Kabir, E., Brown, R.J.: A review of airborne polycyclic aromatic hydrocarbons (PAHs) and their human health effects. Environ. Int. **60**, 71–80 (2013)

Kipling, J.J.: Adsorption from Solutions of Non-electrolytes. Academic Press (2013)

Kong, H., He, J., Gao, Y., Han, J., Zhu, X.: Removal of polycyclic aromatic hydrocarbons from aqueous solution on soybean stalk–based carbon. J. Environ. Qual. **40**(6), 1737–1744 (2011)

Krzyszczak, A., Czech, B.: Occurrence and toxicity of polycyclic aromatic hydrocarbons derivatives in environmental matrices. Sci. Total. Environ. **788**, 147738 (2021)

Kumar, M., Bolan, N.S., Hoang, S.A., Sawarkar, A.D., Jasemizad, T., Gao, B., Keerthanan, S., Padhye, L.P., Singh, L., Kumar, S., Vithanage, M.: Remediation of soils and sediments polluted with polycyclic aromatic hydrocarbons: to immobilize, mobilize, or degrade? J. Hazard. Mater. **420**, 126534 (2021)

Lamichhane, S., Krishna, K.B., Sarukkalige, R.: Polycyclic aromatic hydrocarbons (PAHs) removal by sorption: a review. Chemosphere **148**, 336–353 (2016)

Lee, P.H., Ong, S.K., Golchin, J., Nelson, G.L.: Extraction method for analysis of PAHs in coal-tar-contaminated soils. Pract. Period. Hazard., Toxic, Radioact. Waste Manag. **3**(4), 155–162 (1999)

Lemić, J., Tomašević-Čanović, M., Adamović, M., Kovačević, D., Milićević, S.: Competitive adsorption of polycyclic aromatic hydrocarbons on organo-zeolites. Microporous Mesoporous Mater. **105**(3), 317–323 (2007)

Li, H., Qu, R., Li, C., Guo, W., Han, X., He, F., Ma, Y., Xing, B.: Selective removal of polycyclic aromatic hydrocarbons (PAHs) from soil washing effluents using biochars produced at different pyrolytic temperatures. Biores. Technol. **163**, 193–198 (2014)

Librando, V., Hutzinger, O., Tringali, G., Aresta, M.: Supercritical fluid extraction of polycyclic aromatic hydrocarbons from marine sediments and soil samples. Chemosphere **54**(8), 1189–1197 (2004)

Liu, X., Li, J., Wang, X., Chen, C., Wang, X.: High performance of phosphate-functionalized graphene oxide for the selective adsorption of U (VI) from acidic solution. J. Nucl. Mater. **466**, 56–64 (2015)

Liu, X., Sun, J., Duan, S., Wang, Y., Hayat, T., Alsaedi, A., Wang, C., Li, J.: A valuable biochar from poplar catkins with high adsorption capacity for both organic pollutants and inorganic heavy metal ions. Sci. Rep. **7**(1), 10033 (2017)

Mahmoodi, N.M., Sadeghi, U., Maleki, A., Hayati, B., Najafi, F.: Synthesis of cationic polymeric adsorbent and dye removal isotherm, kinetic and thermodynamic. J. Ind. Eng. Chem. **20**(5), 2745–2753 (2014)

Majd, M.M., Kordzadeh-Kermani, V., Ghalandari, V., Askari, A., Sillanpää, M.: Adsorption isotherm models: a comprehensive and systematic review (2010–2020). Sci. Total. Environ. **812**, 151334 (2022)

Mallah, M.A., Changxing, L., Mallah, M.A., Noreen, S., Liu, Y., Saeed, M., Xi, H., Ahmed, B., Feng, F., Mirjat, A.A., Wang, W. (2022) Polycyclic aromatic hydrocarbon and its effects on human health: an overeview. Chemosphere **296**, 133948

Manoli, E., Samara, C.: Occurrence and mass balance of polycyclic aromatic hydrocarbons in the Thessaloniki sewage treatment plant, vol. 28, no. 1, pp. 176–187 (1999)

Meador, J.P.: Bioaccumulation of PAHs in marine invertebrates. PAHs: An Ecotoxicological Perspect. 147–171 (2003)

Mojiri, A., Zhou, J.L., Ohashi, A., Ozaki, N., Kindaichi, T.: Comprehensive review of polycyclic aromatic hydrocarbons in water sources, their effects and treatments. Sci. Total. Environ. **696**, 133971 (2019)

Mumtaz, M., George, J.D.: Toxicological profile for polycyclic aromatic hydrocarbons. US Department of Health and Human Services, Public Health Service, Agency for Toxic Substances and Disease Registry (1995)

Musah, M., Azeh, Y., Mathew, J., Umar, M., Abdulhamid, Z., Muhammad, A.: Adsorption kinetics and isotherm models: a review. Caliphate J. Sci. Technol. **4**(1), 20–26 (2022)

Nowacka, A., Włodarczyk-Makuła, M.: Effectiveness of priority PAH removal in a water coagulation process. Water Sci. Technol.: Water Supply **15**(4), 683–692 (2015)

Nyström, F., Nordqvist, K., Herrmann, I., Hedström, A., Viklander, M.: Removal of metals and hydrocarbons from stormwater using coagulation and flocculation. Water Res. **182**, 115919 (2020)

Oleszczuk, P., Hale, S.E., Lehmann, J., Cornelissen, G.: Activated carbon and biochar amendments decrease pore-water concentrations of polycyclic aromatic hydrocarbons (PAHs) in sewage sludge. Biores. Technol. **111**, 84–91 (2012)

Pathak, S., Sakhiya, A.K., Anand, A., Pant, K.K., Kaushal, P.: A state-of-the-art review of various adsorption media employed for the removal of toxic Polycyclic aromatic hydrocarbons (PAHs): an approach towards a cleaner environment. J. Water Process. Eng. **47**, 102674 (2022)

Puccia, V., Avena, M.J.: On the use of the Dubinin-Radushkevich equation to distinguish between physical and chemical adsorption at the solid-water interface. Colloid Interface Sci. Commun. **41**, 100376 (2021)

Queiroz, R.N., Prediger, P., Vieira, M.G.A.: Adsorption of polycyclic aromatic hydrocarbons from wastewater using graphene-based nanomaterials synthesized by conventional chemistry and green synthesis: a critical review. J. Hazard. Mater. **422**, 126904 (2022)

Rasheed, A., Farooq, F., Rafique, U., Nasreen, S., Aqeel Ashraf, M.: Analysis of sorption efficiency of activated carbon for removal of anthracene and pyrene for wastewater treatment. Desalin. Water Treat. **57**(1), 145–150 (2016)

Rashid, R., Shafiq, I., Akhter, P., Iqbal, M.J., Hussain, M.: A state-of-the-art review on wastewater treatment techniques: the effectiveness of adsorption method. Environ. Sci. Pollut. Res. **28**, 9050–9066 (2021)

Ravanbakhsh, M., Yousefi, H., Lak, E., Ansari, M.J., Suksatan, W., Qasim, Q.A., Asban, P., Kianizadeh, M., Mohammadi, M.J.: Effect of polycyclic aromatic hydrocarbons (PAHs) on respiratory diseases and the risk factors related to cancer. Polycyclic Aromat. Compd. **43**(9), 8371–8387 (2023)

Ravindra, K., Wauters, E., Van Grieken, R.: Variation in particulate PAHs levels and their relation with the transboundary movement of the air masses. Sci. Total. Environ. **396**(2-3), 100–110 (2008)

Rinawati, Hidayat, D., Supriyanto, R., Permana, D.F., Yunita.: Adsorption of polycyclic aromatic hydrocarbons using low-cost activated carbon derived from rice husk. In Journal of Physics: Conference Series, vol. 1338, no. 1, p. 012005. IOP Publishing (2019)

Sahoo, B.M., Ravi Kumar, B.V., Banik, B.K., Borah, P.: Polyaromatic hydrocarbons (PAHs): structures, synthesis and their biological profile. Curr. Org. Synth. **17**(8), 625–640 (2020)

Saichek, R.E., Reddy, K.R.: Electrokinetically enhanced remediation of hydrophobic organic compounds in soils: a review. Crit. Rev. Environ. Sci. Technol. **35**(2), 115–192 (2005)

Selvaraj, M., Hai, A., Banat, F., Haija, M.A.: Application and prospects of carbon nanostructured materials in water treatment: a review. J Water Process Eng **33**, 100996 (2020)

Shabeer, T.A., Saha, A., Gajbhiye, V.T., Gupta, S., Manjaiah, K.M., Varghese, E.: Removal of poly aromatic hydrocarbons (PAHs) from water: effect of nano and modified nano-clays as a flocculation aid and adsorbent in coagulation-flocculation process. Polycyclic Aromat. Compd. **34**(4), 452–467 (2014)

Silva, M.J., Soares, S.A., Santos, I.D., Pepe, I.M., Teixeira, L.R., Pereira, L.G., Silva, L.B., Celino, J.J.: Optimization of the photocatalytic degradation process of aromatic organic compounds applied to mangrove sediment. Heliyon **6**(10) (2020)

Smol, M., Włodarczyk-Makuła, M., Mielczarek, K., Bohdziewicz, J., Włóka, D.: The use of reverse osmosis in the removal of PAHs from municipal landfill leachate. Polycyclic Aromat. Compd. **36**(1), 20–39 (2016)

Solano, R.A., De León, L.D., De Ávila, G., Herrera, A.P.: Polycyclic aromatic hydrocarbons (PAHs) adsorption from aqueous solution using chitosan beads modified with thiourea, TiO_2 and Fe_3O_4 nanoparticles. Environ. Technol. Innov. **21**, 101378 (2021)

Sri Shalini, S., Palanivelu, K., Ramachandran, A., Raghavan, V.: Biochar from biomass waste as a renewable carbon material for climate change mitigation in reducing greenhouse gas emissions—a review. Biomass Convers. Biorefinery **11**(5), 2247–2267 (2021)

Srogi, K.: Monitoring of environmental exposure to polycyclic aromatic hydrocarbons: a review. Environ. Chem. Lett. **5**, 169–195 (2007)

St. Helen, G., Goniewicz, M.L., Dempsey, D., Wilson, M., Jacob III, P., Benowitz, N.L.: Exposure and kinetics of polycyclic aromatic hydrocarbons (PAHs) in cigarette smokers. Chem. Res. Toxicol. **25**(4), 952–964 (2012)

Stogiannidis, E., Laane, R.: Source characterization of polycyclic aromatic hydrocarbons by using their molecular indices: an overview of possibilities. Rev. Environ. Contam. Toxicol. 49–133 (2014)

Straif, K., Baan, R., Grosse, Y., Secretan, B., El Ghissassi, F., Cogliano, V.: Carcinogenicity of polycyclic aromatic hydrocarbons. Lancet Oncol. **6**(12), 931–932 (2005)

Sun, H., Wu, W., Wang, L.: Phenanthrene partitioning in sediment–surfactant–fresh/saline water systems. Environ. Pollut. **157**(8–9), 2520–2528 (2009)

Sun, K., Song, Y., He, F., Jing, M., Tang, J., Liu, R.: A review of human and animals exposure to polycyclic aromatic hydrocarbons: health risk and adverse effects, photo-induced toxicity and regulating effect of microplastics. Sci. Total. Environ. **773**, 145403 (2021)

Szczeklik, A., Szczeklik, J., Galuszka, Z., Musial, J., Kolarzyk, E., Targosz, D.: Humoral immunosuppression in men exposed to polycyclic aromatic hydrocarbons and related carcinogens in polluted environments. Environ. Health Perspect. **102**(3), 302–304 (1994)

Tang, L., Wang, P., Yu, C., Jiang, N., Hou, J., Cui, J., Xin, S., Xin, Y., Li, M.: Adsorption of polycyclic aromatic hydrocarbons (PAHs) in soil and water on pyrochars: a review. J. Environ. Chem. Eng. 116081 (2025)

Tien, C., Ramarao, B.V.: On the significance and utility of the Lagergren model and the pseudo second-order model of batch adsorption. Sep. Sci. Technol. **52**(6), 975–986 (2017)

Valderrama, C., Gamisans, X., De las Heras, X., Farran, A., Cortina, J.L.: Sorption kinetics of polycyclic aromatic hydrocarbons removal using granular activated carbon: intraparticle diffusion coefficients. J. Hazard. Mater. **157**(2–3), 386–396 (2008)

Vigdorowitsch, M., Pchelintsev, A., Tsygankova, L., Tanygina, E.: Freundlich isotherm: an adsorption model complete framework. Appl. Sci. **11**(17), 8078 (2021)

Vijayanand, M., Ramakrishnan, A., Subramanian, R., Issac, P.K., Nasr, M., Khoo, K.S., Rajagopal, R., Greff, B., Azelee, N.I.W., Jeon, B.H., Chang, S.W.: Polyaromatic hydrocarbons (PAHs) in the water environment: a review on toxicity, microbial biodegradation, systematic biological advancements, and environmental fate. Environ. Res. **227**, 115716 (2023)

Von Lau, E., Gan, S., Ng, H.K., Poh, P.E.: Extraction agents for the removal of polycyclic aromatic hydrocarbons (PAHs) from soil in soil washing technologies. Environ. Pollut. **184**, 640–649 (2014)

Walters, R.W., Luthy, R.G.: Equilibrium adsorption of polycyclic aromatic hydrocarbons from water onto activated carbon. Environ. Sci. Technol. **18**(6), 395–403 (1984)

Wang, J., Chen, Z., Chen, B.: Adsorption of polycyclic aromatic hydrocarbons by graphene and graphene oxide nanosheets. Environ. Sci. Technol. **48**(9), 4817–4825 (2014)

Wang, J., Wang, C., Huang, Q., Ding, F., He, X.: Adsorption of PAHs on the sediments from the yellow river delta as a function of particle size and salinity. Soil Sediment Contam.: Int. J. **24**(2), 103–115 (2015)

Wang, S., Zhang, X., Zhou, Q., Liu, Z., Xu, Z.: Magnetic hypercrosslinked polymer microspheres for the detection, spatial distribution, source identification and potential risks assessment of five polycyclic aromatic hydrocarbons in city river of plateau lake, Southwest China. J. Hazard. Mater. **476**, 135026 (2024)

Wang, X., Liu, Y., Zhang, H., Shen, X., Cai, F., Zhang, M., Gao, Q., Chen, W., Wang, B., Tao, S.: The impact of carbon nanotubes on bioaccumulation and translocation of phenanthrene, 3-CH 3-phenanthrene and 9-NO 2-phenanthrene in maize (Zea mays) seedlings. Environ. Sci. Nano **3**(4), 818–829 (2016)

Wijayaratne, R.D., Means, J.C.: Sorption of polycyclic aromatic hydrocarbons by natural estuarine colloids. Mar. Environ. Res. **11**(2), 77–89 (1984)

Wise, S.A., Sander, L.C., Schantz, M.M.: Analytical methods for determination of polycyclic aromatic hydrocarbons (PAHs)—a historical perspective on the 16 US EPA priority pollutant PAHs. Polycyclic Aromat. Compd. **35**(2–4), 187–247 (2015)

Xue, W., Warshawsky, D.: Metabolic activation of polycyclic and heterocyclic aromatic hydrocarbons and DNA damage: a review. Toxicol. Appl. Pharmacol. **206**(1), 73–93 (2005)

Yakout, S.M., Daifullah, A.A.M., El-Reefy, S.A.: Adsorption of naphthalene, phenanthrene and pyrene from aqueous solution using low-cost activated carbon derived from agricultural wastes. Adsorpt. Sci. Technol. **31**(4), 293–302 (2013)

Yan, H., Wu, H., Li, K., Wang, Y., Tao, X., Yang, H., Li, A., Cheng, R.: Influence of the surface structure of graphene oxide on the adsorption of aromatic organic compounds from water. ACS Appl. Mater. Interfaces **7**(12), 6690–6697 (2015)

Yim, S.: Influence of inorganic salts on aqueous solubilities of polycyclic aromatic hydrocarbons. J. Soil Groundw. Environ. **8**(3), 23–29 (2003)

Yin, Z., Cui, C., Chen, H., Duoni, Yu, X., Qian, W.: The application of carbon nanotube/graphene-based nanomaterials in wastewater treatment. Small **16**(15), 1902301 (2020)

Zango, Z.U., Sambudi, N.S., Jumbri, K., Ramli, A., Abu Bakar, N.H.H., Saad, B., Rozaini, M.N.H., Isiyaka, H.A., Osman, A.M., Sulieman, A.: An overview and evaluation of highly porous adsorbent materials for polycyclic aromatic hydrocarbons and phenols removal from wastewater. Water **12**(10), 2921 (2020)

Zhang, C., Wu, L., Cai, D., Zhang, C., Wang, N., Zhang, J., Wu, Z.: Adsorption of polycyclic aromatic hydrocarbons (fluoranthene and anthracenemethanol) by functional graphene oxide and removal by pH and temperature-sensitive coagulation. ACS Appl. Mater. Interfaces **5**(11), 4783–4790 (2013)

Zhang, J., Li, R., Ding, G., Wang, Y., Wang, C.: Sorptive removal of phenanthrene from water by magnetic carbon nanomaterials. J. Mol. Liq. **293**, 111540 (2019a)

Zhang, Y., Zhu, C., Liu, F., Yuan, Y., Wu, H., Li, A.: Effects of ionic strength on removal of toxic pollutants from aqueous media with multifarious adsorbents: a review. Sci. Total. Environ. **646**, 265–279 (2019b)

Ziyaei, K., Mokhtari, M., Hashemi, M., Rezaei, K., Abdi, F.: Association between exposure to water sources contaminated with polycyclic aromatic hydrocarbons and cancer risk: a systematic review. Sci. Total. Environ. 171261 (2024)

Bioremedial Treatment for the Decontamination of Polycyclic Aromatic Hydrocarbons for Sustainable Environment

12

Rashmi Verma, Richa Saxena, Shreya Kotnala, Shalini Tiwari, B. S. Rawat, and Latif Arifudin

Abstract

Polycyclic aromatic hydrocarbons (PAHs), along with their byproducts, are the widespread contaminants that require efficient remediation techniques. Many anthropogenic and natural factors contribute to the spread, dissemination, and accumulation of PAHs in aquatic and soil systems. PAHs are gaining worldwide attention due to their harmful effects, ability to persist in the environment, and potential for bioaccumulation. The contamination of PAHs primarily occurs from human activities related to the disposal of industrial and household run-off or atmospheric emissions from vehicles and industrial facilities. Once PAHs enter the environment, they experience various processes of transformation, which can be physical, chemical, or biological, known collectively as weathering processes. Treating sites contaminated with PAHs is often a challenging and costly endeavour, so it would be wise to pursue a method that is adaptive, sustainable, environmentally friendly, efficient, and cost-effective. The use of

R. Verma
Department of Biotechnology, School of Basic and Applied Sciences, Shri Guru Ram Rai University, Dehradun, India

R. Saxena (✉)
Department of Physics, School of Basic and Applied Sciences, Shri Guru Ram Rai University, Dehradun, India
e-mail: saxena.richa23@gmail.com

S. Kotnala · S. Tiwari
Department of Chemistry, School of Basic and Applied Sciences, Shri Guru Ram Rai University, Dehradun, India

B. S. Rawat
Department of Physics, FUIT, Uttaranchal University, Dehradun, Uttarakhand, India

L. Arifudin
Department of Mechanical Engineering, King Saud University, Riyadh, Saudi Arabia

H. Gupta et al. (eds.), *Ubiquitous Polycyclic Aromatic Hydrocarbon Contamination*, Emerging Contaminants and Associated Treatment Technologies,
https://doi.org/10.1007/978-3-032-11764-9_12

green technologies can enhance remediation efficiency with minimal negative environmental impacts. This chapter explains the sources and properties of polycyclic aromatic hydrocarbons. Different types of remediation methods for the elimination of PAHs are discussed. This chapter highlights the challenges, limitations, and latest advancements in technologies for removing PAHs, including their mechanism. Combining bioremediation could serve as an effective method for sustainable environmental cleanup.

Keywords

Polycyclic aromatic hydrocarbons • Properties of PAHs • Decontamination • Sustainable environment

Introduction

Environmentally destructive organic pollutants known as hydrocarbons exist throughout ecosystems, thus requiring multifaceted evaluations for assessment purposes. Hydrocarbons, primarily composed of the essential components of crude oil include carbon and hydrogen, while the complex mixture also contains other elements. The mixture consists of oxygen, nitrogen, sulphur, trace metals together with hydrocarbons. Post-refinement, petroleum product derivatives display modified physicochemical characteristics as a result of which their complexity increases and their degradation becomes more challenging. Industrial petroleum waste sludge contributes contaminants such as hydrocarbons, sulphides, and ammonia into soil ecosystems. Hydrocarbon compounds constitute a diverse and extensive category of chemical molecules. Every hydrocarbon molecule possesses a distinct chemical structure and set of properties. The variety of hydrocarbon molecules indicates that distinct bacteria employ various strategies for their degradation (Logeshwaran et al. 2018). The transformation process of hydrocarbons is dependent upon microbial communities and several environmental conditions, including temperature, along with oxygen content and pH, and nutrient accessibility levels (Abdullah et al. 2020).

Saturated hydrocarbons belong to the aliphatic hydrocarbon category, polycyclic aromatic hydrocarbons (PAHs) and monocyclic aromatic hydrocarbons (MAHs) are the two aromatic types of hydrocarbons. The petroleum industry stores both petroleum and natural gas deposits full of paraffin molecules known as aliphatic hydrocarbons that form straight chains or circular arrangements. These hydrocarbons resist double bonds and demonstrate limited polar character, leading to their poor water solubility and low reactivity when the temperature is at room level (Logeshwaran et al. 2018). The aromaticity of these hydrocarbon compounds varies because they contain benzene ring structures (Mahjoubi et al. 2018).

Lipophilic pollutants, PAHs, formed from biological systems and incomplete combustion, generate environmental importance because they persist within air as well as soil and water and produce harmful consequences for human health and ecological systems (Pandolfo et al. 2023). The pollutant group consists of naphthalene and fluorene among lower-molecular-weight compounds, which contain 2–3 benzene rings. The atmosphere receives the former pollutant as it evaporates from its surface, yet the latter instruments itself as solids or liquids before possibly dissolving into water bodies (Varjani 2017; Gupta and Gupta 2016; Gupta and Kumar 2020).

Bioremediation utilizes the inherent metabolic functions of microbes and plants to decompose, detoxify, or convert PAHs into innocuous byproducts. Techniques like bioaugmentation, biostimulation (boosting the activity of indigenous microorganisms with nutrients or oxygen), and phytoremediation are progressively employed to expedite the degradation process (Huang and Batterman 2014).

The efficacy of bioremediation is contingent upon several aspects, including the bioavailability of PAHs, the existence of proficient degraders, and various parameters related to environment. The enhancement of degradation efficacy can be achieved by latest breakthroughs in biotechnology and environmental genomics, enabling researchers for manipulation of microbial communities. These advancements closely adhere to the tenets of sustainable development, seeking to rehabilitate natural ecosystems while reducing ecological disturbance. Focus is directed on microbiological mechanisms, innovative technologies, and comprehensive remediation strategies that promotes a healthy and eco-friendly environment (Zhang et al. 2021).

The research focus concerning bioavailability plays the most vital role in bioremediation methods (Shahsavari et al. 2013). The method of breakdown of PAH in soil remains intricate since sediments and soils contain numerous diverse microorganisms that use different metabolic pathways and show varying substrate preferences. The efficiency of elimination of chemical compounds is relatively low in the environment (Abdel-Fatah 2018; Dendooven et al. 2011). The degradation research uses PHE as an indicator of three-ring PAH compounds (Gupta 2016) and PYR as an indicator of four-ring PAH compounds. For LMW PAHs, the degradation process may either enhance the breakdown of HMW PAHs or inhibit it because of enzymatic activation by intermediates or because of competitive inhibition mechanisms. Multiple investigations that studied indigenous microorganisms' effects on PAHs degradation rates and pollutant survival times were conducted in diverse natural environments during the past twenty years (Khiadani Hajian et al. 2013). Microbial enhancement can be applied for removal of oil polluted soil, including native microbial stimulation along with the modification of oxygen content and nutrient availability, and supply of assorted microbial consortia. Multiple bacterial species take petroleum-based molecules to obtain carbon as well as energy, thus breaking down petroleum compounds (Dendooven et al. 2011; Abdel-Shafy and Mansour 2016). Most remediation methods that use chemicals or physical approaches prove either too expensive or prove to be ineffective in soil clean-up. The integration of compost leads to three main benefits for soil remediation: it grows organic matter and boosts fertility while assisting biological soil clean-up efforts and proving itself as one of the most economical remediation approaches. Soil bioremediation takes advantage of composting as a universal biological waste stabilizing method through which microorganisms decompose organic materials in agricultural and municipal waste to create a practical solution for cost-efficient remediation. Vermicompost contains numerous microorganisms consisting of earthworms together with compost, which enhances the characteristics of soil. In another study, Davis et al. reported the different mitigation techniques over the past years (Ye et al. 2019). Population expansion, global industrialization, escalating energy demand, and the pervasive application of pesticides result in a rising influx of diverse contaminants into aquatic and soil ecosystems, leading to their accumulation. In contemporary society, the primary energy source for industrial and home requirements is petroleum-derived goods. Petroleum hydrocarbon spills, along with leaks, emerge during every stage of petroleum product prospecting, extraction and processing, transportation, and storage operations. Soil contamination by petroleum products causes two detrimental effects on soils due to lowering strength and permeability while increasing drought levels, which ultimately reduce agricultural production (García-Díaz et al. 2013). The identification of soil pollution as significant occurs at pollutant concentrations that reach 1% (w/w) in the environment (Kotoky et al. 2017). Pollutants, including hydrocarbons among other substances and heavy metals, together with organic dyes and solvents, cut down global water purity and damage marine ecosystems severely (Singh et al. 2018, 2022; Gupta 2018). Studies for pollutant degradation and environmental restoration expanded due to the universal spread of pollution, which harms both humans and creatures. Environmental pollutants derived from petroleum hydrocarbon origin have recently gained prominence as xenobiotic aliphatic and aromatic compounds. The group of notable pollution chemicals includes polycyclic aromatic hydrocarbons (PAHs) (Fig. 12.1). The hydrophobic compounds known as PAHs contain at least two fused aromatic rings.

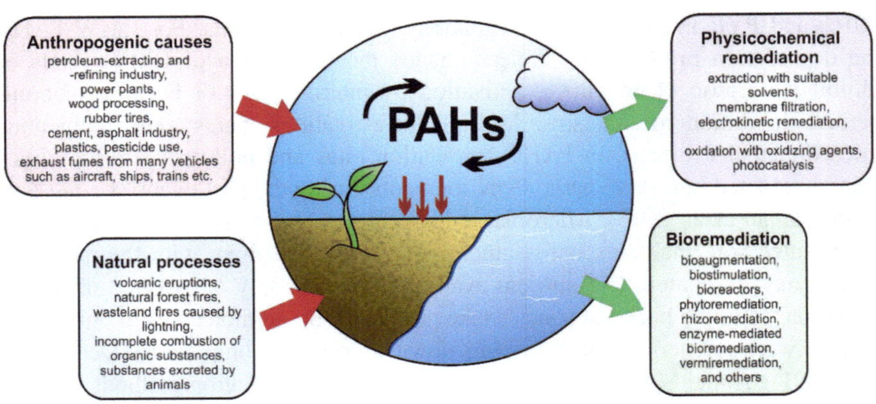

Fig. 12.1 Remediation strategies of polycyclic aromatic hydrocarbons (PAHs) in the environment **(published in an open open-access journal).** (https://creativecommons.org/licenses/by/4.0/) (Davletgildeeva and Kuznetsov 2024)

Sources, Properties, Environmental Behavior, and Toxicological Impacts of Polycyclic Aromatic Hydrocarbons (PAHs)

Polycyclic aromatic hydrocarbons (PAHs) have two or more than that organic pollutants that are fused with benzene rings with atoms of carbon and hydrogen atoms. By natural occurrence, these compounds are ejected into the environment, predominantly through anthropogenic sources (Ravindra et al. 2008). The perseverance of these is likely to occur because of their hydrophobic, lipophilic, and chemically stable nature. These are depicted in Fig. 12.2.

In Table 12.1 PAHs vary in molecular size and structure, which significantly affects their physicochemical behavior. As compared to high molecular weight, the voltality and are easy to degradation resistance of low molecular weight PAHs are more from industrial sources (Kästner and Miltner 2016). Their strong affinity for organic matter causes accumulation of them in soil and sediments, causing pollution. PAHs results in serious ecological risks, due to their lipophilicity, they can bioaccumulate in the fatty tissues of the organisms and move through tropical levels (Dehnavi and Ebrahimipour 2024; Ou et al. 2024). Toxicologically, benzo [a]pyrene are well-known Group 1 carcinogen, is a PAHs capable of forming DNA-reactive diol epoxides that induce mutations, cancer, and developmental

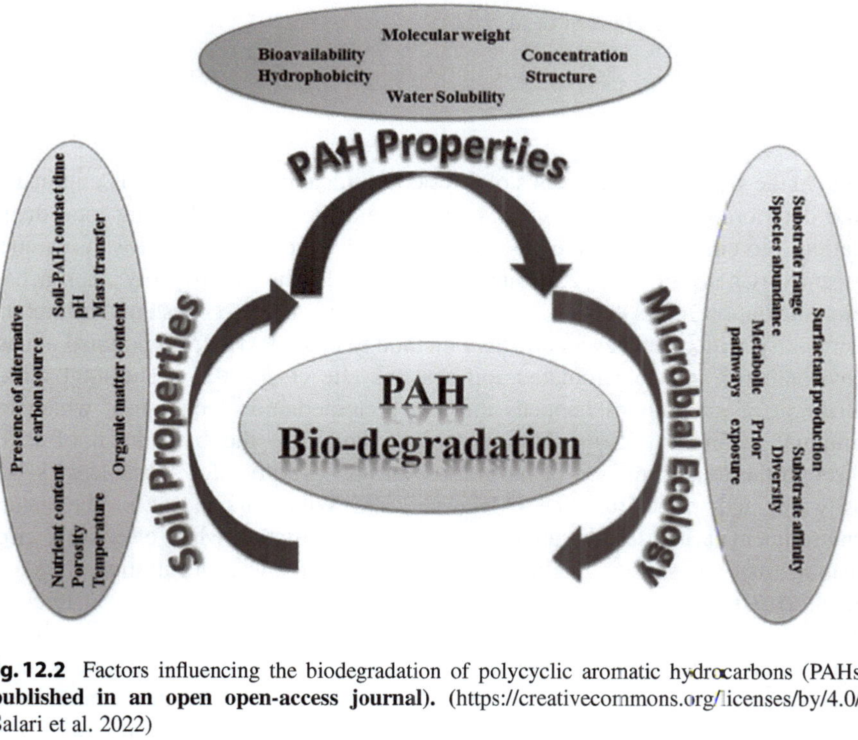

Fig. 12.2 Factors influencing the biodegradation of polycyclic aromatic hydrocarbons (PAHs) **(published in an open open-access journal).** (https://creativecommons.org/licenses/by/4.0/) (Salari et al. 2022)

Table 12.1 Physicochemical properties of polycyclic aromatic hydrocarbons (PAHs)

Property	Description
Molecular weight	Increases with number of rings; ranges from ~ 128 g/Mol (naphthalene) to > 278 g/Mol (benzo[ghi]perylene)
Solubility	Poorly soluble in water (decreases with increasing molecular weight); soluble in organic solvents
Volatility	Low to moderate; lower for higher molecular weight PAHs
Melting/boiling points	Generally, increase with molecular weight; PAHs with fewer rings are semi-volatile
Hydrophobicity	High; tends to adsorb to organic matter in soils and sediments
Stability	Chemically stable and resistant to degradation under ambient environmental conditions

disorders (Boström et al. 2002; Kim et al. 2013; Perera et al. 2006; Gupta and Gupta 2015). They are also associated with endocrine disruption and oxidative stress (de Souza Pohren et al. 2019).

In Table 12.2, due to their stability and toxicity, PAHs require effective remediation from contaminated soils and sediments. Conventional decontamination techniques are primarily classified into physical, chemical, and physicochemical methods. While these strategies may show high removal efficiency, they often face challenges like high costs, secondary pollution, and ecological disruption (Sinha et al. 2010; Medina-Herrera et al. 2024).

Bioremediation

Due to the significant drawbacks of physicochemical methods for PAH elimination and deterioration, such as high costs and the difficulty in achieving total degradation, bioremediation strategies have garnered increased interest by researchers recently (Cheng et al. 2022; Gupta et al. 2025). The eco-friendly approach of bioremediation offers economic advantages while it helps ecosystems restore sustainability, which enables an effective method to clean PAHs from polluted aquatic environments and contaminated soil systems (Jie et al. 2023). The total mineralization of xenobiotics happens through bioremediation techniques, which use minimal technology costs. Microorganisms complete the breakdown of PAHs through the transformation of hazardous substances into safer substances as well as water and carbon dioxide (Malawska and Wilkomirski 1999). The natural environment faces little disruption through bioremediation while effectively eliminating toxic substances, including those present at low concentrations (Ukaogo et al. 2020). The procedure serves mainly to counteract limitations in microorganism multiplication that break down PAHs. The adaptation of microbes to contaminants represents a method for achieving bio-stimulation. Research facilities primarily use endemic crops to build up their level of specific contaminants while choosing the best microbial strains intentionally (Salimnezhad et al. 2021).

Table 12.2 Remediation methods for PAH-contaminated soils (Sinha et al. 2010)

Method	Description	Effectiveness	Drawbacks
Soil washing	Uses water or chemical solvents to physically separate PAHs from soil particles	Efficient for lighter PAHs and soils with low organic content	Produces contaminated wash water requiring further treatment. Less effective for high molecular weight PAHs
Thermal desorption	Heats contaminated soils (200–600 °C) to volatilize and remove PAHs	High removal efficiency, especially for volatile PAHs	High energy consumption. Risk of air pollution from volatilized contaminants. Destruction of soil structure and nutrients
Incineration	High-temperature combustion (> 800 °C) of contaminated media	Complete destruction of organic pollutants	Air emissions (dioxins, furans). High cost and carbon footprint. Loss of soil fertility
Chemical oxidation	Uses oxidants (e.g., hydrogen peroxide, ozone, permanganate) to break down PAHs into less harmful compounds	Rapid degradation of a wide range of PAHs	Risk of forming toxic intermediate compounds. Potential harm to soil microbial communities. Requires careful control and handling
Solidification/ stabilization (S/S)	Immobilizes PAHs using binders like cement or lime, reducing their mobility	Prevents leaching and environmental spread	Does not degrade the pollutant. Increases waste volume. Long-term stability concerns
Solvent extraction	Uses organic solvents to dissolve and extract PAHs from soil or sediment	Useful for compact or clay-rich soils	Toxic solvents pose additional hazards. Solvent recovery and disposal issues

Utilizing live organisms, mostly microbes, fungi, or plants, bioremediation is a biological treatment method that breaks down, detoxifies, or eliminates dangerous chemicals from the environment. It is seen as an affordable, sustainable, and environmentally beneficial substitute for traditional physical and chemical restoration techniques. The biodegradation of complex aromatic chemicals into simpler, non-toxic end products like carbon dioxide (CO_2), water (H_2O), and in the field of PAH bioremediation, the process of converting biomass receives the name bioremediation.

Bioremediation Strategies and Mechanisms

The core principle of bioremediation is to encourage the microbes natural metabolic process for making use of PAHs (Salimnezhad et al. 2021). An estimated 7–9 billion tonnes of trash are produced annually by various businesses in the majority of industrialised and developing nations (Chen et al. 2020), with organic waste accounting for around 80% of the total. Currently, the food production industry in the UK is responsible for a significant portion of waste. In most of the African countries, the majority of these wastes are organic in origin and have no recorded history (Omoni et al. 2025).

In situ and ex-situ methods are important techniques for remediation of PAHs. Contaminants are applied directly on the site without excavation in ex-situ process, employing techniques such as bioventing and biosparging to stimulate native microbial activity through the controlled supply of oxygen and nutrients (Vidali 2001). In constrast, off-site method is used in ex-situ bioremediation for elimination of soil and water contaminants by using technologies like bio piles and land farming, which allow for enhanced environmental control and microbial management (Ghosal et al. 2016).

The mechanism of PAH bioremediation typically involves four key steps. First, contaminant bioavailability is critical, as PAHs must be accessible to microbes for effective degradation. This bioavailability can be improved through the addition of chemical surfactants or naturally derived biosurfactants that increase solubility. Second, microorganisms uptake the PAHs via transport across their cell membranes. In the third step, enzymatic degradation occurs through the action of oxygenase enzymes, such as monooxygenases and dioxygenases, which catalyze the initial oxidation of PAHs, leading to ring cleavage and conversion into simpler intermediates like catechol. Finally, these intermediates undergo either mineralization complete breakdown into carbon dioxide, water, and biomass or transformation into less toxic compounds via microbial metabolic pathways (Ausuri et al. 2021). This integrated biochemical process highlights the potential of microorganisms to detoxify and eliminate PAHs from contaminated environments.

Microorganisms Involved in PAH Bioremediation

The most practical bioremediation techniques include Verm remediation and enzyme-mediated bioremediation, along with phytoremediation/rhizoremediation and bioreactors and bioaugmentation, and biostimulation according to Abdel-Shafy and Mansour (2016). The purpose of bioaugmentation is to enhance degradation efficiency while potentially extending the catabolic abilities of native bacteria using inoculated microorganisms or microorganism groups that degrade pollutants (Abdel-Shafy and Mansour 2016).

Researchers have extensively studied how both bacterial and fungal organisms decompose PAHs. Scientists operate in present times on developing methods to bioremediate PAHs through fungal processes (Abdel-Shafy and Mansour 2016;

Lee 2010). The two primary types of fungi used in PAH bioremediation are white-rot fungi, known as lignolytic fungi, and non-lignolytic fungi (Haritash and Kaushik 2009). A diverse range of microorganisms, including bacteria, fungi, and actinomycetes, play a crucial role in the biodegradation of polycyclic aromatic hydrocarbons (PAH) (Haritash and Kaushik 2009). Among bacteria, *Pseudomonas* spp. are highly efficient in degrading compounds like naphthalene, phenanthrene, and pyrene, while *Mycobacterium* spp. specialize in breaking down high-molecular-weight PAHs such as benzo[a]pyrene (BaP). *Sphingomonas* spp. can metabolize a wide array of aromatic hydrocarbons, and *Rhodococcus* spp. are particularly effective under cold and nutrient-deficient conditions. Additionally, *Acinetobacter* spp. are commonly employed in bioreactor-based PAH treatment systems. Fungal species also contribute significantly; *Phanerochaete chrysosporium*, a white rot fungus, utilizes ligninolytic enzymes like laccase and manganese peroxidase to degrade complex PAHs, whereas *Aspergillus niger* is more adept at degrading low-molecular-weight PAHs (Johnsen et al. 2005). *Trametes versicolor* exhibits strong potential in cleaving stable aromatic ring structures. Furthermore, actinomycetes such as *Streptomyces* spp. produce extracellular enzymes that are effective in degrading high-ring PAHs (Sudaryanto et al. 2007).

PAHs convert into biphenolic intermediates through the action of laccases, manganese peroxidases, and lignin peroxidases that peroxidases that degrade lignin in fungi (Lemaire et al. 2013).

PAH Remediation Strategies

In situ, solid phase, and bioreactor operations are the fundamental techniques for treatment of PAHs polluted soil and groundwater (Breckenridge et al. 1994; Bates et al. 2000). Each PAH bioremediation method possesses its distinct features. It can be affirmed that bioslurry reactors are the most efficient and widely used method for PAH biodegradation, followed by solid phase applications and in situ strategies. PAH bioremediation applications are accessible (Sahle-Demessie et al. 2000).

Solid Phase Bioremediation

Standard soil management techniques are used in a contained, aboveground system to treat contaminated soil or other solid material and improve the microbial degradation of organic contaminants through solid-phase bioremediation.

The composting and landfarming methods are combined to create the designed soil cell. In essence, engineered soil cells are made to resemble aerated compost piles. Some technical constraints to solid-phase PAH bioremediation are present.

Bioreactor Operations

In reactors, bioremediation is the process of treating contaminated water or solids (soil, silt, or sludge) using a contained, designed system. Bioreactors can be fabricated for maximizing biological degradation and reducing the abiotic losses, by modifying physical and chemical features of the contaminated matrix. Soil slurry reactors are essentially the only method available for treating PAH-contaminated soil in a reactor setting. Last but not least, bioreactors can be more successfully inoculated with carefully chosen microorganisms that are frequently better able to quickly and thoroughly degrade specific contaminants due to this enclosed environment.

In Situ Remediation

In situ bioremediation generally seeks to supply necessary co-reagents to the native microorganisms to boost the breakdown of organic substances. This system will guarantee the efficient breakdown of organic pollutants while also containing polluted groundwater. Bioremediation techniques is depicted in Fig. 12.3.

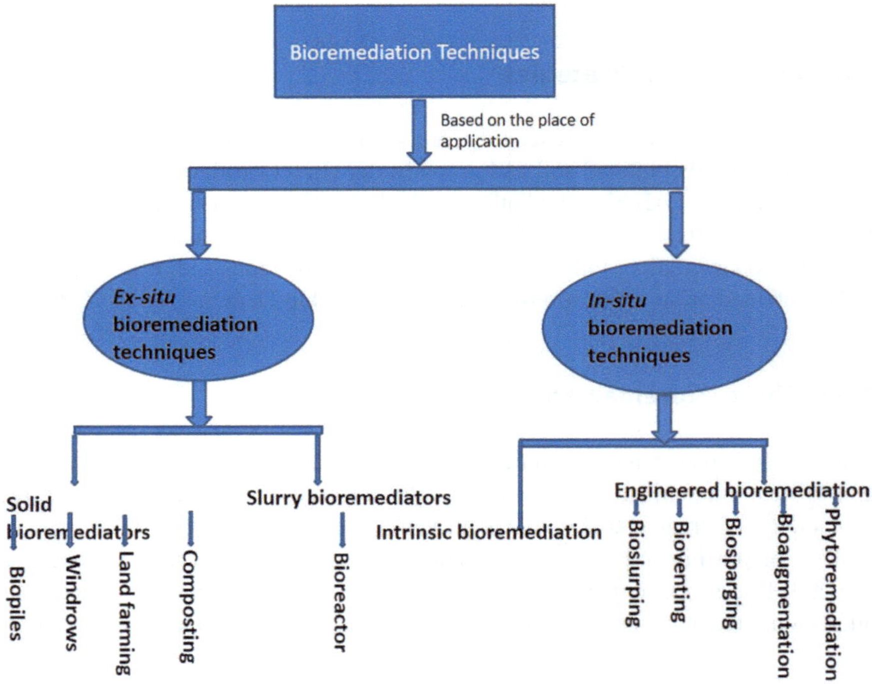

Fig. 12.3 Bioremediation treatments (**published in an open open-access journal).** (https://creativecommons.org/licenses/by/4.0/) (Alori et al. 2022)

Norris Robert (2017) recent reviews of in situ bioremediation technology are great resources for references and case studies for a wide range of in situ bioremediation applications. Early bioventing reports showed how well the procedure stimulated the removal of hydrocarbons from refined petroleum (Chen et al. 2015).

Thermal Management Technologies

For the remediation of PAHs, the thermal method is a fast and effective technique. The foundation of thermal remediation strategies is the mobilisation of PAHs into gaseous flow through the use of heat. This process either destroys the pollutants (i.e., incineration), breaks them down into simpler compounds (i.e., pyrolysis), or immobilises them through stabilisation or vitrification (Vidonish et al. 2016). High elimination efficiency of total petroleum hydrocarbons is achieved by the thermal process (Fig. 12.4). Despite these benefits, high temperatures are necessary for thermal treatment, which can lead to several issues. According to Vidonish et al. heating contaminated locations to high temperatures requires a lot of energy and is therefore expensive. In order to remove organic pollutants, mostly through thermal desorption, microwave irradiation works by dispersing the electromagnetic field energy into the irradiated substances that have absorbed the energy and then converting it into heat (Eldos et al. 2020). In microwave heating, the water dipoles and other polar materials, such as PAHs, rotate due to the electromagnetic field penetrating the contaminated matrix. During the microwave heating, soil moisture is progressively converted into water vapour, resulting in pollution desorption by condensation that helps in pollutant degradation processes.

The key role in elimination of pollutants is moisture level, polarity of matrix, mineral constituents (Vidonish et al. 2016; Eldos et al. 2020). At the microwave heating sediment moisture is progressively converted to water vapour. Table 12.3 presents the summary of various PAH bioremediation techniques (Mohan et al. 2006).

Factors Affecting the Bioremediation of PAHS

Bio-piling and composting are approaches that have successfully applied bioremediation (Reddy et al. 2010) technologies for contaminated site.

Temperature

To cultivate bacteria that break down hydrocarbons, an ideal temperature environment is needed. Temperatures between 20 and 300 degrees Celsius are said to be the most efficient for the breakdown of hydrocarbons [99]. Temperature has an impact on not just the soil's moisture content but also its capacity to retain moisture. Rise of temperature is inversely proportional to biodegradation

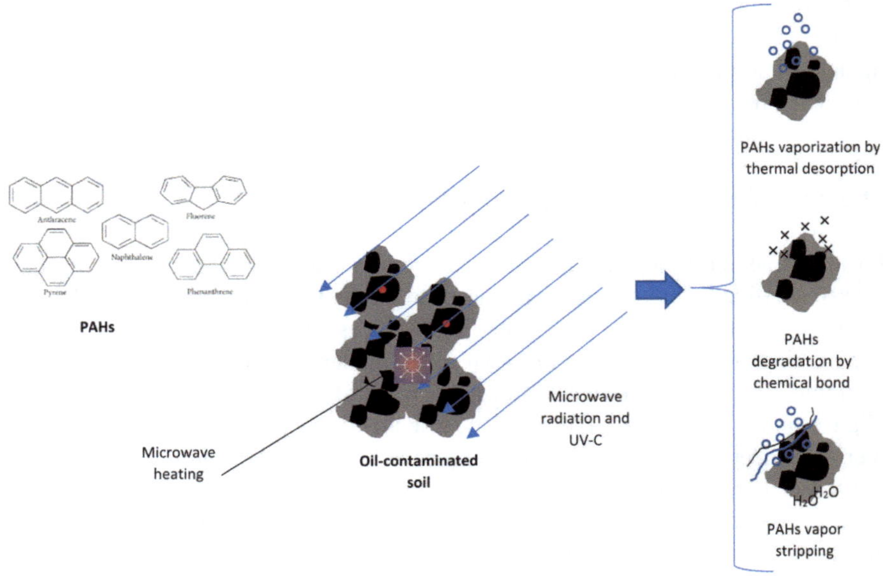

Fig. 12.4 Degradation of PAHs mechanism by microwave irradiation heating **(published in an open open-access journal).** (https://creativecommons.org/licenses/by/4.0/) (Eldos et al. 2020)

Table 12.3 Technologies used for bioremediation of PAHs

Origin of contaminants	Procedure	Process	Reference
PAH-contaminated soil	Ex situ composting with spent mushroom	Composting	(Lau et al. 2003)
PAH-contaminated alkaline soil	Ex situ, composting with soot waste		
Coal tar-contaminated soil	Ex situ, soil slurry bioreactor, solvent pre-treatment	Bioreactors	(Lee et al. 2003)
	Ex situ, soil slurry bioreactor, solvent pre-treatment		(Lee et al. 2001)
PAHs-contaminated sand	Ex situ, ozone oxidation integrated bioremediation	Combined treatment	(Choi et al. 2001)
Pyrene-contaminated soil	Phytoremediation	In situ, nine plant species	(Liste and Alexander 2000)

(Tudararao-Aherobo and Mesogboriwon 2020). Furthermore, as the temperature rises, oxygen solubility falls, lowering aerobic microbes metabolic activity. Most of the literature is available for mesophilic temperature in contrast different range of temperatures (Siron et al. 1995).

Oxygen

The most of the bioremediation literature is found for aerobic metabolism of PAHs in contrast to anaerobic metabolism. Perchlorate (Coates et al. 1999), sodium nitrate, and hydrogen peroxide (Pardieck et al. 1992) have been used. Aerobic biodegradation is found to occur faster in contrast to anaerobic biodegradation. Furthermore, anaerobic PAH breakdown rates under denitrifying circumstances, however, have also been reported to be similar to aerobic settings (McNally et al. 1998).

pH

The pH of many PAH-contaminated locations is not ideal for bioremediation. For instance, large amounts of demolition debris, such as brick and concrete, are frequently found on abandoned gasworks sites. An acidic nature is developed by sulphides oxidation by the leaching of coal spoil by the emission and oxidation of sulphides. For the variation of pH, lime can be added (Alexander 1995). Phenanthrene breakdown in liquid culture has been studied in pH range 5.5–7.5, using Burkholderia cocovenenae, a bacterium isolated by polluted soil from petroleum. Phenanthrene elimination was 40% only at pH 5.5 after a duration of 16 days, whereas it was 80% at circum-neutral pH levels, even though bacterial growth was not substantially impacted by pH. Sensitivity of Sphingomonas paucomibilis is highly sensitive to pH media with breakdown of PAHs (Kästner et al. 1998). For a period of 28 days, pH degradation is found to be 2 in acetic soil with decreased concentration of naphthalene, phenanthrene, and anthracene (Stapleton et al. 1998). PAHs from both acidic and alkaline environments might be beneficial for future research.

Salinity

Ward and Brock (1978) observed that in a series of hypersaline evaporation ponds at the Great Salt Lake in Utah, USA, the rate of hydrocarbon metabolism decreased as salinity levels increased from 3.3 to 28.4%. Another study by Minai-Tehrani et al. (2009) found that the breakdown of PAHs was more effective in a medium with 0% NaCl compared to one with 5% NaCl. Common strategies to address high salinity include dilution to lower salinity, reverse osmosis, ion exchange, or electrodialysis before biological treatment, though these methods can be costly (Margesin and Schinner 2001).

Bioavailability

One of the most important features in bioremediation (Davletgildeeva and Kuznetsov 2024) is found to be bioavailability, which has an impact of physicochemical and microbiological parameters on the pace and degree of freedom. PAH chemicals are categorized as hydrophobic organic pollutants and have poor bioavailability (Fewson 1988). These substances are resistant to chemical, biological, and photolytic degradation, and have less water solubility.

The PAH's solubility decreases with increasing molecular weight, which lowers the PAH's accessibility for microbial cell metabolism. Furthermore, organic materials and mineral surfaces can be rapidly absorbed by PAHs in soil (Miller and Bartha 1989).

The desorption is more irreversible, and the contaminant's (Hatzinger and Alexander 1995) chemical and biological extractability decreases, with more in touch of soil with PAH. Surfactants produced by microbes helps in fixation of soil structure (Miller and Bartha 1989). Harmful effect on the environment is caused by new contaminants are fabricated by synthetic surfactants (Wang and Brusseau 1993; Makkar and Rockne 2003; Pacwa-Płociniczak et al. 2011). Application of some surfactants enhances the PAHs compounds solubility, desorption in liquid medium, but this may not always increase the rate and effectiveness of degradation. (Makkar and Rockne 2003).

Hatzinger and Alexander (1995) however, took a more representational technique, extracting the pollutants using mild organic solvents to represent the soil's bioavailable fraction. Water-based solvents are also being employed since microorganisms can typically only reach pollutants that are in the aqueous phase.to more precisely forecast the portion of organic pollution in bioavailable soil. Furthermore, HPCD does not seem to inhibit lux-type microbes, bioavailability can be accessed chemically and biologically (Reid et al. 2000).

Biosurfactants

Amphiphilic compounds known as biosurfactants can be utilised in industrial, agricultural, therapeutic, and environmental settings to reduce surface tension. They can help improve petroleum bioremediation (Ahmed et al. 2022; Adeoye et al. 2024; Issaka 2024).

In a study, Das et al. (2008) and Pacwa-Płociniczak et al. (2011), they are highly reactive, biodegradable, safe, environment friendly, and active at extremes of salinity and temperature. According to Nguyen et al. (2008) and Nievas et al. (2008), biosurfactants can be to classified on the basis of molar mass, action method, physical and chemical properties, and microbiological derived (Bezza and Nkhalambayausi Chirwa 2016; Pacwa-Płociniczak et al. 2011). Susceptibility for the environment enhances by bioavailability (Maier and Soberón-Chávez 2000). By using two methods, the encouragement of biodegradation of hydrocarbons can be made (Pacwa-Płociniczak et al. 2011).

Pseudomonas sp. Generated biodegradative features of the biosurfactants. The LP1 strain was examined on diesel and crude oil (Pacwa-Płociniczak et al. 2014). In another study, Reddy et al. explained the biodegenerative properties of Brevibacterium sp. (Reddy et al. 2010) Additional PAHs as anthracene and fluorene, can be broken down by strain and can also break down 93.92% of phenanthrene. There are several ways to apply biosurfactants, that is favourable for in situ process. In other cases, because native microorganisms usually do not show moderate efficiency, introducing appropriate microbes might lead to the production of biosurfactants (Perfumo et al. 2017).

Limitations and Future Prospects

The characteristics of the contaminants, the poor performance of microbial communities in specific situations, and the low bioavailability of the contaminants are some of the drawbacks of biodegradation programmes. Certain sectors are vulnerable to contamination by PAHs. When used on a field scale, these therapeutic techniques are highly effective. This is because certain techniques are not viable in locations with limited permeability or mixed contaminants. The selection of mitigation techniques is also affected by advantages, limitations, in contrast to cost of other mitigation techniques, viability, and potential environmental effects. Therefore, choice of most efficient mitigation technique for field application for elimination of PAHs from the polluted locations is crucial.

Future research must focus on understanding and identifying the PAHs degradation method in soil and marine environment. Black-clayish carbon particles and coal tar frequently trap the emitted PAHs beneath the soil environment, greatly lowering their bioavailability. Consequently, it is among the primary obstacles to the effective bioremediation of PAHs. The PAHs mitigation techniques outcomes can be measured by the ratio of available nutrients, concentration of dissolved oxygen, co-contaminants, and abiotic factors. The degradation process of synergistic hydrocarbon is important to understand for proper application of process of in situ biostimulation and other bioaugmentation techniques.

Conclusion

A large amount of trash about 7–9 billion tonnes is produced by industrialised and developed countries along with 80% of organic waste. In African countries the majority of waste is organic. Hydrocarbon pollutants are important because they are difficult to break down due to their low reactivity and require solvents to be hydrophobic. These pollutants effects human health and ecosystem. Reducing the impact of petroleum hydrocarbons in contaminated environments requires biodegradability. These pollutants effect the environment in different ways and breakdown slowly. It has been suggested that bioremediation is a useful, affordable, and flexible treatment method. The technique of bioremediation is interdisciplinary that is affected by professional from different fields, including chemistry, engineering, ecology, microbiology, and geology. For the better understanding of physiology, genetics, evolution, microbial ecology, and biochemistry, the recovery of microbial metabolites comes before the bioremediation. Oleophilic bacteria can be applied single or in groups for eliminating pollutants from environment. The anaerobic and aerobic catabolic pathways involved in biodegradation provide a way for creation of efficient bioremediation techniques of environment contaminants originating from petroleum hydrocarbons.

References

Abdel-Fatah, M.A.: Nanofiltration systems and applications in wastewater treatment: review article. Ain Shams Eng. J. **9**(4), 3077–3092 (2018). https://doi.org/10.1016/j.asej.2018.08.001

Abdel-Shafy, H.I., Mansour, M.S.M.: A review on polycyclic aromatic hydrocarbons: source, environmental impact, effect on human health and remediation. Egypt. J. Pet. **25**(1), 107–123 (2016). https://doi.org/10.1016/j.ejpe.2015.03.011

Abdullah, S.R.S., Al-Baldawi, I.A., Almansoory, A.F., Purwanti, I.F., Al-Sbani, N.H., Sharuddin, S.S.N.: Plant-assisted remediation of hydrocarbons in water and soil: application, mechanisms, challenges and opportunities. Chemosphere. **247**, 125932 (2020). https://doi.org/10.1016/j.chemosphere.2020.125932

Adeoye, J.B., Tan, Y.H., Lau, S.Y., Tan, Y.Y., Chiong, T., Mubarak, N.M., Khalid, M.: Advanced oxidation and biological integrated processes for pharmaceutical wastewater treatment: A review. J. Environ. Manag. **353**, 120170 (2024). https://doi.org/10.1016/j.jenvman.2024.120170

Ahmed, M., et al.: Recent developments in hazardous pollutants removal from wastewater and water reuse within a circular economy. npj Clean Water. **5**(1), 12 (2022). https://doi.org/10.1038/s41545-022-00154-5

Alexander, M.: How toxic are toxic Chemicals in Soil? Environ. Sci. Technol. **29**(11), 2713–2717 (1995). https://doi.org/10.1021/es00011a003

Alori, E.T., Gabasawa, A.I., Elenwo, C.E., Agbeyegbe, O.O.: Bioremediation techniques as affected by limiting factors in soil environment. Front. Soil Sci. **2** (2022). https://doi.org/10.3389/fsoil.2022.937186

Ausuri, J., Vitale, G.A., Coppola, D., Palma Esposito, F., Buonocore, C., de Pascale, D.: Assessment of the degradation potential and genomic insights towards Phenanthrene by Dietzia psychralcaliphila JI1D. Microorganisms. **9**(6), 1327 (2021). https://doi.org/10.3390/microorganisms9061327

Bates, E.R., Grosse, D.W., Sahle-Demessie, E.: Treatment technology for remediation of wood preserving sites: overview. Remediat. J. **10**(3), 35–49 (2000). https://doi.org/10.1002/rem.3440100305

Bezza, F.A., Nkhalambayausi Chirwa, E.M.: Biosurfactant-enhanced bioremediation of aged polycyclic aromatic hydrocarbons (PAHs) in creosote contaminated soil. Chemosphere. **144**, 635–644 (2016). https://doi.org/10.1016/j.chemosphere.2015.08.027

Boström, C.-E., et al.: Cancer risk assessment, indicators, and guidelines for polycyclic aromatic hydrocarbons in the ambient air. Environ. Health Perspect. **110**(suppl 3), 451–488 (2002). https://doi.org/10.1289/ehp.110-1241197

Breckenridge, R.P., Keck, J.F., Williams, J.R.: Characterizing soils for hazardous waste site assessments. Environ. Monit. Assess. **30**(2), 113–138 (1994). https://doi.org/10.1007/BF00545618

Chen, D.M.-C., Bodirsky, B.L., Krueger, T., Mishra, A., Popp, A.: The world's growing municipal solid waste: trends and impacts. Environ. Res. Lett. **15**(7), 074021 (2020). https://doi.org/10.1088/1748-9326/ab8659

Chen, M., Xu, P., Zeng, G., Yang, C., Huang, D., Zhang, J.: Bioremediation of soils contaminated with polycyclic aromatic hydrocarbons, petroleum, pesticides, chlorophenols and heavy metals by composting: applications, microbes and future research needs. Biotechnol. Adv. **33**(6), 745–755 (2015). https://doi.org/10.1016/j.biotechadv.2015.05.003

Cheng, M., Chen, D., Parales, R.E., Jiang, J.: Oxygenases as powerful weapons in the microbial degradation of pesticides. Ann. Rev. Microbiol. **76**(1), 325–348 (2022). https://doi.org/10.1146/annurev-micro-041320-091758

Choi, H., Kim, Y.-Y., Lim, H., Cho, J., Kang, J.-W., Kim, K.-S.: Oxidation of polycyclic aromatic hydrocarbons by ozone in the presence of sand. Water Sci. Technol. **43**(5), 349–356 (2001). https://doi.org/10.2166/wst.2001.0323

Coates, J.D., Michaelidou, U., Bruce, R.A., O'Connor, S.M., Crespi, J.N., Achenbach, L.A.: Ubiquity and diversity of dissimilatory (per) chlorate-reducing bacteria. Appl. Environ. Microbiol. **65**(12), 5234–5241 (1999). https://doi.org/10.1128/AEM.65.12.5234-5241.1999

Das, P., Mukherjee, S., Sen, R.: Improved bioavailability and biodegradation of a model polyaromatic hydrocarbon by a biosurfactant producing bacterium of marine origin. Chemosphere. **72**(9), 1229–1234 (2008). https://doi.org/10.1016/j.chemosphere.2008.05.015

Davletgildeeva, A.T., Kuznetsov, N.A.: Bioremediation of polycyclic aromatic hydrocarbons by means of bacteria and bacterial enzymes. Microorganisms. **12**(9), 1814 (2024). https://doi.org/10.3390/microorganisms12091814

Dehnavi, S.M., Ebrahimipour, G.: Biostimulation of petroleum-contaminated soils with synthetic and natural sources of NPK fertilizer. Soil Sediment Contam. Int. J. **33**(4), 416–429 (2024). https://doi.org/10.1080/15320383.2023.2209186

Dendooven, L., Alvarez-Bernal, D., Contreras-Ramos, S.M.: Earthworms, a means to accelerate removal of hydrocarbons (PAHs) from soil? A mini-review. Pedobiologia. **54**, S187–S192 (2011). https://doi.org/10.1016/j.pedobi.2011.08.006

Eldos, H.I., Ashfaq, M.Y., Al-Ghouti, M.A.: Rapid assessment of the impact of microwave heating coupled with UV-C radiation on the degradation of PAHs from contaminated soil using FTIR and multivariate analysis. Arab. J. Chem. **13**(11), 7609–7625 (2020). https://doi.org/10.1016/j.arabjc.2020.08.031

Fewson, C.: Biodegradation of xenobiotic and other persistent compounds: the causes of recalcitrance. Trends Biotechnol. **6**(7), 148–153 (1988). https://doi.org/10.1016/0167-7799(88)90084-4

García-Díaz, C., Ponce-Noyola, Ma.T., Esparza-García, F., Rivera-Orduña, F. and Barrera-Cortés, J.: PAH removal of high molecular weight by characterized bacterial strains from different organic sources. Int. Biodeterior. Biodegradation. **85**, 311–322 (2013). https://doi.org/10.1016/j.ibiod.2013.08.016

Ghosal, D., Ghosh, S., Dutta, T.K., Ahn, Y.: Current state of knowledge in microbial degradation of polycyclic aromatic hydrocarbons (PAHs): a review. Front. Microbiol. **7** (2016). https://doi.org/10.3389/fmicb.2016.01369

Gupta, H.: Photocatalytic degradation of phenanthrene in the presence of akaganeite nano-rods and the identification of degradation products. RSC Adv. **6**, 112721–112727 (2016). https://doi.org/10.1039/C6RA24602H

Gupta, H.: Anthracene removal from water onto activated carbon derived from vehicular Tyre. Sep. Sci. Technol. **53**(4), 613–625 (2018). https://doi.org/10.1080/01496395.2017.1405038

Gupta, H., Dhiman, S., Kumar, A., Paliwal, S.K., Khan, G., Singh, H., Mishra, A., Mishra, A.K.: Photo-and bio-degradation of selected persistent organic pollutants. Sep. Purif. Rev. 54(1), 3–16 (2025). https://doi.org/10.1080/15422119.2024.2332953

Gupta, H., Gupta, B.: Photocatalytic degradation of polycyclic aromatic hydrocarbon benzo[a]pyrene by iron oxides and identification of degradation products. Chemosphere 138, 924–931 (2015)

Gupta, H., Gupta, B.: Adsorption of polycyclic aromatic hydrocarbons on banana peel activated carbon. Desalin. Water Treat. **57**, 9498–9509 (2016). https://doi.org/10.1080/19443994.2015.1029007

Gupta, H., Kumar, R.: Distribution of selected polycyclic aromatic hydrocarbons in urban soils of Delhi India. Environ. Technol. Innov. **17**, 100500 (2020). https://doi.org/10.1016/j.eti.2019.100500

Haritash, A.K., Kaushik, C.P.: Biodegradation aspects of polycyclic aromatic hydrocarbons (PAHs): A review. J. Hazard. Mater. **169**(1–3), 1–15 (2009). https://doi.org/10.1016/j.jhazmat.2009.03.137

Hatzinger, P.B., Alexander, M.: Effect of aging of chemicals in soil on their biodegradability and extractability. Environ. Sci. Technol. **29**(2), 537–545 (1995). https://doi.org/10.1021/es00002a033

Huang, L., Batterman, S.A.: Multimedia model for polycyclic aromatic hydrocarbons (PAHs) and nitro-PAHs in Lake Michigan. Environ. Sci. Technol. **48**(23), 13817–13825 (2014). https://doi.org/10.1021/es503137b

Issaka, E.: From complex molecules to harmless byproducts: electrocoagulation process for water contaminants degradation. Desalin. Water Treat. **319**, 100532 (2024). https://doi.org/10.1016/j.dwt.2024.100532

Jie, H., Khan, I., Alharthi, M., Zafar, M.W., Saeed, A.: Sustainable energy policy, socio-economic development, and ecological footprint: the economic significance of natural resources, population growth, and industrial development. Util. Policy. **81**, 101490 (2023). https://doi.org/10.1016/j.jup.2023.101490

Johnsen, A.R., Wick, L.Y., Harms, H.: Principles of microbial PAH-degradation in soil. Environ. Pollut. **133**(1), 71–84 (2005). https://doi.org/10.1016/j.envpol.2004.04.015

Kästner, M., Breuer-Jammali, M., Mahro, B.: Impact of inoculation protocols, salinity, and pH on the degradation of polycyclic aromatic hydrocarbons (PAHs) and survival of PAH-degrading bacteria introduced into soil. Appl. Environ. Microbiol. **64**(1), 359–362 (1998). https://doi.org/10.1128/AEM.64.1.359-362.1998

Kästner, M., Miltner, A.: Application of compost for effective bioremediation of organic contaminants and pollutants in soil. Appl. Microbiol. Biotechnol. **100**(8), 3433–3449 (2016). https://doi.org/10.1007/s00253-016-7378-y

Khiadani Hajian, M., Amin, M., Beik, F., Ebrahimi, A., Farhadkhani, M., Mohammadi-Moghadam, F.: Determination of polycyclic aromatic hydrocarbons concentration in eight brands of black tea which are used more in Iran. Int. J. Environ. Health Eng. **2**(1), 40 (2013). https://doi.org/10.4103/2277-9183.122427

Kim, K.-H., Jahan, S.A., Kabir, E., Brown, R.J.C.: A review of airborne polycyclic aromatic hydrocarbons (PAHs) and their human health effects. Environ. Int. **60**, 71–80 (2013). https://doi.org/10.1016/j.envint.2013.07.019

Kotoky, R., Singha, L.P., Pandey, P.: Draft genome sequence of polyaromatic hydrocarbon-degrading bacterium Bacillus subtilis SR1, which has plant growth-promoting attributes. Genome Announc. **5**(49) (2017). https://doi.org/10.1128/genomeA.01339-17

Lau, K.L., Tsang, Y.Y., Chiu, S.W.: Use of spent mushroom compost to bioremediate PAH-contaminated samples. Chemosphere. **52**(9), 1539–1546 (2003). https://doi.org/10.1016/S0045-6535(03)00493-4

Lee, B.-K.: Sources, distribution and toxicity of Polyaromatic hydrocarbons (PAHs) in particulate matter. In: Air Pollution Sciyo (2010). https://doi.org/10.5772/10045

Lee, P.H., Doick, K.J., Semple, K.T.: The development of phenanthrene catabolism in soil amended with transformer oil. FEMS Microbiol. Lett. **228**(2), 217–223 (2003). https://doi.org/10.1016/S0378-1097(03)00751-1

Lee, P.-H., Ong, S.K., Golchin, J., Sam Nelson, G.L.: Use of solvents to enhance PAH biodegradation of coal tar. Water Res. **35**(16), 3941–3949 (2001). https://doi.org/10.1016/S0043-1354(01)00115-4

Lemaire, J., Buès, M., Kabeche, T., Hanna, K., Simonnot, M.-O.: Oxidant selection to treat an aged PAH contaminated soil by in situ chemical oxidation. J. Environ. Chem. Eng. 1(4), 1261–1268 (2013). https://doi.org/10.1016/j.jece.2013.09.018

Liste, H.-H., Alexander, M.: Plant-promoted pyrene degradation in soil. Chemosphere. **40**(1), 7–10 (2000). https://doi.org/10.1016/S0045-6535(99)00216-7

Logeshwaran, P., Megharaj, M., Chadalavada, S., Bowman, M., Naidu, R.: Petroleum hydrocarbons (PH) in groundwater aquifers: an overview of environmental fate, toxicity, microbial degradation and risk-based remediation approaches. Environ. Technol. Innovation. **10**, 175–193 (2018). https://doi.org/10.1016/j.eti.2018.02.001

Mahjoubi, M., Cappello, S., Souissi, Y., Jaouani, A., Cherif, A.: Microbial bioremediation of petroleum hydrocarbon–contaminated marine environments. In: Recent Insights in Petroleum Science and Engineering. InTech (2018). https://doi.org/10.5772/intechopen.72207

Maier, R.M., Soberón-Chávez, G.: Pseudomonas aeruginosa rhamnolipids: biosynthesis and potential applications. Appl. Microbiol. Biotechnol. **54**(5), 625–633 (2000). https://doi.org/10.1007/s002530000443

Makkar, R.S., Rockne, K.J.: Comparison of synthetic surfactants and biosurfactants in enhancing biodegradation of polycyclic aromatic hydrocarbons. Environ. Toxicol. Chem. **22**(10), 2280–2292 (2003). https://doi.org/10.1897/02-472

Malawska, M., Wilkomirski, B.: An analysis of polychlorinated biphenyls (PCBs) content in soil and plant leaves *(taraxacum officinale)* in the area of the railway junction Iława Główna. Toxicol. Environ. Chem. **70**(3–4), 509–515 (1999). https://doi.org/10.1080/02772249909358773

Margesin, R., Schinner, F.: Biodegradation and bioremediation of hydrocarbons in extreme environments. Appl. Microbiol. Biotechnol. 56, 650–663 (2001). https://doi.org/10.1007/s002530100701

McNally, D.L., Mihelcic, J.R., Lueking, D.R.: Biodegradation of three-and four-ring polycyclic aromatic hydrocarbons under aerobic and denitrifying conditions. Environ. Sci. Technol. **32**(17), 2633–2639 (1998). https://doi.org/10.1021/es980006c

Medina-Herrera, N., Martínez-Ávila, G.C.G., Robledo-Jiménez, C.L., Rojas, R., Orozco-Zamora, B.S.: From citrus waste to valuable resources: a biorefinery approach. Biomass. **4**(3), 784–808 (2024). https://doi.org/10.3390/biomass4030044

Miller, R.M., Bartha, R.: Evidence from liposome encapsulation for transport-limited microbial metabolism of solid alkanes. Appl. Environ. Microbiol. **55**(2), 269–274 (1989). https://doi.org/10.1128/aem.55.2.269-274.1989

Minai-Tehrani, D., Minoui, S., Herfatmanesh, A.: Effect of Salinity on Biodegradation of Polycyclic Aromatic Hydrocarbons (PAHs) of Heavy Crude Oil in Soil. Bull. Environ. Contam. Toxicol. 82, 179–184 (2009). https://doi.org/10.1007/s00128-008-9548-9

Mohan, S.V., Kisa, T., Ohkuma, T., Kanaly, R.A., Shimizu, Y.: Bioremediation technologies for treatment of PAH-contaminated soil and strategies to enhance process efficiency. Rev. Environ. Sci. Biotechnol. **5**(4), 347–374 (2006). https://doi.org/10.1007/s11157-006-0004-1

Nguyen, T.T., Youssef, N.H., McInerney, M.J., Sabatini, D.A.: Rhamnolipid biosurfactant mixtures for environmental remediation. Water Res. **42**(6–7), 1735–1743 (2008). https://doi.org/10.1016/j.watres.2007.10.038

Nievas, M.L., Commendatore, M.G., Esteves, J.L., Bucalá, V.: Biodegradation pattern of hydrocarbons from a fuel oil-type complex residue by an emulsifier-producing microbial consortium. J. Hazard. Mater. **154**(1–3), 96–104 (2008). https://doi.org/10.1016/j.jhazmat.2007.09.112

Norris, R.D.: Handbook of bioremediation (1993). CRC Press, Boca Raton (2017). https://doi.org/10.1201/9780203712764

Omoni, V.T., Bankole, P.O., Semple, K.T., Ojo, A.S., Ibeto, C., Okekporo, S.E., Harrison, I.A.: Enhanced remediation of polycyclic aromatic hydrocarbons in soil through fungal delignification strategy and organic waste amendment: a review. Indian J. Microbiol. **65**(2), 775–797 (2025). https://doi.org/10.1007/s12088-024-01383-7

Ou, Y., Wu, M., Yu, Y., Liu, Z., Zhang, T., Zhang, X.: Low dose phosphorus supplementation is conducive to remediation of heavily petroleum-contaminated soil-from the perspective of hydrocarbon removal and ecotoxicity risk control. Sci. Total Environ. **929**, 172478 (2024). https://doi.org/10.1016/j.scitotenv.2024.172478

Pacwa-Płociniczak, M., Płaza, G.A., Piotrowska-Seget, Z., Cameotra, S.S.: Environmental applications of biosurfactants: recent advances. Int. J. Mol. Sci. **12**(1), 633–654 (2011). https://doi.org/10.3390/ijms12010633

Pacwa-Płociniczak, M., Płaza, G.A., Poliwoda, A., Piotrowska-Seget, Z.: Characterization of hydrocarbon-degrading and biosurfactant-producing pseudomonas sp. P-1 strain as a potential tool for bioremediation of petroleum-contaminated soil. Environ. Sci. Pollut. Res. **21**(15), 9385–9395 (2014). https://doi.org/10.1007/s11356-014-2872-1

Pandolfo, E., Barra Caracciolo, A., Rolando, L.: Recent advances in bacterial degradation of hydrocarbons. Water. **15**(2), 375 (2023). https://doi.org/10.3390/w15020375

Pardieck, D.L., Bouwer, E.J., Stone, A.T.: Hydrogen peroxide use to increase oxidant capacity for in situ bioremediation of contaminated soils and aquifers: a review. J. Contam. Hydrol. **9**(3), 221–242 (1992). https://doi.org/10.1016/0169-7722(92)90006-Z

Perera, F.P., et al.: Effect of prenatal exposure to airborne polycyclic aromatic hydrocarbonson neurodevelopment in the first 3 years of life among Inner-City children. Environ. Health Perspect. **114**(8), 1287–1292 (2006). https://doi.org/10.1289/ehp.9084

Perfumo, A., Rudden, M., Marchant, R., Banat, I.M.: Biodiversity of biosurfactants and roles in enhancing the (bio) availability of hydrophobic substrates. In: Cellular Ecophysiology of Microbe, pp. 1–29. Springer International Publishing, Cham (2017). https://doi.org/10.1007/978-3-319-20796-4_35-1

Ravindra, K., Sokhi, R., Van Grieken, R.: Atmospheric polycyclic aromatic hydrocarbons: source attribution, emission factors and regulation. Atmospheric Environment, 42, 2895–2921 (2008). https://doi.org/10.1016/j.atmosenv.2007.12.010

Reddy, M.S., Naresh, B., Leela, T., Prashanthi, M., Madhusudhan, N.C., Dhanasri, G., Devi, P.: Biodegradation of phenanthrene with biosurfactant production by a new strain of brevibacillus sp. Bioresour. Technol. **101**(20), 7980–7983 (2010). https://doi.org/10.1016/j.biortech.2010.04.054

Reid, B.J., Jones, K.C., Semple, K.T.: Bioavailability of persistent organic pollutants in soils and sediments—a perspective on mechanisms, consequences and assessment. Environ. Pollut. **108**(1), 103–112 (2000). https://doi.org/10.1016/S0269-7491(99)00206-7

Sahle-Demessie, E., Grosse, D.W., Bates, E.R.: Solvent extraction and soil washing treatment of contaminated soils from wood preserving sites: bench-scale studies. Remediat. J. **10**(3), 85–109 (2000). https://doi.org/10.1002/rem.3440100308

Salari, M., Rahmanian, V., Hashemi, S.A., Chiang, W.-H., Lai, C.W., Mousavi, S.M., Gholami, A.: Bioremediation treatment of Polyaromatic hydrocarbons for environmental sustainability. Water. **14**(23), 3980 (2022). https://doi.org/10.3390/w14233980

Salimnezhad, A., Soltani-Jigheh, H., Soorki, A.A.: Effects of oil contamination and bioremediation on geotechnical properties of highly plastic clayey soil. J. Rock Mech. Geotech. Eng. **13**(3), 653–670 (2021). https://doi.org/10.1016/j.jrmge.2020.11.011

Shahsavari, E., Adetutu, E.M., Anderson, P.A., Ball, A.S.: Necrophytoremediation of phenanthrene and pyrene in contaminated soil. J. Environ. Manag. **122**, 105–112 (2013). https://doi.org/10.1016/j.jenvman.2013.02.050

Singh, S., Parveen, N., Gupta, H.: Adsorptive decontamination of rhodamine-B from water using banana peel powder: a biosorbent. Environ. Technol. Innovation. **12**, 189–195 (2018). https://doi.org/10.1016/j.eti.2018.09.001

Singh, S., Gupta, H., Dhiman, S., Sahu, N.K.: Decontamination of cationic dye brilliant green from the aqueous media. Appl Water Sci. **12**, 61 (2022). https://doi.org/10.1007/s13201-022-01596-5

Sinha, R.K., Agarwal, S., Chauhan, K., Chandran, V., Soni, B.K.: Vermiculture technology: reviving the dreams of sir Charles Darwin for scientific use of earthworms in sustainable development programs. Technol. Invest. **01**(03), 155–172 (2010). https://doi.org/10.4236/ti.2010.13019

Siron, R., Pelletier, E., Brochu, C.: Environmental factors influencing the biodegradation of petroleum hydrocarbons in cold seawater. Arch. Environ. Contam. Toxicol. **28**(4) (1995). https://doi.org/10.1007/BF00211621

de Souza Pohren, R., Rocha, J.A.V., Horn, K.A., Vargas, V.M.F.: Bioremediation of soils contaminated by PAHs: mutagenicity as a tool to validate environmental quality. Chemosphere. **214**, 659–668 (2019). https://doi.org/10.1016/j.chemosphere.2018.08.020

Stapleton, R.D., Savage, D.C., Sayler, G.S., Stacey, G.: Biodegradation of aromatic hydrocarbons in an extremely acidic environment. Appl. Environ. Microbiol. **64**(11), 4180–4184 (1998). https://doi.org/10.1128/AEM.64.11.4180-4184.1998

Sudaryanto, A., et al.: Levels and distribution of organochlorines in fish from Indonesia. Environ. Int. **33**(6), 750–758 (2007). https://doi.org/10.1016/j.envint.2007.02.009

Tudararao-Aherobo, L., Mesogboriwon, S.: Bioremediation of spent engine oil contaminated soils using indigenous fungi species. Int. J. Sci. Res. Sci. Technol., 445–461 (2020). https://doi.org/10.32628/IJSRST207156

Ukaogo, P.O., Ewuzie, U., Onwuka, C.V.: Environmental pollution: causes, effects, and the remedies. In: Microorganisms for Sustainable Environment and Health, pp. 419–429. Elsevier, Amsterdam, Netherlands (2020). https://doi.org/10.1016/B978-0-12-819001-2.00021-8

Varjani, S.J.: Microbial degradation of petroleum hydrocarbons. Bioresour. Technol. **223**, 277–286 (2017). https://doi.org/10.1016/j.biortech.2016.10.037

Vidali, M.: Bioremediation an overview. Pure Appl. Chem. **73**(7), 1163–1172 (2001). https://doi.org/10.1351/pac200173071163

Vidonish, J.E., Zygourakis, K., Masiello, C.A., Sabadell, G., Alvarez, P.J.J.: Thermal treatment of hydrocarbon-impacted soils: a review of technology innovation for sustainable remediation. Engineering. **2**(4), 426–437 (2016). https://doi.org/10.1016/J.ENG.2016.04.005

Wang, X., Brusseau, M.L.: Solubilization of some low-polarity organic compounds by hydroxypropyl-.Beta.-cyclodextrin. Environ. Sci. Technol. **27**(13), 2821–2825 (1993). https://doi.org/10.1021/es00049a023

Ward, D.M., Brock, T.D.: Hydrocarbon biodegradation in hypersaline environments. Appl. Environ. Microbiol. 35(2), 353–359 (1978). https://doi.org/10.1128/aem.35.2.353-359.1978

Ye, S., et al.: The effects of activated biochar addition on remediation efficiency of co-composting with contaminated wetland soil. Resour. Conserv. Recycl. **140**, 278–285 (2019) https://doi.org/10.1016/j.resconrec.2018.10.004

Zhang, X., et al.: Assessing approaches of human inhalation exposure to polycyclic aromatic hydrocarbons: a review. Int. J. Environ. Res. Public Health. **18**(6), 3124 (2021). https://doi.org/10.3390/ijerph18063124

Green Technologies for the Decontamination of Polycyclic Aromatic Hydrocarbons

13

Shubhangee Agarwal, Priyanka Rani, and Himanshu Gupta

Abstract

Polycyclic aromatic hydrocarbons (PAHs) are dangerous microscopic contaminants which are organic in nature and characterized by the presence of two or more phenyl rings, which are notably resistant to degradation. PAHs originate from man-made as well as natural sources including volcanic eruptions forest fires, petroleum, coal, and oil-based products such as gasoline, as well as improper disposal of industrial and domestic waste, crude oil extraction, and spills. They are potentially a threat to human health, plants, animals, and the environment as a whole. Environmental scientists are exploring numerous methods for treating PAH-contaminated water and soil, which encompass biological, chemical, physical, and thermal technologies. The conventional approach to PAH decontamination typically involves the use of harsh chemicals and a process that consumes significant energy, leading to a push for more sustainable and eco-friendly alternatives. This chapter reviews recent progress in green technologies for PAH decontamination, including bioremediation, electrokinetic remediation, phytoremediation, adsorption (utilizing adsorbents like zeolites, biomass derivatives, geosorbents, silica, and graphene-based materials), advanced oxidation processes, enhanced remediation through biocatalysts, and nanotechnology.

Keywords

Polycyclic aromatic hydrocarbons (PAHs) • Environment • Remediation • Green • Nanomaterials

S. Agarwal (✉) · P. Rani · H. Gupta
Department of Chemistry, School of Sciences, IFTM University, Moradabad, Uttar Pradesh, India
e-mail: shubhangeeagarwal79@gmail.com

H. Gupta et al. (eds.), *Ubiquitous Polycyclic Aromatic Hydrocarbon Contamination*, Emerging Contaminants and Associated Treatment Technologies,
https://doi.org/10.1007/978-3-032-11764-9_13

Introduction

In recent years, the environmental existence of the pervasive organic contaminants PAH has come to an attention. PAHs are widely distributed, resilient, and hazardous organic pollutants, consisting of two or more fused phenyl rings arranged in clustered, angular or linear formations, and characterized by less solubility in water (Johnsen and Karlson, 2005; Gupta and Gupta, 2016). They primarily originate from the incomplete burning of carbon-based materials. PAHs can be present in variety of environmental matrix structures, including air, water, and soil (Ansari et al., 2023; Gupta and Dhiman, 2023). PAHs are categorized into two groups: lower molecular weight (LMW) compounds having two or three ring structures that are generally less hazardous, such as phenanthrene, naphthalene, anthracene, and fluorene, and higher molecular weight (HMW) compounds with four to seven ring structures that are less susceptible to deterioration, mutagenic, and capable of causing tumors, such as pyrene, benz[a]pyrene, and chrysene (Kuppusamy et al., 2016) (Fig. 13.1). The potential for carcinogenicity in persistent organic compound increases with their molecular weight (Eisler, 1987). PAHs that are bound to particles pose significant health risks to humans. Benz[a]pyrene contributes significantly to the total carcinogenic potential, accounting for 51–64% (Ohura et al., 2004). While they exhibit less solubility in water, PAHs are more soluble in organic solvents. Their partial volatile and lipophilic characteristics lead to a strong tendency for absorption and capture by biological matter and soil minerals, resulting in reduced mobility for microorganisms and other degrading agents in the upper layer of soil (Shanker et al., 2017). When PAHs are released into the atmosphere, they experience a variety of mechanisms including physical, chemical, and biological reactions through metabolic and microbial pathways, which are commonly referred to as weathering or degradation processes (Ossai, 2020). These weathering processes are significantly influenced by two non-biotic factors: the ecological characteristics and the composition of the contaminants. These persistent pollutants can adsorb on the upper layer of the soil, volatilize into the surroundings, and dissolve in water (Schwarzenbach et al., 2016). The physicochemical components of each organic pollutant affect their fate, behavior, and movement within the environment. The distribution of these organic hydrocarbons across different environmental matrices will differ substantially based on their molecular characteristics and the surrounding environmental conditions (e.g., temperature). Consequently, the amount of PAHs in the environment changes depending on the place of origin (Chen et al., 2020).

Sources of PAHs

PAHs are common environmental contaminants that are produced when carbon-based materials burn at high temperatures. Both naturally occurring and man-made sources emit PAHs. PAHs from natural sources are negligible in comparison to those produced by humans (Charlesworth et al. 2002; Ravindra et al. 2008; Witt

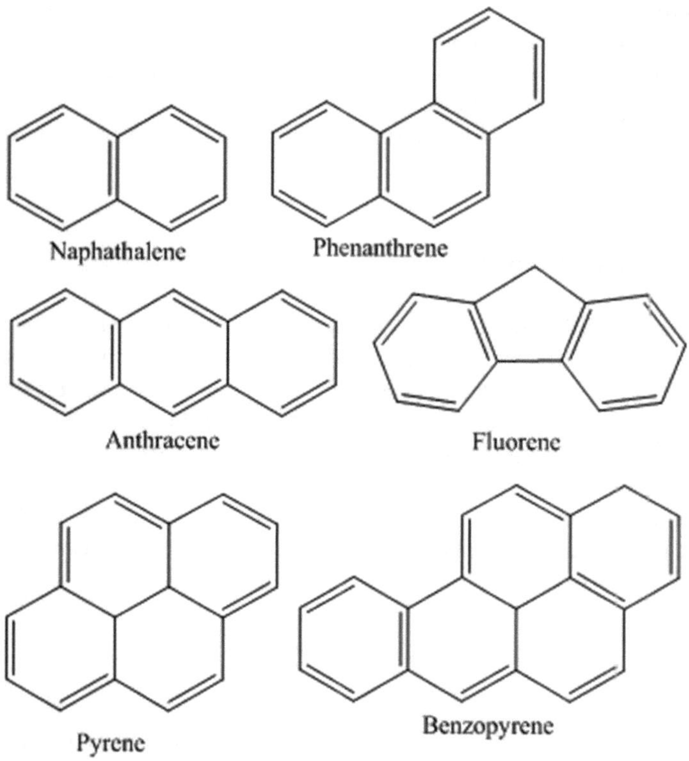

Fig. 13.1 Chemical structure of few PAHs with lower and higher molecular weights

1995). Oil seeps, volcanic eruptions, thermal geological reactions, and fires in natural forests and prairies are examples of natural sources (Wright and Welbourn 2002; Hidayat et al. 2019; Hall et al. 2009; Gupta, 2018). Household waste burning, processing of agricultural products, automobile exhaust, armed forces operations, fossil fuel conversion, emissions from coal tar, and shale formation, petroleum based products processing, manufacturing of steel and iron, aluminum manufacturing, coke production, and partial combustion are all examples of human-induced sources of PAH emissions (Gupta and Gupta, 2018). These contaminants are extensively utilized in the manufacturing of pigments, dyes, insecticides, resins, wood maintenance, grease, and pharmaceutical and agricultural products (Abdel-Shafy and Mansour, 2016). The main ways that PAHs reach the aquatic system are by polluted debris, precipitation absorption, and accidental leakage. They are present in tar sands, coal, and oil (Blanchard et al. 2004). Humans come in contact to these persistent aromatic hydrocarbons through a variety of routes, including atmospheric deposition, and even at low quantities, they can cause cancer. Inhalation, water (from home and recreational activities), food, and work environments are among the exposures. One of the main ways that humans are exposed is through contaminated water (Wang et al., 2009; Wu et al., 2011). The harmful consequences of PAHs necessitate constant and careful monitoring of human intake. According to their sources, PAHs can also be categorized as biogenic (originating from biological sources), pyrogenic (originating from combustion processes), or petrogenic (originating from petroleum products) (Merhaby et al., 2019).

Pathways of Exposure

PAHs are a class of chronic toxic substances that linger in many environments, such as the air, soil, and water, for a considerable amount of time. Their endurance is explained by their high thermal stability and poor water solubility (Na et al., 2021; Kim et al., 2013). Humans are exposed to PAH through a variety of routes, through breathing in particles that are in air and re-suspended contaminants, ingesting and drinking, and getting into contact with dirt and other substances through the bare skin (Menzie et al., 1992). The formation and destiny of these pollutants have become a major concern because to their mutagenic, carcinogenic, and teratogenic properties.

Air

Natural events such as volcanic eruptions and bushfires send PAHs into the surrounding air. However, the main causes of these volatile organic pollutants in the surrounding air are activities brought on by high temperatures, such as partial combustion of biomass from wood, garbage from municipalities, and combustion of petroleum based products (Lammel, 2015). Additionally, chewing tobacco or employing high-temperature cooking techniques like grilling, deep-frying, roasting, or baking near or directly over a flame might cause it (Wróbel et al., 2022).

Soil

In terms of soil quality, agricultural productivity, and other aspects, the ecosystem has suffered from the wet or dry accumulation of atmospheric PAHs on upper layer of soil (Balmer et al., 2019). Forest fires, volcanic eruptions, the degradation of sedimentary rocks that contain petroleum based chemicals, and the production of bacteria and algae are all natural sources of organic pollutants. PAHs level in upper layer of soil is also influenced by human activities including the pyrolysis of organic matter and inadequate burning of carbon-based fuels. Because of PAHs' hydrophobic properties and stable chemical composition, which lead to their build up in soil, they have mutagenic, teratogenic, and carcinogenic qualities. Additionally, surface soil in crowded urban areas contains a significant concentration of PAHs (Bortey-Sam et al., 2014).

Water

PAHs primarily enter water sources through leaching from creosote-impregnated wood, road drainage, waste water from factories, burning petroleum and other fossil fuels, and wet or dry accumulation (Karyab et al., 2013). Groundwater, wastewater, potable water, water bodies like lakes and rivers, and wreckages have all found to contain volatile organic compounds (Mojiri et al., 2019).

Effect of PAHs on Health

Both short-term and long-term health impacts are possible with PAHs. Signs of asthma, respiratory difficulties, pulmonary problems, kidney and liver impairment, glaucoma, and immune system suppression are among the long-term complications (Abdel-Shafy and Mansour, 2016). Short-term or immediate side effects include gastrointestinal stress, diarrhoea, vomiting, discomfort and pain in eyes, anxiety, and more (Abdel-Shafy and Mansour, 2016). Air, plants, animal tissues, water reservoirs, soil, and rocks have also been found to contain a number of PAHs (Chang et al. 2004a, 2004b; Balati et al. 2015; Liu et al. 2014; Vane et al. 2014). Due to their persistence and lack of biodegradability, PAHs can penetrate into the food chain through the soil and reach humans through the respiratory, food, and skin systems (Abdel-Shafy and Mansour, 2016). The immune system transforms PAH into epoxide hydrolase and diolepoxides, which react with genetic material and stop its synthesis, when they reach the human body and may successfully pass through the plasma membrane and be absorbed by the cells (Kaya et al., 2013) (Fig. 13.2).

PAH Derivatives and Their Fate

PAHs penetrate into the environment through the air and water, travel through the marine environment and eventually make their way into the food chain of humans through marine foods. The amount of time PAHs stay in the atmosphere depends on a number of factors, including the chemical makeup, pollutant-degrading microorganisms, their concentration, dispersion, bioavailability, pH, oxygen levels, soil type, temperature and the availability of nutrients. PAHs are pervasiveorganic pollutants that have serious negative impact on human beings and environment (Cai et al., 2017). Inspite of their greater mobility and substantial risk of harm, polar PAHs are less known and are not often studied. Many types of cancer, including those of the skin, bladder, and lungs, are brought on by these carcinogenic

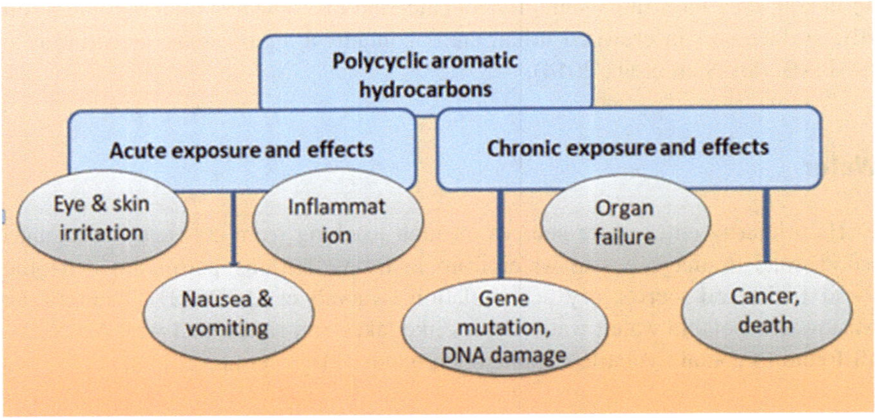

Fig. 13.2 Schematic representation showing the acute and chronic exposure to PAHs

compounds. They can also cause genotoxicity (damaging non-transmissible DNA), nephrotoxicity (damaging kidney), cytotoxicity (damaging cells), ocular toxicity (damaging eye disorders), cardiotoxicity (damaging heart muscles), hepatotoxicity (damaging liver), hemotoxicity (destroying RBCs), teratogenicity (causing malformation of the embryo or fetus), neurotoxicity (damaging brain and nervous system), immunotoxicity (damaging immune system), and mutagenicity (causing transmissible genetic alterations) in both animals and human beings (Ossai et al., 2020). Researchers are very interested in the destination and removal of all those ubiquitous substances from the atmosphere. Since their presence increases the potential impact of health risk. According to numerous research factors including population density, economic development, and meteorological conditions affect both the concentration and geographical distribution of PAH in the surrounding atmosphere (Peng et al., 2016; Zhang and Chen, 2017). The concentration of PAH varies by area, just as these factors do. Therefore, effective measures are needed to limit the negative effects of prolonged exposure on local residents due to the elevated PAH burden in diverse parts.

Decontamination Techniques of PAH

Concern over the immediate and long-term hazards linked with PAH emissions has spread around the world due to the fact that their presence might result in irreparable harm to the environment and living things. In last few decades, numerous decontamination techniques have been created to detect and handle these environmental contaminants. To maintain a sustainable ecosystem, the cleaning methods should be both inexpensive and effective (Abdel-Shafy and Mansour, 2016). Both the qualitative and quantitative content, as well as the degree of contamination, influence the technique (Cunningham, 1996). Chemical and physical procedures are among the different approaches used for decontamination. However, these methods are expensive and introduce a number of harmful substances into the chemical, physical and biological treatments are among the various methods used to degrade PAH (Ossai et al., 2020). Every treatment has pros and cons, but choosing the best one for a contaminated site requires a thorough case analysis that takes into account the features of the site, the pollutants' characteristics, and the possible treatment method. In 2018, Gitipour et al. Therefore, the best course of action is to employ green technologies, which are more appropriate, sustainable, effective, and economical. Physical therapy techniques such as electrokinetic remediation and solvent extraction are simple to use and typically require non-toxic materials. However, the requirement for follow-up care is the primary drawback of this kind of treatment. Degradation of high molecular weight (HMW), persistent and hazardous pollutants is the primary benefit of chemical interventions like catalytic degradation and oxidation via chemicals. The most environmentally friendly method that produces no trash is biological treatment. These methods are the most effective, but they take longer. This chapter goes into great length into a number of green and traditional (physical, thermal, and chemical) methods, including advanced oxidation processes (AOPs), bioremediation, phytoremediation, and hybrid green technologies.

Physical Degradation

Pollutant transfer throughout geochemical cycles and pollutant accumulation on soil or water is examples of physical processes (Ansari et al., 2023). While pollutant transit can happen by condensation or dispersion, retention can happen through adsorption or aggregation in the intergranular area. Numerous variables, such as soil media, temperature, moisture content, soil salinity, or water pH, influence these interactions.

Thermal Degradation

Because of their speed and effectiveness, thermal technologies and incineration are regarded as excellent remediation methods for PAH-contaminated areas. High total petroleum hydrocarbon elimination efficiency is the outcome of the thermal process. Despite these benefits, thermal treatment necessitates elevated temperatures, which may result in several problems. High temperatures at contaminated sites need a lot of energy, which is expensive (Vidonish et al., 2016). High temperatures have the capacity to completely destroy organic matter or site composition and leading to structural degradation that makes it difficult to return the environment to its initial state. Although it's evident that thermal approaches are capable of efficiently eliminating pollution, little is known about how high temperatures affect ecosystems, including soil and sediment organisms, and restoration initiatives (Buttress et al., 2016). As a result, the necessity to develop a approach to maximize thermal treatments with sustainable objectives like ecosystem preservation and energy saving is developing. A thorough approach to thermal treatment is necessary to guarantee that environmental cleanup initiative programs do not unnecessarily damage our planet but rather support worldwide endeavors to enable a sustainable development in the future. According to research, one of the most effective thermal techniques for quickly and affordably cleaning up PAH-contaminated areas is microwave irradiation (Falciglia et al.. 2013; Robinson et al., 2012; Buttress et al., 2016).

Chemical Oxidation Degradation

Another popular technique for breaking down resistant organic contaminants (such PAHs) is chemical oxidation. PAHs can be eliminated from contaminated water using standard chemical oxidation techniques such ozonation, photolysis, and chlorination. More harmful by-products, however, may result from the actual treatment. For instance, natural organic matter in water can react with the classic oxidant chlorine to form harmful halogenated hydrocarbons (Amor et al., 2019). Ozonation has been proposed by a number of authors as a way to eliminate PAHs from water (Ates and Argun, 2018). Ozone can oxidize the pollutant in two ways: (a) the pollutant is directly oxidized by ozone via chemical bonds between electrons or atomic nuclei. (b) By generating radicals that oxidize pollutants to minuscule molecules, metal oxides or organic materials indirectly destroy ozone. pH plays a major role in these degradation processes. Basic pH facilitates the indirect pathway for the breakdown of O_3 and the generation of OH radicals. A direct approach is preferred because ozone becomes more stable and deteriorates the contaminants in an acidic pH. The benzene ring is cleaved by chemical reactions between ozone and PAHs, producing intermediate molecules that are like carboxylate and hydroxylated benzene. Subsequently, intermediate chemicals may directly or indirectly compete with the pollutants being targeted for O_3 and OH^- radicals (Rubio-Clemente et al., 2014). Oxidation limits, such as the existence of OH^- scavengers in water that is contaminated and the generation of hazardous products, may adversely impact PAH decomposition (Miller and Olejnik 2001).

Green Technologies for Decontamination of PAHs

Biodegradation

Biodegradation is a remediation technique that enables organic contaminants to be eliminated through the substrate's natural biological activity (Ghosal et al., 2016; Ansari and Rafatullah, 2023; Sivasamy et al., 2025; Gupta et al., 2025). Microorganisms, plants, and enzymatic reactions are used to detoxify or degrade environmental pollutants (Ansari and Rafatullah, 2023; Kaur et al., 2023). The microorganisms used for biodegradation are bacteria, algae, fungi and yeast (Al-Hawash et al., 2018). The microbes utilized for biological degradation ought to be native to the site or contaminated area (Das and Chandran, 2011). In order to accelerate the biodegradation process, it is crucial to consider the bioavailability of the contaminant to microbial attack, the availability of microbial consortia can degrade the pollutant, and particular environmental factors (oxygen level, soil type, soil nutrient content, pH, electron acceptor agents and temperature) that encourage microbial growth (Moghaddam et al., 2024). Many scientists believe that certain bacterial isolates can degrade PAHs. Under aerobic conditions, PAHs can be broken down by microorganisms belonging to the genera Mycobacterium and Pseudomonas (Motteran et al., 2024). A recent study found that Serratiamarcescens, Bacillus megaterium, Raoultellaornithinolytica, and Aeromonashydrophila decomposed PAHs with above 90% efficiency for acenaphthene and fluorene. The effectiveness of bioremediation methods was found to be significantly impacted by a few environmental factors, including pH and temperature. According to Moscoso et al. (2013), temperatures between 30 and 38 °C, which are higher than room temperature, were found to be ideal for the breakdown of PAHs due to increased microbial proliferation. Nevertheless, as demonstrated by a study that was carried out at a slightly higher temperature between 40 and 45 °C and showed a notable decrease in PAH degradation from 92% (37 °C) to 74%, bioremediation is not a rigorous and reliable method. As a result of decreased oxygen solubility brought on by rising temperatures, bioreactors have less dissolved oxygen available for microbial Restoring the soil's capacity to operate as intended should be the ultimate goal of any remediation operation, in addition to eliminating contaminants from contaminated substrates (Punnapayak et al., 2009). The application of laccase, an enzyme obtained from a Genodermal ucidum isolate that causes white rot. It was found that, with or without the use of a redox mediator, G. lucidum totally breaks down anthracene. Additionally, it was discovered that *Thiobactersubterraneus*, *Alcoligenes* sp., and *Escherichia coli* were effective isolates used for breaking down anthracene and phenanthrene (Abd-Elsalam et al., 2009).

Three main elements influence the pace of the bioremediation process: the contaminants being degraded, the form of microorganisms, and the natural habitat. Inherent bioremediation may result from natural decrease or bioattenuation, however this process may take years (Carolin et al., 2021). Therefore, by altering various parameters, a number of approaches have been proposed to improve the natural microbial digestion of all these kind of contaminants. To boost microbial metabolism of PAHs, biostimulation or bioaugmentation have been employed (Mohapatra and Phale, 2021). According to Zhou et al. (2022), biostimulation is the process of increasing the metabolic activities of endogenous microorganisms by electron acceptors, adding nutrients or biosurfactants. In order to remediate oil-contaminated soil, bioaugmentation technology—which involves introducing specific functional microorganisms into the intended environment—has been used repeatedly (Wu et al., 2016). In order to enhance the breakdown of PAHs by making them more soluble in water and more accessible to microorganisms, many researchers employ biosurfactants, which are surface-active molecules made by microorganisms (Cameotra, 2008). Amphiphilic compounds with both water-resistant and water friendly components are known as biosurfactants. While the hydrophilic portion comes in a variety of forms, the hydrophobic portion is usually a hydrocarbon chain. Molecular weight is used to categorize the microbial surfactants into two groups. Polysaccharides and lipoproteins are classified as high molecular weight (HMW), whilst lipopeptides and glycolipids are classified as low molecular weight (LMW). Biosurfactants' structural variations also affect how functional they are (Johnson et al., 2021). LMW biosurfactants are utilized far more often than HMW biosurfactants due to their significant reduction in surface tension (Carolin et al., 2021). Because of their numerous functional properties, such as dispersion, surface activity, foaming, emulsification, wetting, cleansing, stability under pH, extreme temperature, and salinity conditions, and more, biosurfactants are a promising choice for biotechnological applications such as biological remediation, emulsion stabilization, and microbial enhanced oil recovery (MEOR) (Adetunji and Olaniran, 2021). Table 13.1 lists the often researched biosurfactants along with the microbes that produce them. Despite the environmental advantages, the main drawback of producing biosurfactants is their high cost in comparison to producing synthetic ones. Since synthetic surfactant synthesis uses proven procedures rather than microbial culture, it is believed to be more cost-effective and efficient (Fig. 13.3).

In situ and ex situ are the two categories of bioremediation approaches. Langenbach (2013) further investigated bioremediation methods including composting, land farming, bioslurry, bioventing, and biosparging. In order to accelerate the in-situ degrading process, air is injected into a contaminated area, a procedure known as bioventing. In the saturated zone, bio-sparging is employed to raise the oxygen content, which improves the biological performance of native microorganisms. In a controlled bioreactor, contaminated locations can be treated using bioslurry. In order to achieve maximum removal effectiveness, these bioreactors are designed to monitor, manage, and adjust a number of parameters, including temperature and nutrients. The practice of land farming involves sprinkling contaminated soil over a region. Because the pollutant's biodegradation takes place at the oil-water interface, oxygen, moisture, and nutrients—such as fertilizers—had to be added.

Table 13.1 List of types of biosurfactants and the microbes that produce them

Microorganism	Biosurfactant type	References
Bacillus licheniformis LRK1	Lipopeptide	Nayak et al., (2020)
Acinetobacter sp.	Alasan	Uzoigwe et al., (2015)
Acinetobacter calcoaceticus PTCC1318	Emulsan	Amani and Kariminezhad, (2016)
Bacillus subtilis- RSL2	Lipopeptide	Sharma et al., (2020)
Bacillus tequilensis MK 729,017	Surfactin	Dutta et al., (2020)
Brevibacillus brevis	Surfactin	Krishnan et al., (2019)

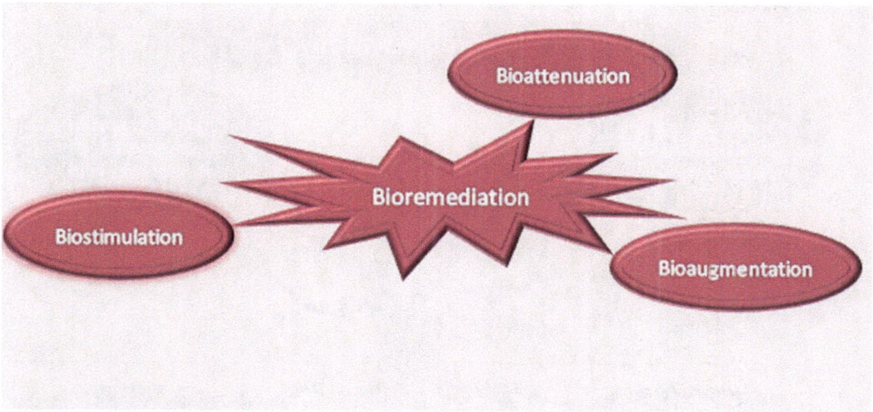

Fig. 13.3 Biodegradation of organic contaminant

Organic wastes can be turned into compost, a humic substance that can be applied as soil fertilizer, through a controlled natural microbial degradation process (Ossai et al., 2020).

Phytoremediation

Phytoremediation also referred to as phytodegradation. It is an ecologically friendly and economical approach that utilizes plants to reduce soil contamination through processes such as removal, digestion, assimilation, or absorption (Wu et al., 2016). The phytoremediation process encompasses various subgroups aimed at extracting organic compounds and metals from contaminated soil. These subgroups include phytoextraction, phytodegradation, phytostabilization, phytofiltration, phytovolatization, and phytostimulation, as shown in Fig. 13.4. Phytoextraction involves using hyper accumulator plants to gather pollutants from the soil (Mani and Kumar, 2014). Phytodegradation serves as an effective secondary treatment for eliminating PAHs from composted soils (Parrish et al., 2004). Pollutants that enter plant cells frequently undergo enzymatic transformations that decrease their toxicity (Kvesitadze et al., 2009). Phytostabilization employs plant cell compartments to sequester toxic substances from critical cellular regions, including the mitochondria, plastids, cytoplasm, and nucleus. Unlike mammals, plants possess different excretory mechanisms and lack a conjugated excretory system. Instead, they store toxic substances within organelles like cell membrane and vacuoles (Kvesitadze et al., 2009). In the process of phytofiltration, the root areas and lipophilic discharges help in translocation, mobilization, or adsorption PAHs in the rhizosphere. Additionally, they promote biodegradation by stimulating microbial activity and metabolic pathways triggered by compounds released from roots, as well as direct plant uptake and metabolism (Hussein et al., 2022). Research indicates that plants are more prone to accumulate low molecular weight

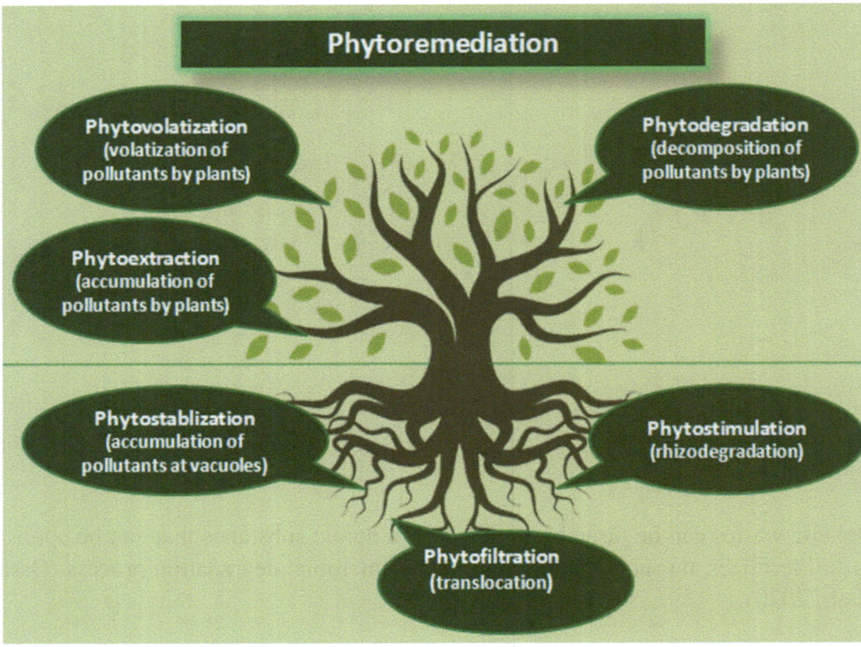

Fig. 13.4 Phytoremediation: the green approach

(LMW) PAHs (Tian et al., 2018). These compounds can exist in vapor form, allowing for easier entry into plants via stomatal openings. This phenomenon explains the higher occurrence of PAHs in the above ground parts of the plant. LMW PAHs may volatilize and travel to the leaves (Tian et al., 2018). The leaves are significant sites for excretion and volatilization processes, contributing to the detoxification of harmful substances, while leaf space index plays a role in biomass growth by affecting photosynthesis (Sun et al., 2011). This method utilizes both plants and soil microbes to remove PAHs at a low cost with least environmental impact. Phytostimulation, or rhizodegradation, is considered the most effective solution for eliminating PAHs from soil. Plants naturally synthesize organic substances such as organic acids, sugars, and secondary metabolites that promote microbial growth. These microorganisms associated with plant roots aid in breaking down PAH compounds (Rohrbacher and St-Arnaud, 2016).

Adsorption via Nanomaterials

The use of nanomaterials in technologies aimed at remediating environmental compartments like air, water, and soil has been explored (Oluwasanu et al., 2019). Various types of nanomaterials have been used as nanosorbents to improve the removal of PAH contamination prior to their reuse (Russo et al., 2020). An effective adsorbent exhibiting a high adsorption capacity possesses certain characteristics related to its surface area and pore size (Nethaji et al., 2013). Adsorbing materials can be divided into natural adsorbents, which include both organic and inorganic types, and synthetic adsorbents. Natural adsorbents can originate from biological or earth-based resources, such as food wastes or materials derived from plants, including vegetable fibers, clay minerals, graphene, straw, sawdust, and date palm. In comparison to other adsorbents, these natural organic options are advantageous due to their low cost and easy availability. They have been utilized to enhance adsorption capabilities and transform waste into valuable substances known as sorbents (Ndimele et al., 2018). However, a significant drawback of their use is the challenge in separating the adsorbent for reuse or disposal after it has been applied in oil spill water (Crini and Lichtfouse, 2019). Synthetic sorbents, frequently produced in laboratories or available commercially, include materials like activated carbon, zeolites, graphene, silica, and those based on nanotechnology. According to the USEPA, one of the best remediation techniques is activated carbon (AC) adsorption. According to Okiel et al. (2011), the use of AC for oil-contaminated water treatment is generally acknowledged as a technically viable technique for eliminating a variety of organic compounds. However, this adsorbent is thought to be costly, particularly in developing nations. The oleophilic and hydrophobic characteristics of these synthetic adsorbents enable them to absorb oil media weighing up to 100 times the weight of media (Ndimele et al., 2018). The diverse scientific applications of graphene and its composites can be attributed to their extensive specific surface area, high tensile strength, conductance, thermal stability, chemical durability, thin film width flexibility, and charge mobility (Novoselov et al., 2012). Owing to their fundamental physicochemical characteristics, including adsorption, high ion exchange capacity, mechanical and thermal stability, and specific surface area—both naturally occurring and synthesized materials exhibit these characteristics (Fletcher et al., 2017). Zeolites play an essential role in the removal of contaminants from soil and water by functioning as molecular sieves, ion exchangers, membranes, and adsorbents. This necessary modified clay substance has proved to be a practical, affordable and readily available sorbent (Lee and Tiwari, 2012). Zeolite and its improved forms were used in a sorption process to extract pyrene, phenanthrene, and benz(a)pyrene from water (Zhang et al., 2017). Benz(a)anthracene, pyrene, fluoranthene, phenanthrene, and fluorene were all removed from water with an average removal rate of 98% using organo-zeolite as an adsorbent (Lemić et al., 2007).

Advanced Oxidation Processes (AOPs)

AOPs are chemical oxidative methods employed for addressing PAH contamination in soil and various types of landfills, including industrial and municipal. These advanced chemical techniques can degrade approximately 90% of PAHs. The oxidising agents such as activated peroxodisulphate (Huling and Pivetz, 2006), hydrogen peroxide (Flotron et al., 2005), Fenton's reagent (Zhang et al., 2019; Gupta and Gupta, 2015; Gupta, 2016), and ozone (Lian et al., 2017) are utilized in this process. These methods generate highly reactive radicals like OH, O_3, SO_4, along with other activated species such as peroxides and persulfate anion, which facilitate the disintergration of the aromatic rings in PAH compounds. AOPs present several advantages over conventional remediation methods, including the in situ degradation of contaminants, a relatively shorter treatment time, and the ability to effectively remediate a diverse array of PAHs due to their highly reactive radicals. Nonetheless, the oxidation of components which are not target such as particulate organic matter or dissolved can complicate the use of these highly reactive oxidizing agents, and their corrosive nature makes them unsuitable for treating potable water (Amor, et al., 2019). AOPs are divided into photochemical and non-photochemical processes, based on the method used to produce OH radicals. The photochemical process employs a source of light—specifically UV light in conjunction with oxidising agents or catalysts. The production of OH^- can be initiated by oxidisers like H_2O_2, O_3, and UV light, as well as by catalysts. The effectiveness of AOPs is affected by various variables, which include the type and amount of oxidant used the intensity and type of light, pH levels, the rate at which free radical are generated, the level of contamination and characteristics of the contaminants, and the composition of the water (Rubio-Clemente et al., 2014). TiO_2 is a unique metal oxide semiconductor which has high photocatalytic decomposition property, high efficiency and extremely low price so, it is selected as photocatalyst. Its catalytic reaction starts in the presence of UV light. When the material absorbs UV light it causes formation of electrons and positive holes in the valence and conduction band. In contrast, electrons can directly oxidize organic pollutants, such as PAHs, by consuming oxygen molecules through a reduction reaction. Conversely, holes can either directly react with PAHs or form an OH radical that can oxidize PAHs. AOPs represent the most promising approach and offer economic advantages when leveraging a solar light source. AOPs are associated with relatively high operating costs due to the expensive nature of chemicals like H_2O_2 and the significant energy utilization is involved. Furthermore, OH radicals are highly non-selective and can react with non-target pollutants found in the contaminated water (Comninellis et al., 2008). Innovative techniques that merge ultraviolet photo-Fenton and Fenton oxidative-coagulation processes have also demonstrated considerable effectiveness in removing PAHs from leachates and solid waste (Li et al., 2016) (Fig. 13.5).

Hybrid Green Technologies

Recent findings indicate that bioremediation and adsorption are currently effective and cost-efficient remediation methods. Nevertheless, bioremediation has its limitations, including the potential accumulation of harmful metabolites or degradation byproducts. For instance, dichlorodiphenyltrichloroethane (DDT) and its metabolic product, dichlorodiphenyldichloroethylene (DDE), can accumulate over time (ATSDR, 2000). In order to overcome these obstacles integrated technologies or hybrid methods have been proposed. In the context of treating drinking water, bioaugmentation is an integrated approach that can be used to clean PAH-polluted water, given that the bacteria employed are safe for human health or can be effectively removed through standard chlorination processes. Phytoremediation, also referred to as plant-aided bioremediation, is one strategy capable of tackling PAH contamination on a large scale. Plants can absorb, store, and chemically

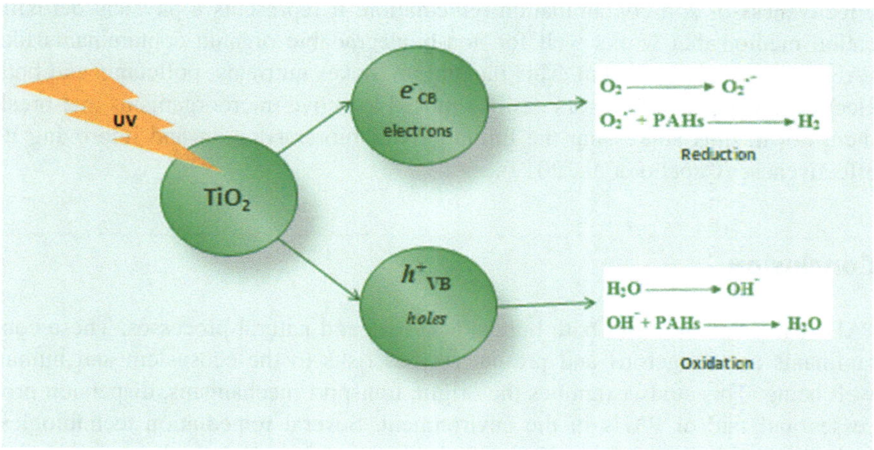

Fig. 13.5 Mechanism of photo catalytic degradation of PAHs using TiO_2

alter pollutants, making research into PAH degradation in water valuable. They can produce enzymes that function as surfactants, which enhance the pollutant's availability in solution (Capello et al., 2019). The electrokinetic approach employs electrical current to decontaminate contaminated soils, and integrating bioremediation with an electrokinetic process enhances the holistic effectiveness of soil contaminant remediation. Nanoremediation is an emerging tool for soil remediation, effectively removing contaminants when combined with other methods such as bioremediation (Ansari et al., 2023). To fully utilize the substantial potential of nanoremediation in the removal of PAHs, research should be done on the creation and application of nanofertilizers (biostimulation and bioaugmentation), nanominerals (biostimulation), or green-synthesized nanooxidizers (for PAH oxidation) (Kuyukina et al., 2021). Furthermore, nanoremediation can be utilized alongside established PAH bioremediation techniques to enhance remediation efficacy and facilitate quicker PAH decomposition in practical applications.

Electrokinetic (EK) Remediation

This method has the ability to eliminate non-biodegradable persistent pollutants like PAHs from soil. It employs electric current to extract contaminants from the affected soil. A small electric current is applied at the contaminated location to create an electric field. The main mechanisms by which the electric field facilitates the movement and mobilization of pollutants towards the electrodes (anode and cathode) during the EK process are electromigration and electroosmosis (Gill et al., 2014). However, the difficulty in dissolving PAHs in water, along with their hydrophobic nature and low dispersal rate, poses challenges to their removal. The use of a combination of surfactants can effectively extract hydrophobic organic compounds from soil with contaminants (Boulakradeche et al., 2015). Integrating the biological restoration process with the electrokinetic process enhances the effectiveness of soil contamination remediation. It represents a physical detoxification method that works well for non-biodegradable organic contaminants like PAHs (Capello et al., 2019). This integration makes nutrients, pollutants and both electron acceptors and donors easily exposed to active microorganisms that break them down, thus addressing the limitations of bioremediation and improving its effectiveness (Capello et al., 2019).

Conclusion

PAHs can originate from both human activities and natural processes. These contaminants are dangerous and present serious risks to the ecosystem and human well being. This study examines the origin, transport mechanisms, dispersion processes and fate of PAHs in the environment. Several remediation technologies, including traditional, advanced, and hybrid or integrated methods, are explored, such as physical, chemical, thermal, electrokinetic, AOPs, phytoremediation, and bioremediation. Bioremediation stands out as a promising method for extracting PAHs. This method evacuates PAHs from the atmosphere by using various kinds of microorganisms, which involves bacteria, fungi, and algae. Microbes like *Serratia marcescens, Bacillus megaterium, Raoultellaornithinolytica,* and *Aeromonashydrophila* decomposed PAHs with above 90% efficiency for acenaphthene and fluorene. Similarly, *Thiobactersubterraneus, Alcoligenes* sp., and *Escherichia coli* are effective isolates used for breaking down anthracene and phenanthrene. Moreover, phytoremediation seems to be among the most economical and environmentally friendly techniques for lowering PAHs in substrates. Recently, there has been a trend toward using integrated or combined treatment technologies to maximize their effectiveness. This approach is a green degradation method that produces no waste, thus preventing further environmental contamination. Zeolites also proved to be very important in a adsorption process. Benz(a)anthracene, pyrene, fluoranthene, phenanthrene, and fluorene were all removed from water with an average removal rate of 98% using organo-zeolite as an adsorbent. Integrated methods represent environmentally friendly green degradation processes with no waste generation that might lead to further pollution. To determine the most appropriate treatment techniques, site characteristics are also critical. The criteria for selection are usually based on site attributes, costs, challenges, advantages, and the type and concentration of pollutants present. Addressing PAH contamination at sites usually presents significant challenges and high costs; therefore, opting for methods that are adaptable, sustainable, eco-friendly, efficient, rapid, and economical is a wise choice.

References

Abd-Elsalam, H.E., Hafez, E.E., Hussain, A.A., Ali, A.G., El-Hanafy, A.A.: Isolation and identification of three-rings polyaromatic hydrocarbons (anthracene and phenanthrene) degrading bacteria. Am. Eurasian J. Agric. Environ. Sci. **5**, 31–38 (2009)

Abdel-Shafy, H.I., Mansour, M.S.: A review on polycyclic aromatic hydrocarbons: source, environmental impact, effect on human health and remediation. Egypt. J. Pet. **25**(1), 107–123 (2016)

Adetunji, A.I., Olaniran, A.O.: Production and potential biotechnological applications of microbial surfactants: An overview. Saudi J. Biol. Sci. **28**(1), 669–679 (2021)

Al-Hawash, A.B., Dragh, M.A., Li, S., Alhujaily, A., Abbood, H.A., Zhang, X., Ma, F.: Principles of microbial degradation of petroleum hydrocarbons in the environment. Egypt. J. Aquat. Res. **44**(2), 71–76 (2018)

Amani, H., Kariminezhad, H.: Study on emulsification of crude oil in water using emulsan biosurfactant for pipeline transportation. Pet. Sci. Technol. **34**(3), 216–222 (2016)

Amor, C., Marchão, L., Lucas, M.S., Peres, J.A.: Application of advanced oxidation processes for the treatment of recalcitrant agro-industrial wastewater: A review. Water. **11**(2), 205 (2019)

Ansari, F., Ahmad, A., Rafatullah, M.: Review on bioremediation technologies of polycyclic aromatic hydrocarbons (PAHs) from soil: Mechanisms and future perspective. Int. Biodeterior. Biodegradation. **179**, 105582 (2023)

Ates, H., Argun, M.E.: Removal of PAHs from leachate using a combination of chemical precipitation and Fenton and ozone oxidation. Water Sci. Technol. **78**(5), 1064–1070 (2018)

ATSDR, T.: ATSDR (Agency for toxic substances and disease registry). Prep. Clement Int. Corp., Contract. **205**, 88–0608 (2000)

Balati, A., Shahbazi, A., Amini, M.M., Hashemi, S.H.: Adsorption of polycyclic aromatic hydrocarbons from wastewater by using silica-based organic–inorganic nanohybrid material. J. Water Reuse Desalination. **5**(1), 50–63 (2015)

Balmer, J.E., Hung, H., Yu, Y., Letcher, R.J., Muir, D.C.: Sources and environmental fate of pyrogenic polycyclic aromatic hydrocarbons (PAHs) in the Arctic. Emerg. Contam. **5**, 128–142 (2019)

Blanchard, M., Teil, M.J., Ollivon, D., Legenti, L., Chevreuil, M.: Polycyclic aromatic hydrocarbons and polychlorobiphenyls in wastewaters and sewage sludges from the Paris area (France). Environ. Res. **95**(2), 184–197 (2004)

Bortey-Sam, N., Ikenaka, Y., Nakayama, S.M., Akoto, O., Yohannes, Y.B., Baidoo, E., Mizukawa, H., Ishizuka, M.: Occurrence, distribution, sources and toxic potential of polycyclic aromatic hydrocarbons (PAHs) in surface soils from the Kumasi Metropolis, Ghana. Sci. Total Environ. **496**, 471–478 (2014)

Boulakradeche, M.O., Akretche, D.E., Cameselle, C., Hamidi, N.: Enhanced electrokinetic remediation of hydrophobic organics contaminated soils by the combination of non-ionic and ionic surfactants. Electrochim. Acta. **174**, 1057–1066 (2015)

Buttress, A.J., Binner, E., Yi, C., Palade, P., Robinson, J.P., Kingman, S.W.: Development and evaluation of a continuous microwave processing system for hydrocarbon removal from solids. Chem. Eng. J. **283**, 215–222 (2016)

Cai, C., Li, J., Wu, D., Wang, X., Tsang, D.C., Li, X., Sun, J., Liu, W.: Spatial distribution, emission source and health risk of parent PAHs and derivatives in surface soils from the Yangtze River Delta, eastern China. Chemosphere. **178**, 301–308 (2017)

Cameotra, S.S.: Biosurfactant—Enhanced bioremediation of hydrophobic pollutants. J. Biotechnol. **136**, S678 (2008)

Cappello, S., Viggi, C.C., Yakimov, M., Rossetti, S., Matturro, B., Molina, L., et al.: Combining electrokinetic transport and bioremediation for enhanced removal of crude oil from contaminated marine sediments: Results of a long-term, mesocosm-scale experiment. Water Res. **157**, 381–395 (2019)

Carolin, C.F., Kumar, P.S., Joshiba, G.J., Madhesh, P., Ramamurthy, R.: Sustainable strategy for the enhancement of hazardous aromatic amine degradation using lipopeptide biosurfactant isolated from Brevibacterium casei. J. Hazard. Mater. **408**, 124943 (2021)

Chang, C.F., Chang, C.Y., Chen, K.H., Tsai, W.T., Shie, J.L., Chen, Y.H.: Adsorption of naphthalene on zeolite from aqueous solution. J. Colloid Interface Sci. **277**(1), 29–34 (2004a)

Chang, C.F., Chang, C.Y., Höll, W., Ulmer, M., Chen, Y.H., Groß, H.J.: Adsorption kinetics of polyethylene glycol from aqueous solution onto activated carbon. Water Res. **38**(10), 2559–2570 (2004b)

Charlesworth, M., Service, M., Gibson, C. E.: PAH contamination of western Irish Sea sediments. Mar. Pollut. Bull.,**44**(12), 1421–1426 (2002)

Chen, C.F., Ju, Y.R., Su, Y.C., Lim, Y.C., Kao, C.M., Chen, C.W., Dong, C.D.: Distribution, sources, and behavior of PAHs in estuarine water systems exemplified by Salt River Taiwan. Mar. Pollut. Bull. **154**, 111029 (2020)

Comninellis, C., Kapalka, A., Malato, S., Parsons, S.A., Poulios, I., Mantzavinos, D.: Advanced oxidation processes for water treatment: advances and trends for R&D. J. Chem. Technol. Biotechnol.: Int. Res. Process., Environ. Clean Technol. **83**(6), 769–776 (2008)

Crini, G., Lichtfouse, E.: Advantages and disadvantages of techniques used for wastewater treatment. Environ. Chem. Lett. **17**, 145–155 (2019)

Cunningham, S.D.: of Soils contaminated with organic pollutants. Adv. Agron. **56**, 55–114 (1996)

Das, N., Chandran, P.: Microbial degradation of petroleum hydrocarbon contaminants: an overview. Biotechnol. Res. Int. **2011**(1), 941810 (2011)

Dutta, P., Tiwari, P., Pandey, L.M.: Oil washing proficiency of biosurfactant produced by isolated Bacillus tequilensis MK 729017 from Assam reservoir soil. J. Pet. Sci. Eng. **195**, 107612 (2020)

Eisler, R.: Polycyclic aromatic hydrocarbon hazards to fish, wildlife, and invertebrates: a synoptic review (No. 11). Fish and Wildlife Service, US Department of the Interior (1987)

Falciglia, P.P., Urso, G., Vagliasindi, F.G.: Microwave heating remediation of soils contaminated with diesel fuel. J. Soils Sediments. **13**, 1396–1407 (2013)

Fletcher, R.E., Ling, S., Slater, B.: Violations of Löwenstein's rule in zeolites. Chem. Sci. **8**(11), 7483–7491 (2017)

Flotron, V., Delteil, C., Bermond, A., Camel, V.: Degradation of polycyclic aromatic hydrocarbons in sewage sludges by Fenton's reagent. In Environmental Chemistry: Green Chemistry and Pollutants in Ecosystems pp. 449–459, Berlin, Heidelberg: Springer Berlin Heidelberg (2005)

Ghosal, D., Ghosh, S., Dutta, T.K., Ahn, Y.: Current state of knowledge in microbial degradation of polycyclic aromatic hydrocarbons (PAHs): a review. Front. Microbiol. **7**, 1369 (2016)

Gill, R.T., Harbottle, M.J., Smith, J.W.N., Thornton, S.F.: Electrokinetic-enhanced bioremediation of organic contaminants: A review of processes and environmental applications. Chemosphere. **107**, 31–42 (2014)

Gupta, H.: Photocatalytic degradation of phenanthrene in the presence of akaganeite nano-rods and the identification of degradation products. RSC Adv. **6**(114), 112721–112727 (2016)

Gupta, H.: Anthracene removal from water onto activated carbon derived from vehicular tyre. Sep. Sci. Technol. **53**(4), 613–625 (2018). https://doi.org/10.1080/01496395.2017.1405038

Gupta, H., Dhiman, S.: Synthetic Nanoparticle-Based Remediation of Soils Contaminated with Polycyclic Aromatic Hydrocarbons. In: Shanker, U., Hussain, C.M., Rani, M. (eds.) Handbook of Green and Sustainable Nanotechnology. Springer, Cham (2023). https://doi.org/10.1007/978-3-031-16101-8_102

Gupta, H., Dhiman, S., Kumar, A., Paliwal, S.K., Khan, G., Singh, H., Mishra, A., Mishra, A.K.: Photo-and bio-degradation of selected persistent organic pollutants. Sep. Purif. Rev. **54**(1), 3–16 (2025). https://doi.org/10.1080/15422119.2024.2332953

Gupta, H., Gupta, B.: Photocatalytic degradation of polycyclic aromatic hydrocarbon benzo[a]pyrene by iron oxides and identification of degradation products. Chemosphere. **138**, 924–931 (2015). https://doi.org/10.1016/j.chemosphere.2014.12.028

Gupta, H., Gupta, B.: Adsorption of polycyclic aromatic hydrocarbons on banana peel activated carbon. Desalin. Water Treat. **57**(20), 9498–9509 (2016). https://doi.org/10.1080/19443994.2015.1029007

Gupta, H., Gupta, B.: Vehicular tire as potential adsorbent for the removal of polycyclic aromatic hydrocarbons. Polycycl. Aromat. Compd. **38**(4), 354–368 (2018). https://doi.org/10.1080/10406638.2016.1217890

Hall, S., Tang, R., Baeyens, J., Dewil, R.: Removing polycyclic aromatic hydrocarbons from water by adsorption on silicagel. Polycycl. Aromat. Compd. **29**(3), 160–183 (2009)

Huling, S.G., Pivetz, B.E.: In-situ chemical oxidation In-situ chemical oxidation.EPA/600th/R-06/072. United States Environmental Protection Agency, Cincinnati, OH (2006)

Hussein, Z.S., Hamido, N., Hegazy, A.K., El-Dessouky, M.A., Mohamed, N.H., Safwat, G.: Phytoremediation of crude petroleum oil pollution: a review. Egypt. J. Bot. **62**(3), 611–640 (2022)

Johnsen, A.R., Karlson, U.: PAH degradation capacity of soil microbial communities—does it depend on PAH exposure? Microb. Ecol. **50**, 488–495 (2005)

Johnson, P., Trybala, A., Starov, V., Pinfield, V.J.: Effect of synthetic surfactants on the environment and the potential for substitution by biosurfactants. Adv. Colloid Interf. Sci. **288**, 102340 (2021)

Karyab, H., Yunesian, M., Nasseri, S., Mahvi, A.H., Ahmadkhaniha, R., Rastkari, N., Nabizadeh, R.: Polycyclic aromatic hydrocarbons in drinking water of Tehran, Iran. J. Environ. Health Sci. Eng. **11**, 1–7 (2013)

Kaur, A., Gupta, H., Dhiman, S.: Biogenic Metallic Nanoparticles: Synthesis and Applications Using Medicinal Plants. In: Shanker, U., Hussain, C.M., Rani, M. (eds.) Handbook of Green and Sustainable Nanotechnology. Springer, Cham (2023). https://doi.org/10.1007/978-3-030-69023-6_101-1

Kaya, E.M.Ö., Özcan, A.S., Gök, Ö., Özcan, A.: Adsorption kinetics and isotherm parameters of naphthalene onto natural-and chemically modified bentonite from aqueous solutions. Adsorption. **19**, 879–888 (2013)

Kim, K.H., Jahan, S.A., Kabir, E., Brown, R.J.: A review of airborne polycyclic aromatic hydrocarbons (PAHs) and their human health effects. Environ. Int. **60**, 71–80 (2013)

Krishnan, N., Velramar, B., Velu, R.K.: Investigation of antifungal activity of surfactin against mycotoxigenic phytopathogenic fungus Fusarium moniliforme and its impact in seed germination and mycotoxicosis. Pestic. Biochem. Physiol. **155**, 101–107 (2019)

Kuppusamy, S., Thavamani, P., Megharaj, M., Naidu, R.: Biodegradation of polycyclic aromatic hydrocarbons (PAHs) by novel bacterial consortia tolerant to diverse physical settings–assessments in liquid-and slurry-phase systems. Int. Biodeterior. Biodegradation **108**, 149–157 (2016)

Kuyukina, M.S., Ivshina, I.B., Krivoruchko, A.V., Peshkur, T.A., Cunningham, C.J.: Improvement of jet fuel contaminated water treatment in a fluidized-bed bioreactor by introducing nickel nanoparticles. Int. Biodeterior. Biodegradation. **164**, 105308 (2021)

Kvesitadze, E., Sadunishvili, T., Kvesitadze, G.: Mechanisms of organic contaminants uptake and degradation in plants. World Acad. Sci. Eng. Technol. **55**(6), 458–468 (2009)

Lammel, G.: Polycyclic aromatic compounds in the atmosphere–a review identifying research needs. Polycycl. Aromat. Compd. **35**(2–4), 316–329 (2015)

Langenbach, T.: Persistence and bioaccumulation of persistent organic pollutants (POPs). Appl. Bioremediation-Act. Passiv. Approaches. **10**, 56418 (2013)

Lee, S.M., Tiwari, D.: Organo and inorgano-organo-modified clays in the remediation of aqueous solutions: An overview. Appl. Clay Sci. **59**, 84–102 (2012)

Lemić, J., Tomašević-Čanović, M., Adamović, M., Kovačević, D., Milićević, S.: Competitive adsorption of polycyclic aromatic hydrocarbons on organo-zeolites. Microporous Mesoporous Mater. **105**(3), 317–323 (2007)

Li, J., Zhao, L., Qin, L., Tian, X., Wang, A., Zhou, Y., Chen, Y.: Removal of refractory organics in nanofiltration concentrates of municipal solid waste leachate treatment plants by combined Fenton oxidative-coagulation with photo–Fenton processes. Chemosphere. **146**, 442–449 (2016)

Lian, L., Yao, B., Hou, S., Fang, J., Yan, S., Song, W.: Kinetic study of hydroxyl and sulfate radical-mediated oxidation of pharmaceuticals in wastewater effluents. Environ. Sci. Technol. **51**(5), 2954–2962 (2017)

Liu, H., Liang, S., Gao, J., Ngo, H.H., Guo, W., Guo, Z., et al.: Enhancement of Cr (VI) removal by modifying activated carbon developed from Zizaniacaduciflora with tartaric acid during phosphoric acid activation. Chem. Eng. J. **246**, 168–174 (2014)

Mani, D., Kumar, C.: Biotechnological advances in bioremediation of heavy metals contaminated ecosystems: an overview with special reference to phytoremediation. Int. J. Environ. Sci. Technol. **11**, 843–872 (2014)

Menzie, C.A., Potocki, B.B., Santodonato, J.: Exposure to carcinogenic PAHs in the environment. Environ. Sci. Technol. **26**(7), 1278–1284 (1992)

Merhaby, D., Rabodonirina, S., Net, S., Ouddane, B., Halwani, J.: Overview of sediments pollution by PAHs and PCBs in mediterranean basin: Transport, fate, occurrence, and distribution. Mar. Pollut. Bull. **149**, 110646 (2019)

Miller, J.S., Olejnik, D.: Photolysis of polycyclic aromatic hydrocarbons in water. Water Res. **35**(1), 233–243 (2001)

Moghaddam, M.B., Hemmati, R., Homaei, A., Vianello, F., Shareghi, B.: Novel cold-adapted lipase from Psychrobacter sp. C18 immobilized on reduced graphene oxide-cellulose nanomatrix with high activity and stability. J. Mol. Liq. **397**, 124210 (2024)

Mohapatra, B., Phale, P.S.: Microbial degradation of naphthalene and substituted naphthalenes: metabolic diversity and genomic insight for bioremediation. Front. Bioeng. Biotechnol. **9**, 602445 (2021)

Mojiri, A., Zhou, J.L., Ohashi, A., Ozaki, N., Kindaichi, T.: Comprehensive review of polycyclic aromatic hydrocarbons in water sources, their effects and treatments. Sci. Total Environ. **696**, 133971 (2019)

Moscoso, F., Ferreira, L., Deive, F.J., Morán, P., Sanromán, M.A.: Viability of phenanthrene biodegradation by an isolated bacterial consortium: optimization and scale-up. Bioprocess Biosyst. Eng. **36**, 133–141 (2013)

Motteran, F., Magalhães, V.E.M., Candido, M.A., da Silva Freitas, D.M., Fernandes, B.S., Gavazza, S.: Taxonomic identification of microbiota from enrichment and isolation of phenanthrene-degrading bacteria in aerobic reactors. Water Air Soil Pollut. **235**(6), 341 (2024)

Na, G., Ye, J., Li, R., Gao, H., Jin, S., Gao, Y., Hou, C., Huang, J.: Fate of polycyclic aromatic hydrocarbons in the Pacific sector of the Arctic Ocean based on a level III fugacity environmental multimedia model. Mar. Pollut. Bull. **166**, 112195 (2021)

Nayak, N.S., Purohit, M.S., Tipre, D.R., Dave, S.R.: Biosurfactant production and engine oil degradation by marine halotolerant Bacillus licheniformis LRK1. Biocatal. Agric. Biotechnol. **29**, 101808 (2020)

Ndimele, P.E., Saba, A.O., Ojo, D.O., Ndimele, C.C., Anetekhai, M.A., Erondu, E.S.: Remediation of crude oil spillage. In: The political ecology of oil and gas activities in the Nigerian aquatic ecosystem, pp. 369–384. Academic (2018)

Nethaji, S., Sivasamy, A., Mandal, A.B.: Adsorption isotherms, kinetics and mechanism for the adsorption of cationic and anionic dyes onto carbonaceous particles prepared from Juglansregia shell biomass. Int. J. Environ. Sci. Technol. **10**, 231–242 (2013)

Novoselov, K.S., Colombo, L., Gellert, P.R., Schwab, M.G., Kim, K.A.J.N.: A roadmap for graphene. Nature. **490**(7419), 192–200 (2012)

Ohura, T., Amagai, T., Fusaya, M., Matsushita, H.: Polycyclic aromatic hydrocarbons in indoor and outdoor environments and factors affecting their concentrations. Environ. Sci. Technol. **38**(1), 77–83 (2004)

Okiel, K., El-Sayed, M., El-Kady, M.Y.: Treatment of oil–water emulsions by adsorption onto activated carbon, bentonite and deposited carbon. Egypt. J. Pet. **20**(2), 9–15 (2011)

Oluwasanu, A.A., Oluwaseun, F.A.P.O.H.U.N.D.A., Teslim, J.A., Isaiah, T.T., Olalekan, I.A., Chris, O.A.: Scientific applications and prospects of nanomaterials: A multidisciplinary review. Afr. J. Biotechnol. **18**(30), 946–961 (2019)

Ossai, I.C., Ahmed, A., Hassan, A., Hamid, F.S.: Remediation of soil and water contaminated with petroleum hydrocarbon: A review. Environ. Technol. Innov. **17**, 100526 (2020)

Parrish, Z.D., Banks, M.K., Schwab, A.P.: Effectiveness of phytoremediation as a secondary treatment for polycyclic aromatic hydrocarbons (PAHs) in composted soil. Int. J. Phytoremediation. **6**(2), 119–137 (2004)

Peng, C., Wang, M., Chen, W.: Spatial Analysis of PAHs in Soils along an Urban–Suburban–Rural Gradient: Scale effect, distribution patterns, diffusion and influencing factors. Sci. Rep. **6**(1), 37185 (2016)

Punnapayak, H., Prasongsuk, S., Messner, K., Danmek, K., Lotrakul, P.: Polycyclic aromatic hydrocarbons (PAHs) degradation by laccase from a tropical white rot fungus Ganodermalucidum. Afr. J. Biotechnol. **8**(21) (2009)

Ravindra, K., Sokhi, R., Van Grieken, R.: Atmospheric polycyclic aromatic hydrocarbons: source attribution, emission factors and regulation. Atmos. Environ. **42**(13), 2895–2921 (2008)

Rinawati Hidayat, D., Supriyanto, R., Permana, D.F., Yunita: Adsorption of polycyclic aromatic hydrocarbons using low-cost activated carbon derived from rice husk. In: Journal of Physics: Conference Series, vol. 1338, No. 1, p. 012005. IOP Publishing (2019), October)

Robinson, J.P., Kingman, S.W., Lester, E.H., Yi, C.: Microwave remediation of hydrocarbon-contaminated soils–Scale-up using batch reactors. Sep. Purif. Technol. **96**, 12–19 (2012)

Rohrbacher, F., St-Arnaud, M.: Root exudation: the ecological driver of hydrocarbon rhizoremediation. Agronomy. **6**(1), 19 (2016)

Rubio-Clemente, A., Torres-Palma, R.A., Peñuela, G.A.: Removal of polycyclic aromatic hydrocarbons in aqueous environment by chemical treatments: a review. Sci. Total Environ. **478**, 201–225 (2014)

Russo, V., Hmoudah, M., Broccoli, F., Iesce, M.R., Jung, O.S., Di Serio, M.: Applications of metal organic frameworks in wastewater treatment: a review on adsorption and photodegradation. Front. Chem. Eng. **2**, 581487 (2020)

Schwarzenbach, R.P., Gschwend, P.M., Imboden, D.M.: Environmental organic chemistry. John Wiley & Sons (2016)

Shanker, U., Jassal, V., Rani, M.: Green synthesis of iron hexacyanoferrate nanoparticles: potential candidate for the degradation of toxic PAHs. J. Environ. Chem. Eng. **5**(4), 4108–4120 (2017)

Sharma, S., Pandey, L.M.: Production of biosurfactant by Bacillus subtilis RSL-2 isolated from sludge and biosurfactant mediated degradation of oil. Bioresour. Technol. **307**, 123261 (2020)

Sivasamy, S., Rajangam, S., Kanagasabai, T., Bisht, D., Prabhakaran, R., Dhandayuthapani, S.: Biocatalytic potential of pseudomonas species in the degradation of polycyclic aromatic hydrocarbons. J. Basic Microbiol. **65**(2), e2400448 (2025)

Sun, R., Zhou, Q., Wei, S.: Cadmium accumulation in relation to organic acids and nonprotein thiols in leaves of the recently found Cd hyperaccumulator Rorippaglobosa and the Cd-accumulating plant Rorippaislandica. J. Plant Growth Regul. **30**, 83–91 (2011)

Tian, K., Bao, H., Liu, X., Wu, F.: Accumulation and distribution of PAHs in winter wheat from areas influenced by coal combustion in China. Environ. Sci. Pollut. Res. **25**, 23780–23790 (2018)

Uzoigwe, C., Burgess, J.G., Ennis, C.J., Rahman, P.K.: Bioemulsifiers are not biosurfactants and require different screening approaches. Front. Microbiol. **6**, 245 (2015)

Vane, C.H., Kim, A.W., Beriro, D.J., Cave, M.R., Knights, K., Moss-Hayes, V., Nathanail, P.C.: Polycyclic aromatic hydrocarbons (PAH) and polychlorinated biphenyls (PCB) in urban soils of Greater London, UK. Appl. Geochem. **51**, 303–314 (2014)

Vidonish, J.E., Zygourakis, K., Masiello, C.A., Sabadell, G., Alvarez, P.J.: Thermal treatment of hydrocarbon-impacted soils: a review of technology innovation for sustainable remediation. Engineering. **2**(4), 426–437 (2016)

Wang, L., Yang, Z., Niu, J., Wang, J.: Characterization, ecological risk assessment and source diagnostics of polycyclic aromatic hydrocarbons in water column of the Yellow River Delta, one of the most plenty biodiversity zones in the world. J. Hazard. Mater. **169**(1–3), 460–465 (2009)

Witt, G.: Polycyclic aromatic hydrocarbons in water and sediment of the Baltic Sea. Mar. Pollut. Bull. **31**(4–12), 237–248 (1995)

Wright, D.A., Welbourn, P.: Environmental toxicology, vol. No. 11. Cambridge University Press (2002)

Wróbel, M., Rybak, J., Rogula-Kozłowska, W., Dwojak, F.: Ocena toksyczności wymywanego miejskiego pyłu drogowego z wykorzystaniem testu Microtox–badanie porównawcze dla Dolnego i Górnego Śąska. Zeszyty Naukowe SGSP/Szkoła Główna Służby Pożarniczej. **81**, 63–76 (2022)

Wu, B., Zhang, Y., Zhang, X.X., Cheng, S.P.: Health risk assessment of polycyclic aromatic hydrocarbons in the source water and drinking water of China: quantitative analysis based on published monitoring data. Sci. Total Environ. **410**, 112–118 (2011)

Wu, M., Dick, W. A., Li, W., Wang, X., Yang, Q., Wang, T., Chen, L.: Bioaugmentation and biostimulation of hydrocarbon degradation and the microbial community in a petroleum-contaminated soil. Int. Biodeterioration Biodegrad.**107**, 158–164 (2016)

Zhang, M.H., Dong, H., Zhao, L., Wang, D.X., Meng, D.: A review on Fenton process for organic wastewater treatment based on optimization perspective. Sci. Total Environ. **670**, 110–121 (2019)

Zhang, P., Chen, Y.: Polycyclic aromatic hydrocarbons contamination in surface soil of China: A review. Sci. Total Environ. **605**, 1011–1020 (2017)

Zhou, N., Guo, H., Liu, Q., Zhang, Z., Sun, J., Wang, H.: Bioaugmentation of polycyclic aromatic hydrocarbon (PAH)-contaminated soil with the nitrate-reducing bacterium PheN7 under anaerobic condition. J. Hazard. Mater. **439**, 129643 (2022)

New Advancements and Multidisciplinary Approaches in the Remediation of Polycyclic Aromatic Hydrocarbons

14

Meghna Rawat, Anjali, Nancy Sharma, and Pawan Kumar

Abstract

Numerous anthropogenic activities caused by rapid industrialization and urbanization release polycyclic aromatic hydrocarbons (PAHs) as toxic environmental pollutants. The PAHs come from unburnt combustion in residential and commercial properties, garbage burning and activities involving fossil fuels. The PAHs pollutants are widely distributed and equally present in the atmosphere, aquatic and terrestrial ecosystems. PAH pollutants directly and indirectly negatively impact human health, well-being, and biodiversity worldwide. The removal of PAHs is a global challenge because of their inherent characteristics, which make them long-lasting pollutants with a broad range of biological and chemical toxicity. The global challenges have been caused by adverse PAH effects, which prompted the remediation strategies that reduce the ecosystem's long-term impact, confirm safety and decrease risks to humans and ecosystems. The sustainable treatment strategy remains challenging to implement for the remediation of PAH. More recently, various integrated advanced and multidisciplinary remediation techniques have been successfully used, including chemical, biological and physical approaches such as nanomaterial-based technologies, microbial bioremediation, electrokinetic and adsorption photocatalysis. Combining these techniques shows a comprehensive and long-term strategy to deal with the problems caused by PAH contamination. The current study primarily examines innovative and multidisciplinary approaches that integrate various

M. Rawat · Anjali · P. Kumar (✉)
Department of Chemistry, Hemvati Nandan Bahuguna Garhwal University, Srinagar, Uttarakhand, India
e-mail: pawankumar@hnbgu.com

N. Sharma
Institut Parisien de Chimie Moleculaire, UMR CNRS-8232, Sorbonne University, Paris, France

H. Gupta et al. (eds.), *Ubiquitous Polycyclic Aromatic Hydrocarbon Contamination*, Emerging Contaminants and Associated Treatment Technologies,
https://doi.org/10.1007/978-3-032-11764-9_14

techniques, which are becoming increasingly popular due to their promise of more effective remediation of PAH in ecological systems.

Keywords

Polycyclic aromatic hydrocarbons (PAHs) • Physiochemical techniques • Hybrid remediation technologies and degradation • Remediation strategies

Introduction

Industrialization has undoubtedly increased the amount of pollutants liberated into the atmosphere, causing the general quality of the aquatic, air, and land to deteriorate, along with adversely impacting human life (Kicińska and Dmytrowski, 2023; Venkatraman et al. 2024). The most common organic pollutants that are present practically everywhere are poly aromatic hydrocarbons (PAHs). These substances seriously affect human and animal health, including mutagenicity, carcinogenicity, immunological suppression, teratogenicity, and metabolic inflammation (Almouallem et al., 2023). They can readily integrate into mammals' digestive tracts because of their lipid-soluble characteristics. To effectively remove these xenobiotics from the atmosphere, it is crucial to prioritize remediation techniques.

Since petroleum is the chief energy resource supporting numerous aspects of everyday life and the global economy, the world's population growth has increased industrial activity (Tornero and Hanke, 2016) (Yim et al., 2007). A number of ecosystems matrices such as air, aquatic, land, sediments and flora & fauna, are getting industrial PAHs contaminants, coinciding with the growth of the oil, gas, and petrochemical industries (Ghosal et al., 2016). Numerous petrochemical industries have been releasing hydrocarbons, grease, lubricants and fuel oils into the environment as refinery wastes. These pollutants are distinguished by their potential to affect and their persistence in the environment. Their natural resistance to oxidation and reduction and chemical stability determine effectiveness of the deposition of the substrate (Chen et al., 2020). PAHs are a set of organic pollutants with various structures and biological toxicity that are widely distributed and regularly found in the environment (Eldos et al., 2020). Sixteen PAHs as priority environmental contaminants due to their widespread presence and toxicity, have been identified by the United States Environmental Protection Agency (EPA) (Dai et al., 2022).

Usually, PAHs are categorized into two groups: Low Molecular Weight (LMW-PAHs) comprising each with less than four rings; acenaphthene, fluorene, phenanthrene, naphthalene, and anthracene, and High Molecular Weight (HMW-PAHs) with four or more rings, including pyrene, chrysene, fluoranthene, benzoanthracene, benzopyrene, benzofluoranthene, dibenzoanthracene, benzoperylene, and indenopyrene (Wilson et al., 2023).

According to (Balmer et al., 2019) (Bright et al., 2023), PAHs are composed of two or more aromatic rings of carbon and hydrogen grouped in clusters, angular or linear, as shown in Fig. 14.1. Depending on where they come from, PAHs can be divided into three classes: biogenic, pyrogenic, and petrogenic. Products of petroleum produce petrogenic PAHs, combustion processes produce pyrogenic PAHs, and biological sources produce biogenic PAHs (Merhaby et al., 2019).

PAHs, which come from manmade and natural sources, primarily build up in road dust and soil (Sun et al., 2022). PAHs in soil sorption and retention, which are fed by soil surface and organic matter—particularly clay and meso- and macropores—are crucial. Human caused sources of PAHs include inadequate waste burning in residential and industrial activities involving fossil fuels (Deka et al., 2020; Hussain et al., 2019; Kicińska and Dmytrowski, 2023).

Due to their entry into marine ecosystems, accumulation, and creation of exposure pathways for living species, the extensive dispersion of environmental PAHs, causes health hazards (Deka et al., 2023) (Banerjee et al., 2022). Remediation efforts that reduce the impact on the ecosystem in the long term, provide a safety

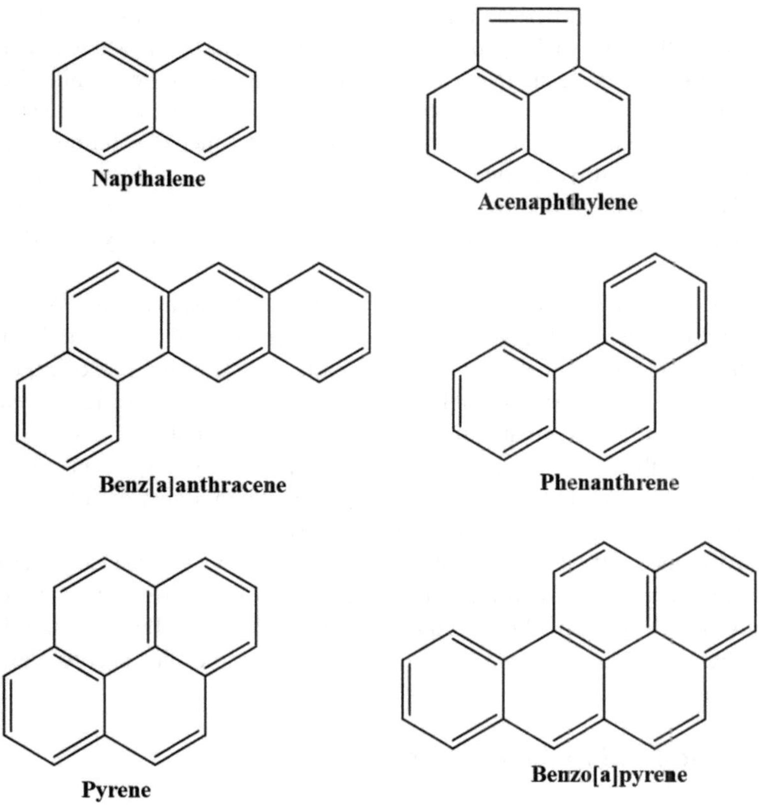

Fig. 14.1 Chemical structure of some common PAHs

guarantee, and lower exposure hazards to the public and flora and fauna have been prompted by the detrimental PAHs effects, which have environmental contamination (Udom et al., 2023; Zhang et al., 2023). They are primarily used as intermediates in the pharmacological, agronomic, photographic, thermoset plastic and lubricating material (Abdel-Shafy and Mansour, 2016).

The study combines several techniques, summarizing conventional and new advanced multidisciplinary strategies based on nanomaterials, genetic plant engineering, and electrochemical sensor detection technology for removing PAH-contaminated soils, water, and air, focusing on hybrid techniques. This study proposes advanced remediation strategies to bridge the gap between environmental sustainability and remediation efficacy. Optimising these technologies, guaranteeing profitability, and bridging knowledge gaps remain challenges. Potential long-term ecological sustainability and innovative remedial expertise may be the main topics of future research.

Characteristics of PAHs

PAHs are characterized by low vapour pressure and low aqueous solubility, along with high boiling and melting points, which make them solid, whereas oxidation and reduction resistance increase as molecular weight increases, while the latter two features tend to decrease (Masih et al., 2012). As the rings increase, polycyclic aromatic hydrocarbons become less soluble in aqueous solutions and more soluble in organic solvents due to their high lipophilicity (Masih et al., 2010). PAHs show various properties, including heat and corrosion resistance, conductivity, physiological action, light sensitivity and emittivity (Akyüz and Çabuk, 2010).

Utmost harmful pollutants usually found in the atmosphere are PAHs (Eldos et al., 2020). These substances are toxic, extremely stable, hydrophobic and lipophilic, which allows them to build up in biota and raises the risk of exposure to humans and other species. As PAHs are toxic, genotoxic, mutagenic, and carcinogenic, they are strictly regulated by law in industrialized and progressed countries. It is significant that their structure has been shown to exhibit high levels of environmental persistency, toxicity and chemical stability due to higher molecular weights and a large number of aromatic rings in PAHs, (Botsou and Hatzianestis, 2012).

Each isomer of PAHs has a remarkably distinct UV absorbance spectrum as a result of every ring structure having a different UV spectrum. It is particularly helpful when identifying PAHs. Additionally, most PAHs are fluorescent, releasing distinct light wavelengths upon excitation (when the molecules absorb light).

Physiochemical Properties of PAH

The properties of individual PAHs are determined by their mobility and distribution in the atmosphere (Mackay and Callcott, 1998). The weathering processes that

PAHs go through after being released into the environment include physiochemical and biological reactions via metabolic and microbial pathways (Latimer and Zheng, 2003; Ossai et al., 2020). The two abiotic elements significantly impacting weathering processes are the pollutant's structure and the environment. The adsorption of resistant pollutants into soil, volatilisation into the atmosphere, and dissolving in water are all examples of weathering processes (Schwarzenbach et al., 2016).

Additional environmental factors and processes, including soil aggregation, moisture levels, volatilization, transformation, leaching and biodegradation by organisms, will all play a significant part in PAHs bioavailability in soil. As a result, PAHs with higher bioavailability and mobility have more detrimental effects overall. Because of their structure, non-polar PAHs have more long-term negative health impacts than polar PAHs.

Route of Transport, Exposure and Toxic Effect of PAHs

In the atmosphere, the PAHs are transported through terrestrial and aquatic environments in the ecosystem due to several processes. The physical characteristics such as chemical stability, hydrophobicity and lipophobicity, allow PAHs to travel great distances through the atmosphere, where they are absorbed earlier by sediments and suspended particles to build up in the biota (Ossai et al., 2020). Dry deposition, wet deposition and the exchange of vapour between air and water are the three primary atmospheric processes that lead to the buildup of PAHs in the land or ocean. Several crucial factors determine the dominant process, such as temperature, wind speed, the pollutant's physicochemical characteristics, salinity, density, precipitation rates, and matrix composition (Latimer and Zheng, 2003) (Golomb et al., 2001) (Hong et al., 2016).

There are two distinct phases of PAHs in the atmosphere after they are released: vapour and solid phase, where they are absorbed by particulate matter (Ravindra et al., 2008; Zhang and Tao, 2009) (Wang et al., 2013). Low-vapour-pressure hydrophobic organic compounds, like PAHs, are more easily sorbed to airborne particles than higher-vapour-pressure chemicals, like benzene. The individual concentrations of PAHs vary in the vapour and other sorbet phases (Kameda, 2011) due to the variation in vapour pressures of distinct PAH compounds.

Since PAHs' non-polar structures do not allow them to dissolve in water, they become slightly immobile once absorbed into sediments. As a result, the low molecular weight of PAHs can dissolve and enter the pore water, making them bioavailable. These PAHs tend to adsorb onto organic colloids present in pore water, which can increase their concentrations beyond their aqueous solubility. Consequently, PAHs are easily transferred across the sediment's pore spaces. This means that the movement and bioavailability in sediments of PAHs can be enhanced by their adsorption to colloids (Dong et al., 2012).

PAHs' toxicity to the ecosystem is affected by photo-oxidation and metabolism, and is more harmful when exposed to UV light. The concentration of PAHs at modest to higher levels is more poisonous to birds and aquatic life. Except in cases where the soil is highly contaminated, PAHs in soil are not likely to harm terrestrial invertebrates. Tumours, immunity, development, and reproduction adversely affect these organisms. Ingestion, skin contact, and inhalation are ways mammals can absorb PAHs (Beyer et al., 2010; Dong et al., 2012; Veltman et al., 2011). It is expected that the PAHs concentration in aquatic species will be significantly higher than the surrounding environment from were collected. However, the PAHs metabolism is adequate to obstruct biomagnification (Inomata et al., 2012). The relative toxicity, as well as the concentration and route of exposure of the PAHs, significantly influence how they affect human health. Symptoms like nausea, vomiting, diarrhoea, eye irritation, irritation and inflammation of the skin have been linked to occupational exposures to pollutant mixtures containing excessive concentrations of PAHs (Unwin et al., 2006). Long-lasting PAHs contact can have several harmful health effects and abnormalities. Skin inflammation and redness can occur due to frequent contact with the skin. Naphthalene, a specific polycyclic aromatic hydrocarbon (PAH), can cause red blood cells to break down and be destroyed if inhaled or ingested in large amounts. The adverse effects of human exposure to PAHs largely depend on how the exposure occurs (Diggs et al., 2011). Most PAHs require metabolism to react with DNA, leading to genotoxic damage by producing diol epoxides. The process of carcinogenicity is significantly influenced by genotoxicity, which may also be present in certain types of developmental toxicity (Gamboa et al., 2008; Lewtas, 2007).

Advancement and Multidisciplinary Approaches for PAH Remediation

Remediation techniques involve reducing environmental pollutants through processes of deterioration or conversion in the soil, water, and air until acceptable levels are reached. Effective and economical treatment of industrial effluents is necessary to develop organizations and a sustainable environment. Several approaches to eliminate PAHs from polluted soils, streams and the environment are imposed to lessen the risk to public wellbeing and the environment (Gupta and Gupta, 2015, 2016; Gupta et al., 2017; Gupta 2018; Kumar and Gupta, 2020). These approaches come into three categories: biological, chemical, or physical. Physical options include electrokinetic therapies and thermal remediation (Y.D. Alazaiva et al., 2022). Ozone, photocatalysis, and oxidation using Fenton's reagent are examples of chemical techniques (Samuel et al., 2023).

The biological approaches use phytoremediation, fungi, and bacteria to cause microbial breakdown (Thacharodi et al., 2023). Despite their effectiveness, physical and chemical approaches might result in secondary pollutants and are often expensive, energy-intensive, or chemical-intensive. Because biological technologies are environmentally friendly and may transform harmful contaminants into non-hazardous molecules, they are becoming increasingly popular as a safe and economical solution to these restrictions. The exported research publications are

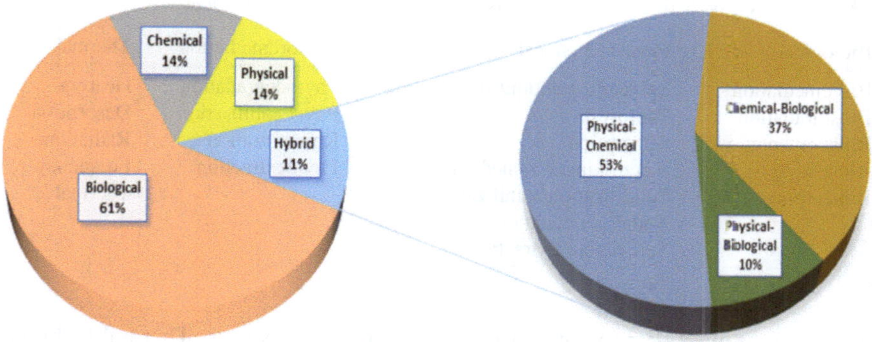

Fig. 14.2 Bibliography analysis of the number of publications on various PAH remediation techniques (Reprinted from original article by Gupta, et al., (2024) after taking permission from Elsevier)

followed by the categorization based on remediation approaches, considering significance screening criteria, operational quality and field application to ensure direct linkage to degradation of PAHs in soil. The current research, for treating PAH-polluted soil, mainly focuses on biological approaches (61%), with hybrid approaches obtaining less attention (11%) of the hybrid approaches, as shown in Fig. 14.2, physical-chemical procedures account for the majority (53%), with chemical-biological approaches following closely behind (37%). The least studied approaches are physical-biological methods (10%).

However, these approaches have several drawbacks, including complicated operating procedures and expensive maintenance and investment expenses (Lamichhane et al., 2016). A sustainable approach considers either an appropriate single method or an integrated approach, integrating several ways to solve the shortcomings of biological, chemical, and physical therapy. These factors are considered to efficiently remediate the PAH-contaminated matrix while reducing the negative environmental consequences.

This study provides a detailed description of several existing and new advancements and multidisciplinary approaches in remediation techniques, including thermal, chemical, electrokinetic, nano adsorption, microbiological treatment, etc. Table 14.1 shows the different methods for the removal of PAHs from the air, soil and water.

Thermal Technologies

Thermal technology is a suitable technique for the relatively high efficiency of PAH compound degradation (Zhang et al., 2017). This method mainly uses an external heat source to remove volatile organic contaminants. Accordingly, the two primary drawbacks of this technology are the potential for air contamination due to

Table 14.1 Methods for removal of PAHs existing in the air, soil and water

Biological	Physicochemical	Phytoremediation	Thermal
Bioremediation Biostimulation Bioaugmentation Bioventing Bioslurping	Solvent extraction/soil washing Photolysis Adsorption Electrokinetic remediation Solidification/stabilization Natural attenuation Soil vapour extraction	Phytodegradation Phytostabilisation Rhizofiltration Phytoextraction	Heating Desorption Radiation Frequency Thermal

organic compound desorption and its high operating expenses. Thermal technologies quick and dependable removal standards make them promising techniques for extracting hydrocarbons from contaminated soils. However, excessive heat and energy use can harm soil quality (Vidonish et al., 2016). The factors that influence the proficiency of thermal technologies include temperature, soil makeup, contaminant amount and treatment time, composition, overall density of soil, contaminant vapour pressure, moisture contents and distribution of particle size (Falciglia et al., 2011). By using thermal technology, the elimination rate of PAHs was improved by 95% (Bates et al., 2008). Thermal technologies effectively removed PAHs from soil under various conditions in a different study. They discovered that reducing heating temperature and residence time led to a reduced elimination rate (Jeon).

Incineration and In-Situ Thermal Desorption (ISTD)

Polyaromatic hydrocarbons are extracted from wood wastes, refinery by-products, coal tar, and contaminated soils using ISTD, also called conductive heating (Gan et al., 2009). In contrast, incineration uses a high-temperature range of 900–1200 °C to effectively destroy PAHs in contaminated plumes. Similar to incineration, ISTD uses heating to separate PAHs, but it is considered reasonably safe as it uses a carrier gas or vacuum for off-site disposal.

Electrokinetic Remediation (EK)

An in-situ method for cleaning up locations tainted with low-permeable PAHs is called EK remediation (Gitipour et al., 2018). This technique affects the PAH-polluted soil due to the usage of low direct electric current between the cathode and anode electrodes. (Lim et al., 2016). This technique is a promising way to remove PAHs, either when used alone or in conjunction with other methods like Fenton or bioremediation. The main challenge to employing traditional EK remediation technology is that PAHs from subterranean environments are hydrophobic and desorb slowly (Pazos et al., 2010). Increasing PAH desorption from the contaminated sites and establishing a conducive atmosphere for the transportation of the electrode chambers are essential for enhancing the EK process.

Furthermore, the increased electroosmotic flow would promote the removal of an organic compound when electrolytes existed in the processing fluid and the pH of electrode chambers was controlled (Alcántara et al., 2008). Since it can be treated in situ, this method has low operational cost with fast response time. The kaolin contaminated with a combination of various PAHs (pyrene, benzanthracene and fluoranthene,) was conducted by EK remediation, According to Alcántara (Alcántara et al., 2010), this technique has an immense potential for remediating polluted soil by 40%.

Solidification/Stabilization (S/S)

S/S is a based on physical change rectification technique that uses the reagent-soil reaction to decrease the mobility of the contaminant. This process involves a chemical reaction between waste from mechanical processes and solidifying reagents. Therefore, this technique can be applied to physical and chemical methods to reduce environmental impacts. While the contaminants form a physical link with the solid matrix during solidification processes, stabilization involves chemical reactions that transform the pollutant into less mobile species. A new technique for PAH-contaminated soil was created by (Ma et al., 2018), with the ideal dosage being 0.02% of sulfonated oil. They discovered that, in comparison to the original sample, the remediated soil had a 25% leaching content of PAHs.

Supercritical Fluid Extraction (SCFE)

SCFE is a suitable and efficient method that uses heated fluid and then applies pressure above the critical point. The basis for SCFE's operation is the high separation process and gaseous-liquid characteristics. PAHs can be removed from solid matrices' binding sites by using organic solvents, surfactants, or plain water. The solid is then rinsed into the extraction fluid (Gan et al., 2009). Furthermore, PAHs require additional biochemical remedies to finish the decontamination process as they desorb in the liquid phase. For instance, a study found that applying a solvent mixture of ethanol and pentanol in three steps resulted in over 95% PAH extraction (Khodadoust et al., 2000). The study also found that the high temperature and pressure caused low viscosity and diffusivity. Prior research indicates that pollutants like PAHs can be quickly removed from various settings. Furthermore, SCFE is regarded as a practical technique for forecasting the bioavailability of organic contaminants based on variables such as molecular weights and concentrations, application versatility, and water solubility (Riding et al., 2013).

Soil Vapour Extraction (SVE)

SVE eliminates dangerous contaminants like vapours from the soil (Speight, 2017). This process involves moving vapours toward extraction wells by applying a negative pressure gradient and progressively volatile pollutants are eliminated from the subsurface. The gathered vapours are then released into the atmosphere. SVE applies to organic compounds with low and high volatility. This method is inappropriate for low air-permeability soils as the induced air prefers to flow down pathways with less resistance. The main drawback of SVE is the downstream contamination and sweep gas treatment for additional downstream treatment (Lim et al., 2016). Numerous scientists have applied the SVE method to remove PAHs from various contaminated locations. For example, SVE effectively removed BTEX from soil contaminated by wet gasoline (Simpanen et al., 2018). SVE was used in one more study (Albergaria et al., 2012) to remediate sandy soil polluted artificially with six different types of PAHs. According to (Soares et al., 2010), contaminants with lower vapour pressures had a 92% removal efficiency rate, but longer remediation times were needed.

Chemical Oxidation

Researchers are interested in treating organic pollutants in wastewater by chemical oxidation. This approach's primary benefits are its efficacy and speed of remediation compared to other methods that require excavation treatments both on and off-site, and its cost is heavily influenced by the chemical agents required. Additionally, this approach can be considered an appropriate remediation option for soil-borne PAH contamination, regardless of the toxicity of polluted sources.

Chemical oxidants are primarily used to introduce gaseous substances or liquids into the ground; this process is straightforward with minimal operating expenses. Additionally, this method can break down oil pollutants without creating harmful compounds; instead, it can produce more biodegradable ones. A few factors affects the PAH oxidation rate include temperature, oxidizing agent used, molecular weight, and the structure of compound (Abdel-Shafy and Mansour, 2016). It is advised to check for ongoing natural reduction processes at the oil-contaminated sites before the use of chemical oxidants (Agarwal and Liu, 2015). Ozone and Fenton techniques are the two popular oxidant methods used to remove PAH from contaminated areas (Wang and Wang, 2021).

Fenton

For PAHs, Fenton oxidation is a practical remediation technique. Numerous experiments have been performed on the use of Fenton oxidation to remove PAHs from polluted soils (Agarwal and Liu, 2015; Gupta and Dhiman, 2023). The primary obstacle for Fenton treatment alone is the requirement for a high oxidant load;

however, Fenton limitations are addressed by integrating the Fenton procedure with other treatment techniques. The main constraints, meanwhile, are the lack of PAHs, the deterioration of the quality of soil, as well as the medium's and the microbial community's acidification. Significant reactions taking place in Fenton's reaction are:

In the presence of ferrous ion (Fe^{2+}), breakdown of hydrogen peroxide (H_2O_2) into hydroxyl radicals ($OH^{\bullet}$):

$$Fe^{2+} + H_2O_2 \rightarrow Fe^{3+} + OH^{\bullet} + OH^{-}$$

Breakdown of organic molecules (RH or R) by the previously generated unstable hydroxyl radicals by either hydroxyl addition or hydrogen elimination.

$$RH + OH^{\bullet} \rightarrow R^{\bullet} + H_2O$$

$$R + OH^{\bullet} \rightarrow R^{\bullet}OH$$

By using hydroxyl radicals produced by hydrogen peroxide catalysis, this method exhibits encouraging potential for rapidly degrading PAHs in contaminated areas. Several chemical reagents, like cyclodextrins, co-solvent, vegetable oil, and surfactant, may make PAHs more accessible in contaminated soils for Fenton oxidation to occur (Yang et al., 2017). For example, co-solvents make PAHs more soluble in soils between the aqueous phase and soil components by reducing the surface tension and increasing the accessibility of PAHs (Usman et al., 2016).

Ozone (O_3)

One of the most effective chemical oxidation technology by ozone for hydrocarbon degradation, as it can be swiftly transported to contaminated areas. According to Lim et al. (2016), natural metal oxides like Al_2O_3, MnO, and geothite are present on the soil surface and help in catalysing the breakdown of ozone into hydroxyl radicals. This technique reduces the time and treatment cost and is used to remove various forms of phenanthrene from contaminated soils by swiftly oxidizing organic molecules (O'Mahony et al., 2006).

Microbial Enzyme-Mediated Bioremediation

The microbial enzyme-mediated bioremediation process uses extracted enzymes to remove PAHs from fungi, bacteria, and other living things. The enzymatic process is incredibly effective and specific due to the ability to catalyse reactions across a broad range of pH and temperature alongwith increased reaction rates. According to the subsections above, the enzymes that cause PAH oxidation are dehydrogenases, oxygenases, phenoloxidases, peroxidase of manganese and lignin. Due to their decreased substrate specificity, the oxidative fungal enzymes are more efficient. The unique manganese peroxidase gene isolated from Cerrena unicolour BBP6 was used to clone Pichia pastoris. This organism exhibited a variety of dye-decolourising abilities and achieved 80% and 91% degradation activity for fluorene and phenanthrene within 24 hours, with the highest recorded recombinant enzyme expression being 154.5 Units/L. The only drawback of this approach is the cost of producing, isolating, and refining enzymes (Kuppusamy et al., 2017).

Phytoremediation

Using plants, phytoremediation detoxifies soil by allowing pollutants to enter the roots causing bioaccumulation or biodegradation in above-ground areas. Phytoremediation provides a financial benefit (4–40 $/ton), compared to chemical treatment (100–500 $/ton), electrokinetic (20–200 $/ton), landfilling (100–500 $/ton), and vitrification (75–425 $/ton), which is significantly less when maintenance, recycling, and transportation costs are excluded (Adeoye et al., 2022). Phytoremediation uses rhizodegradation, phytoextraction, and plant-fungal/bacteria interactions to eliminate PAHs while increasing plant growth and tolerance (Fall et al., 2021). Natural non-ionic surfactants, biochar-amended surfactants, chelating agents, and bacterial strains all facilitated the breakdown of PAHs.

Bacteria-Mediated PAHs Degradation

Utilizing naturally occurring scavenging bacteria, biodegradation is an economical and environmentally acceptable method that cleans the environment by consuming organic contaminants. As a biotechnological approach, they have been used for remediating environmental pollutants. Microbes can use PAHs as solitary carbon sources or through co-metabolism (Chattopadhyay and Ghosh Sachan, 2023). They undergo biodegradation to become simpler metabolites, which are then mineralized into H_2O and CO_2 or CH_4 (anaerobic). With species including Enterobacter, Pseudomonas, Arthrobacter, Rhodococcus, Bacillus, Mycobacterium, Vibrio, Stenotrophomaonas, Corynebacterium, Sphingomonas, and Gordonia, among others, multiple investigations emphasize how bacteria breakdown PAHs.

Fungi-Mediated PAHs Degradation

Strategies for fungi-mediated remediation, including cytochrome P450-dependent and ligninolytic-dependent mechanisms, are essential for combating environmental contamination. Ligninolytic fungi are necessary to break down PAHs because they produce extracellular and intracellular enzymes. Applications for this process include pulp and paper, bioremediation, and biofuel generation. It is essential for the cycling of nutrients.

The breakdown and metabolism of PAHs depend on intracellular enzymes, which are only produced by non-ligninolytic fungi. Compared to bacterial cleanup, fungal cleanup exhibits more promise, especially in mycelial environments, because its hyphal projections and quick substrate expansion provide better accessibility to pollutants (Omoni et al., 2023). Although decomposition by microbes has many benefits, including being less costly, environmentally friendly, versatile, and sustainable, it also has some disadvantages, such as a slow degradation rate, the possibility of incomplete degradation, and the requirement for sporadic maintenance and monitoring.

Emerging Trends for PAHs Remediation

Some limitations of physicochemical techniques include monitoring issues and expensive expenses. On the other hand, combining biological techniques with nanotechnology provides a sustainable and effective way to mitigate PAH contamination in the surroundings. The new advanced and multidisciplinary approaches are based on the different characteristics of nanomaterials that contribute to their effectiveness in PAH bioremediation. This study also covered the methods of nano-bioremediation, including enzyme-enhanced nanomaterials and nanomaterial-assisted microbial degradation, microbiological activity, microbial cells immobilized by nanomaterials, electron transfer assisted by nanomaterials, and even specific environmentally friendly methods of PAH remediation, such as biogenic nanomaterial for PAHs.

Recently, there has been an increase in interest in using tools based on nanomaterials as an emerging technology to remediate different contaminants in polluted locations. Based on this, a diverse range of nanoparticulate materials have been released onto the nanosorbant sector market, providing improved capacities for purifying PAH-polluted water, finally rendering it appropriate for reclamation. These materials are distinguished from their bulk counterparts by their special properties, which include a markedly higher surface-to-volume ratio and improved properties of magnetic and enzymatic (Gupta et al., 2024).

Nano Enzymes for PAHs Remediation

By cross-linking real enzymes, inorganic nanoparticles identified as "nanoenzymes," with size ranging from 0 to 100 nm, may replicate the enzyme properties in a redox process. These particles also show catalytic activity in addition to the unique properties of a nanomaterial.

Nanomaterials' straightforward production process, excellent firmness, budget friendly, together with potent catalytic activity are benefits of nanozymes over natural enzymes. It has been discovered that natural enzymes are a very effective way to degrade PAHs. However, many factors, including these enzymes' low stability, lack of adaptability, and expensive manufacturing, make their widespread use challenging. Nanozymes emerge as a viable remedy to address these problems. For a commercial application, nano-enzymes are favoured over microorganisms because they offer greater operational conditions and flexibility. Unlike bacteria, these nanozymes do not need a particular environment or conditions to grow. The medium of nutrients is unnecessary for nano-enzymes to significantly lower operating costs over microorganisms. Similarly, in comparison with microbial treatments, the mass transfer constraints of enzyme treatment are substantially lower. Above all, it is more suitable for a widespread application for PAHs because of their low costs, easy scaling up, and lack of harmful byproduct creation from nano-enzymes. Nanozymes with peroxidase-like catalytic activity are widely employed for the catalytic breakdown of xenobiotic contaminants. When breaking down organic pollutants in water bodies, nanozymes have demonstrated the best outcomes. (Ramya et al., 2023).

Nanomaterial-Aided Microbial PAHs Degradation Strategies

A suitable technique for PAHs in contaminated soil is provided by nano-bioremediation technology, which benefits the environment. Improving soil and water quality promotes vegetation growth, microbiological survival, and agricultural productivity (Rajput et al., 2023). In Fig. 14.3, several studies have shown to clarify the possible PAHs degradation mechanisms, which vary depending on the microbial strain and the structure of the nanoparticle.

Microbial Cell Immobilized NPs

Biofunctionalized NPs with a higher affinity for PAHs are produced by adding microorganisms, enzymes, or biostabilizers to their surfaces. After PAHs are first adsorbed onto the surface of NPs, π–π interactions take place, which propel disintegration. With successful degradation, NPs can be capably recycled for further PAHs remediation processes. The best removal rates of grease and total petroleum hydrocarbons at 83.9% and 85% respectively by Exogenous bacterial consortia (Enterobacter cloacae and P. otitidis) bound with Fe_3O_4 NPs were achieved in just over four hours (Bestawy et al., 2020).

Microbial Enzyme-Immobilized NPs

Bacteria may effectively break down PAHs outside of cells, although using NPs might increase this effectiveness. The kind and intensity of interactions between NPs and microorganisms are probably essential variables affecting the improved biodegradation of PAHs. Although the enzymes used as biocatalysts in bioremediation are very selective and efficient, instability and short lives limit their practical use (Shahi et al., 2021). Enzyme stability and efficiency are decreased by oxidation, restricting their use as affordable substitutes for synthetic catalysts. By

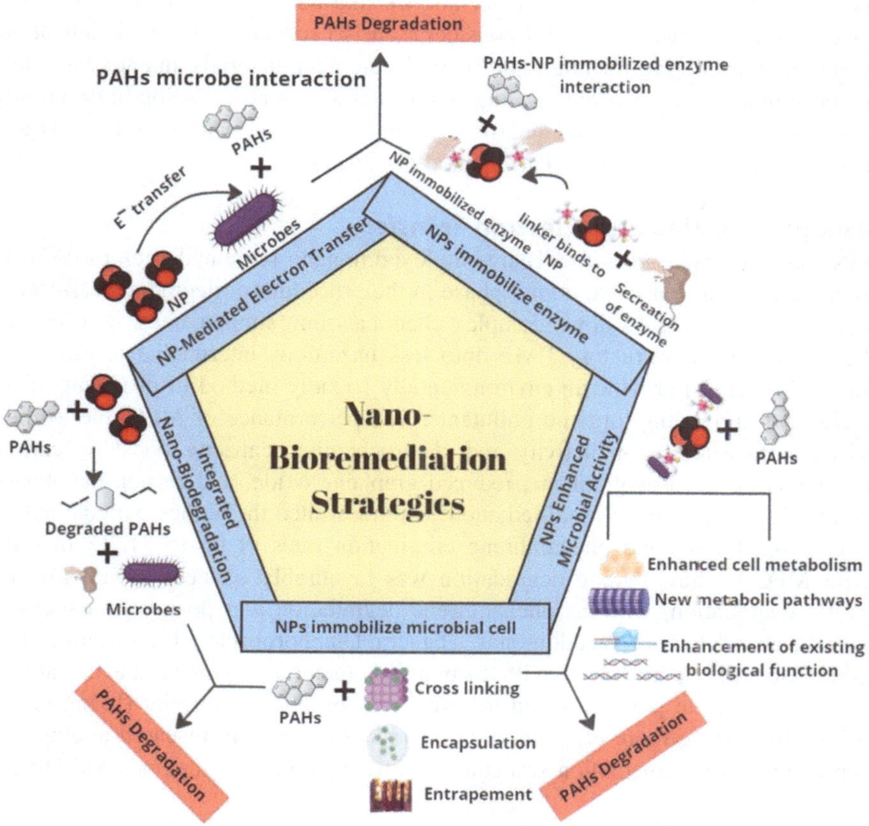

Fig. 14.3 Nano-bioremediation strategies for enhanced PAH degradation (Reprinted from original article by Gupta et al., (2024) after taking permission from Elsevier)

enhancing the stability, lifespan, and reusability of enzymes, NPs provide a remedy for this problem. Enzymes can be readily separated using a magnetic field when bound to these magnetic nanoparticles, increasing their activity significantly from hours to weeks.

Integrated Nano-biodegradation

Bioremediation has limits, particularly in severely contaminated areas containing HMW-PAHs, due to its high expense, limited bioavailability, and time dependence. To overcome these obstacles, integrated techniques, including chemical oxidation, solvent extraction, and bioremediation, can be combined in physical-chemical or physical-biological systems to achieve more successful remediation. The use of nano-sized materials to alter the physicochemical characteristics of pollutants has greatly improved integrated remediation techniques thanks to recent developments in nanotechnology. This approach works with biological processes like biodegradation and chemical techniques like adding surfactants to enhance bioavailability. However, careful consideration is necessary for functionalization, which includes opting for non-toxic biomolecule nanomaterials and steering clear of hazardous reducing agents during nanomaterial synthesis. Adopting naturally and ecofriendly generated nanoparticles (NPs) is a viable way to solve this and lessen environmental impact (Parthipan et al., 2021).

Nanoparticle-Mediated Electron Transfer

NPs and zero-valent iron (nZVI) are employed in nano-bioremediation of PAHs to improve electron transport, a vital phase in the procedure of degradation. Because electron transfer breaks down complex chemicals into simpler ones, it speeds up the conversion of refractory PAHs into less hazardous intermediates. Microbial fuel cells (MFCs) provide an environmentally friendly method of producing electricity and eliminating organic pollutants. The performance of MFC due to their raised surface area, conductivity and electrochemical capacity boost by carbon nanomaterials such as graphene, reduced graphene oxide, and carbon nanotubes. Reduced graphene oxide-modified anodes demonstrated the highest voltage output (30.60–48.61 mV) and phenanthrene elimination rates of up to 71.2% in sediment MFC studies. Pyrene degradation was favourably connected with loss on ignitio degradation, whereas phenanthrene degradation was positively associated with the abundances of Pseudomonas, Thauera, Diaphorobacter, Tumebacillus, and Lysobacter. MFCs enhanced with carbon nanomaterials demonstrate the ability to remove organic pollutants and to effectively produce electricity (Liang et al., 2020). To achieve more efficient PAH degradation, ongoing research attempts to improve nano-bioremediation strategies by investigating new methods and NPs.

NPs Enhanced Microbial Activity

By supporting an environment favourable to microorganisms, the ability to boost microbial activity has been demonstrated by NPs while PAHs are being biodegraded. Significantly accelerating PAHs breakdown, have been shown to improve cell adherence by magnetic iron nanoparticles, boost nutritional availability and ease metabolic processes. In contaminated soil, bioaugmentation using graphene oxide-immobilized bacterial pellets (GOBP) enhances the breakdown of PAHs. High-efficiency bacteria that break down Compared to conventional pellets, Paracoccus aminovorans embedded in graphene oxide alginate-Luria-Bertani composites exhibit an 18.51% increase in PAH elimination (62.86% after 35 days). High-molecular-weight PAHs are the target of GOBP, which also enriches promising native degraders like Pseudarthrobacter and Arthrobacter and increases the number of embedded bacteria. This novel strategy provides a cutting-edge bioaugmentation-based organic pollutant remediation method in difficult soil settings (Ren et al., 2022). Graphene oxide promotes the growth of bacteria and gene expression related to the two-component system, i.e. microbial mobility (flagellar assembly) and microbial chemotaxis and the phosphotransferase system in soil, which helps microorganisms break down PAHs. Graphene oxide speeds up the degradation of PAHs by increasing the number of degrading microorganisms during brief exposure. On the other hand, deterioration of saturation could result from prolonged exposure. In addition to highlighting the significance of microbial mobility and associated genetic processes, this study shows how graphene oxide affects the breakdown of microbial PAHs, offering insights into efficient environmental remediation.

Genetic Plant Engineering PAHs Remediation

The purpose of genetic plant engineering is to improve particular characteristics in microorganisms or plants, increasing their capacity to remove pollutants and adapt to complex environments. This involves causing gene expression to enhance plant phytoextraction and increase the capacity of plant-associated bacteria, including rhizospheric and endophytic bacteria, to degrade petroleum hydrocarbons. Superorganisms with enhanced capacities to combat pollutants in areas that are co-contaminated, such as heavy metals and PAHs, have been created due to recent developments. According to Ashkanani et al. (2024), modified microbial strains are more effective in bioremediation and enhance the bioavailability of pollutants.

Still, there are concerns with applying genetic engineering to co-contaminated soil treatment, including unexpected gene interactions, microbial diversity risks, and uneven lab and field results, which necessitate further research to guarantee safe and effective use.

Electrochemical Sensor Detection Technology

Considering the significant risks and difficulties presented by AHCs class pollutants and the concentrations of μg and ng in the aquatic environment, a quick, delicate, trace detection method is needed. Unlike traditional analytical methods like chromatography titration and chromatography-mass spectrometry, which are complex and time-consuming, electrochemical sensors are quick and straightforward. Moving toward the direction of electrochemical identification of PAHs, researchers have conducted in-depth investigations and analyzed information from numerous electromechanical sensors.

A variety of electrochemical analytical techniques were used, including linear sweep voltammetry (LSV), square wave voltammetry (SWV), amperometry, cyclic voltammetry (CV), differential pulse voltammetry (DPV), and stripping voltammetry (SV). Researchers have looked into developing rapid, sensitive, field-usable, and convenient electrochemical sensors using a variety of materials. Its primary components are metal, metal oxide nanoparticles, graphene, carbon nanomaterials (CNs), carbon nanotubes (CNTs), and nanoparticles (NPs) (Qu et al., 2024).

Conclusion

Many technologies for clearing polluted sites with polycyclic aromatic hydrocarbons (PAHs) are still in the prototype stage, while others have been implemented in practice. However, these technologies are not sufficient to fully address the PAHs problem. This is because each contaminated site is unique, and managing these sites requires careful consideration of various relevant factors. Additionally, the process must take into account the constraints imposed by remedy policies, existing financial resources and public approval.

Several traditional physicochemical techniques, such as thermal, chemical, and electrokinetic treatment, have been reported as relevant findings, but have some limitations, including monitoring issues and high expenses. Compared to the physical, chemical and biological approaches, the hybrid mode of remediation has a greater choice as the higher number of publications, with physical-chemical approaches accounting for the majority (53%) and biological-chemical methods are closely followed by chemical-biological methods (37%) and biological-physical methods are the least researched (10%). The United Arab Emirates has a 100% successful removal rate of the phenolic compounds, such as PAHs, from the water by the electrochemical techniques.

Effective methods for improving the removal of polycyclic aromatic hydrocarbons (PAHs) include bioremediation, microbial degradation assisted by nanomaterials, the immobilization of microbial cells on nanomaterials and the enhancement of microbial activity through enzymes and electron transfer. These methods demonstrate the nanomaterials has versatile role in boosting PAHs removal processes, thereby improving the efficiency of bioremediation techniques. It is advisable to pursue treatment options that are adaptive, sustainable, eco-friendly, efficient, quick, and cost-effective, as addressing PAH-contaminated sites can often be a challenging and expensive undertaking.

Acknowledgement The authors would like to thank Hemvati Nandan Bahuguna Garhwal University, Srinagar Garhwal (Uttarakhand), and Institut Parisien de Chimie Moleculaire, UMR CNRS-8232, Sorbonne University, Paris -75006, France.

References

Abdel-Shafy, H.I., Mansour, M.S.: A review on polycyclic aromatic hydrocarbons: source, environmental impact, effect on human health and remediation. Egypt. J. Pet. **25**(1), 107–123 (2016)

Adeoye, A.O., Adebayo, I.A., Afodun, A.M., Ajijolakewu, K.A.: Benefits and limitations of phytoremediation: heavy metal remediation review. In: Phytoremediation, pp. 227–238. Academic (2022)

Agarwal, A., Liu, Y.: Remediation technologies for oil-contaminated sediments. Mar. Pollut. Bull. **101**(2), 483–490 (2015)

Akyüz, M., Çabuk, H.: Gas–particle partitioning and seasonal variation of polycyclic aromatic hydrocarbons in the atmosphere of Zonguldak, Turkey. Sci. Total. Environ. **408**(22), 5550–5558 (2010)

Alazaiva, M.Y., Albahnasawi, A., Copty, N.K., Ali, G.A., Bashir, M.J., Al Maskari, T., Amr, S.S.A., Abujazar, M.S.S., Nassani, D.E.: Thermal based remediation technologies for soil and groundwater: a review. Desalin. Water Treat. **259**, 206–220 (2022)

Albergaria, J.T., Maria da Conceição, M., Delerue-Matos, C.: Remediation of sandy soils contaminated with hydrocarbons and halogenated hydrocarbons by soil vapour extraction. J. Environ. Manag. **104**, 195–201 (2012)

Alcántara, M.T., Gómez, J., Pazos, M., Sanromán, M.A.: Electrokinetic remediation of PAH mixtures from kaolin. J. Hazard. Mater. **179**(1–3), 1156–1160 (2010)

Alcántara, T., Pazos, M., Cameselle, C., Sanromán, M.A.: Electrochemical remediation of phenanthrene from contaminated kaolinite. Environ. Geochem. Health. **30**(2), 89–94 (2008)

Almouallem, W., Michel, J., Dorge, S., Joyeux, C., Trouvé, G., Le Nouen, D.: A comparative study of the sorption of O-PAHs and PAHs onto soils to understand their transport in soils and groundwater. J. Environ. Sci. **124**, 61–75 (2023)

Ashkanani, Z., Mohtar, R., Al-Enezi, S., Smith, P.K., Calabrese, S., Ma, X., Abdullah, M.: AI-assisted systematic review on remediation of contaminated soils with PAHs and heavy metals. J. Hazard. Mater. **468**, 133813 (2024)

Balmer, J.E., Hung, H., Yu, Y., Letcher, R.J., Muir, D.C.: Sources and environmental fate of pyrogenic polycyclic aromatic hydrocarbons (PAHs) in the Arctic. Emerg. Contam. **5**, 128–142 (2019)

Banerjee, S., Maiti, T.K., Roy, R.N.: Enzyme producing insect gut microbes: an unexplored biotechnological aspect. Crit. Rev. Biotechnol. **42**(3), 384–402 (2022)

Bates, M., Bruno, P., Caputi, M., Caselli, M., de Gennaro, G., Tutino, M.: Analysis of polycyclic aromatic hydrocarbons (PAHs) in airborne particles by direct sample introduction thermal desorption GC/MS. Atmos. Environ. **42**(24), 6144–6151 (2008)

Bestawy, E.E., El-Shatby, B.F., Eltaweil, A.S.: Integration between bacterial consortium and magnetite (Fe3O4) nanoparticles for the treatment of oily industrial wastewater. World J. Microbiol. Biotechnol. **36**(9), 141 (2020)

Beyer, J., Jonsson, G., Porte, C., Krahn, M.M., Ariese, F.: Analytical methods for determining metabolites of polycyclic aromatic hydrocarbon (PAH) pollutants in fish bile: a review. Environ. Toxicol. Pharmacol. **30**(3), 224–244 (2010)

Botsou, F., Hatzianestis, I.: Polycyclic aromatic hydrocarbons (PAHs) in marine sediments of the Hellenic coastal zone, eastern Mediterranean: levels, sources and toxicological significance. J. Soils Sediments. **12**(2), 265–277 (2012)

Bright, A., Li, F., Movahed, M., Shi, H., Xue, B.: Chronic exposure to low-molecular-weight polycyclic aromatic hydrocarbons promotes lipid accumulation and metabolic inflammation. Biomolecules. **13**(2), 196 (2023)

Chattopadhyay, S., Ghosh Sachan, S.: Biodegradation of polycyclic aromatic hydrocarbons and the impact of various genes for their enhanced degradation. Lett. Appl. Microbiol. **76**(2), ovac062 (2023)

Chen, C.F., Ju, Y.R., Su, Y.C., Lim, Y.C., Kao, C.M., Chen, C.W., Dong, C.D.: Distribution, sources, and behavior of PAHs in estuarine water systems exemplified by Salt River, Taiwan. Mar. Pollut. Bull. **154**, 111029 (2020)

Dai, C., Han, Y., Duan, Y., Lai, X., Fu, R., Liu, S., Leong, K.H., Tu, Y., Zhou, L.: Review on the contamination and remediation of polycyclic aromatic hydrocarbons (PAHs) in coastal soil and sediments. Environ. Res. **205**, 112423 (2022)

Deka, J., Baul, N., Bharali, P., Sarma, K.P., Hoque, R.R.: Soil PAHs against varied land use of a small city (Tezpur) of middle Brahmaputra Valley: seasonality, sources, and long-range transport. Environ. Monit. Assess. **192**(6), 357 (2020)

Deka, J., Sarma, K.P., Gupta, N., Ahmed, M.S., Mazumder, M.J., Hoque, R.R.: Polycyclic aromatic hydrocarbons in groundwater of oil-rich regions of upper Brahmaputra Valley, India: linkages of colloidal transport. Arab. J. Geosci. **16**(1), 66 (2023)

Diggs, D.L., Huderson, A.C., Harris, K.L., Myers, J.N., Banks, L.D., Rekhadevi, P.V., Niaz, M.S., Ramesh, A.: Polycyclic aromatic hydrocarbons and digestive tract cancers: a perspective. J. Environ. Sci. Health C. **29**(4), 324–357 (2011)

Dong, C.D., Chen, C.F., Chen, C.W.: Determination of polycyclic aromatic hydrocarbons in industrial harbor sediments by GC-MS. Int. J. Environ. Res. Public Health. **9**(6), 2175–2188 (2012)
Eldos, H.I., Ashfaq, M.Y., Al-Ghouti, M.A.: Rapid assessment of the impact of microwave heating coupled with UV-C radiation on the degradation of PAHs from contaminated soil using FTIR and multivariate analysis. Arab. J. Chem. **13**(11), 7609–7625 (2020)
Falciglia, P.P., Giustra, M.G., Vagliasindi, F.G.A.: Low-temperature thermal desorption of diesel polluted soil: influence of temperature and soil texture on contaminant removal kinetics. J. Hazard. Mater. **185**(1), 392–400 (2011)
Fall, A.F., Nakabonge, G., Ssekandi, J., Founoune-Mboup, H., Apori, S.O., Ndiaye, A., Badji, A. and Ngom, K., 2021. Roles of Arbuscular Mycorrhizal Fungi on Soil Fertility and its Importance in Phytoremediation of Polluted Soil.
Gamboa, R.T., Gamboa, A.R., Bravo, A.H., Ostrosky, W.P.: Genotoxicity in child populations exposed to polycyclic aromatic hydrocarbons (PAHs) in the air from Tabasco, Mexico. Int. J. Environ. Res. Public Health. **5**(5), 349–355 (2008)
Gan, S., Lau, E.V., Ng, H.K.: Remediation of soils contaminated with polycyclic aromatic hydrocarbons (PAHs). J. Hazard. Mater. **172**(2–3), 532–549 (2009)
Ghosal, D., Ghosh, S., Dutta, T.K., Ahn, Y.: Current state of knowledge in microbial degradation of polycyclic aromatic hydrocarbons (PAHs): a review. Front. Microbiol. **7**, 1369 (2016)
Gitipour, S., Sorial, G.A., Ghasemi, S., Bazyari, M.: Treatment technologies for PAH-contaminated sites: a critical review. Environ. Monit. Assess. **190**(9), 546 (2018)
Golomb, D., Barry, E., Fisher, G., Varanusupakul, P., Koleda, M., Rooney, T.: Atmospheric deposition of polycyclic aromatic hydrocarbons near New England coastal waters. Atmos. Environ. **35**(36), 6245–6258 (2001)
Gupta, H.: Anthracene removal from water onto activated carbon derived from vehicular Tyre. Sep. Sci. Technol. **53**(4), 613–625 (2018)
Gupta, H., Dhiman, S.: Synthetic nanoparticle-based remediation of soils contaminated with polycyclic aromatic hydrocarbons. In: Shanker, U., Hussain, C.M., Rani, M. (eds.) Handbook of Green and Sustainable Nanotechnology. Springer, Cham.doi:10.1007/978-3-030-69023-6_102–1 (2023)
Gupta, H., Gupta, B.: Photocatalytic degradation of polycyclic aromatic hydrocarbon benzo [a] pyrene by iron oxides and identification of degradation products. Chemosphere. **138**, 924–931 (2015)
Gupta, H., Gupta, B.: Adsorption of polycyclic aromatic hydrocarbons on banana peel activated carbon. Desalin. Water Treat. **57**(20), 9498–9509 (2016)
Gupta, H., Kumar, R., Park, H.-S., Jeon, B.-H.: Photocatalytic efficiency of iron oxide nanoparticles for the degradation of priority pollutant anthracene. Geosyst. Eng. **20**(1), 21–27 (2017)
Gupta, N., Koley, A., Banerjee, S., Ghosh, A., Hoque, R.R., Balachandran, S.: Nanomaterial-mediated strategies for enhancing bioremediation of polycyclic aromatic hydrocarbons: a systematic review. Hybrid advances. **7**, 100315 (2024)
Hong, W.J., Jia, H., Li, Y.F., Sun, Y., Liu, X., Wang, L.: Polycyclic aromatic hydrocarbons (PAHs) and alkylated PAHs in the coastal seawater, surface sediment and oyster from Dalian, Northeast China. Ecotoxicol. Environ. Saf. **128**, 11–20 (2016)
Hussain, K., Hoque, R.R., Balachandran, S., Medhi, S., Idris, M.G., Rahman, M., Hussain, F.L.: Monitoring and risk analysis of PAHs in the environment. In: Handbook of Environmental Materials Management, pp. 973–1007. Springer, Cham (2019)
Inomata, Y., Kajino, M., Sato, K., Ohara, T., Kurokawa, J.I., Ueda, H., Tang, N., Hayakawa, K., Ohizumi, T., Akimoto, H.: Emission and atmospheric transport of particulate PAHs in Northeast Asia. Environ. Sci. Technol. **46**(9), 4941–4949 (2012)
Kameda, T.: Atmospheric chemistry of polycyclic aromatic hydrocarbons and related compounds. J. Health Sci. **57**(6), 504–511 (2011)
Khodadoust, A.P., Bagchi, R., Suidan, M.T., Brenner, R.C., Sellers, N.G.: Removal of PAHs from highly contaminated soils found at prior manufactured gas operations. J. Hazard. Mater. **80**(1–3), 159–174 (2000)

Kicińska, A., Dmytrowski, P.: Anthropogenic impact on soils of protected areas—example of PAHs. Sci. Rep. **13**(1), 1524 (2023)
Kumar, A., Gupta, H.: Activated carbon from sawdust for naphthalene removal from contaminated water. Environ. Technol. Innovation. **20**, 101080 (2020)
Kuppusamy, S., Thavamani, P., Venkateswarlu, K., Lee, Y.B., Naidu, R., Megharaj, M.: Remediation approaches for polycyclic aromatic hydrocarbons (PAHs) contaminated soils: technological constraints, emerging trends and future directions. Chemosphere. **168**, 944–968 (2017)
Lamichhane, S., Krishna, K.B., Sarukkalige, R.: Polycyclic aromatic hydrocarbons (PAHs) removal by sorption: a review. Chemosphere. **148**, 336–353 (2016)
Latimer, J.S. and Zheng, J., 2003. The sources, transport, and fate of PAHs in the marine environment. PAHs: an ecotoxicological perspective, pp. 7–33
Lewtas, J.: Air pollution combustion emissions: characterization of causative agents and mechanisms associated with cancer, reproductive, and cardiovascular effects. Mutat. Res./Rev. Mutat. Res. **636**(1–3), 95–133 (2007)
Liang, Y., Zhai, H., Liu, B., Ji, M., Li, J.: Carbon nanomaterial-modified graphite felt as an anode enhanced the power production and polycyclic aromatic hydrocarbon removal in sediment microbial fuel cells. Sci. Total Environ. **713**, 136483 (2020)
Lim, M.W., Von Lau, E., Poh, P.E.: A comprehensive guide of remediation technologies for oil contaminated soil—present works and future directions. Mar. Pollut. Bull. **109**(1), 14–45 (2016)
Ma, F., Wu, B., Zhang, Q., Cui, D., Liu, Q., Peng, C., Li, F., Gu, Q.: An innovative method for the solidification/stabilization of PAHs-contaminated soil using sulfonated oil. J. Hazard. Mater. **344**, 742–748 (2018)
Mackay, D. and Callcott, D., 1998. Partitioning and physical chemical properties of PAHs. In PAHs and Related Compounds: Chemistry pp. 325–345. Berlin, Heidelberg: Springer Berlin Heidelberg
Masih, J., Masih, A., Kulshrestha, A., Singhvi, R., Taneja, A.: Characteristics of polycyclic aromatic hydrocarbons in indoor and outdoor atmosphere in the north central part of India. J. Hazard. Mater. **177**(1–3), 190–198 (2010)
Masih, J., Singhvi, R., Kumar, K., Jain, V.K., Taneja, A.: Seasonal variation and sources of polycyclic aromatic hydrocarbons (PAHs) in indoor and outdoor air in a semi arid tract of northern India. Aerosol Air Qual. Res. **12**(4), 515–525 (2012)
Merhaby, D., Rabodonirina, S., Net, S., Ouddane, B., Halwani, J.: Overview of sediments pollution by PAHs and PCBs in mediterranean basin: transport, fate, occurrence, and distribution. Mar. Pollut. Bull. **149**, 110646 (2019)
O'Mahony, M.M., Dobson, A.D., Barnes, J.D., Singleton, I.: The use of ozone in the remediation of polycyclic aromatic hydrocarbon contaminated soil. Chemosphere. **63**(2), 307–314 (2006)
Omoni, V.T., Bankole, P.O., Nwosu, T.F.X., Tennison-Omovoh, C.A., Ojo, A.S., Semple, K.T.: Impact of different nitrogen amendments on the biodegradation of 14C-phenanthrene by endophytic fungal strains in liquid culture. Microbiol. Res. **266**, 127223 (2023)
Ossai, I.C., Ahmed, A., Hassan, A., Hamid, F.S.: Remediation of soil and water contaminated with petroleum hydrocarbon: a review. Environ. Technol. Innovation. **17**, 100526 (2020)
Parthipan, P., Prakash, C., Perumal, D., Elumalai, P., Rajasekar, A., Cheng, L.: Biogenic nanoparticles and strategies of nano-bioremediation to remediate PAHs for a sustainable future. In: Biotechnology for Sustainable Environment, pp. 317–337. Singapore, Springer Singapore (2021)
Pazos, M., Rosales, E., Alcántara, T., Gómez, J., Sanromán, M.A.: Decontamination of soils containing PAHs by electroremediation: a review. J. Hazard. Mater. **177**(1–3), 1–11 (2010)
Qu, G., Liu, G., Zhao, C., Yuan, Z., Yang, Y., Xiang, K.: Detection and treatment of mono and polycyclic aromatic hydrocarbon pollutants in aqueous environments based on electrochemical technology: recent advances. Environ. Sci. Pollut. Res. **31**(16), 23334–23352 (2024)
Rajput, V.D., Kumari, S., Minkina, T., Sushkova, S., Mandzhieva, S.: Nano-enhanced microbial remediation of PAHs contaminated soil. Air, Soil Water Res. **16**, 11786221231170099 (2023)
Ramya, R.K., Theraka, K., Ramprasadh, S.V., Bharathi, S.V., Srinivasan, S., Jacob, S., Kuila, A.: Pragmatic treatment strategies for polyaromatic hydrocarbon remediation and anti-biofouling

from surfaces using nano-enzymes: a review. Appl. Biochem. Biotechnol. **195**(9), 5479–5496 (2023)

Ravindra, K., Sokhi, R., Van Grieken, R.: Atmospheric polycyclic aromatic hydrocarbons: source attribution, emission factors and regulation. Atmos. Environ. **42**(13), 2895–2921 (2008)

Ren, W., Liu, H., Mao, T., Teng, Y., Zhao, R., Luo, Y.: Enhanced remediation of PAHs-contaminated site soil by bioaugmentation with graphene oxide immobilized bacterial pellets. J. Hazard. Mater. **433**, 128793 (2022)

Riding, M.J., Doick, K.J., Martin, F.L., Jones, K.C., Semple, K.T.: Chemical measures of bioavailability/bioaccessibility of PAHs in soil: fundamentals to application. J. Hazard. Mater. **261**, 687–700 (2013)

Samuel, O., Othman, M.H.D., Kamaludin, R., Dzinun, H., Imtiaz, A., Li, T., El-badawy, T., Khan, A.U., Puteh, M.H., Yuliwati, E., Kurniawan, T.A.: Photocatalytic degradation of recalcitrant aromatic hydrocarbon compounds in oilfield-produced water: a critical review. J. Clean. Prod. **415**, 137567 (2023)

Schwarzenbach, R.P., Gschwend, P.M., Imboden, D.M.: Environmental Organic Chemistry. John Wiley & Sons (2016)

Shahi, M.P., Kumari, P., Mahobiya, D., Shahi, S.K.: Nano-bioremediation of environmental contaminants: applications, challenges, and future prospects. Bioremediation Environ. Sustain., 83–98 (2021)

Simpanen, S., Yu, D., Mäkelä, R., Talvenmäki, H., Sinkkonen, A., Silvennoinen, H., Romantschuk, M.: Soil vapor extraction of wet gasoline-contaminated soil made possible by electroosmotic dewatering–lab simulations applied at a field site. J. Soils Sediments. **18**(11), 3303–3309 (2018)

Soares, A.A., Albergaria, J.T., Domingues, V.F., Alvim-Ferraz, M.D.C.M., Delerue-Matos, C.: Remediation of soils combining soil vapor extraction and bioremediation: benzene. Chemosphere. **80**(8), 823–828 (2010)

Speight, J.G.: Removal of organic compounds from the environment. Environ. Org. Chem. Eng, 387–432 (2017)

Sun, J., Shen, Z., Zhang, T., Kong, S., Zhang, H., Zhang, Q., Niu, X., Huang, S., Xu, H., Ho, K.F., Cao, J.: A comprehensive evaluation of PM2. 5-bound PAHs and their derivative in winter from six megacities in China: insight the source-dependent health risk and secondary reactions. Environ. Int. **165**, 107344 (2022)

Thacharodi, A., Hassan, S., Singh, T., Mandal, R., Chinnadurai, J., Khan, H.A., Hussain, M.A., Brindhadevi, K., Pugazhendhi, A.: RETRACTED: bioremediation of polycyclic aromatic hydrocarbons: An updated microbiological review. Chemosphere. **328**, 138498 (2023)

Tornero, V., Hanke, G.: Chemical contaminants entering the marine environment from sea-based sources: a review with a focus on European seas. Mar. Pollut. Bull. **112**(1–2), 17–38 (2016)

Udom, G.J., Frazzoli, C., Ekhator, O.C., Onyena, A.P., Bocca, B., Orisakwe, O.E.: Pervasiveness, bioaccumulation and subduing environmental health challenges posed by polycyclic aromatic hydrocarbons (PAHs): a systematic review to settle a one health strategy in Niger Delta, Nigeria. Environ. Res. **226**, 115620 (2023)

Unwin, J., Cocker, J., Scobbie, E., Chambers, H.: An assessment of occupational exposure to polycyclic aromatic hydrocarbons in the UK. Ann. Occup. Hyg. **50**(4), 395–403 (2006)

Usman, M., Hanna, K., Haderlein, S.: Fenton oxidation to remediate PAHs in contaminated soils: a critical review of major limitations and counter-strategies. Sci. Total Environ. **569**, 179–190 (2016)

Veltman, K., Huijbregts, M.A., Rye, H., Hertwich, E.G.: Including impacts of particulate emissions on marine ecosystems in life cycle assessment: the case of offshore oil and gas production. Integr. Environ. Assess. Manag. **7**(4), 678–686 (2011)

Venkatraman, G., Giribabu, N., Mohan, P.S., Muttiah, B., Govindarajan, V.K., Alagiri, M., Rahman, P.S.A., Karsani, S.A.: Environmental impact and human health effects of polycyclic aromatic hydrocarbons and remedial strategies: a detailed review. Chemosphere. **351**, 141227 (2024)

Vidonish, J.E., Zygourakis, K., Masiello, C.A., Sabadell, G., Alvarez, P.J.: Thermal treatment of hydrocarbon-impacted soils: a review of technology innovation for sustainable remediation. Engineering. **2**(4), 426–437 (2016)

Wang, J., Wang, S.: Effect of inorganic anions on the performance of advanced oxidation processes for degradation of organic contaminants. Chem. Eng. J. **411**, 128392 (2021)

Wang, Z., Ren, P., Sun, Y., Ma, X., Liu, X., Na, G., Yao, Z.: Gas/particle partitioning of polycyclic aromatic hydrocarbons in coastal atmosphere of the North Yellow Sea, China. Environ. Sci. Pollut. Res. **20**(8), 5753–5763 (2013)

Wilson, L.B., Moran, I.L., Anderson, K.A., Tanguay, R.L.: Advances in PAH mixture toxicology enabled by zebrafish. Curr. Opin. Toxicol. **34**, 100392 (2023)

Yang, S., Li, J., Song, Y.: Application of surfactant tween 80 to enhance Fenton oxidation of polycyclic aromatic hydrocarbons (PAHs) in soil pre-treated with Fenton reagents. Geol., Ecol., Landsc. **1**(3), 197–204 (2017)

Yim, U.H., Hong, S.H., Shim, W.J.: Distribution and characteristics of PAHs in sediments from the marine environment of Korea. Chemosphere. **68**(1), 85–92 (2007)

Zhang, H., Ma, D., Qiu, R., Tang, Y., Du, C.: Non-thermal plasma technology for organic contaminated soil remediation: a review. Chem. Eng. J. **313**, 157–170 (2017)

Zhang, S., Li, H., He, R., Deng, W., Ma, S., Zhang, X., Li, G., An, T.: Spatial distribution, source identification, and human health risk assessment of PAHs and their derivatives in soils nearby the coke plants. Sci. Total Environ. **861**, 160588 (2023)

Zhang, Y., Tao, S.: Global atmospheric emission inventory of polycyclic aromatic hydrocarbons (PAHs) for 2004. Atmos. Environ. **43**(4), 812–819 (2009)

MIX
Papier aus verantwortungsvollen Quellen
Paper from responsible sources
FSC® C105338
www.fsc.org

If you have any concerns about our products,
you can contact us on
ProductSafety@springernature.com

In case Publisher is established outside the EU,
the EU authorized representative is:
Springer Nature Customer Service Center GmbH
Europaplatz 3, 69115 Heidelberg, Germany

Printed by Libri Plureos GmbH
in Hamburg, Germany